44099
AF844051

Jordan's arid Badia: deepening our understanding

The Badia (root: بدا, *bada,* meaning: to be clear, come to light) is the place where the Bedu (بدو) live, and is the name commonly given to the arid lands of the Middle East. Though arid, these lands receive some rainfall, with occasional storms that can lead to wadi flow and even flash floods. In most places there is thus sufficient moisture to sustain a sparse vegetation of drought-tolerant plants and, therefore, an associated fauna of plant/seed eaters and larger animals that prey on them. Typically there is also great variation between seasonal and day-night temperatures, together with hours of intense sunshine and occasional strong winds. These harsh environmental conditions in the Badia traditionally have given sanctuary to the Bedu whose livelihoods were (and in places remain) sustained by camels, sheep and goats and by trading between urban centres beyond the Badia borders.

The Jordanian Badia forms up to 85% of the country, bordering with Syria, Iraq and Saudi Arabia. In the past eighty years or so the modern world has penetrated the Badia (settlements, roads, vehicles, tele-communications, security services, etc) while many of the Bedu have adopted a village or urban life style in the Badia or elsewhere. The authorities in Jordan refer to the Badia as Jordan's strategic reserve. Some resources from this reserve (phosphates, basalt rock, clay minerals, small quantities of oil and gas, etc) are today mined, refined and used in Amman or exported. More intensive exploitation of these resources, particularly for renewable energy (wind and solar) and earth energy (oil shale, deep gas and uranium), is set to grow rapidly, which will raise questions of environmental protection and the appropriate division of employment and income with the Bedu.

Smith-Gordon has a special interest in community, public health and education.
To discuss new publications relating to these themes contact publisher@smithgordon.com

Jordan's arid Badia: deepening our understanding

Foreword by
HRH Prince El Hassan bin Talal of Jordan

Editors
Roderic W. Dutton and **Mohammad I. Shahbaz**

SMITH-GORDON

Published by Smith-Gordon,
In cooperation with Nishimura

Sales enquiries except for the countries listed below may be directed to
cvenditti@smith-gordon.com
Postal address: Smith-Gordon, CLE, Media House, Burrel Road, St Ives, Cambs, PE27 3LE, United Kingdom
For special editions contact publisher@smithgordon.com

In India contact Ravindra Saxena, Sara Books Pvt Ltd, 4832/24 Ansari Road, Daryaganj, New Delhi 110002, India
Tel: +91 11 6567 6781; (direct): +91 11 2326 6107; Fax: +91 11 2326 6102
email: sarabooks@eth.net

In North America contact Linda P. Jones, Enfield Publishing and Distribution Co Inc, PO Box 699, Enfield, NH 03748, USA
Tel: +1 603 632 7377, Fax: 603 632 5611
email: info@EnfieldDistribution.com

In Australia contact Gavin Crawford, All Things Medical, Blackburn Building, University of Sydney, Sydney, NSW 2006, Australia
email: gavinc1@mac.com

In Japan contact Nishimura Company
Nishimura Company Limited
1-754-39 Asihimashi-dori, Niigata-shi 950 Japan
Tel: +81 03-3239-7671 Tokyo and +81 025-223-2388 Niigata
info@nishimurashoten.co.jp
www.nishimurashoten.co.jp

In South East Asia contact Alkem Company Pte Ltd.
1 Penjuru Close, Singapore 608617, Singapore
Tel: +65 6265 6666; Fax: + 65 6261 7875
chubong@alkem.com.sg

©2008 Smith-Gordon and Company
First published in Great Britain 2008
Smith-Gordon and Company
Media House, Burrel Road, St Ives, Cambs, PE27 3LE, United Kingdom

Copyright. All rights reserved
Unauthorized duplication contravenes applicable laws
British Library Cataloguing in Publication Data
A catalogue record of this book is available from the British Library
ISBN 978-1-85463-227-2
Typeset and printed by CLE Print Ltd, St Ives, Cambridgeshire, UK

Patron
HRH Prince El-Hassan Bin Talal

Organization
The Jordan Badia Research and Development Centre (BRDC)
at the Higher Council for Science and Technology, Jordan

President of BRDC
Mohammad I. Shahbaz
Higher Council for Science and Technology

Focal point in the UK
Roderic W. Dutton
University of Durham

The Jordan Badia Research and Development Centre
acknowledges the financial and in-kind contributions
made by the following organizations

Major financial contributors, Jordan and UK

Ministry of Finance, Jordan
Ministry of Planning and International Cooperation, Jordan
Higher Council for Science and Technology, Jordan
Department for International Development, UK
Department for Environment Food and Rural Affairs, UK

Major in-kind contributors

Jordan Armed Forces
Royal Jordanian Air Force
Public Security
Royal Badia Force
Ministry of Agriculture
Ministry of Water and Irrigation
Ministry of Municipal and Rural Affairs and Environment
Ministry of Public Works and Housing
Ministry of Health
Department of Statistics
Department of Meteorology
Natural Resources Authority
Royal Jordanian Geographic Centre
Jordan Television (JTV)
The Jordan Cement Factories Co.Ltd
Hussein & Ali Al-Manasier Co., Eagle for Mining

BRDC acknowledges with gratitude the contributions to publication
provided by the following organisations:

Ministry of Culture, Jordan
IALC's Sustainable Development of Drylands Project as funded by USAID
Hussein & Ali Al-Manasier Co., Eagle for Mining
Abdul Hameed Shoman Fund For Supporting Scientific Research
Greater Amman Municipality

Badia Research and Development Programme (BRDP) Steering Committee Members

(From the given start date usually until May 2005 when BRDP was succeeded by BRDC)

Dr. Hani El Mulki, co-Chairman (Secretary-General, HCST), 1994-
Dr. John Hemming, co-Chairman for RGS-IBG, 1994-
Dr. Khalid El-Shuraydeh, deputised for Dr.El Mulki, 1995-
Dr. Bassam Snober, Member and Chairman of the Administration Committee, BRDP (Dean of Agriculture, University of Jordan), 1994-
Dr. Abdulla Al Musa, Member and Chairman of the Administration Committee, BRDP (Dean of Agriculture, University of Jordan), 1997-
Prof. John I.Clarke, Member and Chairman of the RGS-IBG Planning Group, 1994-
Brigadier Salim Khalifeh, Member (DG, Royal Jordanian Geographic Centre), 1994-
Prof. Mahmud Duwayri, Member (DG, National Centre for Agricultural Research and Technology Transfer), 1994-
Dr. Awni Taimeh, Member (DG, National Centre for Agricultural Research and Technology Transfer), 1995-
Dr. Fakhruddin Daghistani, Member (DG, Natural Resources Authority), 1994-
Khalid Al-Shiab, Member (DG, Natural Resources Authority), 1996-
Dr. Usama El Khaldi, co-Chairman (Secretary-General, HCST), 1997-
Qais Al Qaissy, Member (DG, Natural Resources Authority), 1997-
HE Mohammad Hamdan, co-Chairman (Secretary-General, HCST), 1998-
Suleiman Abu Alim (DG, Natural Resources Authority), 1998-
Maan Al Hiyari (DG, Natural Resources Authority), 1999-
Dr. Mohammad Shatanawi, Chairman of the Administration Committee (Dean of Agriculture, University of Jordan), 1999-
HE Dr. Munther Al Masri, co-Chairman (Secretary-General, HCST), 2001-
Dr. Abdel Nabi Fardous (DG, National Centre for Agricultural Research and Technology Transfer), 2001-
Prof. Robert Allison (Durham University), 2001-
Mahmoud Al Kurdy (Deputy DG, Natural Resources Authority), 2003-
Dr. Samir Khalifah (Deputy DG, NCARTT), 2003
HE Dr. Tahir Kanaan, co-Chairman (Secretary–General, HCST), 2003
Mohammad Al Gharaybeh (DG, Natural Resources Authority), 2004

BRDP Administration Committee Members

(From the given start date usually until May 2005 when BRDP was succeeded by BRDC)

Dr. Bassam Snobar, Chairman (Dean of Agriculture, University of Jordan), 1994-
Dr. Abdulla Al Musa, Chairman (Dean of Agriculture, University of Jordan), July 1997-
H.H. Sharifa Zein bint Nasser, Finance and Public Relations Director, BRDP, 1994-
Dr. Zahir Rawajfeh, Member (Dean of Agriculture, JUST University), 1994-
Dr. Abdur-Rahman Al-Fataftah, Programme Director, 1994-
Mohammed Shahbaz, Programme, Director BRDP, 1995-
Dr. Roderic Dutton, co-Director BRDP (Durham University, UK), 1994-
Dr. Wajeh Awais, (Dean Faculty of Science, Yarmuk University), 1994-
Dr. Salman Al-Bdour, (Representing al Al-Bayt University), 1994-
Brigadier Ma'arouf Al-Bakheet (Jordan Armed Forces), 1994-

Hassan Al-Mehaisen, (Jordan Armed Forces), 1994-
Brigadier Hamdan Smairan, (Jordan Armed Forces), 1996-
Colonel Walid Kreshan, (Jordan Armed Forces), 1995-
Rouhe Al-Kaied, (Ministry of Interior), 1995-
Saaed Qawqazeh, (Ministry of Interior), 1997-
Captain Kriem Ashoshan, (Royal Badia Force), 1997-
Dr. Saied Khatari, (Dean of Agriculture, Mut'ah University), 1997-
Dr. Khaled Al Tarawneh, (Dean of Agriculture, Mut'ah University), 1997-
Dr. Marwan Muala, (Dean of Agriculture, JUST University), 1996-
Shaykh Sualem Al Sufian, (Local Community), 1995-
Zaied Al-Bnaian, (Local Community), 1996-
Falha Al-Rshoud, (Local Community), 1997-
Shaykh Rakad Al-Shabib (Local Community), 1997-
Colonel Harb Al Quay (Jordan Armed Forces), 1998-
Dr. Darius Campbell (UK representative), 1998-
Talal Seetan Al Madi (Mafraq Governorate Representative), 1998-
Captain Jamal Al Jamaan (Royal Badia Force), 1998-
Colonel Yahya Otiabi (Jordan Armed Forces), 1998-
Major Ghazi Shudaifat (Royal Badia force), 1998-
Shaykh Khalaf Lwein Al Seibeh (Local Community), 1998-
Dr. Sami Mahmoud (Dean of Science, Yarmouk University), 1998-
Captain Noufan Al Jazzi (Royal Badia Force), 1999-
Prof. Mohammad Al Shatanawi (Dean of Agriculture, Jordan University), 1999-
Dr. Ratib Al Oran (Dean of Agriculture, Mutáh University), 1999-
Dr. Hashem Al Massaed (Local Community), 1999-
Lieutenant Colonel Issam Al Mitlaq (Jordan Armed Forces), 2000-
Dr. Nazeer Al Ansari (Al Al-Biet University), 2000-
Dr. Hani Ghousheh (Dean of Agriculture, JUST), 2000-
Mrs. Nadia Falah Al Masarheh (Local Community), 2000-
Lieutenant Salamah Al Kassasbeh (Royal Badia Force), 2001-
Dr. Ayed Al Omari (Dean of Agriculture, Mutah University, 2001-
Captain Bakheet Al Khalayleh (Royal Badia Force), 2001-
Captain Mohammad Hamid Al Haddid (Royal Badia Force), 2001-
Captain Fawaz Al Hawawsheh (Royal Badia Force), 2001-
Captain Hussein Abu Tayeh (Royal Badia Force), 2001-
Eid Ewied Al Adamat (Local community), 2002-
Captain Khalid Sweilim Al Huwaitat (Royal Badia Force), 2002-
Lieutenant Mohammad Ghlailat (Royal Badia Force), 2002-
Dr. Khalid Al Shuraydeh (Assistant Secretary General, HCST), 2003-
Mrs. Maha Al Khatib (Director, Jordan River Foundation), 2003-
HE Dr. Abdul Latif Arabiat (President, Jordan Society for Combating Desertification and Badia Development), 2003-
Khalid Al Irani (General Director, RSCN), 2003-
Dr. Nasri Haddad (Jordan University), 2003-
Dr. Munir Al Rosan (Hashemite University), 2003-
Dr. Nameq Al Tal (Ministry of Agriculture), 2003-
Salamah Al Sfayyan (Local Community), 2003-
Dr. Radi al Rashidy (Hashemite University), 2003-

Badia Research and Development Programme (BRDC)
Board Members

(From May 2005 to present unless otherwise stated)

Prof. Anwar Battikhi, (Secretary-General), HCST), Chairman, 2008-
Dr. Khalid Al Shuraydeh (Secretary-General, HCST), Chairman, 2005-2007
Dr. Walid Al Turk (Secretary-General, HCST), Chairman, 2007-2008
HE Mohammad Odeh Al Njadat (Local Community), 2005-
Retired Brigadier General Salim Khlifeh (DG, Royal Jordanian Geographic Center), 2005-
Dr. Maher Hijazeen (DG, Natural Resources Authority), 2005-
Late Abdul Muniem Samarah Al Zubi (DG, Lands and Survey Department), 2005-
Dr. Abdel Nabi Ferdous (DG, NCARTT), 2005-2007
Mahmoud Al Haddid (DG, Hashemite Fund for Badia Development), 2005-
Dr. Abdulla Al Musa (Jordan University), 2005-
Dr. Nasri Haddad (Jordan University), 2005-
Dr. Hashem Al Massaid (Local Community), 2005-
Shaykh Sami Mithgal Al Fayez (Local Community), 2005-
Mazen Shoter (DG, Lands and Survey Department), 2007-
Faris Al Junaidy (Secretary General, Ministry of the Environment), 2007-
Dr. Faisal Awawdeh (DG, NCARE)
Mohammad Shahbaz (President, BRDC), 2005-
Dr. Adnan Badwan, General Director, The Jordanian Pharmaceutical Manufacturing Company, 2005-

Arabic-English Transliteration

Transliterating between Arabic and English is made problematic because there is a partial mismatch between the alphabets of the two languages; not all the sounds in one of the languages have an equivalent, or an exact equivalent, in the other.

In this volume the transliteration of authors' and other names uses the lettering chosen by the authors themselves.

The transliteration of place names has required finding a reasonable balance between accuracy, consistency (particularly being consistent with the transliterations used in the previous book on the work of the Badia Programme, published in 1998) and simplicity. Place names that carry the definite article in Arabic (al-) but are commonly used in English speech by members of the Badia team without the 'al-', for example Safawi, are left unadorned in this book. But settlements – usually small settlements – whose names are much more rarely used in discussions in English retain the 'al-', which is written to reflect the elision with the first letter of the name in the cases where this applies in Arabic, such as ar-Ranibat an-Na'imat.

A few common nouns applied to geographical features (and a few other items) for which there is no good English equivalent, such as *marab*, are given in the singular form whether the feature being referred to is singular or plural. Such common nouns are written in italics, except when they are used as part of a proper noun, such as Marab Hassan.

CONTENTS

Bedu society and the livestock economy: continuity and change

Diseases and the health service

Aspects of resource management

المجلس الأعلى للعلوم والتكنولوجيا

مركز بحوث وتطوير البادية الأردنية

Jordan Badia Research and Development Centre

FOREWORD

Jordan's Badia, that arid eighty-five percent of the country's land surface stretching east from the highlands to the borders of Syria, Iraq and Saudi Arabia, is a place of awesome and challenging beauty. From the basalt-black extinct volcanoes near the Syrian border to the dramatic cliffs, sands and mountain peaks around the fabled Wadi Rum, the Badia is a unique and integral part of Jordan's natural and social heritage. It has been said that whatever else may change "there will always be a Badia". This is true, but it is also the case that the Badia itself is undergoing change and that the pace of change, which has been gathering speed in recent decades, is set to accelerate. The exploration for, and perhaps the exploitation of, reserves of oil shale and uranium ore will have a dramatic effect on a traditional way of life. A century ago the Badia was a land of peace and tranquillity; no noise was heard apart from the padding feet of the camel and no light pollution at night obscured the brilliance of the stars and the softer reflections of the waxing and waning moon. The Badia was a divide, separating the western and eastern arcs of the Fertile Crescent, and a trading bridge, traversed and guarded by the Bedouin, linking the centres of population. The Bedouin were not only traders but also, of course, herders of camels, sheep and goats. They understood the desert, how to live in it and how to sustain those resources on which their livelihoods depended and which their limited technologies could exploit.

But as the twentieth century progressed, the way of life of the Badia changed, usually through the actions of organisations and people who lived elsewhere or with the use of technologies and ideas provided by them. The changes have been many and varied: most of the Bedouin of the north-east Badia left their camels, their sheep and their tents, and took up residence in the thirty or more villages south of the Syrian border and in what has become the major 'Bedouin city' of Mafraq; roads, oil pipelines and government offices were constructed; people made increasing use of the expanding education and health services; the security forces became an important source of Bedouin employment; the Bedouin started to travel by truck and pick-up; phosphates and other minerals were mined; surface water was dammed and groundwater was pumped, some as far as Amman; people began to see the wind and the sun not just as problems but as sources of power; and the cities of the highlands began to use the desert as a dumping ground for waste.

In 1992 we initiated the Badia Research and Development Programme (BRDP) in order to monitor the resources, assess the impact of change and, wherever possible, to help shape these changes to benefit the people of the Badia. Many of the findings of the first years of this work have found their way into reports to the government and into the international literature, not least into the first Badia book ('Arid lands resources and their management: Jordan's desert margin', 1998) and it is a great pleasure to see another set of papers, covering so many aspects of the Badia, in this second volume. It also makes me proud to see that in the great majority of the papers the lead author is or has until recently been a member of the BRDC field team or has been given a different career start by support from BRDC. Their work, and that of colleagues from the UK and the USA and from other institutions in Jordan, is making Jordan's Badia better known and better understood throughout the region, and throughout the world.

The Programme started as a joint venture between the Higher Council for Science and Technology (HCST) in Jordan and the Royal Geographical Society (RGS) in the UK, and including the Centre for Overseas Research and Development (CORD) at the University of Durham. A few years ago the RGS (quite properly) left us to forge ahead without them but I am delighted to take this opportunity to thank John Hemming, then Director of the RGS, for the unstinting assistance he gave to BRDP and for his continuing enthusiasm for its aims and achievements.

The successes of BRDP, both in research and in R&D, led to it expanding its remit in May 2005 (when we enjoyed a memorable party in the Badia in celebration!) to cover the whole Badia. At that same time the Programme became a National Centre (BRDC) and the then Director, Mohammed Shahbaz, who had so excellently shepherded the Programme towards its heightened status as a Centre, became President. In addition to our base at Safawi we now have a second base in the south, at Ma'an. The dynamic young staff of BRDC, with high qualifications and years of field experience to their credit, will continue to expand their efforts to use our now deeper understanding of Badia systems to improve the livelihoods of the people for whom the Badia is still their home.

HRH Prince El-Hassan bin Talal

CHAPTER 1

The Jordan Badia: deepening our understanding

Roderic W. Dutton
Wolfson Research Institute, University of Durham, Stockton on Tees, TS17 6BH, UK.
E-mail: r.w.dutton@durham.ac.uk

Mohammad I. Shahbaz
Jordan Badia Research and Development Centre, P. O. Box 902, Jubeiha 11941, Jordan.
E-mail: shahbaz@hcst.gov.jo

To most people in the world the word 'desert' carries strong negative connotations. It is associated with the word 'deserted' which not only implies 'empty' but also 'abandoned'. Even worse, today we frequently use the word 'desertified' which implies a land that was alive and active but which has become abandoned and empty as a result of the predatory activities of mankind. The word Badia, on the other hand, has more positive overtones. The word is linguistically associated with the word 'Bedu', implying that the Badia is the place where the Bedu live. The root meaning of Badia (and Bedu) is to do with wide open spaces and therefore with clean air to breathe and no light pollution to interfere with the brilliance of the sky at night. Whereas the word 'desert' implies a barrenness, nothing, the word 'Badia' hints at a potential of life and resources which, indeed, is what the Bedu have known since the dawn of history and which it has been the excitement and pleasure of all those involved in the Badia programme to reveal and explore. Our research in the Badia has yielded some 20 PhD theses, many Masters dissertations and numerous other surveys and reports. The early years of this work led to the publication of a multi-authored book, 'Arid land resources and their management: Jordan's desert margin', in 1998 (Dutton et al 1998) and to a special edition of the journal Applied Geography in 1999 (Millington et al 1999) as well as to many individual papers published in other books and journals. The current volume has created the opportunity for 26 papers, based on more recent theses and field programmes, to be presented to a wider public.

Since the publication of the first book there have been some significant changes in the structures, remit and objectives of the work in the Badia. After a lengthy period of careful preparation the title of the enterprise was changed, in May 2005, from Badia Research and Development Programme (BRDP) to Badia Research and Development Centre (BRDC[1]); a national centre, implying for it and its staff a higher status and much more permanency. The Director of the Programme, Mohammad Shahbaz, became the President of the Centre, and the group of Jordanian staff who had gained higher qualifications and years of field experience became recognized as a much more cohesive team, perhaps unique in the region. At the same time the remit of BRDC was extended to include all the dry areas of Jordan. Thus a new office has been opened in the southern town of Ma'an and projects initiated in this region, as well as around Petra.

The shift in objectives was a gradual process during the late 1990s and early 2000s. It was a partial shift from research to R&D. Whilst today there are still opportunities to

[1] Most of the field work upon which the chapters in this volume are based was undertaken when BRDC was still BRDP with its only field base at Safawi in the geographical area of the western half of the north-east panhandle of Jordan. Thus the two acronyms, BRDP and BRDC, are used by different authors.

undertake fundamental research in the Badia, more of the research is targeted towards collecting and analyzing data that are likely to lead to development work of potential social and economic benefit to the indigenous population, and to Jordan as a whole. In addition, pilot development projects have been undertaken. These are perhaps the most interesting parts of all the work being undertaken by BRDC today because they combine the technical with the social.

We have divided the 25 subsequent chapters of the book into six sections: resources, processes and modelling; aspects of biodiversity; livestock breeding and health; Bedu society and the livestock economy, continuity and change; diseases and the health service; aspects of resource management. From these headings it can be seen that the book in part is divided by research discipline but we have also created a section, to open the book, which focuses on techniques to locate, analyze and quantify aspects of the resource base, to model geomorphological processes and to use modelling approaches to define the nature and extent of problems associated with man's misuse of the Badia by excessive use of fertilizers and for dumping solid waste. The final section of the book suggests ways in which some of the resources could be better husbanded in the long-term interests of the environment and of sustainable resource use.

Of the Badia's physical resources the Bedu used some of the natural vegetation as forage for their livestock and they exploited a small proportion of the surface water for domestic purposes, for watering the livestock and, in more recent times, for very small-scale agriculture in some of the *marab*. They made no use of the solar or wind energy, neither did they use groundwater (because they did not have the technology to reach it) or minerals; their tent-dwelling and mobile way of life meant that they would have had very little use for minerals even if they had been able to access them. The Badia should be a natural place in which to exploit solar and wind energy; there is very little cloud cover, the flat steppe lands can, in places, be swept by strong winds, and access to conventional energy supplies tends to be very expensive because of distance and the dispersed nature of the small Badia settlements. Up until the very recent past Jordan has enjoyed very cheap energy, first from Saudi Arabia and, until March 2003, from Iraq. Today, however, Jordan pays world prices for its energy requirements and those prices, for oil, rose very sharply to over US$100 per barrel in the latter part of 2007. Therefore Jordan now has every incentive to explore sources of renewable and alternative energies and to revisit previous oil and gas exploration sites to see whether, under today's conditions, they are economically exploitable. As the first substantive paper in this volume, Chapter 2, makes clear, Jordan has good levels of solar radiation and a moderate wind speed, which are suitable for the deployment of solar photovoltaic panels and wind turbines to generate electrical power in remote areas. The potential of utilizing hybrid (wind / solar / battery) power systems to meet the load requirements for a typical rural house (with an annual electrical energy demand of 1 175 kWh) is tested but it seems that the stand-alone photovoltaic/battery system is the most suitable option for the Badia according to performance-system cost relationships.

The comprehensive overview of the physical resources in the Azraq basin (Chapter 3) has the aim of achieving the optimum utilization of these resource. There are huge deposits of industrial minerals in the Azraq basin which are a better source of regional wealth for the country than the scarcer metal deposits. There are promising sites for oil and gas exploration, which is why the research concentrated on the evaluation of the hydrocarbon potential in the basin and on providing a national overview of the hydrocarbon potential of individual fields. The study puts these into a wider context of the potential of both agricultural and range resources and of the environment.

Clay minerals from the same part of the north-east Badia, the Azraq basin, are the subject of a special study in Chapter 4. They were formed from alteration of volcanic ash deposited in shallow water lakes. Two forms of commercially significant clay minerals are present in the area. They consist of smectite and palygorskite (attapulgite) -rich sediments. These Jordanian clay minerals display high bleaching capacities of edible oil after acid and thermal activation, giving results similar to those obtained with commercial industrial products.

A future task of BRDC will be to work with other government agencies and private sector companies, seeking to ensure that exploitation of the mineral, clay and other resources is rational and brings maximum benefit to the people of the Badia at least environmental cost. Landscapes that are highly disturbed by mining and related activities will then be subject to reshaping by geomorphological processes, particularly rain and wind. Chapter 5 demonstrates that although infrequent, rainfall events appear to be a critical control in the evolution of the clast-mantled arid slopes found in parts of the Badia. Critical to this modification is the interplay between the surface configuration, rainfall and runoff generation. Results of field experiments suggest that a close synergy exists between surface character, whole slope form and hydrological response to rainfall. If the surface character is severely altered, by mining activity for example, the response to rainfall will be changed with possible undesirable consequences. Part of the preparation for mineral exploitation in the Badia should be additional landform response modelling exercises, which will turn to practical advantage the gain in our understanding of climate and landform processes that are indicated in Chapter 5.

Chapter 6 provides us with information, from a related modelling exercise, on the relationship between rainfall and runoff. The focus of this study is not on how rainfall affects landforms but rather that the most important factors affecting runoff and infiltration rates are the rainfall intensity and surface moisture rather than ground cover characteristics; medium to high intensity rainfall generates 79-91 % of precipitation as runoff; low intensity storms still generate 59-84 % of the storm as local runoff. These runoff values are very high and reveal the tremendous potential for water harvesting if the water can be collected locally in small structures (as was done by the Bedu using *mahafir*) rather than at a distance in big structures. Losses through runoff are high due to seepage and evaporation, which both increase with catchment size. A problem to be faced when conducting analyses such as those in Chapters 5-6 and by Kirk (1998) previously, is that rainfall in the Badia is typically collected or reported on a monthly basis; there is need for a network of modern weather stations if the storm duration / intensity data required to undertake this work and, for example, to justify and locate 'modern' *mahafir* are to be systematically obtained.

Actually, insufficient or inappropriate collection of data by national agencies in the Badia is but one symptom of the relative lack of attention given to Badia concerns by the wider national community because, no doubt, the Badia is far from the main cities and is thinly populated. A more significant example concerns the disposal of solid waste (Chapter 7). Although this is a national issue it is also a Badia issue because, historically, it has been assumed that dumping waste in the wide Badia lands of eastern Jordan will be problem-free. However, it is increasingly being realized that there is a great range of associated problems, from soil and groundwater pollution to dust, bad odours and danger to employees and scavengers. Communities near landfill sites are becoming more aware of the problems and less tolerant of them. For all these reasons, establishing criteria and, thence finding locations, for safe waste disposal sites, is urgently required. Only in the most recent landfill site, al-Khabawi new site, have modern standards of site selection, preparation and management been applied. BRDC will need to play a role in ensuring that the Badia is not used as a dumping ground and that best practice for waste disposal is employed on all future occasions.

But if helping to prevent bad practice in waste disposal is an important task for BRDC, then helping to protect the groundwater resource is a task of greater importance.

As Chapter 8 makes clear most of the groundwater basins in Jordan are already exploited beyond their estimated safe yield. Total safe yield for all basins was estimated to be about $418.5 \times 10^6\ m^3\ yr^{-1}$ yet the consumed water from these basins was $479 \times 10^6\ m^3\ yr^{-1}$. Not only is the groundwater being over-exploited but its quality is being threatened by nitrate pollution. In another modelling exercise Chapter 8 reports on the creation of a GIS-based model to estimate nitrate leaching from cesspools and agricultural land. It was found that nitrate leaching from agricultural sources was much higher than that derived from cesspools. It was estimated that while about 0.3-0.7 kg $household^{-1}\ yr^{-1}$ could leach to groundwater from cesspools in the study area, the estimated nitrate leaching from agricultural sources could be as high as 483 281 kg yr^{-1}. As an extension to the model, within the GIS environment, several management scenarios were tested to minimize nitrate leaching from both cesspools and agricultural fields. Regularly emptying cesspools could minimize or eliminate nitrate leaching from this source and if farmers followed instructions on fertilizer usage they could reduce nitrate leaching by up to 99 %.

Still on the theme of environmental problems and how to define and quantify them, Chapter 9 is an exemplar of how to combine international experience (obtained from published papers and national and international regulations) with Jordanian expertise and community opinions to arrive at a set of prioritized criteria for landfill site selection. High priority was given by all groups to distance from built-up areas and surface water and groundwater bodies, and to avoiding placing the site upwind of a settlement. The experts showed more awareness, than people from the local communities, of the importance of such technical issues as structural geology (eg avoiding fault lines). But important as the paper is for what it has to say about the specific problem of landfill site selection, it is perhaps even more useful as an example of combining research studies and sets of accepted regulations with the views of both local experts and local communities to reach an understanding of how best to prioritize criteria for action to resolve problems.

The next set of three papers (under the heading 'aspects of biodiversity') reminds us that 'life' in the Badia embraces much more than just human beings and that each form of life requires its own environment. Usefully the three papers discuss contrasting life forms: reptiles and mammals; parasitic plants and fungi; and soil bacteria. The first (Chapter 10) describes the macro-habitats of the Badia with particular reference to those factors that condition the presence of reptiles and mammals and their survival in what would normally be considered to be a hostile environment. A full list of amphibians, reptiles and mammals reported from the Badia is given and the unique or special adaptive features of some of them are recorded.

The parasitic plants broomrape (Halouk in Arabic) (*Orobanche* spp) and dodder (Hamoul in Arabic) (*Cuscuta* spp) are found in the Badia together with fungi associated with the root systems of the natural vegetation (Chapter 11). Representative soil samples from a number of sites were carefully collected from the root zones of the natural vegetation cover and bare ground. Soils were studied for their fungal population by serial dilution and plating on acidified potato dextrose agar (PDA) and malt agar (MA). Segments of plant roots were plated for the recovery of intimately associated fungi and soil samples were processed for their VA fungal populations by extracting VA spores. Several genera of soil-borne fungi were observed, but four, *Penicillium* spp, *Puluria* spp, *Dactylospors* spp, two *Asperigillus* spp, and yeasts were the most abundant. A number of *Actinomycetes* were also observed. It was revealed that the Badia soil is very rich in mycorrhizae fungi and that *Glomus* spp and *Aculospora* spp were the most abundant.

These studies are not only of interest as fundamental biodiversity research. Extended work on *Orobanche* substantiated the validity of olive pulp (the solid by product from olive mills or *jift*) as a bioherbicide against *Orobanche* and a wide range of weed plants. *Jift* application *in vitro* and *in vivo* proved to have a potential herbicidal effect against

Orobanche seeds, broad leaf and grass weeds. Several soil-borne fungi were investigated as potential agents for the biological control of *Orobanche*. The very special environments found in the Badia create particular plant and animal adaptations that fully justify ongoing research to isolate compounds that will have practical functionality.

Interesting properties were also revealed in the study of three genera of bacteria (Chapter 12), *Bacillus*, *Streptomyces* and *Pseudomonas*. The low nutrient levels and harsh environmental conditions in the Badia region explain the low soil bacterial count and diversity and could trigger or favour unusual metabolite biosynthesis by the isolated bacteria. Prevalence of members of the genus *Bacillus* and their wide distribution in soils of the Badia could indicate their potential to withstand unfavourable or harsh conditions as they harbour the endospores that are resistant to inactivation by heat and desiccation. Screening results revealed the characterization of four *Streptomyces* isolates (I_7, AC_{32}, Bb_{36} and Z_{11}) which inhibited the growth of all tested pathogens. The unusual antibiotic profile of these isolates stressed their potential as a source of novel antibiotics. Potential hydrocarbon degrader bacteria, belonging mainly to the genus *Pseudomonas*, were isolated from soil samples that had been exposed to crude petroleum oil spills.

But interesting as these microscopic representatives of the living world may be, the animals that are synonymous with Bedu traditions and way of life, the camel and the sheep, still play a very significant, if reduced role, today. Thus is it very appropriate that this book includes not only a section, with two papers, on 'livestock breeding and health', but also a section which we have entitled 'Bedu society and the livestock economy; continuity and change' which looks at the impact of changing government policies on the role and economic significance of the sheep economy in recent years, on the changing habitations of the Bedu and on the quietly changing roles of women in Bedu households.

If sheep and camels are to have an economically sustainable position in Bedu society in the longer term future their modes of management, including both breeding and health, will need to be modernized. Globalisation of livestock markets has meant that Jordan is fully exposed to the importation of live sheep and red meat from Australia and Eastern Europe, amongst other places. Therefore, for example, the Awassi sheep of the Badia has to be bred so that it is not only tolerant of the harsh climatic conditions of the region but also to have early lambing, improved lambing rates and rapid growth rates per day and per unit of feed. If the Bedu want to respond to globalisation pressures they will have to become livestock farmers as opposed to being simply livestock keepers. It is excellent, therefore, that BRDC had the opportunity to support work designed to improve lambing rates in the north-east Badia (Chapter 13). The study was conducted to evaluate the effect of two different levels of eCG (equine chorionic gonadotropin) dosage on reproductive parameters of Awassi ewes in different flocks synchronized using intravaginal progestagen sponges. Oestrus expression was similar between the groups but the interval from eCG administration to onset of oestrus was shorter ($P < 0.05$) in the group receiving the higher level and the lambing rate was greater. Pregnancy rate was similar between treatments while the number of multiple births tended to be greater ($P < 0.10$) in ewes receiving the higher level of eCG, as was the number of lambs born per served ewe and the number of lambs born per lambed ewe.

Camels are generally reckoned to suffer less from parasites and other diseases than sheep and goats. However, a study (Chapter 14) of blood, faecal, skin and organ samples of 178 camels slaughtered at Ramtha abattoir 1999-2005 revealed the presence of many types of parasities, both internal and external. Of the latter, tick problems due to *Hyalomma dromedarii* and *Hyalomma anatolicum* were seen in 174 (97.7 %) cases and mange due to *Sarcoptes scabie* var *cameli* was found in 31 (17.4 %) of the skin scrapings examined. Camels need greater veterinary protection, if they are to achieve their maximum potential, than is typically provided for them.

In stages, by 1996/7, after some one and a half decades, a major sheep concentrate feed subsidy was removed which had a very significant impact on reducing the number of sheep in the Badia, in a trend strengthened by a set of drought years in the latter 1990s. Chapter 15 shows that although sheep continued to decline up until about 2002, numbers subsequently recovered, at least up to 2005 (latest figures available), which presumably indicates a renewal of confidence in the sheep industry and, perhaps, some adjustment in the efficiency with which they are produced. But while the paper does compare the major sheep production statistics from the 1990s and today it focuses on field work undertaken in the mid 1990s before the removal of the feed subsidy. It thus not only provides an understanding of the many factors that governed sheep marketing at that time but also acts as a baseline for any similar study today, a decade after the removal of the subsidy. A large number of farmers responded to questionnaires and traders were interviewed. The relationship between producers and traders is explored and it is clear that sales were made for a complex mix of commercial and non-commercial reasons. The paper appeals to the government to better understand Bedu society and requirements before embarking on future interventions that affect the livestock industry. A second paper (Chapter 16) is based on fieldwork undertaken in 1994/5, before the removal of the feed subsidy, and in 1998, about two years afterwards, and thus provides us with some information about the short-term responses of sheep owners to the subsidy removal. Initial interviews were conducted in the households of 109 Bedu sheep-owning families, and many of the same families were interviewed again in 1998. The interviews centred around milk production, consumption, processing and sales. The families were divided into three groups or clusters based on variables that described the degree of family dependence on livestock and milk production. In 1998 it was seen that the producers in the subsistence cluster were least affected by subsidy removal while the cluster that had been most dependent on their livestock for income had been most severely affected by the change; there was less money for family expenditure and significant numbers of sheep had been sold. But the degree of overall change had also been limited by a somewhat desperate hope that circumstances would improve. However, following 1998 livestock numbers were reduced, as we have seen above, in response to drought as well as continuing high feed prices, but returns from the sale of milk remained much the same as before.

There is no doubt that the twentieth century was a time of major change for the Bedu in Jordan. To some extent change was forced upon them by, for example, the creation of national borders across their territories as part of the settlement following the First World War. To some extent they were swept along by changes that affected the whole Jordanian population, and indeed the whole region, as new types of occupation (such as army service) and new technologies (such as motor vehicles and television sets) became accessible by them. And to some extent the Bedu changed aspects of their life styles, including the ownership of livestock, as they began to perceive the advantages of new services such as health, education and piped water, and new materials, such as stone and brick for house construction. This found expression in a continual drift away from mobility and dwelling in hair tents, the traditional *bayt ash-sha'r*, towards living in houses (near school, clinics and hospitals) in villages and towns. This movement is carefully and sensitively analyzed in Chapter 17. The authors interviewed 317 people from 34 villages in the north-east Badia and observed and photographed the entire evolving range of Bedu 'homes' from the *bayt ash-sha'r* through transitional village homes to large modern houses, or villas. In spite of using entirely different materials for the construction of these various homes the spatial separation, shape, internal divisions and functions of the original hair tent were strongly reflected in the more recent transitional forms, made of mud and basalt rocks, and even in the more sophisticated villas, though the growing wealth of some families coupled with their increasing exposure to external influences (as well as the intrinsic qualities of the new materials) have gradually changed external shape and internal divisions of space and furnishings in the newest houses. However, some traditional public and cultural functions of the original *bayt ash-sha'r* seem still to require its presence, a dilemma often resolved by erecting the *bayt ash-sha'r* beside the new house. The structure and usage of the different forms of the *bayt ash-sha'r* are described in meticulous detail, making a permanent record which will also help preserve its ancient Arabic nomenclature for a time when, perhaps, hair tents are no longer used.

At the same time that the Bedu have been modifying their thinking about their habitations and their sources of income, and no doubt significantly related to both of these, there has been a ferment of discussion within Bedu households about the division of responsibilities and activities between household members, with particular reference to the roles of women. Research was undertaken to investigate gender divisions, roles and decision-making processes in Bedu society in three villages in part of the north-east Badia of Jordan (Chapter 18). Bedu women were found to contribute significantly to the household economy. Their roles can be divided into unpaid domestic and livestock-related work, and paid work which, typically, is agricultural, seasonal and poorly paid. Unpaid work or work that does not directly yield an income for the family (including work with sheep; men sell the sheep and thus are seen as responsible for their income) is not regarded as productive work. A 'power game' showed that the women are aware of power differentials, in favour of men, but also showed that power relationships are changing and that women foresee, or would like, more change to occur though not to the extent of reversing tradition. Younger and unmarried women are more restricted in their movements outside the family home; they are more bound by tradition and the local interpretation of religious mores. The economic and social jolt of the removal of the livestock subsidy in 1996-7 followed by three years of drought not only undermined the economic viability of sheep keeping but also encouraged some women to speak out (and be listened to) about not keeping sheep if they merely added to women's work while losing money. With income from sheep becoming unreliable and with more educational opportunities being taken up by women, more women are securing paid, even salaried, work, essential to maintain household security, and in consequence are seen as making productive contributions to the household, which is promoting an increase in their influence in family discussions about the household economy.

Access to health services has already been mentioned as one reason for the Bedu adopting a more settled way of life. Work supported by BRDC has been able determine the common infectious diseases prevalent in the area by identifying the predominant causative pathogens, their antibiotic susceptibilities, clinical outcomes, and the risk factors (Chapter 19). The study was conducted over a 2-year period in the north-eastern Jordan Badia, a rural area inhabited mostly by recently settled Bedu and small-scale farmers of low socioeconomic status. Diarrhoeal diseases caused by a variety of enteric pathogens, brucellosis, community-acquired bacteraemia and urinary tract infections are prevalent in that region and were the focus of this study. The reporting of newly emerging parasites, eg *Cryptosporidium parvum* and *Cyclospora cayetanensis*, in this rural area increases awareness of them. The recovered pathogenic bacterial species showed a wide range of resistance to commonly used antimicrobial agents and reflects an uncontrolled use of antibiotics. The work has demonstrated the close links between health, wellbeing and knowledge. Risk factors that correlated significantly with contracting certain diseases included socioeconomic status, level of education, drinking unpasteurized milk, drinking unchlorinated well or tank water, contact with animals, and a low level of personal hygiene. The findings emphasize the need for further longitudinal epidemiological studies in which the issue of indigenous infections and modes of transmission should be further investigated.

Chapter 20 examines the delivery of primary health care in the Badia. The research is based on a sociological study that involved interviewing semi-nomadic and sedentarized Bedu communities of the north-east Jordan Badia. The chapter explores the use of healthcare services by these groups, including the interface between traditional and formal medicine. The most commonly used Arab medicines are reviewed and contemporary patterns of the use of healthcare services are described. Changes in communities' health and illness beliefs are explored, with a focus on parents' accounts of their children's health, including the ways economic activities are linked to perceptions of health and illness and the extent to which families perceive they can exert control over their children's health. Key geographical, organizational, economic and social factors mediating access to healthcare services experienced by settled and semi-nomadic communities are identified and the particular problems of healthcare accessibility among households practising pastoral migration are discussed. The study demonstrates the importance of healthcare planners developing detailed understandings of what are complex and interlinked factors of accessibility, and the motivations behind households' willingness and ability to use health services, if they are effectively to deliver appropriate healthcare interventions in the Badia.

What has been written in the pages above shows that much of the work on which it is based has implications for management if the physical resources are to be exploited rationally and sustainably and if the human resources are to be developed appropriately. In the final section of the book, Chapters 21-26, which we have entitled 'aspects of resource management' some different management approaches are discussed. Chapter 21 (which needs to be read in conjunction with Chapters 7 and 9) combines the careful collection of field date from surveys of Jordanian experts and local communities and other sources integrated with international regulations and best practice to develop a decision-support tool, based on GIS technology, as an aid to improving the management of solid waste in Jordan. The analysis makes it clear that there are only relatively small parts of the Badia suited to solid waste disposal. However, there has been a tendency in the past falsely to assume that waste can be disposed of with ease and safety almost anywhere. In consequence, today, there are examples of many types of problems stemming from lack of planning care. The approach used in this paper can be used to develop other planning tools so long as the appropriate data sources are available.

Using a rather different approach another GIS-based management tool was developed (Chapter 22) that includes various steps of simulation used to estimate the runoff volume for catchments in the Badia. First the map data were accessed and displayed on the screen so the user can locate the catchment for further calculation. Other data files, which are the hydrological database that describe the hydrology of the project area, were also loaded and displayed on screen. The significance of the work rests on the fact that with very high demand on the ground water resources, the limited volume of surface water becomes a promising option for new and innovative water supplies, as also discussed in Chapter 6. Rainfall and runoff were recorded under differing conditions and used in the model to calculate runoff. The simulation tool calculates the peak discharge and runoff volume with acceptable accuracy for the whole catchment. This is an important step towards defining the location, type and size of structures that can be used to control and dam surface water flow. This modelling tool would need, in practice, to be employed in conjunction with other kinds of data gathered from local people and agricultural and environmental experts to show what the water could be used for, how much would be needed and exactly where it would be useful to collect and store it.

A related paper (Chapter 23) focuses specifically on the use of surface water for agriculture on the *marab* in the north-east Badia. In a remarkable example of the need to employ global thinking and knowledge the paper notes that the authors' recent research has recognised important correlations between global climatic indices and the total amount and duration of winter rains in Jordan. In particular the effects of the El-Niño Southern Oscillation (ENSO) on Jordanian rainfall have been studied. The results of the study showed that there is a significant correlation between the global climatic indices (ENSO SST and SOI) and precipitation. The correlations have been used to develop a rainfall prediction model for Jordanian rainfall. The developed rainfall prediction model has been linked with the traditional approaches that have been practised by the local people of the north-east Jordanian Badia. The link shows that rainfall predictability in Jordan could be enhanced. The combined prediction model also offers the potential for forecasts to be made and confirmed at intervals during summer and the early months of autumn, generating forecasts with high reliability and thus contributing to water resources management in Jordan, and, indeed, using the same approach, elsewhere in the Middle East. As the paper points out, pro-active management of rain-fed agriculture and natural pastures on the desert margin offers disproportionate benefits to the subsistence farmers currently struggling to survive in these inhospitable areas.

Chapter 24 examines Jordan's experience of designing and managing rural development and conservation projects including weaknesses in project execution and outcomes that may have resulted from failures of the projects' professional teams to engage effectively, and meaningfully, with the Bedu communities that the projects are supposed to serve. The implementation of conservation projects can cause particular problems because, understandably, the main priority of the Bedu is to protect their own livelihoods and they have little interest in the 'rights' of plants and animals if they feel that such interests may threaten the future of their own families. If conservation work is to succeed, the Bedu have to be convinced that new income-earning opportunities will be created, through tourism, for example, over which they exert some effective influence.

Through the Anaqeed Al-Khair project (Chapter 25) BRDC now has several years of direct small-scale project management experience, working with a so-called 'cluster' of villages. One of the aims has been to encourage the cluster of villages to develop a common set of objectives and activities, and overcome traditional tribal and other cultural divisions by building a shared management structure and approach to work in the interests of all parties, resolving problems and formulating strategies through on-going discussion. The Anaqeed Al-Khair project is an agricultural development project in the north-east Badia of Jordan, focusing on a cluster of 14 villages with a population of 7,800 people and five project activities (sheep selection and management, honey production, sun-dried tomatoes, forage production, improved sheep wool and goat hair processing). Detailed discussions led to the adoption of a cooperative management structure. This was a new concept for the villagers who have been on a very steep learning curve supported by BRDC, which, by providing encouragement and technical training, has, at the same time, demonstrated that the very act of working effectively together can lead to greater benefits being achieved by the cooperative and by its active members. In practical terms, the main obstacles faced by the cooperative have been lack of experience in identifying desirable markets, insufficient attention to added value, and lack of a well-thought-out business plan. BRDC has helped overcome practical difficulties, in a number of ways: assistance in finding buyers who offer better terms, establishment of market information systems, facilitating the identification of value-adding technologies and products, analyzing ways to reduce costs of production and marketing, and offering guidelines and methods for the establishment of improved business plans in part by careful financial auditing, which has led to two projects being dropped (forage production and tent manufacture), leaving three which seem to be sustainable (sheep, honey and sun dried tomatoes) because of their positive income generation, with potential for improvements in marketing. Experiences to date suggest that an expansion of activities using the same community-driven methodology is possible, with great potential for replication in other areas of the Badia.

The final paper (Chapter 26) reports on a project that has had to combine great technical expertise with extremely sensitive interactions with the local community and enormous patience. The project concerns the greatly increased production of drought-tolerant nutritious shrubs in what could be called 'forage shrub reserves.' The main technical challenge has been to optimize the usage of the limited rainfall resources, through different techniques of rainfall harvesting, to ensure that the plantations of shrubs not only remain alive but produce worthwhile quantities of forage for the sheep. But the even greater socio-cultural challenge has been to gain the trust of the local sheep owners, encouraging them to be patient and to think long-term, and to follow expert advice about the safe number of sheep and number of days of permissible grazing each season. The project aims to enhance the awareness of the Badia people about the importance of rangeland management and to learn how certain range management practices can be of direct benefit to them and the resources on which they depend. The local community was involved in the project from the starting point of problem definition, possible solutions, site selection, and choice of plant species suitable to the area and the intended use by livestock owners. A 50 ha demonstration site was prepared, using a variety of water harvesting structures and digging holes for shrub planting. Careful survey and statistical analysis has indicated not only that nutritious plant biomass production has increased but also that after four years of protection to 2006, the number of plant species and families found in the project area were much higher. The project has been able to demonstrate to the government that, by achieving the cooperation of the local Bedu population, rehabilitation of large areas of the Badia steppe is possible. The focus of future project activities will be on selected rangeland site management improvement and rehabilitation, with the selected areas managed by Bedu organizations.

We trust that this volume gives a balanced sample of the full range of the physical and social science research undertaken by members of BRDC's R&D staff, and by colleagues in other institutions in Jordan, the UK and elsewhere. The two major themes concern the changing lives, lifestyles and economies of the Bedu population and the quantity, management, use and misuse of water. Two other papers discuss energy from the Badia – largely untapped resources today that will definitely be exploited in the coming years. Given the environmental dangers and social disruption always inherent in such exploitation, we have also given some priority to aspects of biodiversity (reminding our readers that this too is precious) and to the absolute requirement of ensuring that the local communities are centrally involved in change and are significant beneficiaries of it. We publish this volume in the belief that it does indeed lead to a deepening of our understanding of Jordan's Badia.

Acknowledgements

The editors would like to acknowledge the many contributions that the authors have made to our understanding of the north-east Badia in Jordan, and thank them for their hard work in producing the papers included in this volume. They would like also to acknowledge the excellent work of Chris Orton, Design and Imaging Unit, Department of Geography, Durham University, who prepared all the illustrations in the volume. They would also like to pay a special tribute to Ms Rima Ras who with immense efficiency, patience and charm has maintained communication not only between the two editors – in two different countries for much of the time – but also between the editors and the many authors of the papers that it has been our pleasure to edit into this monograph.

References

Dutton RW, Clarke JI and Battikhi AM ed (1998). Arid land resources and their management: Jordan's desert margin. London: Kegan Paul International.

Kirk AJ (1998). A synthesis of climatic data with specific interest in the precipitation record. In: Dutton RW, Clarke JI and Battikhi A (eds), Arid land resources and their management: Jordan's desert margin. London: Kegan Paul International, pp. 47-66.

Millington A and Dutton RW ed (1999). Socio-economic change and natural resource management in a marginal environment: Jordan's Badia Research and Development Programme. London: Pergamon.

CHAPTER 2

Sizing and techno-economical optimization for photovoltaic-wind hybrid energy systems in comparison to single photovoltaic and wind systems with battery storage in the Badia in north-east Jordan

Mohammad Al-Smairan
Jordan Badia Research and Development Centre, P.O.Box 902 Amman 11941, Jordan.
E-mail: wesam93@yahoo.com

Peter White
Faculty of Engineering and Computing, Coventry University, UK

Les Duckers
Faculty of Business Environment and Society, Coventry University, UK

Most of the energy consumption of the world depends on fossil fuel, which is exhaustible and is being used extensively due to continuous growth of the world population and economic development. Wind and solar energy are becoming increasingly attractive and are being widely used for substitution of oil-produced energy, and eventually to reduce air pollution. Jordan has good levels of solar radiation and a moderate wind speed, which are suitable for the deployment of solar photovoltaic panels and wind turbines to generate electrical power in remote areas. Records at Azraq South (north-east Jordan Badia) show that the monthly average wind speed varies 2.12-5.79 m/s while monthly average daily values of solar radiation range 2.81-7.87 kWh/m^2. In this study, hourly mean wind speed and solar radiation data recorded by the Jordan Meteorological Department have been analyzed to investigate the potential of utilizing hybrid (wind / solar / battery) power systems to meet the load requirements for a typical rural house (with an annual electrical energy demand of 1 175 kWh). In the Badia the analysis indicated that the wind energy calculations should be carried out using a bimodal probability density function rather than the typical Weibull probability density function. Otherwise, the energy output and the capacity factor of the wind turbine will be underestimated by not less than 9.6 %. The analysis indicated that the stand-alone photovoltaic/battery system is the most suitable option for the Badia according to the performance-system cost relationship.

Keywords: Jordan Badia; solar radiation; wind speed; photovoltaic system; wind energy.

1. Introduction

The rapid depletion of fossil fuel resources, the limited reserves and their unstable prices on a worldwide basis have necessitated an urgent search for alternative energy and significantly increased the interest in renewable energy sources. Of the many alternatives, photovoltaic and wind energy have been considered as promising towards meeting the continually increasing demand for energy (Borowy et al 1996, Chedid et al 1997). The wind and solar energy sources are non-polluting, free in their availability, site dependent, non-depletable and potential sources of alternative energy options. These facts make the alternative resources attractive for many applications. Today, wind and photovoltaic generators are utilized in such applications as water pumping, lighting, electrification of remote areas and telecommunications (Borowy et al 1994, 1996, Elhadidy 2002, Elhadidy et al 2000, Tay et al).

Jordan has no significant fossil fuel energy resources of its own and must rely on neighbouring Arab oil producing countries. It imports almost 95-97 % of its energy needs in the form of oil and petroleum products (Ammari et al 2003, Hrayshat Eyad 2005, Kablan 2004, MEMR 2003, NERC 2003). Minimizing reliance on imported energy to meet its current and future energy demand is being given policy priority. The expectations show that the demand for energy will increase at a faster rate than the growth in the Jordan economy. For this reason, Jordanian authorities actively encourage intensive examinations of the possibilities for developing indigenous energy sources, as well as energy conservation and more efficient energy use. The future expectations of Jordan's primary energy sources in the year 2010 include 5 % renewable energy, 5 % oil shale, 15 % natural gas, and 75 % crude oil and oil products (Hrayshat Eyad 2005, NERC 2003, RSS 1983a).

In this study an attempt was made to investigate the possibility of harnessing solar and wind energy to supply part of the energy requirements of Jordan's remote villages. In order to establish the possibility of utilizing solar and wind energy, it was necessary to carry out a comprehensive investigation on their basic human energy needs (RSS 1983b). These villages are usually located far from the national electrical grid at a distance that would make national grid connection very expensive (RSS 1988).

Average monthly wind speed and solar radiation data for the Badia have been analyzed to examine the potential of utilizing hybrid wind / solar power systems to meet the load requirements of a typical rural house in the Badia (with an average electrical energy demand of 3.22 kWh/day and 0.625 kW peak load demand). The measured daily average energy consumption of a typical rural house in the Badia has been considered as the predefined daily load in this study (Shaahid et al 2004).

The selection of a combined photovoltaic/wind power system with battery storage for an autonomous application should have certain economic and technical advantages over a single renewable power system configuration. As system reliability is the first priority in such cases, an oversized system is usually designed in order to keep the loss-of-load probability to extremely low. System economics then play a secondary role in the overall system design method. Therefore, the essential content of this paper refers to the sizing problem of an autonomous renewable system when the reliability factor is to be taken into account (Protogeropoulos 1992).

The sizing procedure will be in the energy:load ratio 1.5-2.0, which is the most appropriate (Celik 1998, 2002). This paper presents the sizing technique using three different sizing scenarios. Scenario 1 considers the yearly average energies produced by the photovoltaic panel and the wind turbine and the yearly average load demand. Scenario 2 is the worst month's scenario. Scenario 3 is the worst month's scenario (Celik 1998, 2002, Morgan 1996, Protogeropoulos 1992, Protogeropoulos et al 1997) and a Hybrid Optimization Model for Electric Renewable (HOMER version 2.09) as the sizing and optimization software tool. Three storage battery capacities have been used in the sizing process. A simulation process has been carried out for all three scenarios using HOMER software in order to rank the systems according to the system cost (HOMER).

Mostly, the wind resource in the Badia is categorized as low. On the other hand, it is expected that photovoltaic technology will become cheaper and practicable in the near future. Therefore, emphases have been given in analyzing the size, cost and performance of a photovoltaic/battery system. At present, a single photovoltaic/battery system is the most suitable solution and cost effective system for the required system autonomy level at remote sites. As the anticipated load increases to the urban consumption (about

6.0 kWh/day) the photovoltaic/battery system remains the optimum choice until the load has increased by at least 272 % when a hybrid photovoltaic/wind/battery system presents a better option.

2. Weather data

2.1. Wind and solar data

The weather resource, measured mainly by solar insulation and wind speed, affects renewable system performances and therefore the size of the components. A low resource would result in a larger system size for the same load demand pattern. The solar insolation is crucial for the photovoltaic power output, while the wind speed distribution determines the power delivery by the wind generator (Protogeropoulos 1992, Protogeropoulos et al 1997). The Badia is known for high solar energy and low wind energy potential. The annual average wind velocity is low, which makes the utilization of wind energy converters surely unfeasible. The 12-year average (1991-2002) indicates that the annual average wind speed was 4.26 m/s. In contrast, utilizing solar energy would be feasible since the daily average over the nine-year period 1991-9 of solar radiation on a horizontal surface was measured to be about 5.58 kWh/m^2/d, while the total annual sunshine hours amount to about 3 300. These figures are very encouraging for the use of photovoltaic generators for electrification of remote villages, based on worldwide success.

3. Energy contribution of the components and system performance definitions

3.1. Photovoltaic and wind fractions

Many researchers (Celik 1998, 2002, Morgan 1996, Protogeropoulos 1992, Protogeropoulos et al 1997) have shown that in an ideal system where, at any time, the total energy produced by the photovoltaic and the wind turbine equals the required load:

$$E_{PV} + E_{WG} = E_L \tag{1}$$

Where E_{PV} is the total energy generated by photovoltaic array; E_{WG} is the total energy generated by wind turbine and E_L the load demand.

Under real operational conditions and due, however, to the variable weather resource and possible non-constant load, for the majority of the time with a hybrid photovoltaic-wind energy system this does not happen. The following two cases arise here (Celik 1998, Morgan 1996).

$$E_{PV} + E_{WG} < E_L \tag{2}$$

$$E_{PV} + E_{WG} > E_L \tag{3}$$

In the first case, equation 2, if there is battery storage in the hybrid system and it is fully charged, the energy deficit is supplied by the battery. If there is no battery storage or the battery is completely discharged the load is not met. Also, in the second case, equation 3, if there is battery storage and the battery is not fully charged, the surplus energy is stored in the battery. If the battery is fully charged or there is no battery the excess energy is dissipated. Equation 2 describes an undersized system whilst equation 3 is a typically oversized system.

According to (Celik 1998, 2002, Morgan 1996, Protogeropoulos 1992, Protogeropoulos et al 1997), the photovoltaic fraction fPV and wind fraction fWG are given by the following:

$$f_{PV} = \frac{E_{PV}}{E_T} \qquad (4)$$

$$f_{WG} = \frac{E_{WG}}{E_T} \qquad (5)$$

The total energy (E_T) is determined from the photovoltaic energy and wind energy (E_{PV} and E_{WG}) respectively.

$$E_T = E_{PV} + E_{WG} \qquad (6)$$

And the following is written: $f_{PV} + f_{WG} = 1$ (7)

Where: $f_{pv} = 1$ corresponds to a single photovoltaic power supply system, in which all the energy is contributed by the photovoltaic power supply system; $f_{pv} = 0$ corresponds to a single wind energy conversion system.

The remaining combinations correspond to a hybrid wind/solar system. f_{PV} solar fraction varies from 1 to 0 (1.0, 0.9, ……..,0.1, 0.0).

3.2 Autonomy definition

The system performance is defined by the term 'autonomy', or the system autonomy level A, which is the ratio of the total number of hours in which loss-of-load (H_{LOL}) occurs to the total hours of operation (H_{tot}), given by:

$$A = 1 - \frac{H_{LOL}}{H_{tot}} \qquad (8)$$

For annual continuous operation, H_{tot} = 8 760 h (Celik 2002, Protogeropoulos et al 1997).

4. Energy generated by wind turbines and photovoltaic panels and their corresponding areas

The wind energy calculations are made by matching the power–wind speed characteristics of commercial wind turbine machines with the hourly wind speed data. The energy calculations were made using bimodal wind speed distribution rather than Weibull distributions, otherwise the energy calculations would have been underestimated. In this study, the data of the year 2002 have been considered as a representative year for analysis (JMD 2003).

In this analysis, the Southwest wind power SW whisper 175 model is considered; it has a rated capacity of 3.0 kW.

The solar energy calculations are made by matching the hourly solar radiation data and a given photovoltaic array area together with the characteristics of photovoltaic modules.

HOMER models the photovoltaic array as a device that produces dc electricity in direct proportion to the global solar radiation incident upon it, independent of its temperature and the voltage to which it is exposed. HOMER calculates the power output of the photovoltaic array using the equation:

$$P_{PV} = f_{PV} Y_{PV} \frac{I_T}{I_S} \qquad (9)$$

Where, f_{PV} is the photovoltaic derating factor; Y_{PV} the rated capacity of the photovoltaic array (kW); I_T the global solar radiation (beam plus diffuse) incident on the surface of the photovoltaic array (kW/m^2); and IS is 1kW/m^2, which is the standard amount of radiation used to rate the capacity of the photovoltaic array (JMD 2003).

For the purposes of energy storage capability, conventional batteries are included in this analysis. Commercially available models, such as the Trojan Battery Company's Trojan L16P models (6V, 360Ah, 1075kWh) are considered in this analysis.

A power electronic inverter is needed to maintain flow of energy between the AC and DC components (JMD 2003, Khan et al 2005). For sizing the inverter, in addition to the average power consumption, peak power consumption needs to be considered. This is estimated by adding the wattages of the largest number of devices that are likely to be operated at the same time. The continuous rating of the inverter should be greater than this figure. Usually one needs 1.5-2 times the continuous rating (JATsAPC 2004, W&SLtd 2005).

The photovoltaic array and the wind turbine are not sized with respect only to the hourly load demand pattern. The monthly energy output from the photovoltaic array and wind turbine over a typical year, the weather data fluctuations and the overall conversion efficiencies are taken into account. In this way, three basic renewable component-sizing scenarios were tested. According to the three sizing scenarios 1, 2 and 3, photovoltaic and wind turbine areas are identified required to satisfy various criteria (Celik 1998, 2002, Morgan 1996, Protogeropoulos 1992, Protogeropoulos et al 1997). Table 1 shows the monthly energies produced by a wind turbine and a photovoltaic panel calculated from the Badia weather data and the required area to meet the load demand to explain the principles of scenarios 1, 2, and 3. The load demand that the system is sized to deliver is 3.22 kWh/d, where this load demand is the load of a typical family house in the Badia, which was measured by the author in the year 2004. The second and the third column show the monthly total specific energy produced by 1 m^2 wind turbines and photovoltaic panels respectively, calculated from the monthly average values of solar radiation and wind speed data for the study area. Column 4 shows the monthly load demand requirement. Considering only using photovoltaic panels in the system, the size of the photovoltaic panels in square metres that meets the load is given in column 6. Also, in the case of using a wind turbine only in the system the size in m^2 for meeting the load requirement is shown in the fifth column.

The total renewable area of the wind turbine and the photovoltaic panel required is given in column 7. Table 1 shows that the yearly average specific energies produced by the wind turbine and the photovoltaic array are 10.6 kWh.m^{-2} and 16.51 kWh.m^{-2} respectively. The corresponding areas of the wind converters and the photovoltaic array required to cover the load demand are 9.25 m^2 and 5.93 m^2 respectively; figures used in scenario 1 to size the system. January is the poorest solar month, producing a total energy of 8.467 kWh.m^{-2}. The poorest wind month, with a 0.96 kWh.m^{-2} total energy, is November. The worst solar month, January, requires a photovoltaic area of 11.8 m^2 to meet the load, while the worst wind month, November, requires a 100.0 m^2 area of wind turbine. Scenario 2, the worst month's scenario, considers these figures (11.8 m^2 photovoltaic areas and 100.0 m^2 wind turbines area) to size the system. The worst month, November, considered by scenario 3, requires the greatest combined photovoltaic and wind turbine area, with a total area of 108.63 m^2, of which 8.63 m^2 is photovoltaic area and 100.0 m^2 is wind turbine area. These figures were used to size the system in scenario 3.

The required monthly photovoltaic and wind turbine areas for the Badia weather data are shown in Figure 1. The 12-month picture shows the largest standard deviation of the wind turbine size and the small standard deviation of the photovoltaic array size. For example, April requires only 4.93 m^2 but November requires 100.0 m^2 wind turbine usage to meet the load. In contrast, the photovoltaic requirement varies very little from one month to the next. Therefore, sizing the system according to the worst month(s) sce-

Table 1. Monthly energies produced by the photovoltaic and wind turbine calculated from Azraq south weather data and the required areas to meet the load demand

Month	Specific energy by wind turbine (kWh/m²)	Specific energy by photovoltaic (kWh/m²)	Load demand (kWh)	Required wind turbine area (m²)	Required photovoltaic area (m²)	Total area (m²)
Jan	4.57	8.467	99.8	21.9	11.8	33.65
Feb	3.36	11.24	90.2	26.9	8.02	34.88
Mar	11.6	15.1	99.8	8.58	6.61	15.19
Apr	19.6	18.6	96.6	4.93	5.19	10.12
May	18.5	21.18	99.8	5.41	4.71	10.12
Jun	18.7	22.95	96.6	5.16	4.21	9.37
Jul	14.1	23.26	99.8	7.07	4.29	11.36
Aug	17.9	22.33	99.8	5.57	4.47	10.04
Sep	8.41	18.78	96.6	11.5	5.14	16.63
Oct	4.5	16.09	99.8	22.2	6.2	28.36
Nov	0.96	11.2	96.6	100	8.63	108.6
Dec	4.83	8.979	99.8	20.7	11.1	31.78
Average	10.6	16.51	97.9	9.25	5.93	15.18

Figure 1. The required areas of photovoltaic and wind turbine for Azraq south weather and load data

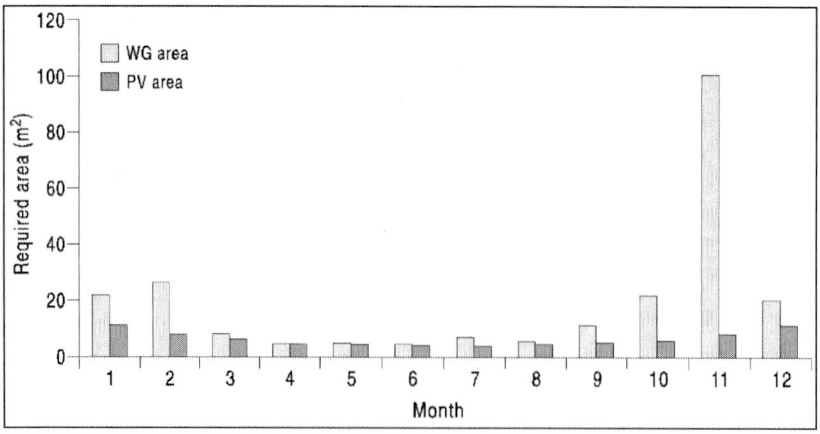

narios, the wind turbine converter is oversized for most of the year and the photovoltaic converter is reasonably sized due to the small monthly variation.

Table 2 shows the required area of photovoltaic and wind turbine corresponding to the three scenarios at different fractions of components. The areas of photovoltaic and wind turbine have been calculated for each solar fraction.

Up to now it was shown that the systems sized using scenario 1 require less renewable photovoltaic and wind turbine areas than those required using scenarios 2 and 3. But the system autonomy (which is the most important criterion for system sizing) has not yet been considered. Therefore, as one of the main design criteria, the system autonomies corresponding to different scenarios must be calculated (Celik 1998, 2002, Mor-

Table 2. Required areas of photovoltaic and wind turbine corresponding to different solar and wind fractions for Azraq south weather and load data

fPV	fWG	PVarea*	WGarea*	PVarea**	WGarea **	PVarea***	WGarea ***
1	0	5.93	0	11.8	0	8.63	0
0.9	0.1	5.337	0.925	10.62	10	7.767	10
0.8	0.2	4.744	1.85	9.44	20	6.904	20
0.7	0.3	4.151	2.775	8.26	30	6.041	30
0.6	0.4	3.558	3.7	7.08	40	5.178	40
0.5	0.5	2.965	4.625	5.9	50	4.315	50
0.4	0.6	2.372	5.55	4.72	60	3.452	60
0.3	0.7	1.779	6.475	3.54	70	2.589	70
0.2	0.8	1.186	7.4	2.36	80	1.726	80
0.1	0.9	0.593	8.325	1.18	90	0.863	90
0	1	0	9.25	0	100	0	100

* scenario 1; ** scenario 2; *** scenario 3.

gan 1996). Simulations by HOMER software have been done using hour-by-hour weather and load data to determine the system autonomies according to the calculated areas for the photovoltaic and wind turbine given in Table 2. Three different battery:load ratios are used; 1.0, 2.0, and 3.0-day storage capacity. HOMER mixes and matches these components to find what combination satisfies the load requirements for the planning area at the highest autonomy level and lowest net present cost (Celik 1998, 2002, Kamel et al 2005, Morgan 1996).

5. Results and discussion

5.1. Annual systems autonomy simulations according to different scenarios

The system autonomies and the coverage of load for three different battery:load ratios according to scenarios 1, 2 and 3 are given in Table 3, 4 and 5. The total coverage of load (CoL) of the above systems was calculated for all solar fractions. It is expected that scenario 2 systems will have the highest CoL figures, >>1.0, followed by scenario 3 systems. Scenario 1 systems are expected to have CoL figures close to unity since they are based on the annual average figures of Table 2.

Figure 2 shows the coverage of load at different solar fractions for the three scenarios. It can be seen from the same figure that there is a large variation in the total CoL value over the year. Scenario 1 CoL figures indicate that at all solar fractions, the CoL show a constant curve, around 1.0, due to sizing according to annual average energy and load demand. Also, as seen from the same figure, the CoL figures increase from solar fraction 1.0 (photovoltaic alone) to solar fraction 0.0 (wind alone) according to scenarios 2 and 3. This is due to sizing using the worst renewable months and worst renewable month, where the wind resource is very poor, making the system oversized in order to cope with the load demand.

It is clear from the trends in Figure 2 that overall the coverage of load is closest to unity for scenario 1 systems. For 45.5 % of the time however, the CoL is less than unity, ie the load is not being met and the system is undersized. This leads to the low autonomy values shown in Table 3. Conversely scenario 2 and 3 systems will, for the majority of the time, be oversized leading to higher autonomy values.

Sizing systems based on scenario 1 give on average a total coverage of load lower

Table 3. Total energy production, energy-to-load ratio and yearly system autonomy for varied solar fractions at different battery:load ratios according to scenario 1

Solar fraction	Total energy production (kWh)	Energy-to-load ratio	Autonomy (%)		
			BLR=1.0	BLR=2.0	BLR=3.0
1	1 160	0.98	67	79	80
0.9	1 266	1.07	76	87	88
0.8	1 150	0.97	73	80	81
0.7	1 255	1.06	80	86	87
0.6	1 139	0.96	76	80	81
0.5	1 245	1.05	79	84	84
0.4	1 129	0.96	75	79	80
0.3	1 235	1.05	75	80	81
0.2	1 119	0.95	71	75	77
0.1	1 225	1.04	70	75	77
0	1 330	1.13	69	75	76

Table 4. Total energy production, energy-to-load ratio and yearly system autonomy for varied solar fraction at different battery:load ratios according to scenario 2

Solar fraction	Total energy production (kWh)	Energy-to-load ratio	Autonomy (%)		
			BLR=1.0	BLR=2.0	BLR=3.0
1	2 320	1.97	81.7	98.6	100
0.9	6 887	5.86	96	99.5	100
0.8	11 455	9.74	96.1	99.4	100
0.7	11 223	9.55	95.6	99.3	100
0.6	15 790	13.43	95.3	99.2	100
0.5	15 558	13.24	94.6	99	99.8
0.4	20 127	17	93.7	98.8	99.7
0.3	24 693	21	92.4	98	99.6
0.2	24 461	20.8	89.9	96.5	99.2
0.1	29 029	24.7	87	94.4	98
0	28 797	24.5	83.3	91.5	96.2

Table 5. Total energy production, energy-to-load ratio and yearly system autonomy for varied solar fractions at different battery:load ratios according to scenario 3

Solar fraction	Total energy production (kWh)	Energy-to-load ratio	Autonomy (%)		
			BLR=1.0	BLR=2.0	BLR=3.0
1	1 524	1.29	73.6	94.6	97.5
0.9	6 167	5.24	94.4	99.1	99.8
0.8	10 817	9.2	94.6	99.1	99.8
0.7	10 668	9	94	98.9	99.7
0.6	15 310	13	93.5	98.7	99.7
0.5	15 161	12.9	92.6	98	99.6
0.4	19 812	16.8	91.4	97.5	99.4
0.3	24 453	20.8	89.8	96.4	99.2
0.2	24 312	20.6	88	95	98.6
0.1	28 946	24.6	85.8	93.4	97.4
0	28 797	24.5	83.3	91.5	96.2

Figure 2. Coverage of load at different solar fractions for the three scenarios

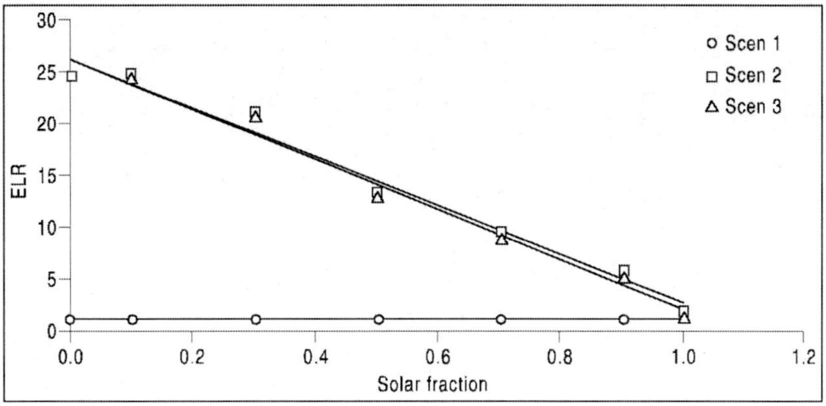

than scenario 2 and 3 systems. This means that less renewable hardware is needed to cover the load energy demand. This preliminary stage of system sizing does not however indicate the loss of load. Therefore, analysis using the yearly system autonomy is the next criterion to determine the best sizing scenario. The yearly system autonomy is treated as a further design criterion in this sizing and optimization technique (Morgan 1996, Protogeropoulos et al 1997). The yearly system autonomies for varied solar fractions at different battery:load ratios according to scenarios 1, 2 and 3 are shown in Figures 3, 4, and 5 which reveal there is a rapid increase in autonomy between the 1-day and 2-day battery capacities and a very small increase between the 2-day and the 3-day battery capacities. As Figure 3 shows for battery:load ratio (BLR) = 1, the achieved system autonomies vary from about 67 % for solar fraction 1 (photovoltaic alone) to about 80 % for solar fraction 0.7 (hybrid system). It is observed that scenario 1 systems are limited to autonomy values of less than 88 % for 3-day storage capacity. Furthermore, additional storage for these systems does not always increase the system autonomy. Also, for the 2- and 3-day storage the system autonomy decreased as the solar fraction moved towards 0 (ie wind turbine alone), because of the low wind resource in Azraq south. In general, if a system were expected to provide an autonomy figure higher than 88 %, then scenario 1 would not be the right sizing methodology. Celik (1998) showed that scenario 1 systems are limited to autonomy values of less than 86.0 %.

Moving to those systems sized according to the worst renewable months, scenario 2, the increase in autonomy is obvious. Figure 4 shows the yearly system autonomy for varied solar fraction at different battery:load ratios according to scenario 2. A comparison between Figures 4 and 3, shows the system autonomy for scenario 2 is higher than that in scenario 1. Because scenario 2 takes the most pessimistic view (the worst solar and wind months), systems sizing based on scenario 2 offer the highest autonomy values. For example, scenario 2 systems provide nearly 99.3 % autonomy on average, peaking at 100 % for solar fraction = 1.0 (photovoltaic alone) and 0.9, 0.8, 0.7 and 0.6 (hybrid system) for the largest battery capacity.

Scenario 3 considers the worst renewable month case. Figure 5 shows there is not a big difference between scenario 2 and scenario 3. The reason is that the combined worst solar and wind month is somewhat the same as the worst solar and wind months in scenario 2. Scenario 3 takes a slightly more optimistic approach compared to scenario 2. From Figure 5 a series of autonomy maxima are observed similar to those observed with scenario 2 systems. Considering the worst combined month instead of the worst months,

Figure 3. Yearly system autonomy for varied solar fractions at different battery:load ratios according to scenario 1

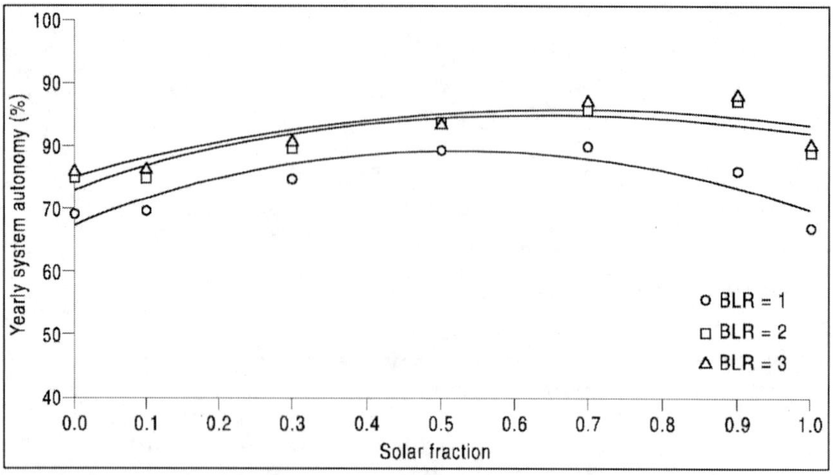

Figure 4. Yearly system autonomy for varied solar fractions at different battery:load ratios according to scenario 2

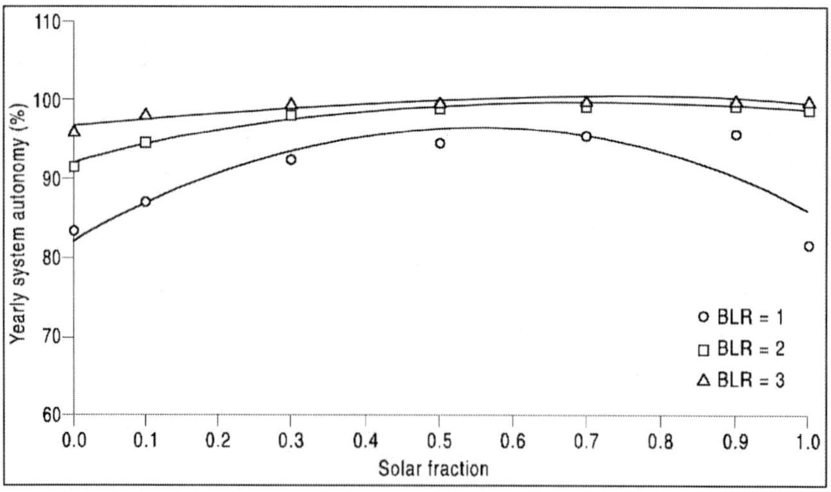

scenario 3 systems provide an average autonomy figure of 98.8 % for the biggest battery capacity, slightly lower than scenario 2. The highest autonomy value of scenario 3 is 99.8 % as seen in Figure 5 for the solar fraction 0.9 (hybrid system).

In a system, if an autonomy level of 75 % is regarded as satisfactory, scenario 1 with a large battery capacity is preferable to scenario 2 and 3 with a small battery capacity. It has been shown that the systems based on scenario 2 give higher autonomies than those based on scenarios 1 and 3. But the system cost (which is another important criterion for system sizing) has not yet been considered. Therefore, as one of the main design criteria, the system costs corresponding to different scenarios must be calculated. Simulations by HOMER software have been done using hour-by-hour weather, load and cost

Figure 5. Yearly system autonomy for varied solar fractions at different battery:load ratios according to scenario 3

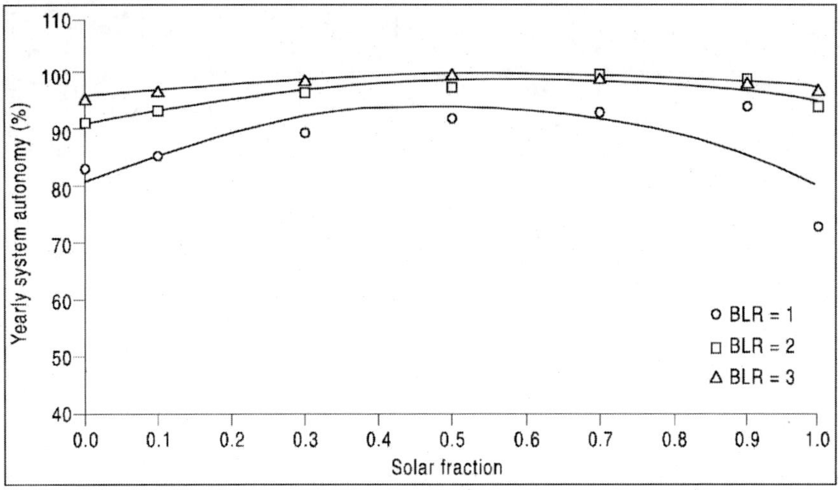

data to determine the system cost according to the calculated areas for the photovoltaic and wind turbine given in Table 2. Three different battery:load ratios are used; 1, 2, and 3-day storage capacity.

5.2. Comparison between different scenarios regarding the system cost

It has been shown that the single photovoltaic energy system in Azraq South, in the Badia, provides a higher autonomy level than the single wind and hybrid energy systems at the optimum design point, which have energy:load ratio (ELR) 1.5-2.0 for any BLR (Celik 1998, 2002). However, the autonomy analysis must be considered with the system cost, the second most important design criteria after autonomy level. As system autonomy is the first priority in most cases, an oversized system is usually designed in order to keep the loss of load probability to extremely low levels. System economics then play a secondary role in the overall system design method (Celik 1998, 2002, Morgan 1996, Protogeropoulos 1992, Protogeropoulos et al 1997). The optimum autonomous system depends on both autonomy and economic considerations and may be defined as the most economic system for a certain autonomy level requirement (Protogeropoulos 1992).

The three different scenarios for sizing renewable systems are compared here with regard to the system cost. As was shown, different scenarios return quite different system performance figures, especially scenario 1. Therefore, a comparison between scenarios regarding system cost would not provide an objective judgment unless the system performance figures are equal for different scenarios (Celik 1998).

The costs per photovoltaic panel, wind generator, inverter and battery charge controller have been given. The characteristic parameters for the lead-acid battery are also added into the cost of the system. The components capital costs refer to 2006 prices. In calculating the cost of the system a 20-year lifetime and a 10 % discount rate are also assumed. By using a capital component cost data base, the total net present costs are calculated for each system size by HOMER for the three different scenarios, with an hourly time step, as the battery size is varied. The results are shown in Tables 6, 7 and 8.

Table 6. Total net present cost and cost of energy for varied solar fractions at different battery: load ratios according to scenario 1

Solar fraction	BLR = 1.0 TNPC ($)	BLR = 1.0 COE ($/kWh)	BLR = 2.0 TNPC ($)	BLR = 2.0 COE ($/kWh)	BLR = 3.0 TNPC ($)	BLR = 3.0 COE ($/kWh)
1	8 478	1.22	8 666	1.08	8 898	1.1
0.9	8 588	1.12	9 044	1.03	9 278	1.04
0.8	7 911	1.07	8 181	1.02	8 536	1.05
0.7	8 301	1.03	8 653	0.995	8 918	1.01
0.6	7 612	0.994	7 917	0.978	8 299	1.01
0.5	7 954	1	8 271	0.983	8 681	1.02
0.4	7 259	0.967	7 554	0.955	8 062	1
0.3	7 598	1	7 907	0.984	8 444	1.03
0.2	6 925	0.973	7 282	0.962	7 825	1.01
0.1	7 307	1.03	7 664	1.01	7 207	1.06
0	7 697	1.11	8 037	1.07	8 580	1.12

Table 7. Total net present cost and cost of energy for varied solar fractions at different battery: load ratios according to scenario 2

Solar fraction	BLR = 1.0 TNPC ($)	BLR = 1.0 COE ($/kWh)	BLR = 2.0 TNPC ($)	BLR = 2.0 COE ($/kWh)	BLR = 3.0 TNPC ($)	BLR = 3.0 COE ($/kWh)
1	14595	1.78	15171	1.53	15392	1.54
0.9	20494	2.13	20784	2.08	21237	2.12
0.8	26790	2.78	27075	2.72	27556	2.75
0.7	25556	2.67	25848	2.6	26327	2.63
0.6	31861	3.34	32159	3.23	32647	3.26
0.5	30623	3.23	30932	3.11	31418	3.14
0.4	36923	3.93	37241	3.76	37738	3.78
0.3	43215	4.67	43545	4.44	44057	4.42
0.2	41953	4.66	42305	4.38	42828	4.31
0.1	48250	5.53	48620	5.14	49148	5
0	47062	5.64	47454	5.18	47919	4.97

Table 8. Total net present cost and cost of energy for varied solar fractions at different battery: load ratios according to scenario 3

Solar fraction	BLR = 1.0 TNPC ($)	BLR = 1.0 COE ($/kWh)	BLR = 2.0 TNPC ($)	BLR = 2.0 COE ($/kWh)	BLR = 3.0 TNPC ($)	BLR = 3.0 COE ($/kWh)
1	10 233	1.38	10 901	1.15	11 165	1.14
0.9	16 657	1.76	16 972	1.71	17 420	1.74
0.8	23 398	2.47	23 707	2.39	24 183	2.42
0.7	22 597	2.4	22 913	2.31	23 389	2.34
0.6	29 297	3.13	29 619	3	30 106	3.01
0.5	28 490	3.07	28 820	2.93	29 312	2.94
0.4	35 227	3.84	35 569	3.64	36 074	3.62
0.3	41 914	4.66	42 266	4.38	42 790	4.3
0.2	41 151	4.67	41 516	4.36	42 043	4.26
0.1	47 824	5.57	48 202	5.15	48 713	5
0	47 062	5.64	47 454	5.18	47 919	4.97

Figure 6. According to scenarios 1, 2, and 3, total system cost for varying solar fractions at battery:load ratio 3

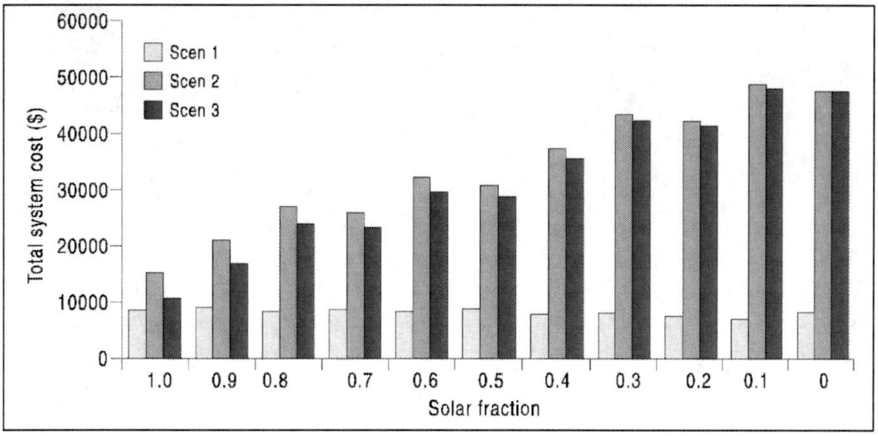

Figure 6 shows the system cost according to three different scenarios for varying solar fractions for the largest battery capacity. Considering only the largest battery capacity figures shown in Figure 6, scenario 1 systems costs are somewhat similar for the single photovoltaic, wind and hybrid energy systems. It may be observed that scenario 2 and 3 systems show a tendency to an increasing system cost from the pure photovoltaic system (solar fraction = 1.0) to the pure wind generator system (solar fraction = 0.0) due to the large extra area required for wind turbine sizing according to scenarios 2, and 3.

Thus, according to scenarios 2, and 3, a pure photovoltaic system costs about $15 392 and $11 165 respectively, while the pure wind turbine system cost $47 919 for the two scenarios. Overall, scenario 1 offers the least costly system at all the solar fractions compared to scenarios 2 and 3, because systems have been sized according to the annual average energy and load demand, while in scenarios 2 and 3 the systems have been sized according to the worst renewable months and worst month respectively. For scenarios 2 and 3 the photovoltaic system alone (solar fraction = 1) is less costly than the wind system alone and the hybrid systems for all the time. For low solar fraction systems the cost of scenario 2 and 3 systems is similar, although the discrepancy increases as the solar fraction increases.

5.3. Techno-economic analysis

5.3.1. Performance-total cost analysis for hybrid and single systems for different scenarios

It has been shown that systems sized according to scenarios 2 and 3 provide more autonomy than scenario 1, and the stand-alone photovoltaic systems provide more autonomy than the stand-alone wind and hybrid systems at the same ratio of energy-to-load. This is, however, an incomplete analysis as it does not take into account the system cost. The basic optimization criterion is the lowest system cost with simultaneous consideration of autonomy requirements as the first and basic design criterion. Figure 7 shows cost versus autonomy for largest battery capacity; each data point, of a total of three, along the lines is determined by a different sizing scenario. The first point is by the yearly average scenario, scenario 1, which sizes the systems at a point where the energy:load ratio (ELR) is equal to unity. The second point (scenario 3) is the worst month scenario. This scenario chooses the worst month in which the largest total area of photovoltaic

Figure 7. Yearly system autonomy versus cost for single photovoltaic, single wind and hybrid energy systems for the BLR of 3.0

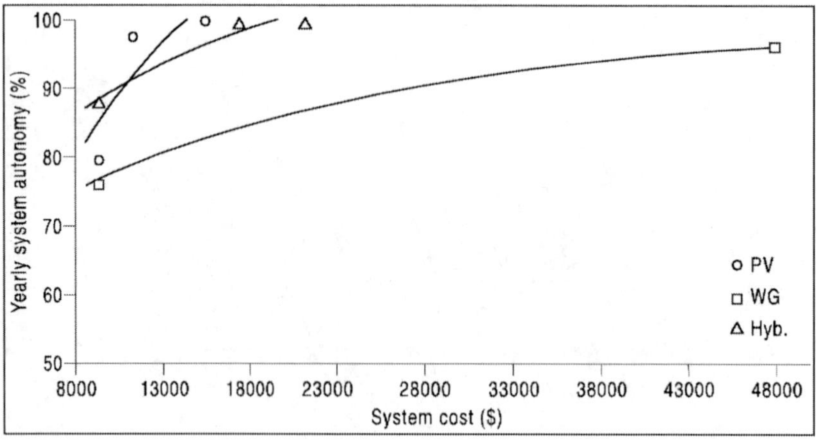

and wind generator is needed. The third point (scenario 2), is the worst month's scenario. The worst months are the ones in a year that require the largest photovoltaic and wind turbine sizes to meet the load (Celik 2002).

Figure 7 shows the yearly autonomy levels and the corresponding system costs for the pure photovoltaic array, pure wind, and for one of the near optimal combinations of the hybrid energy systems, for the battery:load ratio of 3.0.

For this battery capacity and at the same system cost, the pure photovoltaic system returns the highest annual autonomy values and proves most optimal in terms of the performance-cost relationship. For example, looking at Figure 7, for the cost of $13 000, the photovoltaic system provides nearly 100 % yearly autonomy, while the hybrid system provides 95 % and the single wind system provides only 82 %. The single photovoltaic system returns higher autonomy figures than the single wind system for the same system cost. For example, the 90 % autonomy level is achieved with a $10 000 system cost for the photovoltaic and hybrid systems, while the same system autonomy can be achieved with $28 000 for the wind system. Very importantly, Figure 7 shows that achieving a further small increase in the system autonomy means a sharp increase in the system cost, especially for the wind systems. For example, the wind system can achieve an 85 % yearly autonomy level with a system cost $18 000, but an extra 10 % autonomy requires an extra $25 000 investment.

6. Conclusions

The potential of wind and solar energy in the north-east part of the Badia, and a developed technique for finding the optimum wind / solar / battery combination of a hybrid system for a typical house in the Badia were investigated.

Solar resources in the north-east Badia bear excellent potential compared to wind energy and utilization of wind resources which might not be cost effective in most cases.

In view of ample monthly average daily wind speed (2.12-5.79 m/s) and global solar radiation (2.81-7.87 kWh/m^2) the study indicates that the Badia is a suitable candidate for development of stand-alone photovoltaic/battery systems for remote applications. It has been observed that the wind energy conversion system alone and the hybrid systems cannot cope with the load on a 24 h basis due to the low wind resource. The study shows that the photovoltaic system alone with battery storage has the highest autonomy level and minimum system cost compared with wind system alone and hybrid energy systems, according to different sizing scenarios. The photovoltaic system alone proves to be techno-economically better than either the single wind system or the hybrid energy systems. The principal reason why the single photovoltaic system is cheaper and has a higher autonomy level than the wind system or the hybrid systems is the fact that the energy generated by the photovoltaic system alone can be matched more closely to the load.

The photovoltaic system sizing according to scenario 2 and 3-day battery storage proves to be techno-economically better than either of the single wind and hybrid energy systems for the 100 % autonomy level required in the Badia.

The work presented here can be used as a planning tool by local and national governments or their agencies to look at levels of investment and the corresponding electrical power autonomy for users.

Acknowledgements

This work was a part of a PhD study at Coventry University supported by the Jordan Badia Research and Development Centre.

References

Ammari HD and Al-Maaitah A (2003). Assessment of wind-generation potentiality in Jordan using the site effectiveness approach. Energy 28:1579-1592.

Borowy BS and Salameh ZM (1994). Optimum photovoltaic array size for a hybrid wind/PV system. IEEE, Transactions on Energy Conversion 9(3):482-488.

Borowy BS and Salameh ZM (1996). Methodology for optimally sizing the combination of a battery bank and PV array in a wind/PV hybrid system. IEEE, Transactions on Energy Conversion 11(2):367-373.

Celik AN (1998). The system performance and sizing of autonomous photovoltaic, wind and the hybrid energy systems. PhD Thesis, Division of Mechanical Engineering and Energy Studies, University of Wales, Cardiff.

Celik AN (2002). Optimization and techno-economic analysis of autonomous photovoltaic-wind hybrid energy systems in comparison to single photovoltaic and wind systems. Energy Conversion and Management 43:2453-2468.

Chedid R, Akiki H and Rahman S (1997). A decision support technique for the design of hybrid solar-wind power systems. IEEE, Transaction on energy conversion 13 (1):76-82.

Elhadidy M (2002). Performance evaluation of hybrid (wind/solar/diesel) power systems. Renewable Energy 26 (3):401-413.

Elhadidy M and Shahid S (2000). Parametric study of hybrid (wind + solar + diesel) power generating systems. Renewable Energy 21 (2):129-139.

HOMER V.2.09, National Renewable Energy Laboratory (NREL), 1617 Cole Boulevard, Golden, Colorado 80404-3393. http://www.nrel.gov/homer, accessed 20/10/2004.

Hrayshat Eyad S (2005). Wind availability and its potential for electricity generation in Tafila, Jordan. Renewable and Sustainable Energy Reviews 9(1):111-117.

JATsAPC, Determining Your System Requirements. http://www.jatsgreenpower.com/sizing.htm, accessed 21/11/2004.

JMD (2003). Internal report, Jordan Meteorological Department, Amman, Jordan.

Kablan MM (2004). Techno economic analysis of the Jordanian solar water heating system. Energy 29(7):1069-1079.

Kamel S and Dahl C (2005). The economics of hybrid power systems for sustainable desert agriculture in Egypt. Energy 30(8):1271-1281.

Khan MJ and Iqbal MT (2005). Pre-feasibility study of stand-alone hybrid energy systems for applications in Newfoundland. Renewable Energy 30(6):835-854.

MEMR (2003). Annual Report, Ministry of Energy and Mineral Resources, Amman, Jordan

Morgan TR (1996). The performance and optimization of autonomous renewable energy systems. PhD Thesis, University of Wales, Cardiff.

NERC (2003). Files, National Energy Research Centre, Amman, Jordan.

Protogeropoulos C (1992). Autonomous Wind/Solar Power Systems with Battery Storage. PhD thesis, University of Wales College of Cardiff.

Protogeropoulos C, Brinkworth BJ and Marshall RH (1997). Sizing and techno-economical optimization for hybrid solar photovoltaic/wind power systems with battery storage. International Journal of Energy Research 21:465-479.

RSS (1983a). Assessment and analysis of available energy resources, vol. (III), in the potential of solar energy application in Jordan. Internal report, Royal Scientific Society, Amman, Jordan.

RSS (1983b). Possible application of solar energy in Jordan, vol. V, in the potential of solar energy application in Jordan. Internal report, Royal Scientific Society, Amman, Jordan.

RSS (1988). Utilization wind energy and solar energy systems to supply the basic energy needs to the rural areas in Jordan. Internal report, Royal Scientific Society, Amman, Jordan.

Shaahid SM and Elhadidy MA (2004). Prospects of autonomous/stand-alone (photovoltaic+diesel+ battery) power systems in commercial applications in hot regions. Renewable Energy 29 (2):165-177.

Tay LH, Keerthipala WWL and Borle LJ. Performance analysis of a wind/diesel/battery hybrid power system. http://www.itee.uq.edu.au/~aupec/aupec01/129_TAY_AUPEC01, accessed 25/3/2004.

W&S Ltd. Hybrid systems. http://www.windandsun.com/wind_sys.htm, accessed 19/5/2005.

CHAPTER 3

Integrated studies of the Azraq basin for optimum utilization of the natural resources

Bassam Sunna'
Private Consultant and National Coordinator of Azraq Project, P.O. Box 950693, Amman 11195, Jordan.
E-mail: bassamsunna@yahoo.com

As an outcome of the Third Scientific Week held in 1995 under the umbrella of the Higher Council for Science and Technology, convened under the theme of 'Earth resources: towards optimal exploitation', the major national institutions involved in earth resources were drawn upon for proposals on vital research areas. The purposes of this were to encourage cooperation, coordination and integration towards problem solving in applied and economically vital fields. The project entitled 'integrated studies of the Azraq basin for optimum utilization of the natural resource' was one of the selected projects.

In order for the project to reach the required goals in support of the sustained and rational planning, development and management of the natural resources, the project was undertaken in accordance with the following principles:

- Integration of disciplinary activities;
- Integration of base mapping with the resources inventory;
- Integration of resource evaluation in order to prepare the thematic and land use maps which are the ultimate goal of the project.

There are huge resources of industrial minerals in the Azraq basin; such minerals are a better source of regional wealth for the country than the scarcer metal deposits. Also in the Azraq basin, groundwater is one of the main sources of water supplying Amman. Detailed assessments of these resources were implemented. The ability to find new reserves of oil and gas is an essential requirement for the country's economic future. The investigation and development of alternative energy resources are also vital. The project concentrated on the evaluation of the hydrocarbon potential in the basin and on providing a national overview of the hydrocarbon potential of individual fields. Agriculture and environment are also covered in the paper.

Keywords: Jordan Badia, Azraq basin, geomorphology, energy, minerals, water, agriculture, maps

1. Aims of the paper

The aims of this paper summarizing the outcomes of the project are:

- To show the value of the integration between various disciplines for land survey and land resources appraisal and for developing these resources in the exploration, evaluation, and exploitation phases;
- To show the methodology applied in order to better evaluate the natural resources;
- To develop new exploration strategies regarding natural resources in the basin as a pilot area for the remaining basins in Jordan.

2. Location of the Azraq basin

The Azraq basin is located within the north-east Badia of Jordan (Figure 1) between the following coordinates: 36°-38° E and 30° 30′ - 33° N (or, using the Palestine grid, 250-400 E and 055-230 N).

The Azraq basin has a surface area of about 17 000 km^2 distributed as follows: 15 745 km^2 in Jordan (92.13 %); 857 km^2 in Syria (5 %); 488 km^2 in Saudi Arabia (2.87 %).

3. Physiography

The highest elevation point in the whole basin is 1 234 m a.s.l. in the Umm al-Quttayn area and the lowest elevation point is 506 m a.s.l close to Azraq.

Two major geomorphological divisions were identified. Each was sub-divided into different zones which were delineated according to their stratigraphy, structural evolution, types of deformation, and drainage pattern, landform, fracture patterns and image tonal characteristics. The two main divisions are: the volcanic rocks (basalt plateau sub-province); and the sedimentary rocks.

The basin is bounded from the north by basalt hills extending to Jabal Druz in Syria, this basalt represents the main recharge area in the basin. The central part of the basin which is locally known as Qa'a Azraq (a playa) represents the collecting area for basin drainage. It is composed of alluvial deposits, which are partially intercalated with the sabkha deposits (Sunna 1999a).

Figure 1. Location map of the Azraq basin (after Sunna 1999d)

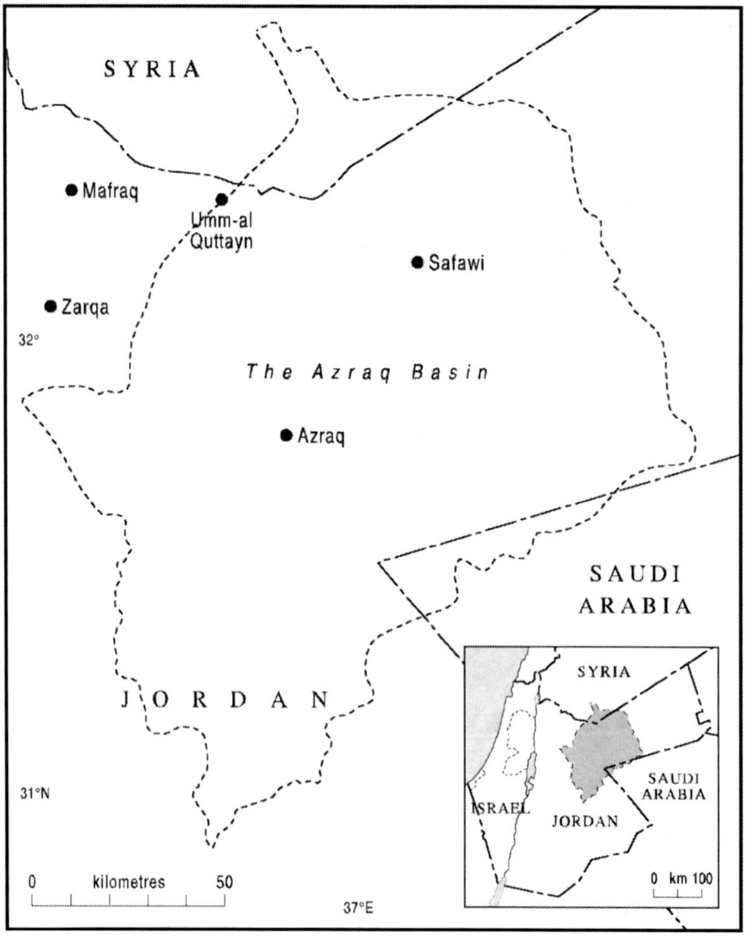

4. Stratigraphy of the Azraq basin

The basin area incorporates exposures of sedimentary rocks and basalts, ranging in age from Cretaceous to Quaternary (Sunna 1999c). In the southern part of the basin and on the surface, the Quaternary deposits and recent sediments cover the underlying Tertiary deposits. The latter are intermittently exposed at the surface in the south, south-west and south-east parts. Eocene and top tertiary sediments lie on the top of the sequence beyond the 'Fuluq' fault to the south and south-east, at Wadi Hazim and Jabal al-Fuluq. The sedimentary sequence includes limestone, chert, marl, chalk, sandstone, clay and evaporites. These rocks are frequently covered with a variably thick sequence of superficial deposits including alluvium, mud-silt flats, chert pavement, Pleistocene gravels, and sand and evaporite incrustations.

To the north and north-east the basalt eruptions of different ages appear on the surface and extend northwards to cover a wide area known as the 'basalt plateau'. This basalt is related to the North Arabian Volcanic province, which extends from Syria across Jordan into Saudi Arabia, covering in Jordan an area of 11 000 km^2.

In the subsurface a thick sedimentary section that varies in thickness as well as varying in the lithostratigraphic and formational units represents the Azraq basin. These sediments range in age from early Palaeozoic to Pleistocene and are primarily composed of carbonates, sandstones and shales. The area was the subject of extensive oil exploration activities, which added a lot of information regarding the stratigraphic sequence and sedimentary section.

The major thickness reduction in the sequence appears towards the south and south-west, while a remarkable increase in thickness is observed east of Azraq village towards the Fuluq fault.

5. Structural set up

The principal and most important structural characteristic of the Azraq basin is the widespread and intense faulting. Fault tectonics have been active in the basin area from the Precambrian to the present. Many prominent regional and local structural features control the Azraq basin. However, the tectonics of a basin are closely related to the regional tectonics of Jordan oriented in four principal directions (Sunna 1999b):

Set A	N 15° E to N 15° W	Generally referred to as N-S
Set B	N 30° E to N 60° E	Generally referred to as NE-SW
Set C	N 75° E to N 105° E	Generally referred to as E-W
Set D	N 30° W to N 60° W	Generally referred to as NW-SE

In discussing the new concept concerning the geological model of the basin it is of great importance to differentiate between two important features, which are: the Azraq basin and the Azraq depression (Sunna 1999a)

Azraq basin: The Azraq basin is a north-west trending topographic and structural depression positioned between the Jordan Arch-central plateau region on the west and the north-eastern plateau-Risha area on the east.

Azraq depression: The depression was formed as a result of four major fault systems and one graben crossing the whole Azraq basin and bordered by stable blocks on the east and west. The faults are:

Fuluq fault (NW-SE trend), from the east;
Sirhan fault system (NW-SE trend), from the west;
Baqa'- Wisad fault system (E-W trend), from the north;
Ghadaf - Makhruq Graben (E-W trend), from the south.

Generally speaking it can be stated that the Azraq depression is a fault-bounded major block with Mesozoic fill. The general characteristics of the depression and its major structural elements include the following (Figure 2):

- Azraq inner lake;
- Azraq central lake or outer lake;
- Rajil monocline;
- Al-Dahikiya structure;
- Hamzeh graben.

The location of this depression within the Azraq basin, with the presence of the major fault systems, resulted in separating the Azraq basin into several blocks surrounding the central depression. They are:

- The eastern block (Fuluq block);
- The southern block (Adh-Dhirwa block);
- The northern block (Safawi block);
- The western block (Muwaqqar block).

Figure 2. Geological model of the Azraq basin

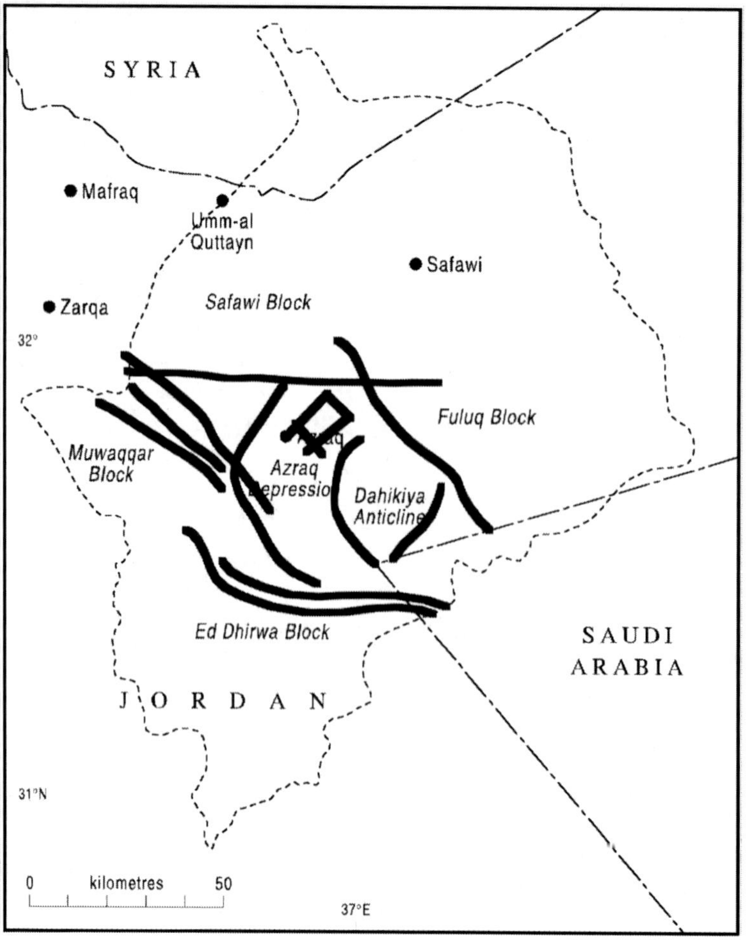

Gentle anticlinal – synclinal folds with various trending axial traces have been observed in the sedimentary rocks and also along some of the seismic lines in the basalt area. Many elliptical, oval and circular features were found representing one or other of: topographic highs; anticlines, synclines or domes; cryptovolcanic activity.

Two major anticlinal structures were identified: the Dahikiya anticline; the Fuluq anticline.

6. Selection of target areas

As a result of the geological studies several target areas for agricultural purposes and for ground water and oil exploration were delineated for further evaluation as shown in Table 1.

There are two types of target areas for further follow up. The first is areas of geomorphological anomalies which suggest that follow-up surveys will be worthwhile. They can be defined as: target areas for oil; or for water; or topographic highs; or cryptovolcanic activity.

The second is areas of opportunity, which are summarized below:

6.1 Target areas for hydrocarbons exploration

These can be categorized as:

- 'Excellent' areas of opportunity;
- 'Very good' (geomorphological anomalies);
- 'Good' (stratigraphic traps along the edge of the deep Jurassic sedimentary rocks);
- 'Fair' (the rest of the basin in particular NW and SW of the basin).

The following target areas were defined by the geological studies:

- Area Oil-1 Ghadaf – Makhruq graben;
- Area Oil-2 Safra area;
- Area Oil-3 Dahikiya area;
- Area Oil-4 Fuluq area.

Table 1. Target areas for resource exploitation

Sheet No	Geomorph. anomalies	Targets for agriculture	Targets for water	Targets for oil
1	7		W1	Oil-1
2	17	AGR1	W1 and W2	Oil-1
4	1			
5	11	AGR2	W3, W4, W5, and W6	Oil-1
6	23	AGR3	W6 and W7	Oil-2
7	15		W7, W8, W9, W10, W11, W12,W13	Oil-3
8	11		W14	Oil-4
9	1		W15	
10	8	AGR4	W15,W16, W17,W18	
11	3		W19, W20, W21, W22, W23, W24	
12	1		W25, W26	
13	1		W22	
14	4		W27, W28, W29	
Total	103	4	29	4

6.2. Target agricultural areas

Site AGR1 (Wadi Makhruq area): The surface is mostly covered with chert and is high in silt. Because of this initial assessment, it was decided not to study this area and instead study a different area as a substitute. However, because this area is a very large one, it is highly recommended that the area be studied in detail especially as there is a high possibility of finding groundwater close to it as reported by the geology team. The area is an ellipsoidal shape located between wadi al-Makhruqadi and wadi adh-Dhirwa Umm Luweiza. The elevation is 625-735m a.s.l.

Site AGR2 (East Muwaqqar area): This is one of the promising areas for potential agricultural development. Although the studied profiles contained soluble salts in their lower horizons, the salts could be leached with good quality water. Soils are deep and contain high clay amounts (Rawajfih et al 2000). It is recommended that this area be studied in detail by digging more profiles at shorter distances, and taking samples with the auger to make sure the samples represent well the total area. If the total area has similar properties, it would be one of the areas that justify investment of resources for sustainable agricultural development. Structurally the area is a synclinal block on controlled longitudinal plains. It is located WSW of Azraq (in the vicinity of Qasr Amra and Qasr al-Harraneh).

Site AGR3 (al-Mafarikh): The al-Mafarikh area has soils containing high gypsum and carbonates and also is very high in soluble salts. Some of the studied profiles contain very high soluble salts in the surface horizons (Rawajfih et al 2000). It is recommended that this area be excluded from any agricultural development. Any attempt to reclaim soils in this area would be too expensive and not feasible economically. It is an ellipsoidal shape SW of Azraq village and NW of Shoumari reserve. The satellite image shows that the area might contain gypsum mixed with clay in sporadically distributed parts.

Site Number AGR4 al-Ariyyatayn-Tell Hassan Area: The al-Ariyyatayn-Tell Hassan area is a very large area and is also very variable in characteristics. Because of that, one cannot generalize about the true potential of this area for agricultural development. It is recommended that the areas with profiles of good soil properties be studied in details to delineate the areas of promising potential for agricultural development. It is an area of plateaus partly covered with basaltic rock pieces and thick soil north and east of Tall Hassan, east of Tall ar-Rimah, and both sides of the tapline track road west and east of Baghdad Highway and Tall ar-Rimah (Rawajfih et al 2000).

7. Industrial rocks and minerals in the basin

The Azraq basin in north-east Jordan is considered rich in its industrial rocks and mineral resources (Khoury 2000). Most of the commodities are concentrated in one area, which is the Azraq mudflat (al-Qa'a) area. Currently some commodities are exploited in a hazardous manner. The results of this unplanned quarrying could be witnessed by the destruction of nature in the desert areas, where limestone concretions and pyroclastic deposits are mined.

Mudflat area: The mineral assemblages in the Azraq mudflat area (the Qa'a) are related in their mode of origin to a closed lake system (Khoury 2000). The lake has been affected by climatic changes (wet and dry), which have controlled the type of the sediments. The source and importance of these minerals are described below.

Bentonite: Smectite-rich beds (bentonite) are widely distributed in the mudflat and Ayn al-Badia areas. The quality of the deposits is below the bentonite specifications. The clays need to be beneficiated by the removal of the non-clay minerals and the separation of the less than two-micron-size fraction. Hydrocylon techniques proved to be an efficient process for the beneficiation of smectites (Khoury 2000).

Homogeneous continuous smectitic clays: Found in the Ayn al-Baida area, they are located beneath the overburden of 2-12 m and total about 105 million tonnes. The evaluation of the clay deposits from Ayn al-Baida as binding material in foundry sand gave acceptable results. The smectite clay from Azraq has proved its importance for use as an adsorbing material of heavy metals and as a feed for farm animals. In the Qa'a area is a smectitic (montmorillonite) rich bed of greater than 1 m thickness. The reserves in Qa al-Azraq total 1 040 million tonnes.

Palygorskite: The newly discovered palygorskite deposits in the Qa'a area proved to be of high quality and the preliminary investigations indicated its possible use in drilling mud, paints and pharmaceuticals in addition to its use as a filtering medium. They are found in two horizons: the first (5 m thick) appears at 5 m depth. The second (10 m thick) appears at 55 m depth. They were located in boreholes two and eight in the Ayn al-Baida and Qa'a areas in boreholes two and eight. Palygorskite is reported in DH-1, HZ-11and WR-4.

Diatomite: The Azraq depression is also characterized by the presence of huge diatomaceous earth deposits, thought to be similar to the Moler type. The use of the material as filter aids or functional fillers needs beneficiation to reach high–grade quality. The raw material however proved to be of great advantage for the treatment of industrial wastewater effluents. Two horizons were identified. The first is 4.5-31 m thick and the second 2-20 m thick. They are located beneath an overburden of, respectively, 11-42.5 m and 37-92.5 m (Khoury 2000). They are located in the Azraq depression, in particular the Azraq Lake. The estimated reserve in the first horizon is 1 042 million tonnes, and in the second 212 million tonnes.

Rock salt: Rock salt is found in most of the boreholes in the mudflat inter-mixed with clay of the Azraq formation. In the mudflat where the water table is shallow, the soluble ions reach a concentration of 30 %. The brine is pumped from shallow wells, less than 12 m deep, into evaporation pans. The development of rock salt production is a matter of production techniques and marketing.

The basalt area: The exposed basaltic areas are of good potential for quarrying to use as decorative stones, crushed aggregates, and roofing granules. Research and development in the fields of rock wool and cast basalt industries are prerequisites for the future planning and development of the Badia region.

Tuff and zeolite: The scoriaceous pyroclastic materials are widely distributed in the Badia area. The cement factories and the private sector in Jabal al-Ariyyatayn, Tall Hassan and Tall ar-Rimah widely and randomly exploit the material. The pyroclastic materials from these volcanoes proved to have excellent properties for use as lightweight aggregate and for cultivation purposes in addition to their use for the production of pozzolanic Portland cement (Khoury 2000). The hazardous mining process needs better organization with the objective of benefiting and developing the Badia region. The use of the pyroclastic material in cultivation needs special attention, and should be the subject of research in the country. Zeolites are widely distributed in two parts of the basin: the northern area, some 65.38 km^2 at 29 volcanic centres; and the north-eastern area (Tulul ash-Shahba), some 66.74 km^2 at 19 volcanic centres. The intensive research work undertaken on the zeolitic tuff from the Tall ar-Rimah and Jabal al-Ariyyatayn volcanoes, proved the possible use of the material for high purity table salt.

Areas of sedimentary rocks: Huge and unlimited reserves of porcelanite and chalk were found. The high purity of the porcelanite layers and the fine particle size of the silica mineral phases enable its use as an abrasive materiel and as filler in the paint, plastics, and rubber industries. The softness and high purity of the chalk from the studied areas enable its use as a filler and in lime production. The evaluation of the Azraq limestone as a decorative and building stone indicated that unlimited reserves are present. The best investigated varieties, which are of international standards for use as building and decorative stone, are the large concretions. Huge gypsum deposits are found in the Azraq mudflat area and extend to the south and south-east. Chert fragments are the undesired impurities so future mining should concentrate on selectivity and screening during the crushing stages to remove the undesired impurities. Calcination of gypsum is highly recommended for future industry.

8. Energy resources

The objectives of the energy sources study were to: review all the previous reports dealing with hydrocarbon possibilities in the Azraq basin, including the Hamzeh field; review the seismic data, reports and previous interpretations; reinterpret and give additional interpretation primarily for detecting structural traps; and recommend a suitable location of favourable structures for further exploration (Sunna et al 1999).

To fulfil these objectives a nationwide study was undertaken to review and evaluate the hydrocarbon evidence in the surface and subsurface horizons, including occurrences in the Hamzeh field, the Risha gas field, and in the Sirhan and Dead Sea regions.

In the Azraq basin the seismic maps provided by the Natural Resources Authority (NRA) were evaluated and the various reflections in the seismic lines were reinterpreted. The fault patterns at the different levels in the subsurface were checked in order to improve understanding of tectonic complexity. Finally, the energy resources in the Azraq basin were reviewed.

The study mainly focused on the structural anomalies and the areas of opportunity. Although it is evident that the other factors which help oil and gas accumulation (that is the source, cap and reservoir rocks in addition to the properties of the fluids) will contribute importantly to defining a target area to be explored, detailed discussion of these factors lay outside the remit of the present study. The locations that we define as the 'areas of opportunity' are those areas having the greatest oil potential, where more detailed study followed by expensive and detailed exploration activity should be focused. Generally speaking it can be stated that the two most important structures found to be attractive in the depression are:

- The Dahikiya double plunged anticline; and
- The Makhruq-Ghadaf graben located between the depression and the southern block.

The Dahikiya anticline (Oil-3 target area located on the Omary sheet) is a faulted single-plunged anticline located between the following coordinates: 105-130 N and 342-360 E (Palestine Grid).

The dip on both sides of the anticline is gentle with a relatively flat top forming a platform-like surface. It is located at the southern margin of the depression and is characterized on the seismic lines by broad anticlinal flanks. All of the anticlinal flanks are associated with faults to some extent.

Manual interpretation of the satellite image showed that the anticlinal shape is very clear with a major lineament along the axis of the anticline. The direction of this axis is north-east to south-west in its south-western side and bending in the centre of the anticline to the north and then to the north-west to south-east direction in the northern side. The Hamzeh field is located at the central part where the axis of the anticline has a north-south trend.

The primary target area is the Ghadaf-Makhruq graben (al-Kharqa graben). Most of the seismic lines named, SW 4-29, cross the Ghadaf-Makhruq Graben except lines 14, 16.

9. Geomorphological anomalies

The study identified several broad positive geomorphological anomalies. Some of them are either anticlines suitable for oil exploration or synclines suitable for water exploration. The anomalies are limited in size and need careful and more detailed studies prior to any drilling. They represent one or other of the following: target areas for oil or for water; topographic high; crypto-volcanic activity or volcanic centres.

Table 2 summarizes the anomalies identified during the course of the present work, including some identified by CoreLab (Sunna et al 1999).

10. The Azraq depression

A total of 34 wells have been drilled in the Azraq basin to evaluate the hydrocarbon potential of the basin. Most of them (17 wells) have been drilled in the Hamzeh field located along the west (down thrown) edge of the Fuluq fault. Only four wells produced oil: HZ-1, HZ-12, HZ-14, and HZ-17. A total of 28 oil wells were drilled in the depression: Hamzeh (Hz 1-17); W. Rajil (WR 1-4); W. Ghadaf (WG1 and 2); Heavy Oil (HO 1 and 2); Azraq (AZ1); Dahikiya (DH1); Adla (AD1).

A petroleum exploration programme undertaken by NRA (1980 to 1996) discovered hydrocarbons in Hamzeh field in the Azraq basin in 1983. Hydrocarbon production from the field commenced in January 1985 from Hz-1 from the Hummar formation. Wells HZ-12 and HZ-14 were brought into production in 1986 from the Shueib and Hummar formations, respectively. In September 1986, well HZ-14 was dually completed in both formations. The seismic data, with 7 174 km lines, cover the Azraq area.

Medium gravity (API 30°), low sulphur oil is produced from four wells (HZ-1, HZ-12, HZ-14 and HZ-17).

Local production of oil during the years 1997-2001 was 1.6-1.9 thousand tonnes (MEMR 2000).

However, evaluation of the data available showed that some of the drilled wells, such as WG1 and WR1, were in unfavourable structural positions, and some of the others, such as WH1, which seem to be in a favourable position, were located very close to the big Jabal Fuluq fault. Thus, many of the wells were not ideally sited. It is therefore important to conduct new drilling tests in the relatively large promising areas situated in the Azraq depression. These may be divided into a number of blocks (Sunna et al 1999):

Table 2. Geomorphological anomalies

Province	Sheet name and no	No of anomalies
Depression	Azraq–6	9
	Omari 7	3
Eastern block	Azraq–6	2
	Azraq–6	12
Northern block	9 Zarqa	1
	Umm al-Quttayn 10	9
	Safawi 11	3
	Tulul ash Shaba 12	1
Southern block	1 Qatrana	7
	2 Qasr Tuba	17
Western block	Sahab	13
	6 Azraq	16

The eastern block: Two oil wells were drilled in this block: W. Hazim (WH1) and Fuluq (F1). The block has been divided into five sub-blocks concerning the hydrocarbon potential: A highly faulted zone is located east of the Fuluq fault. The width of this zone is 3-4 km. The only two wells drilled in this block are located in this highly faulted zone. F1 well was drilled close to SP 490 along seismic line AZ69. There are no indications of a major clear anticline.

- Area of 'very good' potential (sub-block A). A series of anticlines and synclines were selected and these need further evaluation and interpretation of the seismic lines crossing the whole block;
- Area of 'good' potential (sub-block B);
- Area of 'moderate' potential (sub-block C);
- Area of lower potential (sub-block D). Most of the area in sub-block D has low potential due to the fact that there is no subsurface deep structural trap. The seismic lines show horizontal bedding or slightly dip to the north-east and east;
- The highly faulted zone east of the Fuluq fault. The width of this zone is 3-4 km. The only two wells drilled in this block are located in this highly faulted zone. F1 well was drilled close to SP 490 along seismic line AZ69. No indications of a major clear anticline.

The northern block: Fina Company performed a most comprehensive study of this sub-block for NRA. Only two oil wells were drilled in this block: Qitar al-Abd (QA1); and Strat H5 (SH5).

Three types of subsurface folds were identified: a series of anticlines and synclines; a separate anticline and syncline formed as a result of faulting (the anticline in the up-thrown side and the syncline in the down-thrown side); and a very distinguished separate anticline or possibly a domal structure.

The southern block: Light oil was discovered in upper Ordovician Dubeidib sandstone in well WS4 located south of the block. The WS4 discovery well and WS5, 6 and 7 step outs were all drilled on a single large horst block. All four wells encountered show light oil in poor quality reservoir rocks. The widespread distribution of these shows confirms that the Sirhan area may contain a considerable quantity of oil. The absence of reasonable quality reservoir rocks is the main hindrance to economic oil production in the WS4 area.

The WS4, 5 and 7 wells all show that anomalously high geothermal gradients of 4.1 °C/100 m are present in south-eastern Sirhan. The gradients in surrounding wells WS2 andWS3 are 2.8 and 2.2 °C/ 100 m respectively. The anomalously high gradients in the WS4 area may be responsible for the abundant oil shows and may have caused a post-Palaeozoic second stage of oil generation in the Hisuk and basal Mudawwara shales.

There is no oil well drilled in this block. Therefore in order to evaluate the subsurface geology of this block and to correlate it with the other blocks the limited number of wells drilled in the area surrounding the block will be discussed. The nearest wells located south of the basin are Wadi Sirhan-2 (WS2): Siwaqa well SQ1 and Qurrayat well. The block has been divided into four sub-blocks (A to D) concerning the hydrocarbon potential:

The main prospect in the southern block is Palaeozoic in age, since the Mesozoic and Tertiary sequences are thin and presumably immature. The available seismic data indicated normal and/or wrench faults of modest size. The geochemical studies and electrical logs indicated that the source rocks for this large Palaeozoic basin are thick and widespread Silurian shales which have been found with a thickness up to 800 m in exploration wells such as WS1 and WS2, and also found to be thickening towards the southern part of the basin. (Sunna et al 1999).

The western block: This block is located between the following coordinates: 255-310 E, 100-155 N (Palestine Grid). It is mostly located within the Sahab sheet and it also covers the western part of the Azraq sheet, scale 1:100 000. Only one oil well was drilled in this block (Safra).

11. Oil shale within the Azraq basin

It seems that there are two types of oil shale horizon within the Azraq basin: near surface or mineable deposits such as those at Attarat, Maghar and Khan Ez-Zabib and the adjacent area; and deep deposits in the centre and north of the basin.

The work carried out on the oil shale within the Azraq basin covers less than 10 % of the total basin area, and does not allow a full assessment of the results. Locally, only the Attarat and Maghar areas were covered with an exploratory campaign. In the remaining areas the oil shale was only descriptively recorded.

Several hundred water wells of different depths have been drilled in the Azraq basin during the last four decades. The lithological description of these wells shows the presence of a thick bed of oil shale, but unfortunately no further studies had been undertaken on other properties.

The minimum thickness of oil shale (70 m) was recorded in borehole NDW11 in the southern part of the basin, and the maximum thickness (655 m) was recorded in borehole NDW1 in the north-eastern part of the basin. The depth to the oil shale is 200-800 m.

The estimated oil shale reserve is:

- Wadi Maghar sheet area 21.6 billion metric tonnes
- Attarat Umm Ghudran sheet area 10.4 billion metric tonnes

Other energy resources include: asphalt bearing rocks; wind energy; solar energy. There is also potential for geothermal energy. Seventeen wells have been drilled in the basin tapping the middle aquifer (B2/A7) which is the source of the thermal water. Most of these wells are discharging thermal water ranging in temperature 30-62 °C.

12. Hydrogeology of the Azraq basin

The Azraq basin (17 000 km^2) is one of the main important basins in Jordan for domestic and agricultural water supply. It has been exploited since the mid 1970s. Consequently, the aquifer balance is disturbed and a major decline in water level has taken place. A groundwater flow model was prepared as a tool to manage the groundwater resources of this basin. The study focused on: reviewing the previous geological and hydrogeological studies of the Azraq basin; re-evaluation of the previous mathematical models; preparation of hydrogeological maps; and a hydrogeological study of target areas (areas with promising water resources having development potential), with the aim of providing information for the sustainable management of the Azraq basin.

A general downward trend of water levels of 0.3-0.8 m/yr can be recognized in all observation wells of the shallow aquifer. This clearly reflects the deficit between inflow (original spring discharge and upward leakage) and outflow (abstraction and evaporation) in the Azraq depression, where the abstraction reached 55 MCM in 1998 (Al Hadidi et al 2001).

Several previous workers have reported the presence of three main aquifers in the basin: the shallow aquifer, which consists of basalt and Rijam Formation (B4) separated in places by the marls and chalk of the Shallala formation (B5); the middle aquifer system which consists of B2/B1/A7 Amman, and Wadi Sir formation; and the lower (deep aquifer) which consists of the Kurnub sandstone formation. But less emphasis has been placed on the middle and lower aquifers.

The groundwater that used to discharge at Azraq (Al Hadidi et al 2001) was one of the major sources of water in northern and eastern Jordan. However, two issues are causing problems in the Azraq basin.

The first problem is extraction of water. In the Azraq basin, the upper aquifer contains substantial volumes of excellent quality water, which is easily exploited. In the last 20 years there has been a significant increase in the number of irrigation wells constructed by the private sector abstracting water. Therefore, on a basin scale, current abstraction exceeds recharge, and groundwater is being 'mined' from storage. The total volume of groundwater removed from storage is 40-45 MCM yr (about 15-25 MCM yr, is being pumped to Greater Amman), while the calculated recharge is 10-35 MCM yr depending on which estimates of recharge and abstraction are adopted. It is clear that the upper aquifer of the Azraq basin is being exploited beyond its safe yield and as such storage depletion and water quality deterioration will continue to occur (Al Hadidi et al 2001).

In the Umm al-Quttayn well in the north-western part of the basin early human settlements in the locality obtained their water by storing flash flood runoff in purpose-built reservoirs (Al Hadidi et al 2001) since at this elevation the groundwater is around 400 m beneath ground level. This method of obtaining water has now been replaced by groundwater abstraction from deep boreholes. The Umm al-Quttayn region is different from Azraq in that it covers a larger area (34 villages), has an older and less sophisticated distribution system and the water is abstracted using a widely spaced network of 11 municipal wells. Umm al-Quttayn and surrounding villages have seen a dramatic increase in groundwater abstraction over the last decade. In 1993 the total abstraction was 6.7 MCM (WAJ 1994), 81 % of which was exported from the region (Noble 1994).

The second problem is the chemistry of the ground water. The water is saline both in the basin and in the mud pan of Qa'a Azraq (Azraq Lake). Because water can leave this closed basin only by evapo-transpiration (always leaving the dissolved matter behind), salt concentrations are highly elevated in the centre of the basin. Consequently, several workers consider that an interface between fresh and saline groundwater has developed. To assess this danger, it is necessary to know the exact location, as well as the gradient of this interface. The integration studies have cleared up the situation and work in progress on the geological model will help re-evaluate the relationship. As for the basin, Drury sampled several wells located north of Azraq town in 1993 for chemical analysis. The results indicate that the waters are all of either sodium chloride or calcium carbonate types. The evolution of the major ion chemistry of the water, increasing from low to high, is explained by the relatively high volume of recharge of rainwater in the north, becoming more saline further south due to the infiltration of smaller volumes of water that are charged with salts derived by dissolution of the evaporitic crust (Khoury 2000).

A three–dimensional groundwater flow model (Processing Modflow) has been conducted in order to calculate the groundwater budget and aquifer characteristics and predict the aquifer response to groundwater withdrawal up to 2025 (Al Hadidi et al 2001).

The hydraulic conductivities of the basalt B4/B5 aquifer system vary 1-75 m/day in the horizontal direction and 0.4 m/day in the vertical direction. The low conductivity value was found at the north-eastern and southern parts of the Azraq basin. The high conductivity value was found at the central and north-western parts of the basin.

The average annual rainfall in the model area varies from less than 50 mm in the south-east to approximately 150 mm in the north-west. The total recharge into the B4/B5 aquifer is around 36 MCM/yr with 11.2 MCM/yr inflow across the northern boundary, 2.9 MCM/yr inflow from the east and north-east, 3 MCM/yr inflow from the south-west, 0.4 MCM/yr inflow across the western boundary, 2.1 MCM/yr upwards from the B2/A7 in the oasis area and 16.4 MCM/yr as direct excess rainfall recharge.

Leakage from the basalt B4/B5 aquifer through the B3 is approximately 9.8 MCM/yr, but is strongly affected by the amount of abstraction during the unsteady state. The spring discharge into the pools was about 20 MCM/yr and the outflow across the eastern boundary reached 6.2 MCM/yr towards the Sirhan basin.

According to the present abstraction (ca 53 MCM/yr), the maximum water level declines in the model area will reach about 13.2 m and 15.3 m by the years 2015 and 2025, respectively. In addition, two cones of depression will appear, one will be in the central part of the basin and the other will be close to the Baghdad road well field (north Badia). The model shows that the optimal use of the groundwater from the Azraq basin should not exceed 30 MCM/yr (Al Hadidi et al 2001).

Based on the maps prepared by the geological and geophysical group in the same project, twenty-nine areas with promising water resources development potential had been allocated for groundwater exploration. The water group studied and analyzed the details of each area using different methods (structural, hydrological and geophysical). Results of the study show that there are areas of higher and lower potential. Despite the mining of the shallow aquifer, demonstrated by many previous studies (Al Hadidi et al 2001) these promising areas could be taken into consideration for future investigations. So, the water group has recommended that the exploitation of the B2/A7 aquifer in the future will be valuable in order to protect the decline of the water level in the shallow aquifer. Also based on this study, several recommendations were made for the future investigation of the Azraq basin.

13. Land cover and potential land use mapping of the Azraq basin

The objectives of the land use map-making study were to survey both present land use in the basin, and land use potential. Special emphasis was placed on agricultural land, and on the oil, mineral and water areas. An actual land use map and a potential land use map (Figure 3) have been prepared based on the results of the work of all the groups in accordance with the developed classification. The maps are an important planning tool in that they quantify the different resources in the basin and suggest what can be done to utilize the natural resources. They will assist planners choosing between alternative ways of using the land. They will also help determine where best to locate new settlements. Finally, they provide responses to many other questions about the management and use, including conservation, of the natural environment.

As a result of this work several projects were proposed concerning mineral, energy and water resources, and promising agricultural areas in addition to infrastructure projects concerning the future development in the basin.

The potential land use map shows target areas of good potential for mineral resources, areas of good potential for energy resources (oil, oil shale and geothermal energy) and areas of good potential for water resources. The map also shows areas of good potential for agriculture, as a 'preliminary study', and proposed rangeland including wadi beds and alluvial fans, and rural-agricultural resources that need further examination, including the *qa'a* and *marab*. Finally, areas of good potential for nature reserves are marked.

14. Proposed projects

Based on the field and laboratory research summarized in this chapter, the following R&D projects have been proposed:

14.1. Proposed minerals projects

- Project No. M1, evaluation of the mineral resources of the Azraq depression (clay minerals, diatomite and gypsum)

Figure 3. Potential land use map of the Azraq basin

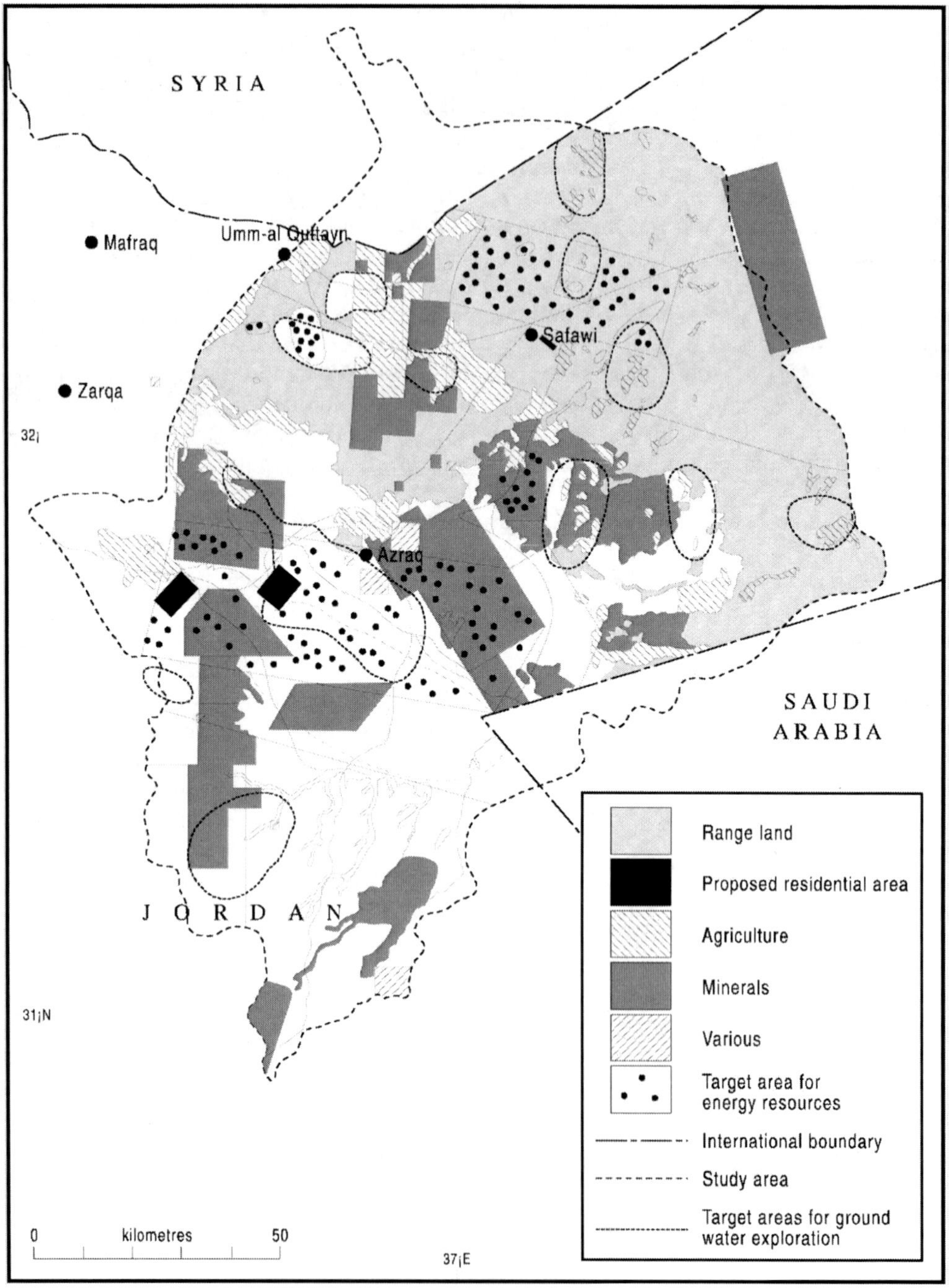

- Project No. M2, evaluation of the tuff and zeolite in the Azraq basin
- Project No. M3, evaluation of porcelanite and chalk in the Azraq basin
- Project No. M4, evaluation of ornamental stone (limestone)
- Project No. M5 evaluation of basalt for industrial application

14.2. Proposed energy projects

- Project No. E1, exploration for oil in the Azraq basin
- Project No. E2, exploration and evaluation of the oil shale deposits in the Azraq basin
- Project No. E3, evaluation and utilization of the geothermal energy resources in the Azraq basin

14.3. Proposed water projects

- Project No W1, testing of the target areas close to agricultural land
- Project No W2, testing of the Ghadaf - Makhruq graben for artificial recharge
- Proposed agricultural projects
- Project No. A1, soil studies in the volcanic terrain of the northern part of the Azraq basin
- Project No. A2, soil mapping in the target areas in the Azraq basin
- Project No. A3, detailed study of the cultivated land east of the Azraq basin

14.4. Proposed infrastructure projects

Proposed settlements: Two areas were proposed for future settlements. Selection was based on the following criteria: close to the agricultural land and good access for both groundwater and geothermal energy for future use in the towns; in a central position and close to Amman; in a central area concerning the mineral resources in the basin; close to the communication and transportation networks; midway between two existing villages, Azraq and Muwaqqar.

Proposed roads: Two roads were proposed to be constructed in the future. The first is a ring road of the Azraq Lake and connects the Azraq-Saudi border road with the Azraq-Ain Baida farm area. The second connects this ring road with the Mafraq-Iraqi border highway between Safawi and Ruwayshid

14.6. Proposed GIS projects

Production of an atlas for the Azraq basin: The digitized map layers constitute a comprehensive GIS database that could be used for a wide range of applications including: resource management and planning, monitoring the impact of any regional plans on the natural environment, and making recommendations on the potential land use for the area. The production of a coloured atlas for all the layers would provide a visual support for the main results of the studies and would be an appropriate means to market them to potential investors (Al Mubarak et al 2002).

Acknowledgements

This writing of this chapter has been dependent on contributions to the field and laboratory work undertaken by the following specialists, in addition to the author Eng Bassam Sunna who was also project leader: Geol. Jolia Sahawneh, Dr. Abdallah Al Zoubi, Geol. Mahamad Sabha, Geol. Izzedeen Waynagh, Geol. Antwanite Shamoun, Geol. Manal Dababneh, Dr. Khalid Tarawneh, Geol. Ahmad Smadi, Geol. Awdeh Al Mashan, Dr. Hani Khoury, Geol. Mohamd Qa'adan, Geol. Hani Malak, Geol. Ayoub Ghrir, Geol. Tariq Nassir Ed Deen, Geol. Ahmad Abu Jiab, Geol. Jamal Al'ali, Abu Snouber, Geol. Mohamad Abu Qudaira, Geol. Samir Mnaizel, Eng Kamal Jreisat, Geol. Ali Swarieh, Geol. Robeir Massarweh, Geol. Ghassan Kilani, Dr Khair Al Hadidi, Eng. Ali Subah,

Geol. Antwanite Shamoun, Geol. Manal Dababneh, Dr Zahir Rawajfih, Dr Sa'eb Khresat, Eng. Mohammed Shahbaz, Dr Alia Hatough Bouran, Ibrahim Al Khader, Eng. Mabarak Al Mubarak, Eng. Rafeh Ashour and Geol. Lufi Abu Sa'd.

Ecological and environmental studies in the Azraq basin were carried out by a team consisting of Eng. Mohammed Shahbaz , Dr.Alia Hatough Bouran and Eng. Ibrahim Al Khader (Shahbaz et al 2000); not discussed in this paper.

References

Al Hadidi K and Subah A (2001). Groundwater resources in the Azraq basin, Jordan. Badia Research & Development Programme, Higher Council for Science and Technology, V4 page 115, fig. 2, table 7.

Al Mubarak M and Ashour R (2002). Final report of the Geographic Information System, Jordan. Badia Research & Development Programme, Higher Council for Science and Technology, V7 page 11.

Khoury H (2000). Industrial rocks and minerals in the Azraq basin, Jordan. Badia Research & Development Programme, Higher Council for Science and Technology, V3, page 447, fig. 231, table 188, annexe 9.

MEMR (2000). Annual report of the Ministry of Energy and Mineral Resources.

Noble PL (1994). Quantification of recharge to the Azraq basin, NE Badia. MSc thesis, University College London,

Rawajfih Z and Khresat S (2000). Agricultural potential of promising areas in the Azraq basin, Jordan. Badia Research & Development Programme, Higher Council for Science and Technology, V5 page 39, fig. 4, table 6.

Shahbaz M, Hatough Bouran A and Al Khader I (2000). Ecological and environmental studies in the Azraq basin, Jordan. Badia Research & Development Programme, Higher Council for Science and Technology, V6 page 92, fig.7, table 3, annexe 4, appendex 1.

Sunna BF (1999a). Geomorphology of the Azraq basin, Jordan. Badia Research & Development Programme, Higher Council for Science and Technology, Vol.1.1 page 102, fig. 7, table 3, plate 14, enclosure 2.

Sunna BF (1999b). Stratigraphy of the Azraq basin, Jordan. Badia Research & Development Programme, Higher Council for Science and Technology, V1.2 page 202, fig. 34, table 42, enclosure 31.

Sunna BF (1999c). Structural geology and geological model of the Azraq basin, Jordan. Badia Research & Development Programme, Higher Council for Science and Technology, V1.3 page 145, fig 34, table 5, plate 49.

Sunna BF (1999d). Target areas for groundwater and hydrocarbon exploration and for agricultural purposes, Jordan. Badia Research & Development Programme, Higher Council for Science and Technology, V1.4 page 168, fig. 6, table 15, plate 4, enclosure 35.

Sunna BF and Jreisat K (1999). Energy resources in the Azraq basin, Jordan. Badia Research & Development Programme, Higher Council for Science and Technology, V3 page 333, fig. 40, table 134, plate 31, appendices 3.

CHAPTER 4

Mineralogical distribution of clay minerals in north-east Jordan and their industrial application as bleaching agents of edible oil

Mohammed K.A. Nawasreh

Natural Resources Authority, P.O. Box 7, Amman-Jordan. E-mail: nawasreh@nra.gov.jo

The clay deposits of the Qa'a al-Azraq area in the north-east Jordan Badia region belong to the Azraq Formation (Pliocene). They were formed from alteration of volcanic ash deposited in shallow water lakes. Two forms of commercially significant clay minerals are present in the area. They consist of smectite and palygorskite (attapulgite) rich sediments. The Jordanian clay mineral types display high bleaching capacities of edible oil after acid and thermal activation reached 99 % and colour changes up to 98 % giving results similar to those obtained with commercial industrial products.

Keywords: Jordan Badia, smectite, palygorskite, bentonite.

1. Introduction

Qa'a al-Azraq represents a closed mudflat that is part of the Azraq Basin. The *qa'a* is characterized by a flat surface and low relief and covers about 350 km^2. It is located about 120 km north-east Amman (Figure 1). The Qa'a is surrounded by basalts and volcanic tuff. It was a lake originally formed in the post-Miocene time. The clastic sediments, evaporites and wadi sediments were accumulated and deposited in the lake during a period of high evaporation rates and low water supply (1956).

The clay minerals of the Azraq Formation were formed by transformation of volcanic ash. The alteration of these materials formed kaolinite, smectite and illite during early diagenetic processes, while palygorskite was formed by transformation of smectite and direct precipitation from solution (Qa'adan 1992). The Azraq Formation consists mainly of a sequence of evaporites, gypsiferous clays, silty clay, marly clay and diatomite intercalation with lacustrine carbonate deposited in a closed fresh water lake with periodic wet and dry cycles (Ibrahim 1997).

Two types of significant clay minerals are present in the area. They consist of smectite (montmorillonite) known as bentonite and palygorskite as major constituents together with mixed layered illite/smectite with traces of illite and kaolinite. The non-clay minerals observed were mica, feldspar and quartz minerals (Nawasreh 2001, Nawasreh et al 1998).

Two types of treatment were applied to the Jordanian clay samples. Bentonite was activated with hydrochloric acid (HCl), whereas palygorskite was thermally activated. Acid activation proceeds with partial dissolution of the smectite structure (octahedral and tetrahedral sheets) with subsequent release of structural cations followed by replacement of the interlayer cations by H^+. This process enhanced the sorptive properties of bentonite by increasing the specific surface area and pore volume (Christidis et al 1997, Odom 1984, Richardson 1978, Rupert et al 1987).

Thermal treatment of palygorskite samples modifies their structure and enhances the sorptive properties. Heating palygorskite in the range 280-400 °C led to the loss of zeolitic water (internal pore water) and hydroxyl water. This dehydration activates the open channels for absorption and adsorption properties which provide the mineral with its decolourizing properties and bleaching capacity (Heivilin et al 1994).

Figure 1. Location map of the study area

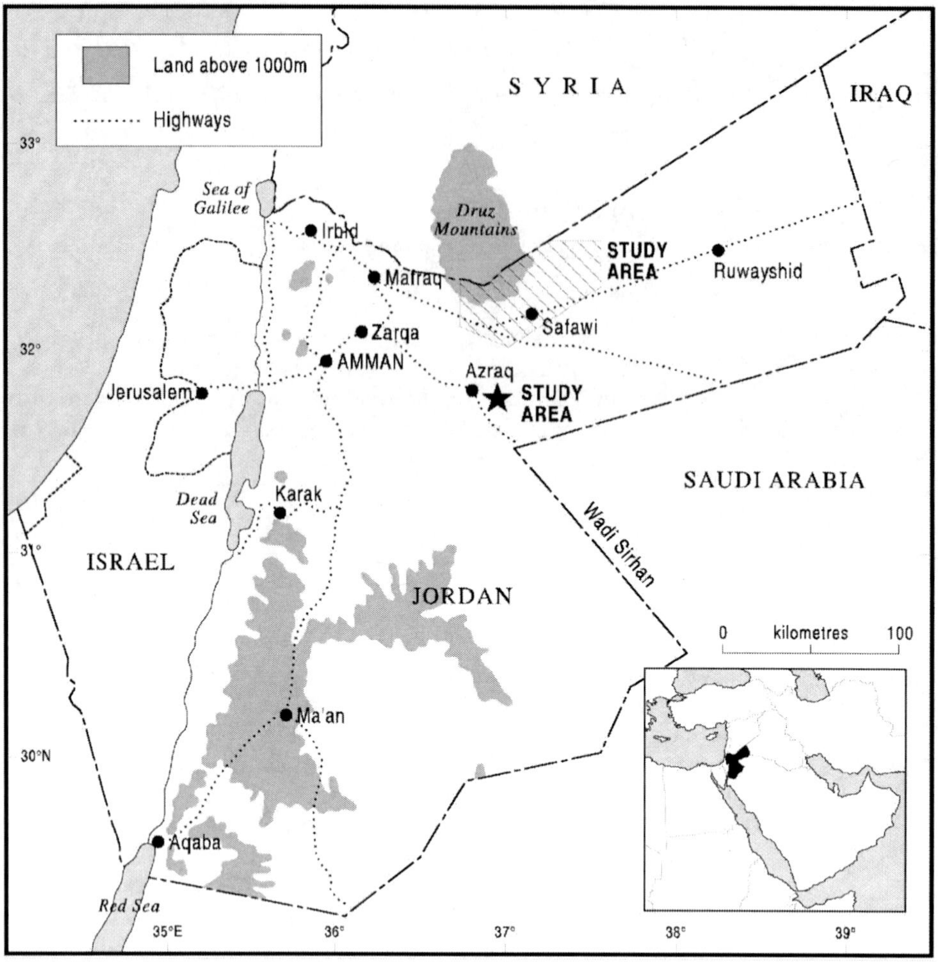

2. Materials and methods

Due to the differences in mineralogy, the cation exchange capacity (CEC) and the specific surface area (Table 1) of bentonite in the area, two bentonite samples were chosen to be representative for most of Qa'a al-Azraq basin. The two samples were subjected to acid activation. Bentonite sample 1 is found to consist of monovalent interlayer cation-saturated smectite, while albite and cristobalite are present as impurities. The clay minerals identified in sample 2 are also dioctahedral smectite, traces of mixed layered illite/smectite and kaolinite that is recognized in trace amounts. The major differences between sample 1 and sample 2 are the presence of calcite, dolomite, quartz, mixed layered illite/smectite and kaolinite with a lesser amount of albite and absence of cristobalite in sample 2 as illustrated by the XRD trace (Figure 2). Palygorskite samples were almost identical in their mineralogy with the presence of trace amounts of kaolinite and/or illite (Figure 3).

The raw bentonite materials were placed in an oven at 80 °C overnight to remove excess moisture content then crushed and ground using a mortar and pestle to pass through a 125 mm sieve (BS 410). The samples were treated with hydrochloric acid (HCl)

Table 1. Cation exchange capacity and specific surface area of the raw and acid activated bentonite

Sample No.1	CEC meq/100 g	Surface area m^2/g	Sample No. 2	CEC meq/100 g	Surface area m^2/g
Raw Clay	102.30	454.02	Raw Clay	73.64	428.13
1 N & 1h	85.59	512.00	2N & 1h	71.36	542.60
1N & 4h	81.24	584.39	2N & 4h	68.43	563.14
1 N & 8h	80.31	522.38	2N & 8h	68.11	596.00
1N & 12h	75.25	497.11	2N & 12h	63.94	615.43
1N & 16h	71.84	413.00	2N & 16h	55.67	431.64
8N & 1h	77.48	567.10	4N & 1h	54.00	569.00
8N & 4h	68.03	593.58	4N & 4h	51.19	509.27
8N & 8h	61.89	503.00	4N & 8h	43.82	407.69
8N & 12h	56.73	461.94	4N &12h	41.45	368.93
8N & 16h	48.57	367.41	4N &16h	41.38	364.84

Figure 2. X-ray diffractometer trace of a Bentonite clay sample

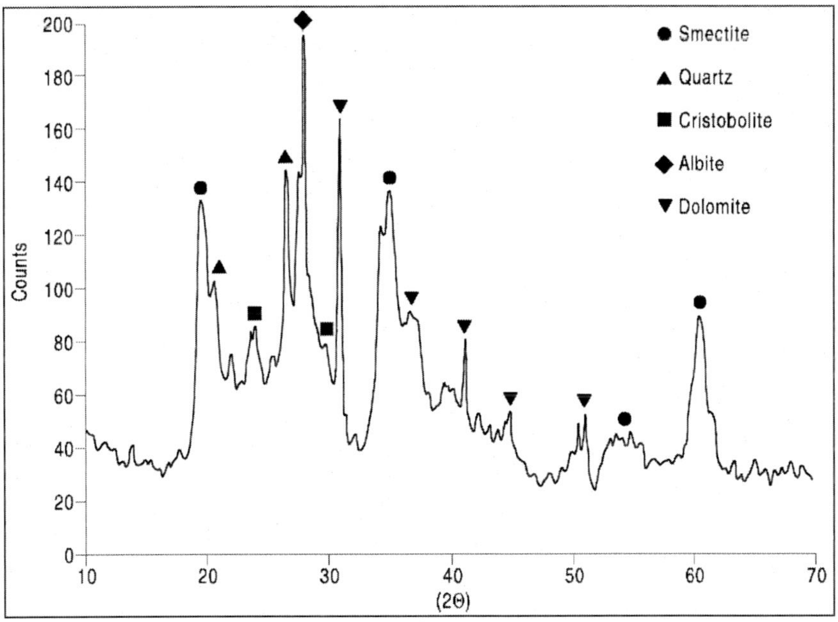

analytical grade, and the strength of acid solution used was 1N, 2N, 4N and 8N. The residence time of acid activation varied 1-16 hr. Ten grams of each bentonite sample were placed in 250 ml Pyrex beakers containing a magnetic stirrer and 150 ml of acid (HCl) was added to the beakers. The beakers were covered with watch-glasses to prevent evaporation during activation. The set were placed on the magnetic hot plate at a constant temperature (70 °C). A combination of different acid strengths and contact time for individual samples were employed to establish optimum treatment conditions. The clay samples were filtered and the clay cake washed several times with distilled water until the water discarded reached pH 6. The activated clay was then dried in the oven at 100 °C for two hours. Finally the clay cake was ground gently, using an agate mortar, in order to pass through 125 μm.

Figure 3. XRD pattern of a Palygorskite clay sample

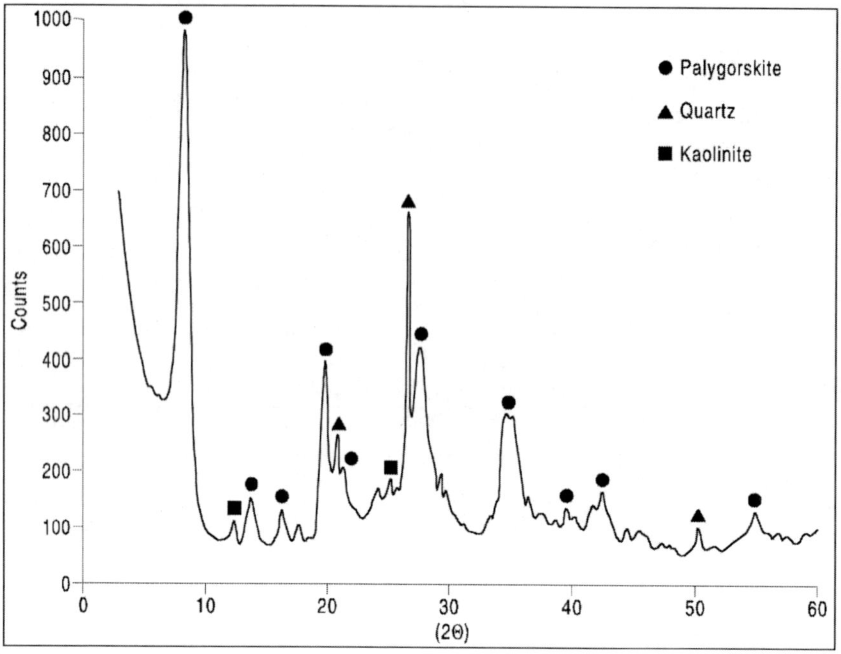

Activation of palygorskite by heating was performed using temperatures of 280 °C and 400 °C. The raw palygorskite samples were crushed and ground in the same manner as bentonite. Palygorskite samples were divided into two equal portions; 5 gm of the first portion were placed in the oven at 280 °C for two hours and 5 gm of the second portion were placed in the oven at 400 °C for two hours in order to choose the optimum temperature for activation.

The surface areas of the raw and activated clay materials were determined by using the 2-Ethoxyethanol (EGME) method as described by Carter et al (1965) and Inglethorpe et al (1993). Mineralogical and structural changes of the raw and acid activated bentonite were examined by XRD in random state (Figures 4 and 5) using a standard Philips PW1710 diffractometer controlled by PC-APD software employing a copper x-ray tube radiation operating at 35 kV and 40 mA within the range 3-60 2θ at a speed of 0.8 2θ/minutes (Van Olphen et al 1979) (Moore et al 1989). Ammonium acetate was used to determine the CEC for the raw and activated clay materials, following the procedure described by Wilson (1987) and Inglethorpe et al (1993).

Crude high erucic acid and refined rapeseed oil provided by Seatons (John L. Seaton and Co Ltd England, Hull) was used for bleaching experiments of thermal activated palygorskite and acid activated smectite. One gram of activated clay was placed in a 50 ml Pyrex beaker with magnetic stirrer and 20 ml of high erucic acid crude oil was added to the clay. The beaker was covered with a watch glass to avoid excess evaporation during the bleaching process. The experiments were performed at atmospheric pressure using a hot plate magnetic stirrer at a constant temperature of 90 °C for a period of 20 minutes. The treated oil was filtered under vacuum and collected in clean glass tubes and stored in labelled polyethylene bottles. The bleaching experiments were operated in the manner adapted from (Christidis et al 1997, Morgan et al 1985, Richardson 1978, Srasra et al 1989).

Figure 4. XRD traces of acid activated sample 1

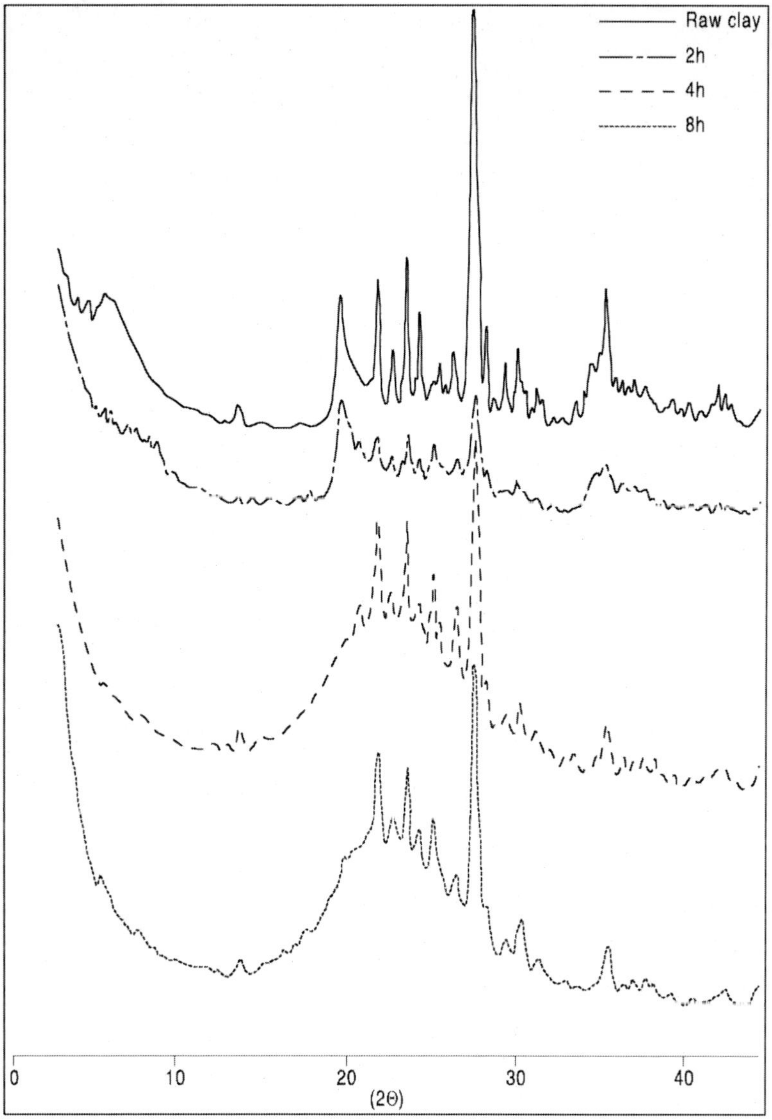

The bleaching capacity of the activated clay was calculated by measuring the removal of β-carotene from the crude oil using a UV-Visible spectrophotometer. β-carotene standards were prepared to establish the calibration curve. This was achieved by dissolving β-carotene in spectroscopic grade acetone (Aldrich) and measuring its absorbance at 450 nm using a Unicam heliosa UV-Visible spectrometer V1.30. The colour changes in the treated oil were also determined spectrophotometrically at 450 nm and 430 nm (using the same equipment) by diluting 5 ml of bleached oil in 15 ml of spectroscopic grade acetone (Figs 6 and 7). The amount of colouring agent adsorbed was calculated by converting the absorbance unit to β-carotene concentration (Christidis et al 1997, Srasra et al 1989). The bleaching capacity was calculated using the following relationship:

Figure 5. XRD traces of acid activated sample 2

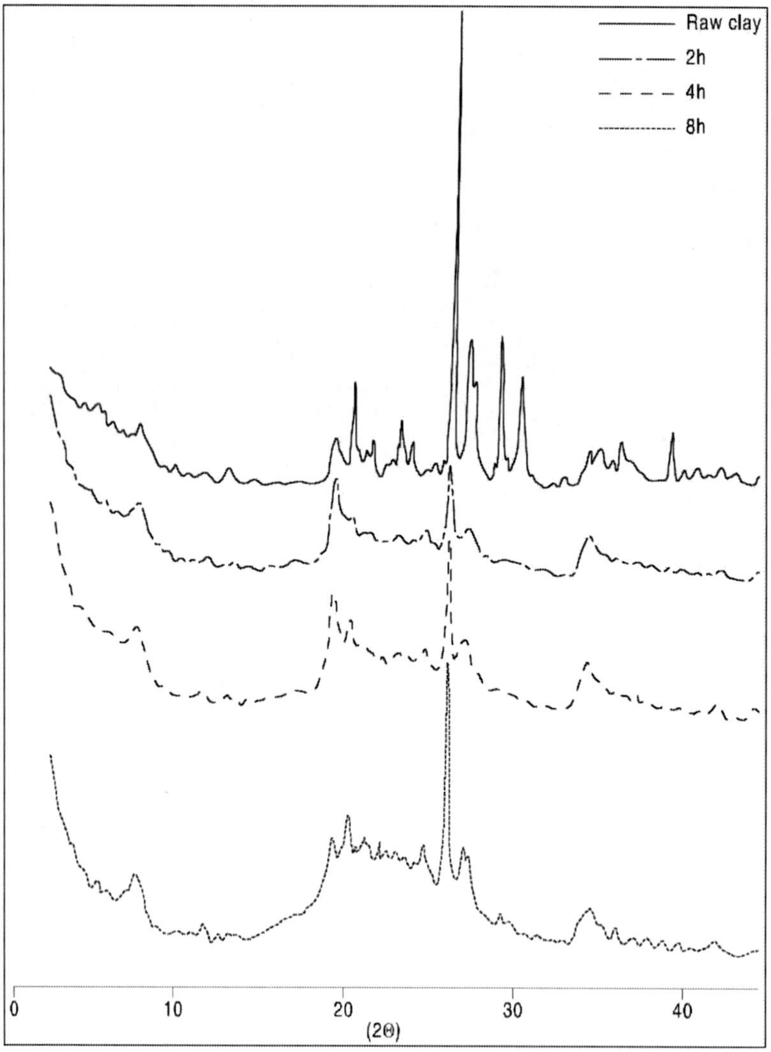

Bleaching capacity = (C-°C/°C) *100.
Where: °C = Concentration of β-carotene in the crude oil μgm/ml; C = Concentration of β-carotene in the bleached oil.

To evaluate the performance of the bleaching capacity of Jordanian clay samples the same experiments were performed using commercial industrial product samples as standard samples. These were Ca-bentonite activated with acid (the strength of acid and the residence time are unknown) provided by Süd Chemie Ltd with the trade name Tonsil optimum 215 FF, Tonsil standard 3151 FF, and attapulgite (palygorskite) activated with Tonsil 350 FF and Tonsil 550 FF (the temperature and the time of thermal activation are unknown). The bleaching capacities for the standard samples were calculated and comparisons between the performance of the commercial bleaching samples and the Jordanian activated clay samples were drawn.

Figure 6. Absorbance spectra of β-carotene obtained from crude rapeseed oil and treated with acid of variable strength and time

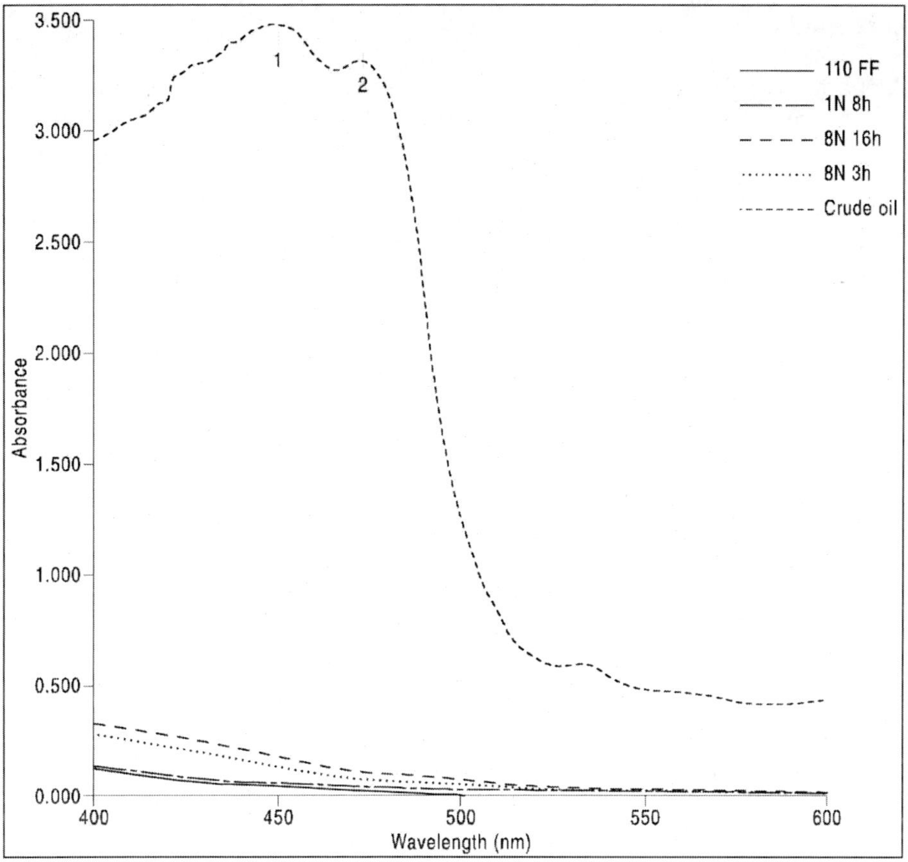

3. Results and discussion

The structural modification of the Qa'a al-Azraq acid treated bentonite sample has been demonstrated by XRD examination (Figures 4 and 5). The changes of the diffraction patterns with increasing time of treatment and acid strength reflect a loss of crystallinity of the treated sample. It was found that by increasing the contact time there was a more drastic change in the diffraction results. Reflection peaks of d001 and d020 became broader after two hours of contact time and then disappeared (complete destruction) with further increase in contact time.

Sample 2 of Qa'a al-Azraq bentonite also displayed a similar behaviour, but calcite peaks in the diffraction pattern were completely removed by acid activation (Figure 5).

The process reduces impurities and enhances the smectite (clay) content. Therefore, the specific surface area and sorptive properties are enhanced and the bleaching capacity are improved. Acid activation also develops the microporosity of bentonite, indicated by increasing the surface area, which is considered an essential factor of bleaching capacity. The CEC was decreased with increasing of acid strength and duration of contact time in spite of increasing the smectite content by dissolving carbonates, particularly in sample 2. This might be as a result of replacement of the cations by H+ and also due to continuous dissolution of smectite flakes and deposition of amorphous silica (Christidis et al 1997, Srasra et al 1989).

It is interesting to note that the background in the XRD traces of the activated samples displayed a noticeable increase in intensity of 20-28 2θ in both samples (Figures 4 and 5). This increase is related to the production of amorphous silica formed due to the attack of acid protons on the octahedral and tetrahedral clay layers. The result obtained from the current research has been also noticed by other researchers (Christidis et al 1997, Rhodes et al 1992, Srasra et al 1989, Taylor et al 1986).

Figure 7. Absorbance spectra of β-carotene obtained from the crude rapeseed oil and oil treated with thermal activation of variable strength for differing times

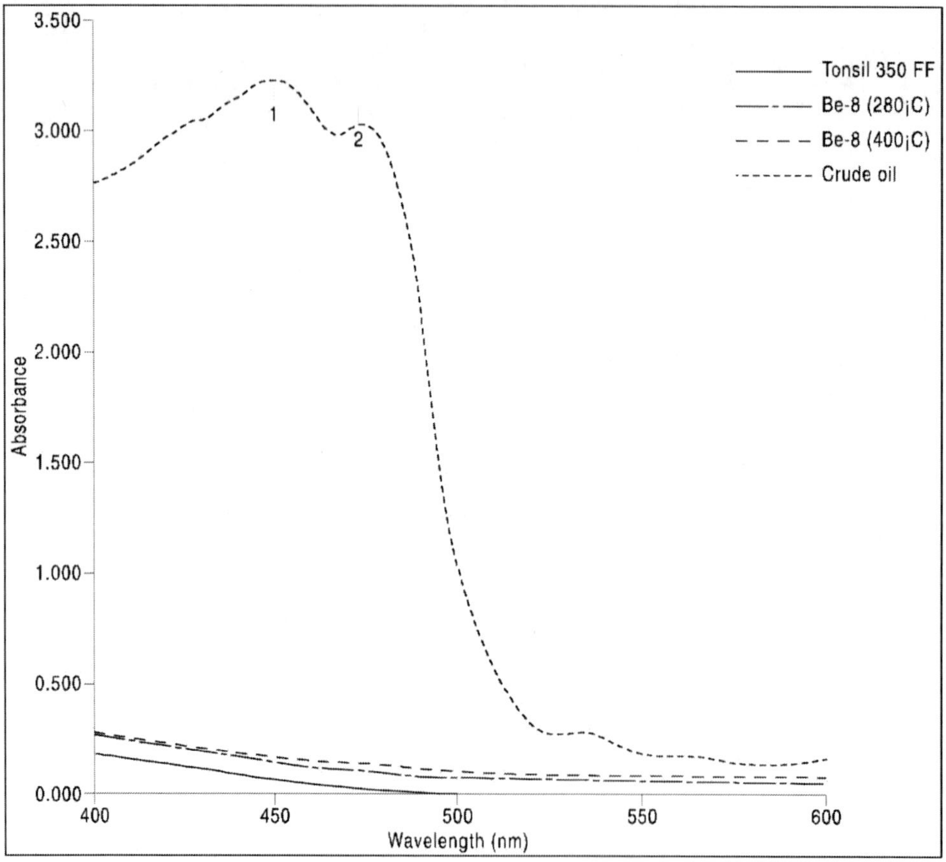

The maximum bleaching capacity obtained for sample 1 was 99.55 % with a 96.52 % colour change (amount of β-carotene removed), which was produced with a treatment at 1 N of acid strength and eight hours of contact time (Table 2). Activation of the same sample with 8 N acid strength for 1 h produced 94.80 % bleaching capacity and 91.88 % colour change. With a further contact time of 12 hours there was a significant loss in the decolourizing properties of the sample. At 16 h of contact time with the same acid strength the bleaching capacity dropped rapidly to about 82 % which might be due to the long period of residence time leading to the collapse of the bentonite structure. Activation of sample 2 with 2 N HCl strength for 1 h contact time produces a bleaching capacity of 97 % and about 96 % change of colour. Activated fractions of sample 2 were found to have higher bleaching capacities and were more rapid than sample 1. Both samples were found to display higher bleaching capacities and colour changes, after activation with dilute acid rather than with the concentrated acid which were considered to be sufficient to develop the maximum bleaching capacity of the sample. The maximum removal of β-carotene obtained with an activated clay of sample 2 was 99.53 %, associated with 97.15 % colour change (Table 3).

Both bentonite samples from the Qa'a al-Azraq area displayed loss of bleaching capacity and decolourizing ability after prolonged acid activation and duration of contact time (Tables 2 and 3). This impairment of bleaching capacity is due to more depletion of the octahedral sheet of smectite followed by destruction of the smectite structure.

A comparison of oil treatment results for the Jordanian bentonite samples with standard commercial Tonsil Bleaching Earth (215 FF and 110 FF) samples produced by Süd Chemie Ltd Company are contained in (Tables 2, 3 and 4). Sample 1 was found to display a higher bleaching capacity after treatment with 1 N HCl for 4 h of contact time than the commer-

Table 2. Absorption of β-carotene and bleaching capacity by acid activated bentonite during the bleaching process of crude rapeseed oil (sample 1)

Sample No.1	Absorption at 450 nm (absorbance units)	Bleaching capacity %	Concentration of β-carotene in solution (μg/ml)	Absorption at 430 nm (absorbance units)	Colour changes %	Concentration of β-carotene in solution (μg/ml)	Concentration of the absorbed β-carotene (10^{-4} moles/100gm)
1 N & 1h	0.239	92.90	1.488	0.223	92.34	1.39	7.255
1N & 4h	0.049	98.54	0.305	0.094	95.07	0.59	7.696
1N & 8h	0.015	99.55	0.093	0.101	96.52	0.63	7.795
1N & 12h	0.036	98.93	0.224	0.069	97.63	0.43	7.726
1N & 16h	0.057	98.31	0.355	0.092	96.83	0.57	7.678
8N & 1h	0.175	94.80	1.090	0.236	91.88	1.47	7.404
8N & 4h	0.041	98.78	0.255	0.039	98.66	0.24	7.715
8N & 8h	0.125	96.29	0.778	0.105	96.39	0.65	7.520
8N & 16h	0.234	93.05	1.457	0.209	92.81	1.30	7.267
8N & 12h	0.617	81.68	3.842	0.482	83.41	3.00	6.379

Table 3. Absorption of β-carotene by acid activated bentonite during the bleaching process of crude rapeseed oil (sample 2)

Sample No.2	Absorption at 450 nm (absorbance units)	Bleaching capacity %	Concentration of β-carotene in solution (μg/ml)	Absorption at 430 nm (absorbance units)	Colour changes %	Concentration of β-carotene in solution (μg/ml)	Concentration of the absorbed β-carotene (10^{-4} moles/100gm)
2N & 1h	0.086	97.45	0.54	0.130	95.53	0.81	7.608
2N & 4h	0.067	98.01	0.42	0.105	96.39	0.65	7.653
2N & 8h	0.044	98.69	0.27	0.093	96.81	0.58	7.709
2N & 12h	0.047	98.60	0.29	0.066	97.77	0.41	7.702
2N & 16h	1.060	68.52	6.60	1.128	61.24	7.02	5.351
4N & 1h	0.048	98.57	0.30	0.065	97.77	0.40	7.698
4N & 4h	0.016	99.53	0.10	0.083	97.15	0.52	7.773
4N & 8h	0.039	98.84	0.24	0.035	98.80	0.22	7.720
4N &12h	0.030	99.11	0.19	0.074	97.46	0.46	7.739
4N &16h	0.044	98.69	0.27	0.059	97.97	0.37	7.709

cial sample 215 FF (95.87 %). It can be seen from the results that acid activated Jordanian bentonite samples produced very similar results to commercially available samples. Treatment of crude rapeseed oil with locally activated bentonite samples gave products with properties identical to those of commercial refined rapeseed oil, as seen in Table 4.

Products of thermal activation palygorskite samples, the bleaching capacity and colour changes of these samples, are listed in Table 5. It can be seen that thermally activated samples of palygorskite performed in a similar manner to industrial products (commercial standard palygorskite samples). It was found the bleaching capacity and colour changes of Jordanian palygorskite samples ranged from 90.70-98.54 % and 87.71-95.01 % respectively, whereas the bleaching capacity of commercial samples ranged from 86.84-93.85 %. The maximum bleaching capacity displayed by thermal activated palygorskite samples was 98.54 %, colour change 95.01 %, recorded for a sample MNJ 13 at a temperature of 280 °C, which dropped to 96.82 % after activation at 400 °C for two hours (Table 5).

Table 4. Absorption of β-carotene and bleaching capacity of tonsil standard commercial products (acid activated bentonite, thermal activated palygorskite and refined oil) during bleaching process of crude rapeseed oil

Sample No.	Absorption at 450 nm (absorbance units)	Bleaching capacity %	Concentration of β-carotene in solution (µg/ml)	Absorption at 430 nm (absorbance units)	Colour changes %	Concentration of β-carotene in solution (µg/ml)	Concentration of the absorbed β-carotene (10^{-4} moles/100gm)
215FF	0.139	95.87	0.866	0.093	96.80	0.58	7.487
110FF	0.033	99.02	0.205	0.082	97.18	0.51	7.733
T350	0.207	93.85	1.289	0.164	94.35	1.02	7.330
T550	0.443	86.84	2.758	0.177	93.91	1.10	6.782
T3151	0.269	92.01	1.67	0.297	89.79	1.85	7.187
Refined oil	0.035	98.96	0.218	0.108	96.28	0.67	7.729

Table 5. Absorption of β-carotene and bleaching capacity of Jordanian thermal activated palygorskite during the bleaching process of crude rapeseed oil

Sample No. plus preparation temp (°C)	Absorption at 450 nm (absorbance units)	Bleaching capacity %	Concentration of β-carotene in solution (µg/ml)	Absorption at 430 nm (absorbance units)	Colour changes %	Concentration of β-carotene in solution (µg/ml)	Concentration of the absorbed β-carotene (10^{-4} moles/100gm)
Crude Oil	3.367	0.00	20.965	2.91	0.00	18.09	0.000
MNJ1 280	0.127	96.23	0.791	0.206	92.91	1.28	7.515
MNJ 1 400	0.204	93.94	1.270	0.29	90.02	1.81	7.337
MNJ9 280	0.149	95.57	0.928	0.259	91.08	1.61	7.464
MNJ9 400	0.313	90.70	1.949	0.357	87.71	2.22	7.084
MNJ10 280	0.052	98.46	0.324	0.098	96.63	0.61	7.689
MNJ10 400	0.101	97.00	0.629	0.192	93.39	1.20	7.575
MNJ11 280	0.114	96.61	0.710	0.178	93.87	1.11	7.545
MNJ13 280	0.049	98.54	0.305	0.145	95.01	0.90	7.696
MNJ13 400	0.107	96.82	0.666	0.217	92.53	1.35	7.562
MNJ16 280	0.110	96.73	0.685	0.197	93.22	1.23	7.555
MNJ16 400	0.155	95.40	0.965	0.223	92.32	1.39	7.450
MNJ5 280	0.880	73.86	5.479	0.834	71.29	5.19	5.71
MNJ3 280	1.116	66.85	6.949	1.115	61.62	6.94	5.221

Activation at a temperature of 280 °C for two hours produced a bleaching capacity slightly higher than that of samples treated at a higher temperature of 400 °C. Nawasreh (2001) indicated that the adsorption properties of palygorskite decreased progressively with increasing activation temperatures dropping to about 7 % after treatment at 600 °C. He also pointed out that the free water and the zeolitic water in the channel structure of the palygorskite were driven off at temperatures of 200-300 °C. The lower temperature (280 °C) was found to be sufficient to remove zeolitic water and the water from open channels that led to make these channels very active to adsorb the undesirable materials. Jordanian palygorskite is characterized by a very fine particle size with short needles and the high bleaching capacity is related to its open channel structure. Micropores along the needle-like crystals and the aggregates of palygorskite fibres form a brush-heap structure.

Two samples of bentonite containing trace amounts of palygorskite, kaolinite and illite were thermally activated to evaluate the ability of these samples to absorb β-carotene from crude rapeseed oil. The result shows that they produce the lowest bleaching values, but it might be that if they were activated with hydrochloric acid they would give a bleaching capacity higher than the thermal activation.

Bleaching capacity and colour changes in oil samples were found to increase substantially with an increase of specific surface area for all local thermally activated palygorskite samples, with a few exceptions. In the case of acid activated bentonite the maximum specific surface area was not associated with the maximum bleaching capacity and colour changes.

4. Conclusion

The performance of bentonite to get optimum bleaching capacity and colour changes can be achieved by using dilute acid and a short period of contact time. This reduces the cost of energy and acid consumption during the activation. The ability of both types of Jordanian clay minerals to eliminate the double bonds of β-carotene, solvents, pigments, soap, phospholipids, organic compounds and trace metals that are associated with edible oil is considered to be the most significant property.

As far as the industrial applications of the clay minerals in the refining of edible oil are concerned, the results obtained from this research show that the bleaching capacity of the Jordanian bentonite and palygorskite samples is comparable with commercial standard samples and their performance was equal to or higher than commercial standard samples.

References

Baker J and Harza E (1956). Princess project: definitive plan Jordan, Amman.

Carter DL, Heilman MD and Gonzalea FL (1965). Ethylene glycol monoethyl ether for determining surface area of silicate minerals. Soil Science 100:356-360.

Christidis G, Scott PW and Dunham AC (1997). Acid activation and bleaching capacity of bentonite from the islands of Milos and Chios, Aegean, Greece, App. Clay Sci, 12:329-347.

Heivilin F and Murray HH (1994). Hormites palygorskite (attapulgite) and sepiolite, In: Carr, D (ed), Industrial minerals and rocks, Colorado: Soc. for Min. Metal. and Exploration.

Ibrahim KM (1997). The regional geology of Al Azraq area map sheet No. 3531; Bulletin 36, Natural Resources Authority, Jordan, Amman

Inglethorpe SD, Morgan DJ, Highley DE and Bloodwoth AJ (1993). Bentonite: industrial minerals laboratory manual; technical report WG/93/20, Keyworth: BGS, Mineralogy and Petrology Series.

Moore DM and Reynolds RC (1989). X-Ray Diffraction and the identification and analysis of clay minerals. Oxford University Press.
Morgan DA, Shaw DB, Sidebottom MJ, Soon TC and Taylor RS (1985). The function of bleaching earth in the processing of palm, palm kernel and coconut oils. JAOCS 62:292-299.
Nawasreh MK (2001). Investigations of Jordanian industrial minerals. PhD Thesis, Cardiff University, UK 331.
Nawasreh MK and Pooley FD (1998). Mineralogical distribution and industrial evaluation of Azraq clay minerals in the North Badia region, Jordan. Appl. Geo. Conf, Leicester.
Odom IE (1984). Smectite clay minerals: properties and uses. Phil. Trans. R. Soc. London A311:391-409.
Qa'adan M (1992). Mineralogy and origin of the recent deposits in the Azraq Depression. Masters thesis, University of Jordan.
Rhodes CN and Brown DR (1992). Structural characterisation and optimisation of acid-treated montmorillonite and high-porosity silica supports for ZnCl2 alkylation catalysts. J. Chem. Faraday Trans 88 (15):2269-2274.
Richardson LL (1978). Use of bleaching clays in processing edible oils. J. Am. Oil chemistry Soc 55:777-780.
Rupert JP, WT, G and Pinnavaia TJ (1987). Catalytic properties of clay minerals. In. editor, In: Newman, ACD (ed), Chemistry of clays and clay minerals. London: Mineralogical Society.
Srasra E, Bergaya F, Van Damme H and Arguib NK (1989). Surface properties of an activated bentonite: decolorization of rape-seed oil. Applied Clay Science 4:411-421.
Taylor DB and Jenkins DB (1986). Acid activated clay. Am. Soc. Min. Eng. of AIME, Transactions 282:1901-1910.
Van Olphen, H and Fripiate JJ (1979). Data handbook for clay materials and other non-metallic minerals. Oxford: Pergamon Press.
Wilson MJ (1987). Handbook of determinative methods in clay mineralogy. New York: Chapman and Hall.

CHAPTER 5

Slope-controlled variations in the hydrological response of clast-mantled arid slopes

N.J. Rosser
Institute for Hazard and Risk Research, Geography Department, Durham University, Science Labs, Durham DH1 3LE, UK. E-mail: n.j.rosser@dur.ac.uk

R.J. Allison
University Office, University of Sussex, Brighton, UK

D.L. Higgitt
Department of Geography, University of Singapore, Singapore

This paper presents evidence from field experiments of feedbacks between ground surface form and process on a series of clast mantled arid slopes. Although infrequent, rainfall events are suggested to be a critical control in the evolution of such landscapes. The relative contribution of events of varying size and the degree to which the ground surface configuration accelerates or restricts modification of the surface is suggested to influence the overall evolution of slope profiles via the erosion, transfer and deposition of both surface clasts and the underlying fine grained sediments. Critical to this modification is the interplay between the surface configuration, rainfall and runoff generation. This paper examines this interaction by focusing on a set of experimental sites at Marab Suway'id, east of Safawi, north-eastern Jordan.

Results from a series of infiltration and rainfall simulation experiments are presented from plots systematically located to assess the variation in surface hydrology and response to variations in ground surface and slope profile form. The experiments presented here are also the first attempt to quantify the influence of clast surface coverage on rainfall partitioning and sediment mobilization in this environment. Results suggest that a close synergy exists between surface character, whole slope form and hydrological response to rainfall. Characteristics commonly suggest a self-regulating type of behaviour, indicated by the development of emergent spatial structures, non-linear spatial variations in surface character and negative feedbacks between slope surface form and hydrology.

Keywords: slopes, hydrology, sediment transfer

1. Background

Extensive research has been undertaken on the hydrology, hydraulics and sediment dynamics of clastic or rock fragment mantled semi-arid surfaces (1994a, Poesen et al 1994), with overland flow and hydraulic processes critical controls on the removal and transfer of sediments on coarse debris mantled slopes (Abrahams et al 1994a, Abrahams et al 1994b, Bryan 1922, Kirkby et al 1974, Melton 1965, Roux et al 1986). Despite this the influence of surface roughness remains a complex and commonly contradictory issue (Lane 2005, Römkens et al 2002). Surface roughness acts to increase the storage capacity of rainwater, reduces the velocity of runoff, influences rain-splash detachment and lowers the erosive power of flow (Huang et al 1990).

Conversely, on rough surfaces flow tends to concentrate, depths and velocities increase, promoting scour, head cut development and an increased tendency for rilling

Figure 1. Map of the location of Marab Suway'id within the basalt flows of the NE Badia

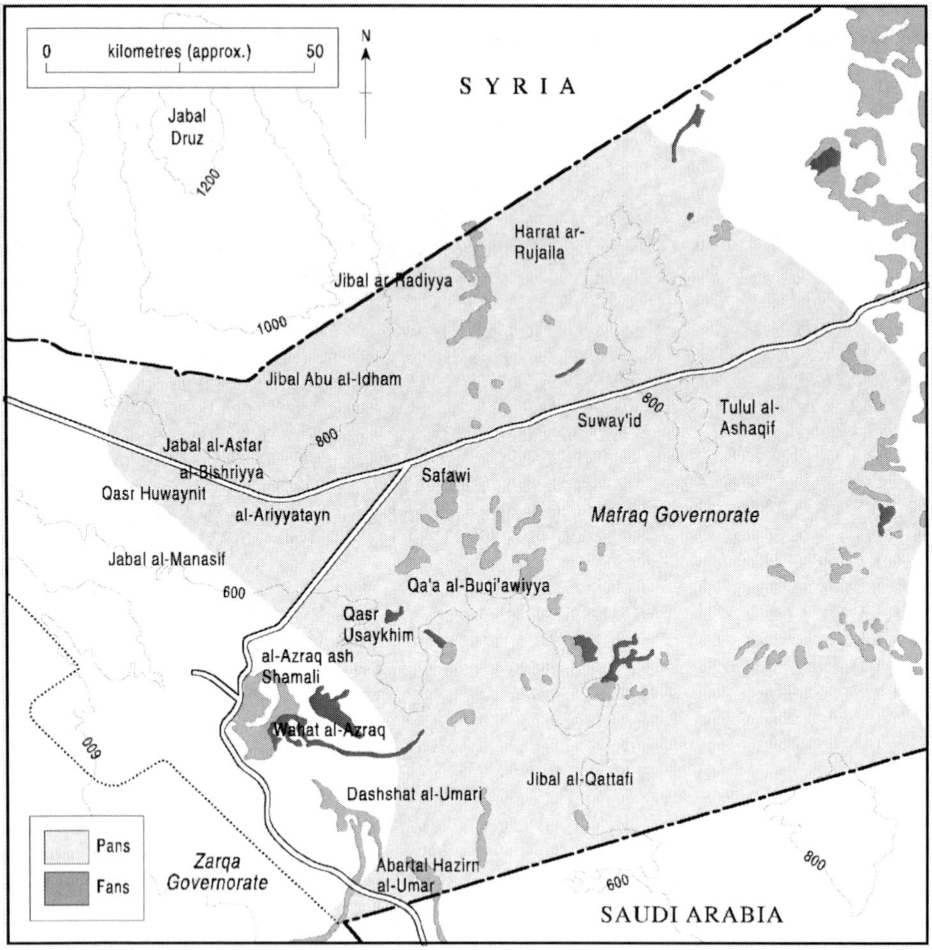

(Abrahams et al 1994b, Helming et al 1997, Puigdefabregas et al 1998). At the micro scale rock fragment cover (Collinet et al 1984), rock fragment size (Wilcox et al 1988, Yair et al 1979), surface roughness geometry (Bunte et al 1993) and soil matrix properties have all been identified as influential controls on flow and sediment dynamics (Poesen et al 1994). What is less well understood is the influence of surface roughness down a whole-slope profile, in particular where the interaction between surface process and surface modification is apparent. It is commonly recognized that arid areas exhibit variability in surface hydrology, specifically infiltration (Blackburn 1975, Johnson et al 1988, Sharma et al 1980). Heterogeneity in infiltration is mirrored in runoff generation from a surface (Puigdefabregas et al 1998). Previous work undertaken in the north-east Badia has identified a clear linkage between slope form and ground surface coverage, with evidence to suggest this relationship is homeostatic (Allison et al 1998, Higgitt et al 1999). The nature of down-slope variation in surface character is characteristic of an increasing intensity of surface modification processes down-slope. This is indicated by a dramatic variation in the ground surface configuration down slope profiles in the Badia (Figure 3). The (relatively) massive nature of the surface clasts compared to the underlying fines, the continuity of the ground surface coverage across the landscape and the systematic variation in ground surface form suggest an iterative type of surface modification. High magnitude rainfall, albeit infrequent in occurrence, is suggested to drive this modification.

2. Study area

The data presented in this paper were collected at Marab Suway'id, which lies on the 8.9 Ma olivine phyric Abed basalts (Tarawneh et al 2000), some 40 km to the east of Safawi (Figure 1). The basin represents a classic infilled wadi network, termed locally marab, with slopes of varying profile form feeding into the wadi channel. This site has been reported on extensively by Allison and Higgitt (1998), and was identified for this study as a suitable location for hydrology experiments due to the good access and the low degree of disturbance to the ground surface clast coverage. Stratified sediments in the marab bed indicate the episodic nature of flow and erosion events in the catchment.

3. Slope surface characteristics

The slope surfaces of the north-east Badia are characterized by large clasts which are commonly varnished (Figure 2). The open clast surface matrix has exposed fine grained underlying sediments, which in places form crusts, vesicular layers and caliche deposits. The fines are clays and silts derived locally from weathering of the basalt, combined with Aeolian deposits from surrounding limestone dominated regions. Deposits com-

Figure 2. Photograph showing the clast mantled slope in the upper reaches of Marab Suway'id

Figure 3. Slope profile and ground cover characteristics

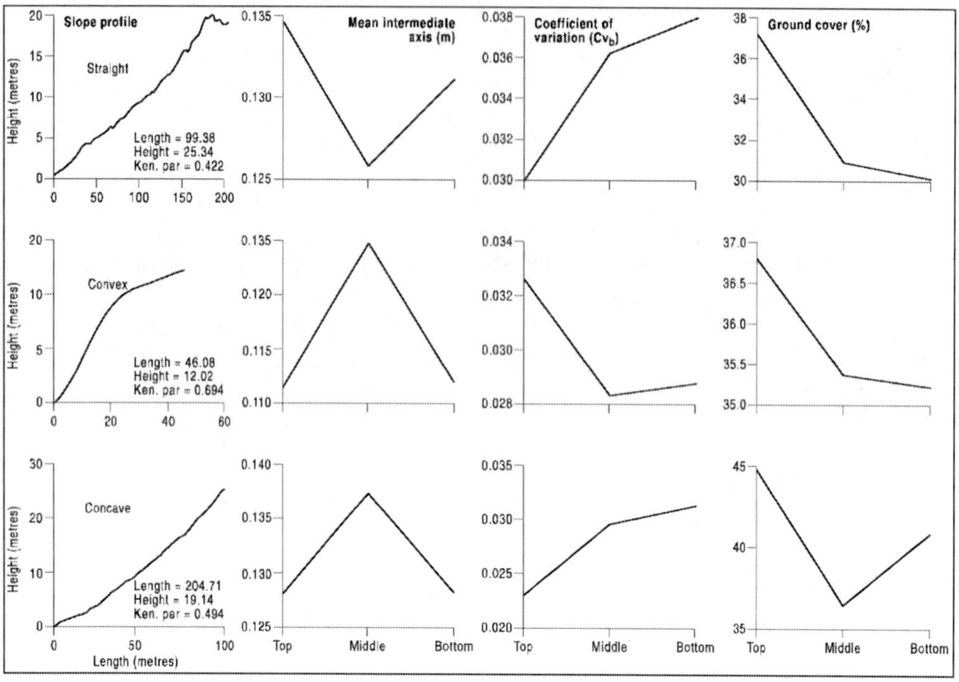

monly thicken in depth towards the toe of the slope profile. Clasts lie upon the surface but become increasingly buried down-slope (Higgitt et al 1999). A general association between slope and the clast surface characteristics was established by Allison and Higgitt (1998). Geology, and hence age dominates clast size distribution, which is in turn modified locally by the position of the sample site upon a slope profile. In a later study Higgitt and Allison (1999) examined further the relationship between slope process activity and the nature of the clast surface, with a clear spatial variation in ground surface form down profile, as illustrated in Figure 4. The degree of clast exposure was suggested to be related to the down-slope increase in transport and deposition of the underlying substrate. Evidence suggests that surface clasts are mobile, with spatial distributions with a departure from complete spatial randomness.

Identifying mechanisms for clast movement under the current rates and magnitudes of processes in this environment remains difficult to directly monitor. Clasts represent discrete objects more massive than those that can theoretically be transported by overland hydraulic action, making associations with literature on rock fragment dominated surfaces difficult. A self-sustained surface position of the clast, suggested by Higgitt and Allison (1999) to be derived via a process of winnowing whereby each clast maintains its surface position and spatial distributions are dictated by a gravity driven down-slope settlement is a possible explanation of this process. There is considerable evidence of pedological processes, with deposits of calcrete commonplace. Gypsum deposits are also common within the weathering horizon, suggesting the presence and movement of water that, combined with the clay content of the sediment, gives the potential for swell or heave processes.

A semi-quantitative method of slope profile selection was employed to increase the significance of the results generated from only a small number of sites. These were de-

Figure 4. Down-slope variation in coarse clastic ground surface coverage, displayed from low level aerial images. The extent of each image is approximately 6 x 9 m

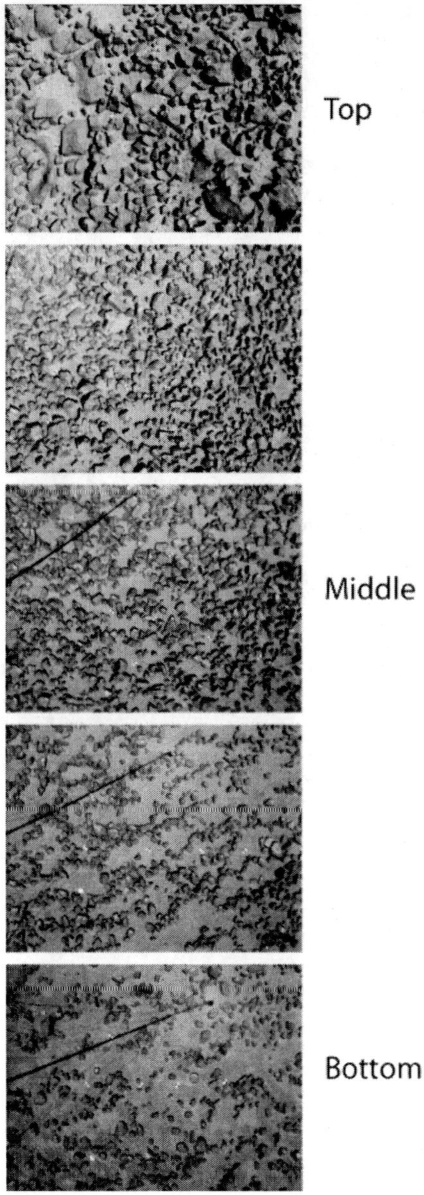

rived from three profiles, the characteristics of which are described in Figure 3, which were considered to be statistically straight, convex and concave. Upon each profile a crest, middle and toe sample area was selected. At each of these locations low-level aerial photography was used to derive measurements of the surface clasts using an image analysis procedure. The images extend across an area of approximately 6 x 9 m. Experimental plots (1 x 0.5 m) were positioned within this image to mirror the ground surface characteristics obtained for the whole image area. Three plots were positioned at each of

Figure 5. Rainfall simulator, with concrete experimental plot and collection tray shown below. The size of the plot is 1 x 0.5 m

the crest, middle and toe locations at each of three slope profiles, which was intended to examine the degree of heterogeneity in surface hydrology. Data from a total of 27 plots are presented at three sites (top, middle and bottom), on three profiles (straight, convex and concave).

4. Experimental methods

4.1 Rainfall simulation and infiltration

Rainfall simulation is employed to study runoff generation, infiltration and sediment dynamics in a diversity of environments, a review of which is beyond the scope of this paper. The technique has been applied to a wide range of pedological, hydrological and geomorphological situations (Foster et al 2000, Holden et al 2002, Wainwright et al 2000). Ultimately the size of the plot dictates the range of processes able to be studied; a small plot (< 2 m^2) commonly does not have a sufficient area to developed flow convergence or rill development.

Large plots have the benefit of spatial representivity but are logistically complex to manage. Given the semi-quantitative method of plot positioning described above and the lack of field evidence of flow convergence and rill development in the Badia, a small (0.5 m^2) plot is deemed adequate for this study.

There is a broad range of design and principles behind rainfall simulation, discussed more widely by, for example, Parsons and Lascelles (2000). This study employs a drop-former based simulator, as this design tends to produce a more tightly constrained distribution of drop sizes, produces a spatially consistent distribution of rainfall across the plot and creates repeatable experimental conditions which are essential when comparing surface character.

Under this method of drop formation the kinetic energy of drops is entirely determined by the rainfall drop-height. A simulator based on the design described by Bowyer-Bower and Burt (1989) and Robinson and Naghizadeh (1992) is employed in this study, with modifications for field site conditions (Figure 5).

Drops formers were constructed from Tygon tubing of 0.0023 m outside diameter and 0.007 m inside diameter, through which a 0.025 m length of 0.006 mm nylon was threaded and crimped secure. In total 627 drop formers evenly covered an area of 1 × 0.5 m beneath which a randomizing 0.004 × 0.004 m galvanized wire mesh was hung freely, which generated a 4 mm theoretical drop size. Drop height from the central point of the drip screen was kept constant at 1.8 m from the randomizing screen to the ground. Calibration tests showed that a given intensity could be recreated within ± 2 mm h^{-1} after the system had stabilized.

The drop zone was shielded from wind and sunlight using a corrugated plastic canopy during the experiments. Plots were constructed the day prior to the experimental run using concrete walls approximately 0.2 m high and 0.075 m wide, such that the inside dimension of the plot was exactly 0.5 m x 1 m. Concrete is moulded over clasts and into cavities which crossed the boundary and the walls were buried to a depth of 0.03 m where they traversed bare sediments to reduce lateral loss of water. Results from the rainfall simulation are compared directly to standard twin ring infiltration experiments undertaken on the exposed sediments in between surface clasts.

4.2 Storm simulation design

The design of the rainfall simulation experiment was based on rainfall characteristics both previously observed and those monitored in the region. In order to assess the levels of spatial variability in surface hydrology, the rainfall simulation was designed to force runoff generation using a storm simulation based upon observed events. Rainfall data from the Badia does not record maximum intensity, drop-size distributions or storm durations (Kirk 1997). Field observations imply much higher intensities than suggested by published data for this area (Kirk 1997). A standard deviation of over 45 % of the mean annual rainfall total is commonplace across the region. Over the longer-term, three gauges studied by Kirk (1997) show a more stable picture of annual rainfall totals, averaging 67.3-76.1 mm. Analysis of the magnitude and frequency of storm events, using data generated from an automatic weather station by Kirk (1997) shows that the majority of storm events total less than 5 mm of rainfall. Conversely in high rainfall years storms may frequently generate more than 10 mm of rainfall in one event. The pattern observed in the Badia is concurrent with observations from other dryland environments, with numerous documented examples of large but infrequent rainstorms and associated floods, several in the Middle East (Greenbaum et al 1998). The role of infrequent cellular convective storms is seen as a critical component of slope hydrology in the Negev (Greenbaum et al 2000). Here, localized convective Mediterranean storms have high intensities, with a means of 60 mm h^{-1} and maximums of 120 mm h^{-1}, often causing major flood events (Greenbaum et al 2000). The highest maximum rainfall intensity is reported by Lange et al (2000), at the Nahal Zin experimental catchment, at 224 mm h^{-1}. Only 5 % of storms in the Nahal Zin catchment generate overland flow and runoff. Small rainfall totals monitored during such storm events, of 12-19 mm do not reflect the magnitude of subsequent flood events. The disparity between the monitored rainfall and the observed floods reflects both the highly spatial and temporal nature of storm events. High intensities rarely last for more than eight minutes in the Negev (Kidron et al 2001) and rainfall totals vary by up to six times over a distance of only 150 m within a single convective storm cell.

Figure 6. Monitored storms in the Badia during the research period

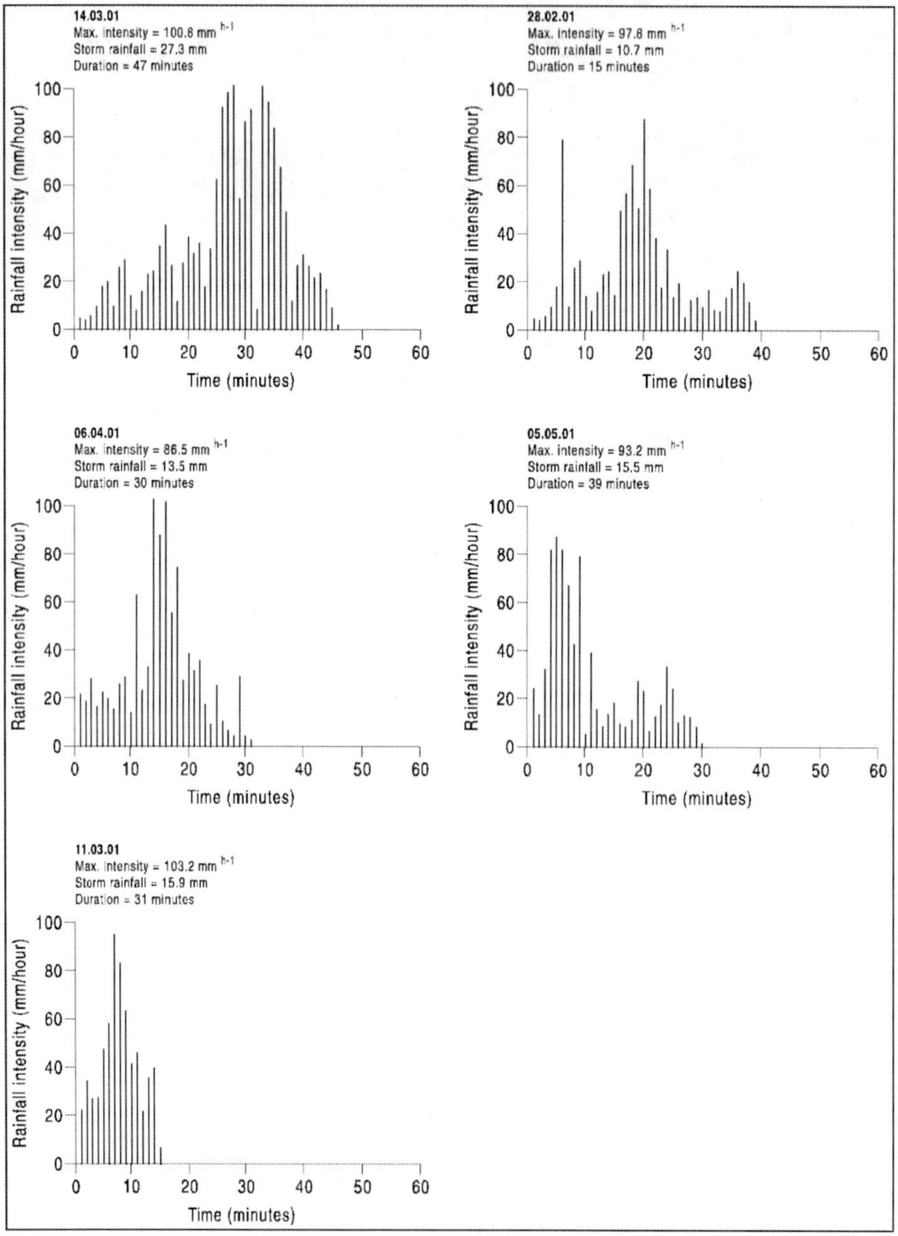

The maximum resolution of storm data previously collected in the Badia is 15 minutes (Kirk 1997); these data are too crude to give a detailed understanding of short, intense rainfall events. Attempts to rectify this were made during the six month field period at the start of 2001, whereby every storm which occurred during the field period was recorded. A sampling interval of one minute was used and a resolution of 0.2 mm achieved. Five events with a rainfall total in excess of 10 mm were monitored between 10.01.01 and 20.06.01 reflecting a wet year in the north-east Badia (Figure 6). Timings of neither storms nor rainfall totals match the data recorded at the permanent monitoring sites in the north-east Badia, suggesting the inadequacy of the spatial cover of the monitoring network to measure intense localized storm events. A series of three temporary rain gauges were set at the field site. The duration of storms ranged from 14-47 minutes and rainfall totals ranged from 10.7-27.3 mm. Within each storm there is a large variation in the intensity of the rainfall. Storms often followed a pattern of initial slight rain, followed by heavy intense rain, and then a final period of less intense rain. Maximum intensities recorded are of the order of 100 mm h^{-1} but this period of intensity only lasted for no longer than 10 minutes.

Figure 7. Top, simulated rainfall at 60 mm h-1 with simulator output overlain; bottom, simulation storm hydrograph, with input rainfall intensities and measured simulator output

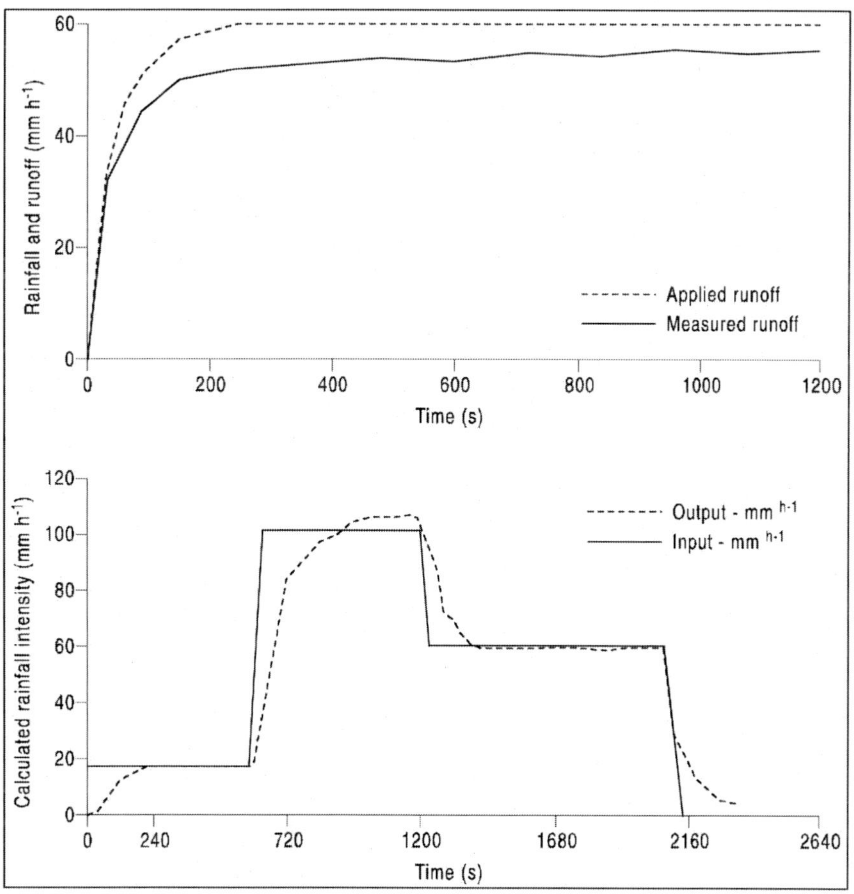

The choice of a single intensity for simulation is often arbitrary with little quantitative justification for the selection of that value. A simulation of a storm event approximated to a recorded event is hence suggested to be of no lesser value in understanding the surface response and variability. A model storm hydrograph was developed (Figure 7). Applied rainfall intensities were selected to be within the ranges observed from the storms monitored by others (Kirk 1997) and those within this study. The duration of the storm is kept constant at 2 100 s, with three distinct periods of rainfall; an initial 600 s of 17.5 mm h^{-1}; 600 s of 100 mm h^{-1}; and a final period of 900 s at 60 mm h^{-1} (Figure 7, solid line). These intensities were found to instigate runoff during the trial experiments after 180 s, 48 s and 112 s respectively.

The actual simulator output was assessed during calibration, and these values are used within the analysis (Figure 7, dashed line), which, is used for all rainfall partitioning calculations. Total rainfall for the simulated storm is 34.6 mm, which over the 0.5 m^2 plot, is approximately 17.3 l of water. Mean drop size increased with intensity, but stabilized after 70 mm h^{-1} to a mean D_{50} = 2.6 mm. Although the mean drop size is less than that measured in the field, the distribution of drop sizes generated by the drip screen is preferential to alternative methods of rainfall simulation.

5. Results

5.1. General observations

Infiltration and runoff generation rates varied considerably between both distal and contiguous plots. Low levels of rainfall instigated runoff, though in general significant flow was only noted shortly after the onset of intense rainfall. Runoff was Hortonian. The nature of surface wetting is dictated initially by the shelter afforded to the surface by clasts and their distribution, whereby overhangs sheltered areas from raindrop impact. Standing water quickly developed in depressions and was often confined or dammed by clasts. Little or no clast surface storage beyond surface wetting was noted due to their rounded, convex and varnished nature. Rain-splash transferred the coarser surface particles (< 2 mm) which concentrated at the sheltered clast perimeters.

Runoff was generated during all 27 experimental plots when standing water breached micro-depressions and flowed down through the plot, or over the lip (Table 1). Runoff was continuous but fluctuated and in some cases mirrored changes in rainfall intensity. The plot area contributing to runoff generation increased during the simulation as areas of standing water coalesced. Sediment yields were generally low. No significant hydraulic erosion of surface sediments was noted. Fine surface organic matter was initially transported from the surface via floatation and then in places some sediment detached as micro flow paths around clast perimeters developed, although this was very limited. Deposition of detached sediments occurred in areas of standing water. Flow direction was dictated by the local slope of the microtopography. When rainfall application halted, runoff quickly ceased. Significant evaporation and limited infiltration left standing surface water for up to two and a half hours after the simulation.

5.2. Rainfall partitioning and infiltration

Aggregated results from all experiments returned a mean runoff coefficient of 0.223, equivalent to 13.3 mm h^{-1} runoff under this storm hydrograph, indicating a mean steady state infiltration rate of 46.1 mm h^{-1}. Surface water storage was estimated immediately after 10 of the 27 experiments. Standing water was commonly observed to accumulate in over 75 % of the exposed sediment areas. Standing water reached a maximum depth of 27 mm, with a mean value of 13 mm.

Table 1. Rainfall simulation plot experiment statistics

Plot position	Plot number	Maximum runoff rate (mm h^{-1})	Mean runoff rate (mm h^{-1})	Runoff standard deviation	Total runoff (mm)	Mean sediment yield (g mm^{-1} h^{-1})	Total sediment yield (g)	Sediment yield standard deviation	Maximum sediment yield (g)	Time to runoff initiation (s)	Time to maximum runoff (s)
STT	1	80.400	42.249	21.389	4929	0.059	1.653	0.059	0.213	11.590	1140
STT	2	66.000	41.249	14.890	3953	0.009	0.212	0.014	0.070	11.300	1080
STT	3	74.640	41.983	18.692	5073	0.223	6.468	0.427	1.496	11.220	1080
STM	1	40.800	31.469	9.331	3278	0.015	0.347	0.046	0.231	14.000	1140
STM	2	54.960	37.173	12.211	4182	0.124	0.335	0.040	0.207	12.130	1140
STM	3	53.760	27.500	10.055	2750	0.006	0.143	0.008	0.029	12.470	1200
STB	1	54.480	27.844	14.839	2996	0.044	0.845	0.102	0.458	11.070	1140
STB	2	28.560	18.400	8.321	2070	0.226	6.113	0.230	1.218	12.400	1350
STB	3	41.280	18.691	12.424	1947	0.048	1.205	0.074	0.312	13.010	1260
CCT	1	69.360	31.324	18.316	3785	0.019	0.558	0.040	0.201	9.180	1200
CCT	2	76.800	40.270	21.307	4027	0.016	0.391	0.044	0.201	15.300	1350
CCT	3	86.160	40.648	26.150	5081	0.047	1.421	0.048	0.131	13.090	1500
CCM	1	46.800	28.720	12.556	3590	0.045	1.347	0.061	0.262	11.300	1350
CCM	2	69.120	36.665	18.479	3972	0.016	0.417	0.436	0.204	12.000	1140
CCM	3	48.960	39.777	12.362	2554	0.084	1.854	0.109	0.581	11.450	1020
CCB	1	54.000	33.757	12.862	4155	0.055	1.751	0.060	0.232	10.110	1020
CCB	2	56.400	29.956	14.930	2746	0.123	2.708	0.089	0.369	10.470	1200
CCB	3	65.520	41.639	16.669	4870	0.041	1.116	0.060	0.264	8.390	1110
CVT	1	75.600	35.881	16.970	4173	0.013	0.403	0.020	0.213	9.110	660
CVT	2	55.200	28.412	13.923	3486	0.050	0.506	0.076	0.213	12.010	1140
CVT	3	103.200	39.545	20.568	3876	0.012	1.105	0.023	0.422	6.300	810
CVM	1	136.000	87.778	25.798	9399	0.153	4.064	0.108	0.181	11.300	930
CVM	2	106.800	71.647	24.504	5561	0.091	1.579	0.107	0.405	12.000	840
CVM	3	72.480	54.707	19.626	4321	0.053	1.043	0.012	0.359	12.000	1290
CVB	1	53.760	27.880	15.005	2091	0.014	0.244	0.012	0.258	11.300	1410
CVB	2	48.000	34.320	9.862	3432	0.025	0.584	0.056	0.201	11.000	1320
CVB	3	65.520	44.271	14.182	4796	0.012	0.299	0.025	0.205	10.560	1260
Mean		66.095	38.287	16.156	3966	0.060	1.434	0.088	0.338	11.336	1151

Aggregated over the mean percentage of exposed sediment cover (52.9 % of plot area for all 27 plots) this represents 19.7 % of the applied rainfall volume. Infiltration, taking account of surface storage, therefore represents 58 % of applied rainfall, with a mean value of 36.2 mm h^{-1}. Infiltration varied considerably between locations, but corresponded well with previous studies in the region. For example, Warburton (1998) suggested a mean infiltration rate of 24 mm h^{-1} for the Badia sediments, the higher rate here attributed to the presence of clasts included within the experimental plot area.

Observation of infiltration processes from the rainfall simulation is complicated by the coalescence of ponded areas, often represented by an oscillation of runoff about a mean steady value. Runoff can therefore only be considered as quasi-stable on these surfaces, whereas infiltration rates maybe stable or steady. Results from a series of 27 twin-ring infiltration experiments are presented to give a better estimation of water exchange at the surface in areas devoid of clasts. Hortonian type infiltration regression curves are fitted to data from each sample site and also aggregated from the slope crest, mid-point and toe (Figure 8).

Figure 8. Infiltration curves: top, for the nine sample positions; bottom, for data aggregated by slope position

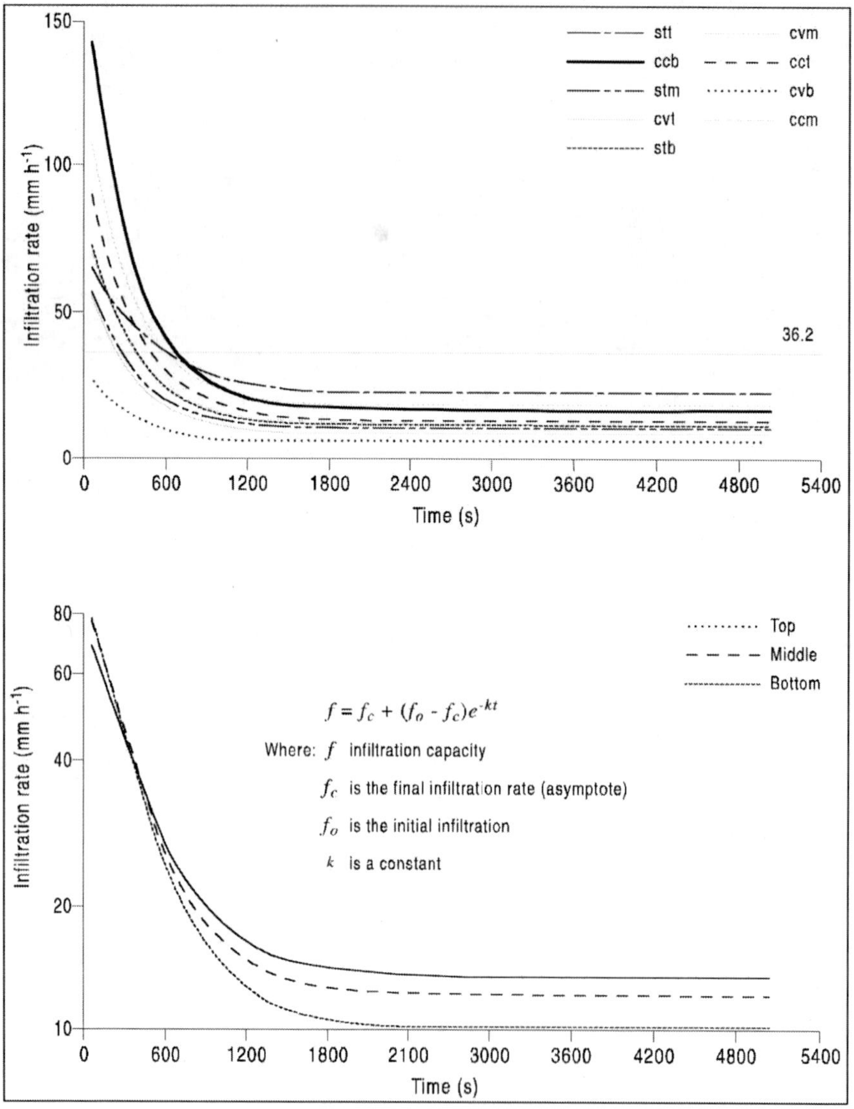

The infiltration experiments show a difference to those results generated from the rainfall simulation experiments, with mean steady state infiltration values of 13 mm h^{-1}, as opposed to a value of 36.2 mm h^{-1} derived from the rainfall simulation. The rainfall simulation experiments encompassed a relatively large plot area (0.5 m^2), which included both clast and exposed sediments; the ring infiltration experiments were undertaken over an area of 0.02 m^2 which did not include clasts. Given the low observed levels of seepage from the plot during the rainfall simulation, the difference in infiltration rates is attributed to the presence and variable nature of clasts in the area under consideration.

Figure 9. Relationship between mean rainfall intensity and mean sediment yield for all 27 experiments

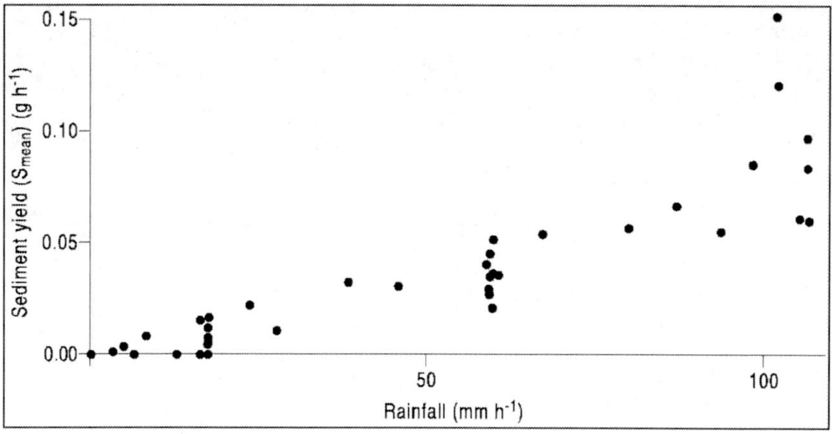

5.3. Sediment transfer

Sediment yield from the experimental plots was low. The mean sediment yield for all 27 experiments was 1.43 g, with a maximum of 6.47 g and a minimum of 0.14 g. As with runoff generation, sediment transfer fluctuated through the experimental period. Peak sediment yields occurred after a mean time of 900 s from the beginning of the experiments which corresponds closely with the onset of peak rainfall intensity. A statistically strong relationship between mean runoff and mean sediment yield ($r^2 = 0.707$, $P = 0.01$), and more significantly rainfall intensity and mean sediment yield is found ($r^2 = 0.789$, $P = 0.01$) (Figure 9). It is interesting to note also the increase in mean sediment yield variability with intensity; higher rainfall intensities exaggerate differences in surface response between sites. The diversity in surface response is attributed to the variable protection afforded to the surface by the clast cover. The generally low levels of sediment removal from the plots can be attributed to several factors. The plot area in itself is relatively small and the surface sediments were commonly crusted and rapidly sealed after the onset of rainfall. Runoff from a plot of 0.5 m^2 will not develop concentrated overland flow, so over a wider area scour or rill development may occur, but there is limited evidence of this, possibly due to the control afforded by the surface clasts.

5.4. Slope hydrology

The experiments show heterogeneity in slope surface hydrology between plot positions down-slope and also significant variation between contiguous plots at single sample sites. This suggests that slope position influences not only rate of infiltration and runoff generation, but also the nature of spatial variability of surface response. The systematic down-slope change in ground surface cover is suggested to be a slope dominated process which acts to modify the surface. Aggregated data by treating plots individually and relating the hydrological response to the local ground surface characteristics yields only insignificant correlations. More promising results are found when the data are considered as ordered down-slope profiles. Here three characteristic profile forms are considered separately.

'Straight' profile

Regression analysis between slope, ground cover and slope hydrology variables returns significant least squares correlations (Table 2a-c), between runoff and ground cover, but not sediment nor timings. This implies that sediment yields are linked to applied rainfall intensity more significantly than ground cover, as demonstrated in Figure 9. Poor relationships are derived from the statistics from only the area of simulation plot (0.5 m^2), including: ground cover percentage within simulation plot (Gcp_{plot}); total clast edge length within simulation plot ($EdgeTot_{plot}$), mean intermediate axis length within simulation plot (B_{plot}). Greater significance is achieved from the statistics for the area of the 6 x 9 m image area, including: ground cover percentage within image analysis plot (Gcpim), total clast edge length within image analysis plot ($EdgeTot_{im}$), coefficient of variation in intermediate axis within image analysis plot (CVB_{im}), and intermediate axis within image analysis plot (B_{im}).

From this arose a curious situation; plot surface hydrology appears to be more sensitive to surface characteristics at a wider scale than those specific characteristics of the plot itself. Slope, ground surface cover and spatial measures of the surface form all return statistically significant values of r^2 presenting a complex system within which many factors appear to influence hydrology, although some collinearity between variables is inevitable.

The results also support the application of *appropriate statistics* that can be conceptually related to process action, as suggested by Dunkerley (1999). Table 2a details the correlations with the weighted variables; including; volume weighted mean intermediate axis within the image analysis plot ($VolwMb_{im}$), area weighted mean intermediate axis within the image analysis plot ($AreawMb_{im}$), and edge weighted mean intermediate axis within the image analysis plot ($EdgewMb_{im}$), each of which shows statistically significant relationships with runoff variables ($\alpha = 0.05$) in excess of those with non-weighted measures. The strength of correlations as the weighted mean approaches a volume approximation of clast form ($VolwMb_{im}$) suggests the importance of edge length, elsewhere suggested to be important as a transition zone for water infiltration (Poesen et al 1992), and clast area, which affords protection from raindrop impact (Poesen et al 1994), in ad-

Table 2a. Slope form, ground cover surface hydrology correlation statistics (r-values): straight slope profiles

	Runoff			Sediment			Timings	
Variable	R_{max}	R_{mean}	R_{Tot}	S_{max}	S_{mean}	S_{Tot}	TR_{Start}	TR_{max}
Gcp_{plot}	0.039	0.019	0.006	0.026	0	0.041	0.087	0.015
$EdgeTot_{plot}$	0.008	0.006	0.010	0.199	0.353	0.200	0.137	0.048
B_{plot}	0.000	0.031	0.026	0.123	0.113	0.134	0.068	0.001
Gcp_{im}	0.731	0.698	0.663	0.017	0.007	0.044	0.313	0.413
$EdgeTot_{im}$	0.681	0.588	0.572	0.039	0.016	0.075	0.372	0.333
CVB_{im}	0.744	0.745	0.699	0.01	0.004	0.031	0.280	0.448
B^{im}	0.281	0.113	0.136	0.157	0.070	0.208	0.443	0.041
PltAngle	0.547	0.797	0.690	0.032	0.016	0.015	0.035	0.544
UpslpAngle	0.580	0.430	0.435	0.073	0.032	0.119	0.427	0.228
PltDist	0.695	0.852	0.765	0.003	0.002	0	0.127	0.551
Curv	0.663	0.555	0.544	0.019	0.084	0.045	0.386	0.310
Fractal	0.738	**0.829**	0.758	0	0.007	0	0.191	0.521
Theissen	0.717	**0.847**	0.767	0.001	0.001	0.002	0.153	0.002
VolwMb	0.732	0.700	0.664	0.017	0.007	0.043	0.312	0.414
AreawMb	0.710	0.645	0.620	0.027	0.011	0.059	0.344	0.374
EdgewMb	0.661	0.552	0.542	0.046	0.020	0.085	0.387	0.085

The bold type in the Table represents $P = 0.05$

dition to clast volume. Here clast volume is suggested to act as a variable for both edge length and area, and as a hydraulic roughness element which acts to impede and channel flow. The implication is that simplistic measures of ground cover, for example mean intermediate axis, may not always be appropriate.

The most significant relationship obtained is between plot distance from the slope crest (PltDist) and mean runoff (R_{mean}) ($r^2 = 0.853$); conventionally plot distance from slope crest is proportional to contributing area and flow volume and hence process intensity; in this experiment flow volume remained constant between sites, so the influence of PltDist was a local characteristic. Down-slope variation is not only apparent, but significantly linked with distance down-slope. Plot angle (PltAngle) can be conceptually linked to process action at a point when steeper increases runoff, as shown by the relationship between PltAngle and R_{mean} ($r^2 = 0.797$). These variables may represent both cause and effect; PltDist is the distance down-slope of the plot, and as such is also the distance over which runoff accumulates, or a measure of the distance-dependant modification process which increases down-slope. The variability in surface hydrology can be seen clearly by examining the mean runoff hydrographs from the rainfall simulations conducted at the three slope locations (Figure 10, top). The hydrographs, although similar, show reduced runoff maxima and reduced runoff totals at the slope middle and bottom plots, suggesting increased infiltration, again implying a negative feedback between the relative position of the plot, its local ground cover and resultant hydrological response. Lag-times reduce down-slope with runoff initiation and maxima occurring more quickly from slope top than bottom.

'Concave' slope

Variation in ground surface cover appears more complex on the convex and concave profiles compared to that on the straight profile; it is perhaps reasonable to assume that

Figure 10. Down-slope variation in runoff response to simulated rainfall

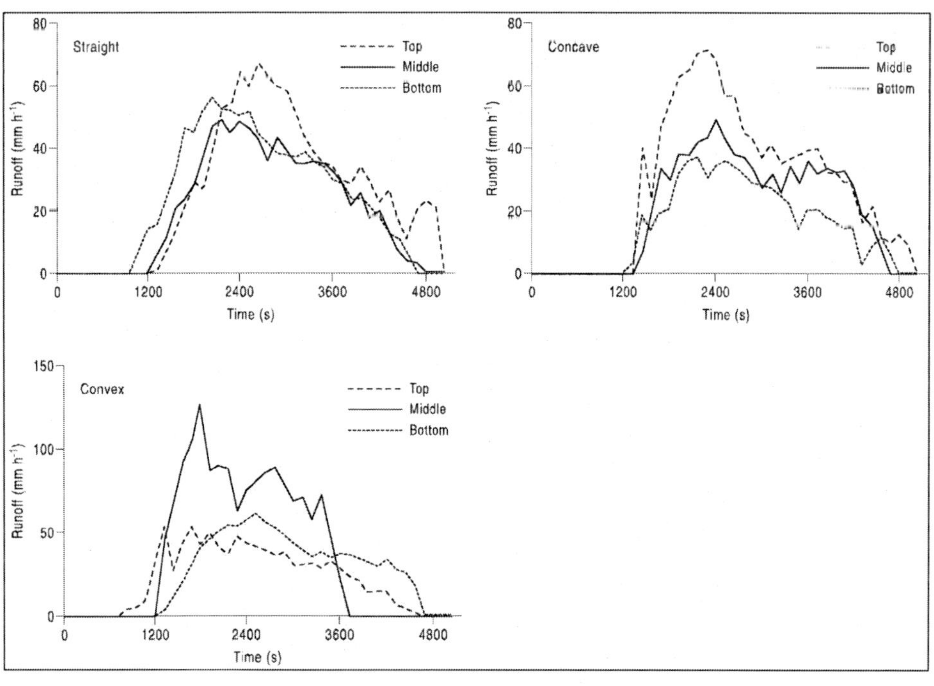

slope hydrology reflects this. Significant correlations are obtained between Runoff maximum (R_{max}) and Ground cover percentage (Gcp), the coefficient of variation in intermediate axis lengths (Cvb_{im}), the angle of the 15 m of slope above the plot (UpslpAngl) ($\alpha = 0.1$), and the weighted mean clast intermediate diameters. There are no significant relationships between Mean runoff (R_{mean}) and Runoff total (R_{Tot}), nor any of the slope, ground cover, sediment yields nor runoff timings. The correlation analysis in Table 2a-c examines linear trends in the relationship between variables. On the straight slope, change in ground cover character commonly appears to be linear with distance down-slope. The suggestion raised is that if the relationship between variables is equivalent to that on the straight profile, then the nature of down-slope change on the concave profile is non-linear. The relationship between ground cover and slope hydrology is influenced by slope form, here qualified by the consistent significance of the profile curvature variable (Curv) and up-slope angle (UpslpAngle). The runoff hydrographs for the three positions down the concave profile (Figure 10, centre) are similar in form. The top location generates the maximum mean runoff and greatest total volume of runoff, and shows a rapid response, which is a pattern that reverses down-slope. The top plot shows a flashy response, closely mirroring the applied rainfall intensity, which is attributed to limited underlying sediments to infiltrate water. The high slope angles at the top plot (top mean plot angle = 11.8°; middle mean plot angle = 9.8°; bottom mean plot angle = 7.1°) promote the rapid flow of water from the plot. Increases in mean sediment yield and total sediment yield correlate with increases in local mean plot angle.

Table 2b. Slope form, ground cover surface hydrology correlation statistics (r-values): concave slope profiles

		Runoff			Sediment		Timings	
Variable	R_{max}	R_{mean}	R_{Tot}	S_{max}	S_{mean}	S_{Tot}	TR_{Start}	TR_{max}
Gcp_{im}	0.535	0.04	0.228	0.059	0.050	0.309	0.033	0.221
$EdgeTot_{im}$	0.035	0.001	0.100	0.07	0.107	0.059	0.13	0.011
Cvb_{im}	0.571	0.052	0.115	0.261	0.286	0.244	0.262	0.441
B_{im}	0.250	0.015	0.194	0.001	0.005	0.193	0.013	0.036
UpslpAngle	0.534	0.040	0.228	0.058	0.049	0.309	0.033	0.219
PltAngle	0.395	0.040	0.041	0.313	0.362	0.137	0.351	0.416
PltDist	0.381	0.039	0.037	0.314	0.365	0.13	0.355	0.410
Curv	0.572	0.052	0.115	0.261	0.285	0.245	0.261	0.440
Theissen	0.008	0.003	0.030	0.175	0.234	0.003	0.257	0.106
Fractal	0.393	0.040	0.040	0.313	0.363	0.136	0.352	0.415
VolwMb	0.578	0.052	0.120	0.256	0.279	0.250	0.254	0.439
AreawMb	0.561	0.051	0.109	0.268	0.294	0.237	0.271	0.443
EdgewMb	0.529	0.050	0.091	0.283	0.315	0.215	0.294	0.444

5.4.3. 'Convex' profile

The top plot at the convex profile was the quickest of all to generate runoff. The degree in variability of runoff appeared to decrease down-slope, with the exception of the middle plot location. The middle plot location showed the highest levels of runoff and also the most rapid response after the onset of intense rainfall. The middle plot was located on the edge of a smooth convex ridge with an extensive low gradient plateau upslope. The local slope was highly convex (Curv = -4.2), with a high up-slope angle (9.0°). The convex profile returns significant correlations between Mean runoff (R_{mean}), mean clast intermediate diameter and its derivatives. There are also significant links with Mean sediment yield (S_{mean}) and Time to maximum runoff (T_{max}). No significant correlations

(α<= 0.1) are apparent between sediment yields, nor timings. Runoff maximum (R_{max}) appears to be poorly related to all variables. Mean runoff correlates well with the edge weighted mean intermediate axis ($EdgewMb_{im}$) and the mean un-weighted intermediate axis (B_{im}). It is interesting to note the reversal of the significance of the weighted mean values in this instance. Both convex and concave profiles fail to return any significant relationships with either slope angle or distance down-slope variables. Slope height and length on the convex profile (12.0 m and 46.1 m) are significantly less than on the straight profile. Overall volumes of runoff generated due to the reduced up-slope catchment area may result in less differentiation in form and process between the extremities of the slope due to the lower distance. When using standardized variables, the convex profile presents a large variation in profile curvature between plot locations. A significant variation in up-slope angle with distance down-slope is apparent. Sediment variables appear to mirror changes in up-slope angle. An increase in up-slope angle reduces the mean sediment load, total sediment yield and to a lesser extent the mean maximum sediment load. Time to runoff increases slightly at the middle plot location, and reduces at the slope bottom. Time to maximum runoff increases down-slope.

Table 2c. Slope form, ground cover surface hydrology correlation statistics (r-values): convex slope profiles

		Runoff			Sediment		Timings	
Variable	R_{max}	R_{mean}	R_{Tot}	S_{max}	S_{mean}	S_{Tot}	TR_{Start}	TR_{max}
Gcp_{im}	0.001	0.149	0.042	0.074	0.04	0.013	0.332	0.375
$EdgeTot_{im}$	0.433	0.289	0.242	0.323	0.279	0.179	0.002	0.452
Cvb_{im}	0.014	0.289	0.111	0.177	0.112	0.001	0.376	0.260
B_{im}	0.401	0.779	0.456	0.657	0.500	0.102	0.227	0.015
UpslpAngle	0.154	0.006	0.025	0.026	0.033	0.097	0.137	0.569
PltDist	0.106	0	0.007	0.006	0.011	0.077	0.178	0.551
PltAngle	0.320	0.125	0.131	0.168	0.156	0.152	0.036	0.543
Curv	0.523	0.592	0.408	0.567	0.459	0.178	0.043	0.232
Fractal	0.071	0.009	0.001	0	0.002	0.061	0.212	0.528
Theissen	0.157	0.007	0.026	0.028	0.035	0.098	0.134	0.570
VolwMb	0.141	0.575	0.281	0.422	0.299	0.018	0.378	0.071
AreawMb	0.235	0.689	0.363	0.536	0.391	0.043	0.339	0.017
EdgewMb	0.337	0.761	0.427	0.622	0.466	0.077	0.278	0

5.4.4 Multiple regression of controls on slope surface hydrology

Feedbacks between form and process complicate the nature of down-slope change across the range of profile forms examined. Exploratory multiple regression has been undertaken to assess the degree to which slope hydrology can be predicted from multiple variables (Table 3). Three successful regression models are presented, relating clast characteristics, local and whole slope form and runoff and sediment dynamics. Again sediment dynamics and runoff timing variables all show poor r^2 values, but from runoff variables may be derived two statistically significant models. Mean and maximum levels of surface runoff can be predicted from slope curvature (Curv), ground cover sorting (CVB) and the plot location on the slope (Pltdist). Analysis of the model, using a Cook-Weisberg test, returns a high probability (0.604) suggesting the residuals are not heteroskedastic. A variance inflation test for collinearity suggests a tolerance of 0.222, and residuals are normally distributed. This apparent independence between variables has interesting implications for understanding the interrelationships between variables, and the relative contribution of micro- and meso- scale influence of surface form / process interaction, discussed further below.

Table 3. Multiple-regression between slope, ground cover and surface hydrology

	Hydrological variable	Curvature	CVB	Theissen	Pltdist	Constant	Prob >F	R2
Runoff	Runoffmean	-2.102	2.518	46.385	-0.094	-75.384	0.000	**0.763**
	Runoffmax	-2.184	2.632	11.198	-0.175	-7.892	0.000	0.591
	Totalrunoff	-149.659	184.052	1784.413	-8.860	-2512.2	0.003	0.496
Sediment	Sedmean	-0.003	0.003	0.046	0.000	-0.066	0.511	0.133
	Sedmax	0.001	-0.009	0.327	0.000	0.264	0.847	0.058
	Sedtotal	-0.060	0.061	-0.149	0.004	-0.282	0.839	0.060
Time	t_{start}	-0.063	0.227	11.432	0.008	-8.571	0.113	0.278
	t_{max}	9.242	23.280	844.696	1.804	-700.78	0.001	0.553

6. Discussion

The rainfall simulations show that splash detachment plays a critical role in detaching sediment mobilization. Hydrological response to applied rainfall is increasingly muted with distance down-slope. The variation in surface hydrology is a reflection of not just ground cover, but a combined influence of the slope and ground cover at any given location in the landscape. Hydrology at a point on the slope is dictated by not only the local conditions, for example ground cover, but also by the situation of that plot within the context of the whole slope profile. The reworking of sediments and modification of the clast mantle is suggested here as a critical control on the variation in ground surface character. The processes identified at the scale of the rainfall simulation experiments are seen to be of insufficient magnitude to modify the clast surfaces over anything other than long timescales through processes of weathering and iterative movement. The apparent control of clast cover on surface hydrology, either directly or indirectly, suggests that flow velocities would never reach the magnitude sufficient to transport clasts by entrainment and there is no evidence in the field of this. This supports the suggestion of Cooke et al (1993) that although overland flow and slope hydrology may maintain the character of contemporary pavement surfaces, they cannot, by definition, be the process primarily responsible for their initial formation (Vincent et al 1995). The development of the surfaces is more complex than purely hydraulic transport of surface clasts, or runoff driven creep (Abrahams et al 1994a). Surfaces are well drained and there appears to be a good degree of connectivity down the slope profile, with only limited evidence of lateral drainage into preferential flow paths. The depth, velocity and routing of overland flow appear to be moderated by the clastic surface, and increasingly so down-slope. Three dominant aspects of surface modification are sediment movement, clast and clast population modification, and clast movement.

6.1. Sediment movement

There is abundant field evidence of down-slope transfer of sediments. Higgitt and Allison (1999) noted the increasing depth of fines down-slope, reflecting an increase in the burial of surface clasts. Sediments are not homogenous and as a result create a very different surface hydrology from slope crest to toe as demonstrated by these experiments. Sediment erosion as examined from the rainfall simulation appears complex, but again directly linked to ground surface character at a plot and the location of the plot on the slope-profile. Sediment transfer at the slope crest appears supply-limited. Slopes with high angles on the top and middle sections show an increase in sediment erosion down-slope, possibly reflecting a higher availability of material to erode and steeper plot angles. The convex profile shows a reduction in sediment transfer, albeit slight. Sediment deposition increases within plots with low local slope angles, which are commonly those with spatially organized surfaces, with low angled inter-clast areas, which promote deposition. Linear micro-dams of clasts are commonly infilled with fine sediments.

6.2. Clast and population modification

The clast population, mean size and sorting all vary systematically down-slope. The number of smaller clasts reduces towards the slope bottom, and the whole population becomes more normalized, suggesting some form of process modification. No process identified from the rainfall simulation experiments appears to directly account for these variations. The nature of infiltration and the development of standing water do however indicate the degree to which overland flow generation may occur, and the variability in this down the slope profile. The variation in sediment dynamics down-slope may also hold influence over the retention of water, particularly with such clay rich sediments, and hence may influence the degree to which weathering occurs. Small rock fragments (< 0.1 m) may be transported down-slope, which would produce the variation in ground cover observed. Increased rounding and smoothing, indicative of increased levels of abrasion down-slope, mirrors a reduction in the infiltration rate down-slope, which may account for the population normalization. The variation in hydrological response observed may therefore explain the variation in clast population; down-slope larger volumes of water as generated by the greater contributing area, are retained by increased sediment depths, promoting weathering and greater levels of clast abrasion via clast iterative movements.

6.3. Mechanisms of clast movement

In the absence of direct observations of the movements of clasts, the understanding of mechanisms of surface modification has to be derived from interpretation of the ground surface form and its hydrological response described above. Features upon the surface indicate that the position of ground surface clasts is most likely modified by small-scale iterative processes acting over a long time period. The case for an iterative process is supported by the relatively continuous clast dispersal across the surface, albeit with notable systematic spatial variations. The level of sediment transfer across the surface is low, even under the intense rainfall intensities simulated here. The apparent control of flow dynamics by the surface roughness afforded by the clasts suggests that flow velocities may not reach sufficient levels to generate scour, and hence levels of sediment removal are likely to remain the same. The massive nature of the surface clasts suggests that they cannot be moved during concentrated overland flow processes. Movement by shrink, swell and resettling of the underlying sediments during wetting, or ice needle action or pedological heave seem to be more feasible driving forces. This is particularly pertinent given the links established above between the spatial variation observed above in water transfer from the surface, the relative occurrence of clast population sorting and the systematic down-slope variation in this, which mirrors down-slope variations in surface organization.

7. Conclusions

Several conclusions can be drawn from the experiments conducted upon the north-east Badia surfaces. Links between surface form, slope and process show a close synergy suggesting that the surface adjusts and responds to process modification, dominated by the presence of surface water and the retention of this into the underlying sediments. It is evident that surface hydrology is highly variable between locations, but can be linked to slope form and ground cover. Statistically significant links exist at the micro-plot scale and the meso-slope scale; suggesting the importance of a consideration of whole slope form. Any generalized understanding of slope hydrology should ideally take into account the degree in variability of slope hydrology as influenced by ground cover and slope form. Splash detachment is a dominant control on sediment entrainment and mobilization and varies with rainfall intensity. An assessment of the spatial nature of surface clast distribution yields significant correlations with surface hydrology, suggesting that the surface clasts act as spatially structured roughness elements.

The experiments undertaken have also suggested the occurrence of a self-regulation in the behaviour of these surfaces. This includes the interaction between the clast mantle and rainfall acting to promote infiltration into the surface, the increasingly muted hydrological response down-slope where process intensity is at its highest and the apparent tendency for the clast mantle to organize into structures which may act to resist further modification. This type of surface organization, at a range of spatial scales, is therefore a vital control on slope hydrology in this environment; a landscape facet commonly attributed to self-regulating behaviour. In turn, the occurrence of clast organization appears to be linked to slope (gravity) controlled processes, but not directly, or solely slope hydrology. Sediment yields, under even high rainfall intensities, are low, suggesting that any processes of surface or profile modification are slow. It is also clear from this study that only minor perturbations in either the ground surface coverage or the site configuration can have a profound effect on the hydrological response of the surface.

Acknowledgements

The authors would like to thank the staff of the Jordan Badia Research and Development Programme, Safawi, in particular Mr Mohammed Shahbaz, Dr Salim Al-Oun and Dr Roderic Dutton, The Centre for British Research in the Levant, Amman for their support of this research, and the technical staff of the Geography Department, Durham University. Valuable comments and suggestions were also gratefully received from Louise Bracken and Andrew Goudie during the preparation of this manuscript.

References

Abrahams AD, Howard AD and Parsons AJ (1994a). Rock-mantled slopes. In: Abrahams, AD and Parsons AJ (eds), Geomorphology of Desert Environments. London: Chapman and Hall pp. 173-212.

Abrahams AD and Parsons AJ (1994b). Hydraulics of overland flow on stone-covered desert surfaces. Catena 23:111-140.

Allison RJ and Higgitt DL (1998). Slope form and associations with ground surface boulder cover in arid environments. Catena 33:47-74.

Blackburn WH (1975). Factors influencing infiltration and sediment production of rangelands in Nevada. Water Resources Research 11:929-937.

Bowyer-Bower TAS and Burt TP (1989). Rainfall simulators for investigating soil response to rainfall. Soil Technology 2:1-16.

Bryan K (1922). Erosion and sedimentation in the Papago Country, Arizona. U.S. Geological Survey Bulletin 730-B:19-90.

Bunte K and Poesen J (1993). Effects of rock fragment covers on erosion and transport of non-cohesive sediment by shallow overland flow. Water Resources Research 29(5):1415-1425.

Collinet J and Valentin C (1984). Evaluation of factors influencing water erosion in West Africa using rainfall simulation. International Association of Hydrological Sciences; IAHS-AISH Publication 144.

Cooke RU, Warren A and Goudie AS (1993). Desert geomorphology. London: UCL Press.

Dunkerley DL (1999). Banded shrublands of arid western New South Wales: responses to interannual rainfall variability. Ecological Modelling 121:127-138.

Foster IDL, Fullen MA, Brandsma RT and Chapman AS (2000). Drip-screen rainfall simulators for hydro- and pedo-geomorphological research: The Coventry experience. Earth Surface Processes and Landforms, The Journal of the British Geomorphological Research Group 25(7):691-707.

Greenbaum N, Margalit A, Shick A, Sharon D and Baker VR (1998). A high magnitude storm and flood in a hyperarid catchment, Nahal Zin, Negev Desert, Israel. Hydrological Processes 12:1-23.

Greenbaum N, Schick AP and Baker VR (2000). The palaeoflood record of a hyperarid catchment, Nahal Zin, Negev Desert, Israel. Earth surface processes and landforms 25(9):951-971.

Helming K, Prasad SN and Romkens MJM (1997). Drainage network characteristics on eroding soil surfaces. In: Wang, SSY, Langendoen, EJ and Shields, J, F.D. (eds). Proceedings of the Conference on Management of Landscapes Disturbed by Channel Incision, Oxford, MS, May 20 - 22.

Higgitt DL and Allison RJ (1999). Characteristics of stone covers on the surface of basalt flows in arid, northeast Jordan. Geomorphology 28:263-280.

Holden J and Burt TP (2002). Infiltration, runoff and sediment production in blanket peat catchments: implications of field rainfall simulation experiments. Hydrological Processes 16(2537-2557).

Huang C and Bradford JM (1990). Depressional storage for Markow-Gaurssian surfaces. Water Resources Research 26 (9):2235-2242.

Johnson CW and Gordon WH (1988). Runoff and erosion from rainfall simulator plots on sagebush rangeland. Transactions of the American Society of Agricultural Engineers 31(421 - 427).

Kidron GJ and Yair A (2001). Runoff-induced sediment yield over dune slopes in the Negev desert. 1: quantity and variability. Earth surface processes and landforms 26(5):461-474.

Kirk AJ (1997). Relationships between sediment, moisture and soil crust characteristics in arid environments. PhD Thesis, University of Durham.

Kirkby A and Kirkby MJ (1974). Surface wash at the semi-arid break in slope. Zeitschrift für Geomorphology, Supplement Band 21:151-176.

Lane (2005). Roughness – time for a re-evaluation? Earth surface processes and landforms 30:251-253.

Lange J, Leibundgut CH and Schick AP (2000). The importance of single events in arid zone rainfall-runoff modelling, (B). Physics and Chemistry of the Earth B 25(7-8):673-677.

Melton MA (1965). Debris-covered hillslopes of the southern Arizona desert – consideration of their stability and sediment contribution. Journal of Geology 73:715-729.

Parsons AJ and Lascelles B (2000). Rainfall simulation in geomorphology. Earth surface processes and landforms 25 (7):679-679.

Poesen J and Ingelmo-Sanchez F (1992). Runoff and sediment yield from topsoils with different porosity as affected by rock fragment cover and position. Catena 19:451 - 474.

Poesen J and Lavee H (1994). Rock fragments in top soils: significance and processes. Catena 24:1-28.

Puigdefabregas J, Gutierrez L, Sole A, Del Barrio G and Boer MM (1998). Differential responses of hillslope and channel elements to rainfall events in a semi-arid area. Geomorphology 23(2 - 4):337-351.

Robinson DA and Naghizadeh R (1992). The impact of cultivation practice and wheelings on runoff generation and soil erosion on the South Downs: some experimental results using simulated rainfall. Soil Use and Management 8(4):151-156.

Römkens MJM, Helming K and SN P (2002). Soil erosion under different rainfall intensities, surface roughness, and soil water regimes. Catena 2 - 3:103-123.

Roux JS and Roos ZN (1986). The relationship between the size of particles in surface wash sediment and rainfall characteristics on a low angle slope in a semi-arid climate. Zietschrift fur Geomorphologie 30(3):357-362.

Sharma ML, Gander GA and Hunt CG (1980). Spatial variability of infiltration in a watershed. Journal of Hydrology 44(1 - 2):101-122.

Tarawneh K, Ilani S, Rabba I, Harlavan Y, Peltz S, Ibraham K, Weinberger R and Steinitz G (2000). Dating of the Harrat Ash Shaam Basalts, northeast Jordan (Phase 1). Jordanian Natural Resources Agency, Geological Survey of Israel Report GSI/2/2000.

Vincent P and Sadah A (1995). Downslope changes in the shape of pediment debris, Saudi Arabia. Sedimentary Geology 95(3-4):207-219.

Wainwright J, Abrahams AD and Parsons AJ (2000). Plot-scale studies of vegetation, erosion and overland flow interactions: Case studies from Arizona and New Mexico. Hydrological Processes 14(16):2921 - 2943.

Warburton J (1998). A reconnaissance survey of surface water infiltration characteristics. In: Dutton RW, Clark JI and Battikhi AM (eds), Arid land resources and their management: Jordan's desert margin. London: Kegan Paul International pp. 157-170.

Wilcox BP, Wood MK and Tromble JT (1988). Factors influencing infiltrability of semi-arid mountain slopes. Journal of Range Management 41:197-206.

Yair A and Lavee H (1979). Runoff generative processes and runoff yield from arid talus mantled slopes. Earth surface processes and landforms 1:235-247.

CHAPTER 6

Assessing runoff potential for water harvesting generation in the Badia

Jarrah Al-Zu'ubi
School of Geography, Manchester University, Oxford Road, Manchester, M13 9PL, UK and PO Box 391, As-Salt-19110, Jordan. E-mail: jarzubi@hotmail.com

Clive Agnew
School of Geography, Manchester University, Oxford Road, Manchester, M13 9PL, UK

Based on simulated rainfall at sites in the north-east Badia and on earlier work that related runoff to total rainfall, rainfall intensity (high or low) and whether or not the antecedent surface condition was wet or dry, it is shown that the most important factors affecting runoff and infiltration rates are the rainfall intensity and surface moisture rather than ground cover characteristics; medium to high intensity rainfall generates 79-91 % of precipitation as runoff; low intensity storms still generate 59-84 % of the storm as local runoff. These values are very high and reveal the tremendous potential for water harvesting if the water can be collected locally. Losses through runoff are high due to seepage and evaporation, which both increase with catchment size. Runoff needs to be collected within small scale catchments located as close as possible to the slopes generating the runoff, as traditionally practised by Bedu by using the *mahafir* technique. A problem is that rainfall is typically collected or reported on a monthly basis; there is need for a network of weather stations if the storm duration / intensity data required to justify and locate 'modern' *mahafir* is to be systematically obtained.

Keywords: Rainfall simulation, water harvesting

1. Introduction

Jordan is facing a water crisis. Population growth is expected to continue at over 3.2 % per annum (Royal Court 1998 pp 40-42) while agricultural intensification is adding to the national water burden. Renewable resources do not meet present water demands (NES 1991 pp. 27-9), and alternative supplies are being sought eg desalination and recycling. Over-abstraction of ground water has resulted in significant falls in the groundwater table in the Azraq basin and domestic supply is only provided two to three days a week in Amman. The situation is exacerbated by possible changes in the seasonal pattern of rainfall. Jordan requires more efficient use of its existing water supplies and this is particularly the case in the drier eastern region known as the Badia. One strategy that has been widely practised in the past in the Middle East is water harvesting. This technique aims to make more effective use of precipitation through capturing and directing surface flows or harvesting rainfall directly.

The Mediterranean climate is characterized by irregular rainfall, droughts, and high-energy storm events. During the last two decades, systematic studies have been carried out in order to ascertain the processes involved and their hydrological response (Martinez-mena et al 1998). There is an extensive amount of literature describing runoff generation in humid regions, but fewer studies have been carried out in hyper-arid areas, especially under a Mediterranean climate (Abrahams et al 1988, Lavabre et al 1993, Yair et al 1976).

Figure 1. Study area and site locations (Source: BRDP, Safawi Centre files)

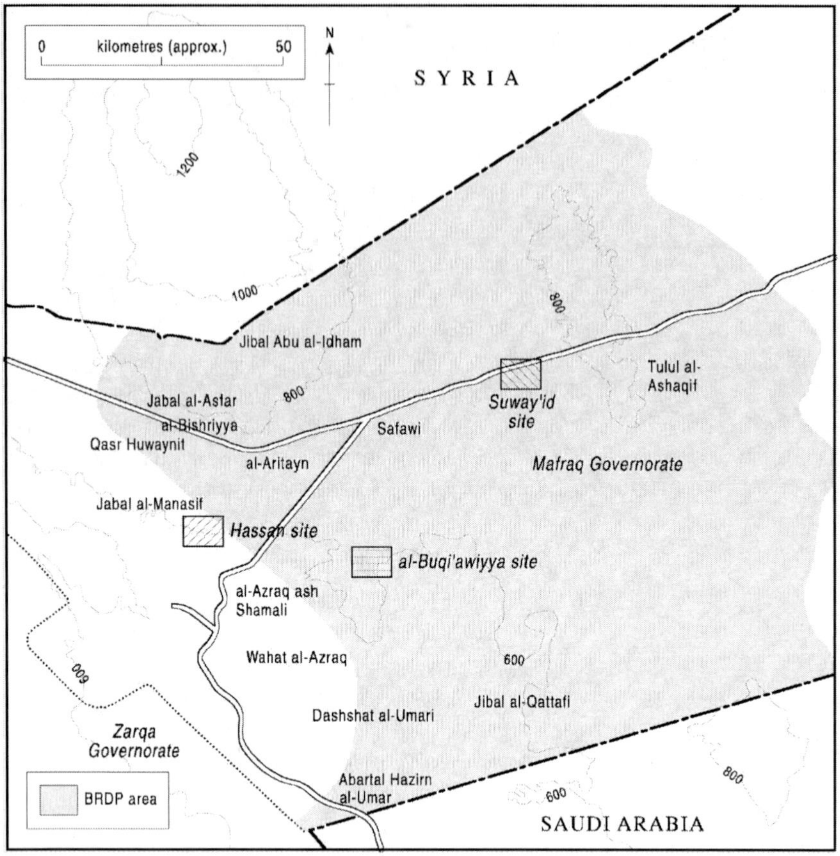

The Jordanian rainy season lasts October-April with most precipitation received in January and February. Only around 1.3 % of Jordan's area receives an average annual precipitation of more than 500 mm, only 1.8 % 300-500 mm, 3.8 % 200-300 mm, and 12.5 % 100-200 mm. The remaining 80.6 % receives less than 100 mm/yr. The above is an inadequate picture of Jordan's water situation. It is clearer to say that only about 3 % of the country receives an average annual rainfall over 300 mm. In addition, 83 % of the total rainfall over the whole country occurs in areas receiving less than 300 mm/yr, which means that only 17 % of rainfall can be used in dry land farming. The remaining 83 % requires irrigation and other technologies to be developed, except for rangeland, where 300-150 mm/yr of precipitation may be adequate. Generally humidity ranges from 15 % to 30 % in summer to 50 % in winter (Table 1).

The potential evaporation rates range from about 2 000 mm/yr in the extreme north-west edge of the country, to more than 4 000 mm/yr in the Aqaba areas. Along the Rift Valley the potential evaporation increases from a minimum of 2 000 mm/yr in the north, to some 2 500 mm/yr in the Dead Sea area and to more than 4 000 mm/yr in Aqaba. These rates are 5-80 times the average amounts of precipitation. Potential evaporation rates on the plateau increase to the east and south, rising from around 3 000 mm/yr at the eastern edge of the highlands to around 4 000 in the centre of the plateau. The southern desert exhibits potential evaporation rates of 3 500-4 400 mm/yr. This high evaporation makes precipitation generally ineffective, with large soil moisture deficits retarding plant growth (DoM 1998).

Table 1. Water situation in Jordan (supply and demand). MCM = million cubic metres

Year	Supply	Demand	Deficit (MCM/yr)
1995	882	1 104	(222)
2000	960	1 257	(297)
2005	1 169	1 407	(238)
2010	1 206	1 457	(251)
2015	1 225	1 550	(325)
2020	1 250	1 658	(408)

Source: Department of Meteorology files, 1998

The study area, located in the north-east Badia region (Figure 1) is a stone-covered arid landscape where precipitation seldom exceeds 200 mm/yr. In the north-west part of the study area, the Druz mountains raise the annual precipitation to a round 350 mm (Higgitt et al 1999). The study site is located within the BRDP (Badia Research and Development Programme) territories, part of the Hamad basin and the Azraq oasis, which is the most important basin in Jordan as it is the main source of potable water of the capital, Amman. Comprising around 80 % of Jordan, the Jordanian Badia, is part of the North Arab Desert. It stretches into Syria, Iraq and Saudi Arabia, with elevations of 600-900 m asl. Climate in the Badia varies widely between day and night, and between summer and winter. Daytime summer temperatures can exceed 40 °C, while winter nights can be very cold, dry and windy. Rainfall is minimal throughout the year, averaging less than 50 mm annually. Although all the regions of the Badia are united by their harsh desert climate, similar vegetation types and sparse concentrations of population, they vary considerably according to their underlying geology. The volcanic formations of the northern basalt desert extend into Syria and Saudi Arabia, and are recognizable by the black basalt boulders, which cover the landscape. East of the basalt desert, the desert is an undulating limestone plateau, which extends to the Iraqi border. There is some grassland in this area, and some dryland agriculture is practised. North-east of Amman, the eastern desert is crossed by a multitude of vegetated wadis, which include the Azraq oasis and the Shomari Wildlife Reserve (MWI files).

Water harvesting systems, *mahafir*, can be found in this region. These are mounds of earth excavated on some *qa'a* but not all and consisting of 'depressions in an arid or semi-arid region that are periodically inundated by surface runoff' (Agnew et al 1995). Agnew mentioned that there are number of hypotheses about the operation of the *mahafir* and it needs more investigation. The main objective of this study is to establish the changes in runoff generation and hydrological responses as influenced by different combinations of land characteristics representative of a Mediterranean arid zone and to increase the knowledge of runoff generation mechanisms under these environmental conditions. Allison (1998) mentioned that there is a potential to harvest water from the desert areas of Jordan, but no data have been collected to corroborate this, nor has any work been done in the BRDP area to examine the likely consequences of water control and management. Also, he emphasized that there are a number of issues that provide focus for much of the surface water development strategy work such as high magnitude runoff events during intense storms, the collection of water, provision of water for irrigation and the provision of water for livestock and groundwater recharge.

2. Water harvesting

Water harvesting is an umbrella term describing a range of techniques for collecting, concentrating and conserving water from various sources for various purposes. Different approaches to classifying water-harvesting methods have been proposed by a number of authors. Generally, these are based on the characteristics of the runoff producing and storage elements of the system. At the simplest level two groups can be identified:

- Within-field methods in which the transfer of water takes place over a short distance (maximum 50-100 m) usually by sheet flow. This category includes micro catchment, natural-counter ridges, furrow dyking, strip planting and stone bunds. (Al-Tabini et al 2007);
- External catchment in which runoff is collected from a catchment area at a considerable distance from the receiving area and is transferred by channel flow or overland flow. This category includes terraced wadi systems, hillside conduit systems, dams used for recession planting, *majaluba* systems and *qa'a* systems (Patrick 2000).

Interest in water harvesting can be traced back to at least the Nabataens (earlier than 300BC). In 1994 FAO (1994) hosted an international workshop on water harvesting in the Middle East and North Africa. UNESCO held a similar workshop for the Arab world (UNESCO 1986) and a smaller scale workshop (UNESCO 1986) as well. UNEP (UNEP 1983) and various other national and international bodies have also funded the meetings of the International Rainwater Catchments Systems Association (IRCSA), which focused on domestic water supply.

Common characteristics of water harvesting include:

- A catchment area: this is the area, which supplies the runoff to the cropped or run-on area. It can be a field, a hillside, a road, a roof, a river, or any other area from which water can be collected and supplied to the cropped area or storage area by, for example, channels and governed by the landscape;
- That water harvesting is particularly well suited to ASAL (arid and semi-arid lands) in that it facilitates taking maximum advantage of rainfall. Usually, the runoff area is larger than the run-on area;
- The ephemeral nature of water input in ASAL and the high rates of evaporation. Water harvesting enables us to store the water and then redistribute it effectively as we need;
- The fact that water harvesting systems involve only small engineering works and do not need too much capital to establish, which makes them suitable for developing countries;
- The fact that water harvesting systems often harvest soil (sediment) as well, depending on the characteristics of the crust, slope of the runoff area and the volume of water (rainfall).

Water harvesting designs can also increase the system efficiency in catchment hydrology by reducing conveyance losses by converting overland flow to channel flow. The nature of the surface materials and associated geology play a role in flow discontinuities in the catchment area. This is affected through changes in surface infiltration where the type and characteristics of the land surface are expected to exert significant influence upon the volume and duration of runoff.

Water harvesting shares many characteristics in common with other soil and water conservation (SWC) techniques. Water harvesting deals with water, but SWC deals with soil conservation and the prevention of erosive flows of water.

Suitability for water harvesting, normally understood to mean environmental and increasingly social suitability, could also be extended to include political suitability, feasibility and desirability. In Jordan, environmentally, water harvesting should be suitable due to the historical evidence of the *mahafir*. The study area contains wadis and hill slopes, which are potential sites for water harvesting. The social suitability increases daily due to the shortage of water in the country; Azraq groundwater basin is one of the main sources of potable water for Amman. Dottridge (1998a) explained the dilemma of the Azraq basin because the upper aquifers there are already over-abstracted beyond the sustainable limit, and abstraction must be reduced if spring flows are to be restored. Also water harvesting is an important issue in the social development of the Bedu who live in the area as it could be a good source of water for their livestock. Politically, the Middle East is one of the critical areas in the world for water supply enhancement due to the conflict between countries about water.

A major international convocation of water harvesting experts took place in Cairo in 1993 (FAO 1994) indicating the widespread interest in the promise of water harvesting. Some of the findings and recommendations are listed below:

- The lack of adequate hydrological data and information is a limiting factor for confident planning;
- National hydrological institutions should be strengthened with regard to the collection of data relevant to water harvesting;
- A regional data bank should be established;
- There is no agreed-upon definition and terminology with regard to water harvesting;
- A working group should be set up by FAO to standardize definitions and terminology and classifications of water harvesting;
- Most water harvesting projects are performing inefficiently;
- More attention needs to be paid to the balance of the water supply and water use;
- Water harvesting projects should be planned, designed and implemented by a multi-disciplinary team;
- Recent development of special equipment should be introduced for large-scale water harvesting to speed up implementation and reduce labour costs.

3. Theory

This study concerns the assessment of runoff potential in a hyper-arid environment. There are many factors affecting runoff behaviour such as rainfall intensity, evaporation, infiltration, interception, rock size, drop size, kinetic energy of rainfall, wind, slope and crust type (Abrahams et al 1991, Anderson et al 1990, De Lima 1988, Emmett 1980, Jean et al 2000). Stone cover is an important factor in determining flow velocity (Jean et al 2000) and in arid environments we can expect a variety of rock fragment distributions and crust types. Flow velocity is of importance for the routing of runoff over the land surface and for calculation of the erosivity and transport capacity of the flow. To calculate the flow velocity a good characterization of the hydraulic resistance is required.

According to current studies and models, the hydraulic resistance of the surface is characterized by a roughness coefficient (Takken et al 2000), usually known as the Manning coefficient (n) or the Darcy-Weisbach friction factor (f).

The Darcy-Weishach frication factor is defined as: $f=8gRS/u^2$ (1)

The Manning roughness coefficient is defined as: $n=R^{2/3}S^{1/2}/u$ (2)

Where: g: is acceleration of gravity (ms^{-2}); R: is the hydraulic radius (m); S: is the sine of the surface slope angle; u: the mean velocity (ms^{-1}).

Takken (2000) mentioned that unit discharge is used in the calculations and assumed the hydraulic radius to be equal to flow depth (d). According to this, flow width is a constant and the surface is thus considered to be fully submerged. Recent research has shown the sensitivity of the hydraulic roughness coefficients used within the models (De Roo et al 1989, Jetten et al 1998). Takken argued that Manning's *n* is generally assumed to be independent of flow parameters and values. But the Darcy-Weisbach friction factor is dependent on flow rate and is expressed in relation to the Reynolds number (*Re*):

$$Re = ud / \upsilon \tag{3}$$

Where: υ: is the kinematics viscosity (m^2s^{-1}).

In hill slope flow, the roughness of the surface is relatively high and the flow depths small, so the roughness can protrude above the flow. In this situation, Jetten mentioned that the *f-Re* relation may vary depending on surface characteristics and inundation of the roughness elements. Gilley and Finkner (Gilley et al 1991) developed two empirical relationships relating respectively the Darcy-Weisbach roughness coefficient (*f*) and the Manning roughness coefficient (*n*) to the random roughness (*RR*) of the surface and Reynolds number (*Re*):

$$f=6.3RR^{1.75}/Re^{0.661} \tag{4}$$

$$n=0.172RR^{0.742}/Re^{0.282} \tag{5}$$

After that, Lawrence (1997) developed a model to predict the Darcy-Weisbach friction factor based on the inundation ratio, which is the ratio between the depth of the flow and a characteristic roughness length (*k*) that represents the size of the roughness elements. Depending on this ratio, she developed three flow regimes:

Well inundated flows ($d/k \geq 10$) the fundamental equation for turbulent flow used, implying gradual decrease in the ratio: $1/\sqrt{f} = 1.64+0.803\ ln(d/k)$

At marginal inundation ($2<d/k<10$): $\sqrt{f} =10\ (d/k)^{-2}$

At partial inundation ($d/k \leq 2$): $f= 8/\iota\pi PCd \min [\pi/4,d/k]$

Where: *Cd*: is the coefficient of drag; *P*: is the fractional surface cover.

3.1 Rock cover

Takken (2000) mentioned that the hydraulic resistance is dependent on flow rate and there is a variation of hydraulic resistance of bare, rough surfaces with discharge and slope. The best way to predict flow velocity could be made when detailed data on surface roughness is available. This study seeks to investigate the impact of different surfaces upon runoff in the Badia area.

A stony surface is a common characteristic of the Jordan Badia. The Mediterranean region is characterized by shallow soils with high rock fragment content due to various factors such as geology, climatology, topography and land use (Poesen et al 1996). Many studies have examined the effect of the slope on the runoff. Experimental work by Abrahams and Parsons (Abrahams et al 1991) indicated that the interaction of surface stone cover with underlying sediment controls the hydrological and erosional response of slopes. Also, they indicated that when the slope is greater than 12°, the surface runoff coefficient decreases suddenly. Poesen (Poesen 1986) found that a 2° slope has a greater amount of surface runoff than a 15° slope. Jean et al (2000) carried out an experimental work to examine the influence of the stone cover, they agreed with the above studies that beyond a threshold the higher the slope gradient, the greater the storage amount.

There are two main types of stone cover placement: loose stones on the surface; stones embedded within the ground. Higgitt et al (1999) mentioned that the variations in clast cover and exposure characteristics in relation to slope position and lithology might have a significant effect in controlling runoff and infiltration. Surface stones have a range of effects on the hydrology of a hillslope. Stones intercept the raindrops and redirect the water onto the intervening regolith surface. This depends on the degree to which the stones are loose on the surface or embedded (Brown et al 1996). It has been argued that loose stones may be associated with a sheltered marginal annulus, which reduces the impact of raindrops that will prevent surface sealing. Such stones may promote local infiltration of water from the upper parts of these stones. On the other hand, embedded stones having no sheltered margins may be completely flanked by sealed regolith that makes the surface impermeable. Embedded stones are associated with reduced infiltration rates and greater runoff amounts (Brown et al 1996). Also, Higgitt et al (1999) mentioned that studies quantifying the relative exposure and burial of stones along the slope profile are very problematic and the topic requires further investigations. Rosser's work (2002) considerably advances our knowledge of these questions.

3.2. Crust

Infiltration and runoff occur at the interface between the atmosphere and the surface. There is a widely reported phenomenon of crusting and sealing of this interface in semi-arid environments, which is of relevance to the issues of infiltration and runoff (Patrick 2000). In drylands the lack of the buffer effect of vegetation can enhance the effects of rainfall kinetic energy. The sudden destruction of the structure of the soil surface under high intensity high-energy bursts of rainfall has an effect on crusting too. Little work has been adapted to account for a sealing effect, due to the complexity of the multiple variables involved and their interactions, which make it very difficult to determine the key controls, but this is changing (Patrick 2000). In addition, the fact that sealing develops under saturated conditions further complicates the objective of modelling.

Bowyer-Bower (1993), based on work in semi-arid Swaziland, found that the surface characteristic rather than subsurface storage controls the infiltration response. Also, the rainfall intensity has an effect on infiltration. She found that, at the start of wetting, lower infiltration was obtained with high rainfall intensities and dry antecedent moisture conditions compared to low rainfall intensities and wet antecedent moisture conditions. The results achieved by Bowyer-Bower (1993) showed strong evidence that the hydrological response of the soil is controlled by successive changes in the properties of the surface soil. The main factor is that the permeability of the soil surface decreases more rapidly than subsurface, which affects the amount of water that will enter the soil; in other words, it will reduce the infiltration rates due to the sealing phenomenon. High rainfall intensities affect the properties of the soil surface more quickly per unit time than low energy and low intensity rainfall. Also, the surface cover, degraded or non-degraded soil, affects the response time to rainfall intensity. There is a faster response by degraded soils than by non-degraded soils (Bowyer-Bower 1993).

Some soils, which contain silt and clay, seal extremely quickly during a rain event and this minimizes the infiltration rates (Brown et al 1996). Such soils have an infiltration depth of just a few millimetres before the soil starts to seal. This will result in increasing the runoff rates due to the low infiltration rates. In dryland soils, it appears that a vesicular phenomenon occurs due to some sort of surface seal (Brown et al 1996). This phenomenon has been reported in many countries such as the United States, Israel, Argentina and Australia. Vesicles appear to be a stable phenomenon protected by the surface crust. The development of surface sealing restricts infiltration for two reasons. First, the seals themselves have low permeability. Second, the regolith beneath the seal develops the vesicular material that reduces the amount of conducting soil matrix available to support infiltrating water and it seems to lower hydraulic conductivity of the regolith (Brown et al 1996). But Poesen (1984) observed that the runoff decreases with increasing slope, which might be a result of the crust. In other words, the infiltration rates here are increasing and the crust does not seal the surface to reduce the permeability of the surface. Also Fox et al (1997) attributed the decrease in infiltration rates with increasing slope angle to greater pond depth on spatially varying seal properties.

Hence the key factors that need to be taken into account for assessing runoff potential are rainfall characteristics, slope, surface rock cover and surface crusts.

4. Fieldwork and methodology

4.1. Field trips

Two field trips were undertaken, an initial trip during the period May-August 2001 (Figure 1), in the dry season and the second during the period October 2001 to July 2002, which contains periods before, during and after the wet season. Both trips took place to the eastern parts of Jordan (Badia) within the Azraq Basin; Tall Hassan, al-Buqi'awiyya and Suway'id (Figure 1). The sites were located 40 km to the east of Safawi and 40 km west and southwest of Safawi (Figure 1). BRDP at Safawi provided logistic support (room, food, and transportation from the centre to sites, water supply and a library). There was a problem with the water supply for experimental work. The only effective storage was 20 li jerry cans, which had to be filled at the centre and taken to the field sites. The shortage of water limited the choice of rainfall simulator and the number of rainfall simulation runs at each site.

4.2. Rainfall simulator

There is a wide range of rainfall simulators. The main advantage of using a rainfall simulator is that it provides greater control. We can produce rainfall on demand whenever we need it and for the duration we require (supply issues not withstanding). If we want to rely on natural rainfall, it will take a long time for a suitable range of events to occur in the research area (Boywer-Bower et al 1989).

The simulator should meet the requirements of the research, especially the variables required for the model, and also meet economic and logistic constraints. Several factors affect the selection of the appropriate simulator such as:

- It must be suitable for use in a semi-arid environment in a developing country with a limited budget;
- It will be used in the field rather than a laboratory;
- It should meet the objective of the research, which is assessing water harvesting potential by characterizing relative and absolute runoff values;
- The use of the output information as a decision support tool for future interventions.

The simulator used in this research was the Amsterdam lap4 simulator. It has been used in many experimental studies since 1984, especially in dryland areas (Boywer-Bower et al 1989, Imeson et al 1986, Patrick 2000). For the present research, there were some modifications in the design to meet the initial requirements of the research. The main concern was that the height of the simulator had to be raised 2 m to increase the kinetic energy of the raindrops by 70 % (Patrick 2000). The design criteria used by Imeson (Patrick 2000) indicated that it should:

- Be usable in a remote, hilly area, having few access roads and limited water supply;
- Be inexpensive;
- Give consistently reproducible results;
- Approximate natural rainfall as much as possible;
- Be simple and light enough for one person to set up, use and dismantle;
- Provide the main scientific requirements including raindrop size, intensity range and stability, height of fall and number of drops per unit time.

The design and manufacturing of the drop former was completed at the University of Manchester using two sheets of 10 mm thick, clear PVC with dimensions of 525 x 525 mm. Located in the upper perspex sheet was a manually controlled air-release valve. The lower PVC sheet (separated by 10 mm from the top sheet) has holes drilled in it through which Tygon tubing (trade name of Cole Palmer) of 0.7 mm internal diameter, 2.3 mm external diameter and 15 mm long were threaded, each with pieces of fishing line (0.65 mm diameter and 25 mm long) inserted. The internal end of the fishing line is crushed to keep it in place when the drop former box is under pressure. On the upper plate of the drop former box there is the main inlet point where the pressurized water from the supply system enters the gap between the PVC sheets.

The water supply system is gravity fed, with an operating pressure of 0-275 kpa, although the pressure required to create the required range of rainfall intensities was only 0-34 kpa. The minimum volume of water required is 5.4 litres, although a reservoir barrel holds 25 litres. The height of the drop-former above the impact surface is 2 m and it is changeable up to 3.5 m. The windshields were fitted to the main frame of the simulator and the height of the simulator was adjustable to 3.5 m. Due to high winds and the difficulty of reaching the height of 3.5 m while ensuring the safety of the researcher, the height was modified to be 2 m.

Several complications arise with the simulator use such as the limited plot size, edge effects at the plot area, the differences in the drop size distribution and the energy characteristics of natural and simulated rainfall. The main potential sources of errors during the experimental work using a drop-type rainfall simulator, were discussed by Bowyer-Bower (1989). First, there is a need for a buffer zone of rainfall around the runoff plots in order to reduce edge effects. Second, the installation of the runoff plot boundaries is difficult in stone-covered ground. Third, there is a need for wind breaks. Finally, the most important potential error is the runoff collection; the main purpose of the rainfall simulation is to collect the runoff to determine the infiltration rates or to calculate the sediment movement. Runoff collection must be as accurate as possible. Usually this can be easily achieved using the simulator by collecting the runoff at the down slope end of the plot in a simple trough. This can be facilitated by digging a trench, placing the trough in it, and sealing the edge of the trough with the plot area to minimize runoff losses.

During the fieldwork some problems occurred including:

- The air outlet of the drop former was located about 8 cm from the edge, so there was a problem to get all of the air out;
- To reduce the effects of the edges there was a need to wet these boundaries but care had to be taken to prevent water infiltrating and moving to inside the plots;
- During the simulation, the area around the legs of the frame becomes saturated and muddy and this could result in movement of the drop former so that it was no longer level;
- Sometimes, the volume of runoff moved the trough.
- The total simulator length of 2.75 m meant there was significant vibration in high winds;
- Sometimes, water went under the trough and could not be collected.

Figure 2a. The drop former, the manometer and the jerry can

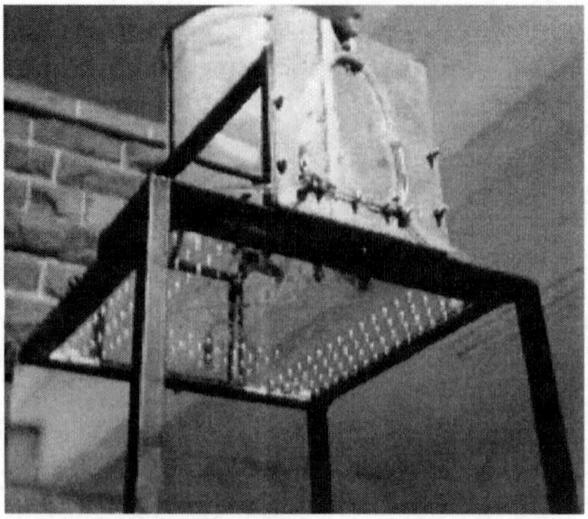

Figure 2b. The actual size of the simulator used in the field study

Figure 2c. The simulator in the field

Preliminary checking of the operation of the simulator, tested rainfall intensities, volume and spatial distribution to ensure the reliability of the system. Drop size was measured to ensure the proper amount of kinetic energy. The simulation of rainfall was not undertaken in high wind speeds and was repeated when the trough was displaced to ensure the data collection was reliable.

4.3. Site locations

During the second field trip a series of rainfall simulations was undertaken at three main sites, Tall Hassan, al-Buqi'awiyya and Suway'id, which are major *qa'a* (Figure1). These sites were chosen because they were: easy of access and within the BRDP area; close to meteorological stations; within the catchment areas of the *marab* and *qa'a*.

Previous researchers conducted experiments in these sites (Shatnawi (2002), Patrick (2000), Rosser (2002) and AlAyyash (1993)) studying ground water, geology, soil properties and actual rainfall.

These sites are very important for local people as they hold water which can be used for their livestock. The problems facing the livestock sector included the reduction of the range area, its low productivity, the rising cost of production due to the higher cost of inputs, such as labour and feed, the limited water resources and the weakness of marketing animal products (Shoshan 1998).

The sites are all within the Azraq basin and are representative of the area.

4.4. Experimental work

The experimental work took place during two different periods including wet and dry seasons. During the winter, the work was difficult due the high wind speed and low temperatures. Two main hypotheses arose from the analysis of the pilot study: can rainfall intensity alone determine runoff; does the surface cover type affect the runoff.

Two main rainfall intensities (RI in Table 2) were applied during the simulation (8-12, 62-70 mm/hr), as they represent the rainfall behaviour during the last 40 years according to rainfall intensity curves produced by the Ministry of Water and Irrigation. Two main soil surfaces were selected: the bare soil and 50 % rock cover surfaces. The selection was based on the preliminary analysis of the pilot study results of the first field trip. Also, these two types of surface cover can be observed as representative of the area. For each plot two simulations took place, a dry and wet run (saturated surfaces). Another two types of soil surface were also used: one, with the surface crust removed and two, ploughed soil. Thus we were able to examine the efficacy of ploughing and preparing the land for planting and agriculture.

Data recorded included discharge, surface cover characteristics, random roughness (RR; measured by a micro soil profile with 22 wood pins; the measurements were taken as a cross-sectional line), temperature and wet perimeter. The data were sorted into two main groups, according to rainfall intensity, for both dry and wet runs, and four categories, according to surface cover type.

5. Results

5.1. Basic data

The total number of rainfall simulation experiments is 720. This very large data set was generated to ensure representative results. Each point of data presented (Table 2) is the mean of 20 runs of the same conditions and criteria, which are time to runoff, time to steady state (illustrated in Figure 3) and the ratio of runoff volume to the total volume of precipitation.

Table 2. Summary of runs has been carried out in all sites

Site		Tall Hassan				Buqi'awiyya				Suway'id			
Surface cover	RI:	Low		High		Low		High		Low		High	
	Condition:	Dry	Wet	Dry	Wet	Dry	Wet	Dry	Wet	Dry	Wet	Dry	Wet
Bare soil		24	24	24	24	12	12	12	12	24	24	24	24
50 % rock		24	24	24	24	12	12	12	12	24	24	24	24
Crust removal		24 (D)		24 (D)		12 (D)		12 (D)		24 (D)		24 (D)	
Loose soil		24 (D)		24 (D)		12 (D)		12 (D)		24 (D)		24 (D)	

5.2. Analysis

Typically, the rainfall simulations generated a time to runoff of 1-8 minutes depending upon antecedent conditions. Time to steady state was influenced by surface conditions. The ratio of runoff to rainfall volume varied from 0.3 to above 0.9 depending upon surface wetness and ground cover. To synthesize and test the significance of relationships the data were analyzed statistically using paired samples t-test analysis, Wilcoxon and ANOVA were carried out to compare the means of the wet and dry runs for each rainfall intensity. Two significant levels were used (5 % and 10 %) for all the analyses. The results presented here are based on the analysis of antecedent conditions and surface characteristics.

5.3. Antecedent conditions

During most of the year the soil conditions are dry, but during the rainy season the soil becomes wet and can become saturated in the upper layer. This condition has its own effects on infiltration rates and runoff generation. The following sections will discuss the effect of high and low rainfall intensities on the ratio between runoff volume and total rainfall volume. This can be the most valuable indicator of runoff generation. In this analysis we have combined all ground cover conditions.

Tables 3 and 4 show the impact of dry and wet surfaces during low and high rainfall intensity; the ratio is higher when the surface is wet. There is strong evidence that the means are not the same in any run ie T-stat > T-critical with very low variability ob-

Figure 3. Typical runoff curve

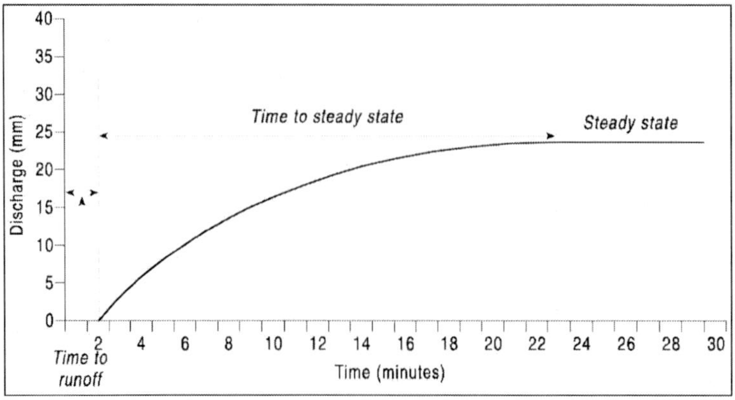

Table 3. Ratio of runoff/ total rainfall volume for high rainfall intensity

RO/ volume	Dry	Surface condition	Wet
Means	0.725		0.913
StdDev	0.065		0.054
T-Critical		1.97	
T-Stat.		24.28	

Table 4. Ratio of runoff/total rainfall volume for low rainfall intensity

RO/ volume	Dry	Surface condition	Wet
Means	0.609		0.8410
StdDev	0.0642		0.114
T-Critical		1.97	
T-Stat.		19.40	

served. The comparison took place between all sites together, ie combing bare surfaces with rock-covered surfaces. The antecedent conditions (wet or dry) affect the runoff generation. Wet surfaces reduce the infiltration rates, which increase the runoff amount. This is not surprising but confirms the importance of climate, ie wetness rather than ground cover, on runoff prediction given the low standard deviations.

5.4. Rainfall intensity

Significantly Table 5 shows that for an initially dry surface the mean ratio of RO/volume is 0.79 for high rainfall intensity, but only 0.58 for low intensity.

The results give strong evidence that rainfall intensity, not just amount, plays a key role in runoff volume. The amount of data variation increases compared to antecedent conditions. Detailed investigation of the data revealed that surface cover (eg proportions of rock fragments) had very little impact on runoff generation under high intensities. Under low rainfall intensities the opposite was found where ground cover became much more important in reducing runoff.

5.5. Ground cover

Two main rainfall intensities were used: high (62-70 mm/hr) and low (8-12 mm/hr). Tables 6-9 show the effects of rainfall intensity for different surface conditions.

Table 5. Ratio of runoff/ total rainfall volume for high and low rainfall intensity

RO/ volume	High RI	Surface condition (dry)	Low RI
Means	0.790		0.585
StdDev	0.0864		0.1616
T-Critical		1.96	
T-Stat.		17.34	

Table 6. Ratio of runoff/ total rainfall volume for high rainfall intensity during dry conditions for bare and rock cover surfaces

RO/ volume	Surface condition (dry) and high intensity rainfall Bare		Rock
Means	0.714		0.736
StdDev	0.0455		0.0792
T-Critical		1.98	
T-Stat.		1.92	

Table 7. Ratio of runoff/ total rainfall volume for high rainfall intensity during wet conditions for bare and rock cover surfaces

RO/ volume	Surface condition (wet) and high intensity rainfall Bare		Rock
Means	0.948		0.878
StdDev	0.0404		0.0421
T-Critical		1.98	
T-Stat.		9.31	

Table 8. Ratio of runoff/ total rainfall volume for low rainfall intensity during dry conditions for bare and rock cover surfaces

RO/ volume	Surface condition (dry) and low intensity rainfall Bare		Rock
Means	0.570		0.648
StdDev	0.0318		0.0648
T-Critical		1.98	
T-Stat.		8.36	

Table 9. Ratio of runoff/ total rainfall volume for low rainfall intensity during wet conditions for bare and rock cover surfaces

RO/ volume	Surface condition (dry) and low intensity rainfall Bare		Rock
Means	0.772		0.909
StdDev	0.1237		0.0367
T-Critical		1.98	
T-Stat.		8.23	

Table 6 shows that under high intensity rainfall ground cover does not influence the amount of runoff; T-stat < T-critical. This is a significant result. Table 7 reveals a statistical difference between the ratio of means of bare and rock-covered surfaces, but only because of the low variability of each data set. In reality, Table 7 shows that the magnitudes of the ratios of runoff volume to rainfall are of similar high values ie 0.95 and 0.88. Tables 8 and 9 show that low intensity rainfall is more likely to be affected by ground cover, under wet or dry antecedent conditions, but the differences between ground cover types is less than those caused by antecedent wetness.

Thus the two key factors emerging are rainfall intensity and previous volumes of rainfall rather than ground cover.

6. Discussion

The key to using the results is that the proportion of runoff to rainfall, dependent also on rainfall amount, can be predicted from just four simple classes based on rainfall intensity (RI) and whether surfaces are wet or dry: high RI/dry; high RI/wet; low RI/dry; low RI/wet.

In order to assess runoff potential from a slope we then need the following data: rainfall amount; rainfall intensity low or high; surface wet or dry.

The rainfall volume (1966-90) for many, but not all, parts of the Azraq basin is known, as is the surface area making it possible to calculate total rainfall (Table 10).

Runoff potential can be applied to individual storms, or can be based upon probabilities of rainfall and calculated across the season. Tables 11, 12 and 13 show how this can be done for the Azraq data, indicating that there is the potential to collect, for example, 12.2 mm of the 15.6 mm falling in October. The probabilities are the estimated

Table 10. Azraq basin stations: mean monthly rainfall, 1966-1990 (mm)

Station name	Oct	Nov	Dec	Jan	Feb	Mar	Apr	May	Total	Area km^2	Volume MCM
Salkhad S1	15.6	38	62.5	62.5	69.5	58.6	20.5	6.9	334.1	150	50.115
Imttan S2	5.8	21.5	40.6	40	39.8	31.4	13.2	5.1	197.4	760	150.024
K.Awad S3	4.5	18.2	36.7	44.5	35	32.1	11.5	2.7	185.2	na	
Umm al-Quttayn F1	12.2	17	32.4	35.3	38.5	37.1	14	3	189.5	210	39.795
Dayr al-Kahf F4	12.8	16.5	28.8	28	30.1	30	24.5	4.1	174.8	460	80.408
Safawi F2	10.5	13.2	16.1	12	14.9	11.2	6.5	2.6	87	650	56.55
AL48	7.4	17.8	23	34.8	27.8	28.8	8.3	1.7	149.6	na	
al-Ariyyatayn F6	11.1	13.5	15.7	33.3	18.8	16.5	11	4.7	124.6	445	55.447
Azraq station F9	5	9.1	12.8	10.3	10.2	12.4	7.4	1	68.2	660	45.012
Muwaqqar CD3	9	16.2	36.8	42	34.3	42	22.9	1.3	204.5	270	55.215
Umary F11	5.9	3.9	7.8	12.5	6.1	8.1	11.1	1.5	56.9	1000	56.9
Mean	9.1	16.8	28.7	31.3	29.5	28	13.7	3.1	160.2		
Std dev	3.7	8.5	15.7	15.9	17.6	15.2	6.2	1.9	78.2		
Coef variation	40.2	50.7	54.8	50.9	59.6	54.2	45.4	60.0	48.8		

Source: Department of Meteorology files, 1998

Table 11. The simulated runoff/rainfall ratio

	Dry surface	Wet surface
Low intensity	59 %	84 %
High intensity	79 %	91 %

Source: Tables 3-5

condition of the surface (0.3 for high RI dry, 0.2 for high RI wet, 0.4 for low RI dry and 0.1 for low RI wet). Thus high intensity rainfall on a wet surface will yield more runoff than on a dry surface (Table 11).

Table 13 shows the predicted runoff volume for the nine stations out of the 19 in the Azraq basin for which data are available.

Table 12. The calculations of predicted runoff for Salkad S1 station

Month	Oct		Nov		Dec		Jan		Feb		Mar		Apr		May	
Rainfall mm	15.6		38.0		62.5		62.5		69.5		58.6		20.5		6.9	
Probability	%	RO	%	RO	%	RO	%	RO	%	RO	%	RO	%	RO	%	RO
% high RI dry	0.4	4.9	0.3	9.0	0.3	14.8	0.3	14.8	0.3	16.5	0.4	18.5	0.4	6.5	0.3	1.6
% high RI wet	0.2	2.8	0.3	10.4	0.3	17.1	0.3	17.1	0.3	19.0	0.2	10.7	0.1	1.9	0.1	0.6
% low RI dry	0.2	1.8	0.1	2.2	0.1	3.7	0.1	3.7	0.1	4.1	0.3	10.4	0.4	4.8	0.5	2.0
% low RI wet	0.2	2.6	0.3	9.6	0.3	15.8	0.3	15.8	0.3	17.5	0.1	4.9	0.1	1.7	0.1	0.6
Total runoff	12.2		31.2		51.3		51.3		57.1		44.5		14.9		4.9	

Table 13. The estimated runoff for 11 stations out of 19 in the Azraq basin

Station name	Oct	Nov	Dec	Jan	Feb	Mar	Apr	May	Total	Area km²	Volume MCM
Salkhad S1	12.23	31.20	51.31	51.31	57.06	44.48	14.90	4.88	267.37	150	40.11
Imttan S2	4.55	17.65	33.33	32.84	32.68	23.83	9.60	3.61	158.08	760	120.14
K.Awad S3	3.53	14.94	30.13	36.53	28.74	24.36	8.36	1.91	148.50	NA	NA
Umm al-Quttayn F1	9.56	13.96	26.60	28.98	31.61	28.16	10.18	2.12	151.17	210	31.75
Dayr al-Kahf F4	10.04	13.55	23.64	22.99	24.71	22.77	17.81	2.90	138.41	460	63.67
Safawi F2	8.23	10.84	13.22	9.85	12.23	8.50	4.73	1.84	69.44	650	45.13
Khalidiyeh AL48	5.80	14.61	18.88	28.57	22.82	21.86	6.03	1.20	119.79	NA	NA
al-Ariyyatayn F6	8.70	11.08	12.89	27.34	15.43	12.52	8.00	3.32	99.29	445	44.19
Azraq station F9	3.92	7.47	10.51	8.46	8.37	9.41	5.38	0.71	54.23	660	35.79
Muwaqqar CD3	7.06	13.30	30.21	34.48	28.16	31.88	16.65	0.92	162.66	270	43.92
Umary F11	4.63	3.20	6.40	10.26	5.01	6.15	8.07	1.06	44.78	1000	44.78

* The total in mm is based on the Table 11-12 calculations. The total in MCM can be calculated for the 9 stations for which the surface area is known.

Table 14. Comparison between previous data and current research of runoff volume generated

Data source	Total area km²	Total rainfall MCM	Total runoff MCM	%
(Ayed, 1996) Azraq basin	12 800	1 119	27.85	2.49
Current researcher-calculated runoff volume (9 stations) (Table 13)	4 605	589	469.5	79.7

Table 14 compares the runoff estimated by Ayed (1996) with the runoff using the model proposed in this research. According to Ayed the proportion of runoff to rainfall volume is much lower and he argued that there were significant losses in seepage and evapotranspiration.

The total runoff volume of the Azraq basin was calculated in Ayed's model by multiplying the average yearly rainfall, which is 88 mm, by total area, which is 12 800 km^2.

The huge difference between Ayed's model and that of the present researcher is the difference between actual runoff, measured locally, and the runoff estimated at a distance; in effect most of the runoff is lost by evaporation as the source is far from collection points in large dams. This underlines the importance of the *mahafir* which were placed close to collecting slope points such that only small slope lengths were required to generate reasonable volumes of water. For example, a 5 m slope with 12.3 mm of runoff generates 61 li per m width, so for a *mahafir* of 10 m length this means 600 li (Agnew et al 1995).

The problem is how to apply the four key outcomes of wet/dry/high/low rainfall intensity as the data are not available to make the calculations correctly over a wide region; we have a method but not the input data. More data need to be gathered and more modern weather stations need to be installed in the basin to collect climatic data correctly. And can we really scale up to whole catchments? From Table 11 we can see that it is unlikely that the water could reach the end of the catchment. This therefore casts doubt on the solution of constructing large dams across wadis. The sustainable solution is to collect water within the catchments, not at the end of them by building big dams. The technology used by the Bedu, the *mahafir*, is a good solution. They relied upon their observation and indigenous knowledge to keep the route of runoff short to the *mahafir* in order to reduce losses by seepage and evaporation. Another key strategy they employed was the technique of spreading the *mahafir* over the catchments to increase the probability of collecting runoff as they could not predict where the rain would fall.

The approach used here can be further developed with rainfall frequency analysis for runoff. The amounts of rainfall have been computed for each frequency duration (Figure 4) and are presented in Table 15 for low rainfall intensity used in this research and in Table 16 for high rainfall intensity. It is noted that low intensity rainfall yields more rainfall because of the longer durations at higher probabilities. Also, there is little difference between wet/dry conditions for high intensity but it does become important for low intensities, hence water harvesting is much less effective at the start and end of the rainy season and better when rain is more frequent, as is to be expected.

Figure 4. Rainfall intensity for Azraq station with frequency and duration

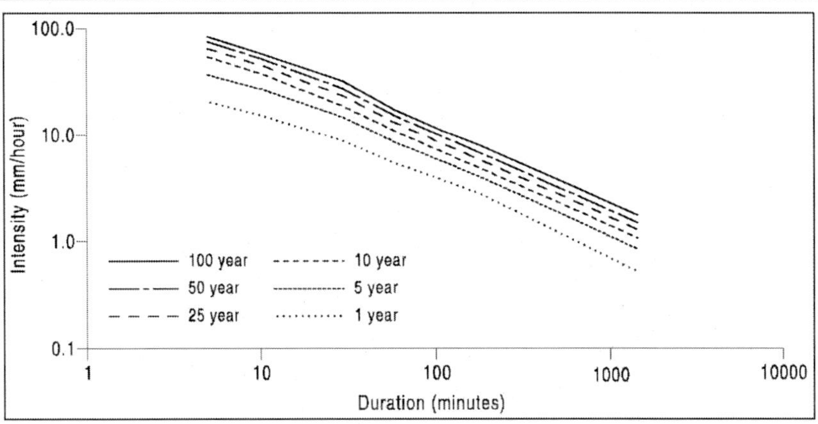

Table 15. Amounts of rainfall calculated from rainfall duration frequency during low rainfall intensity

Rainfall intensity (mm/hr)	Duration (mins)	Frequency (yr)	Amount*	Wet runoff coefficient 0.84 Low intensity wet runoff	Dry runoff coefficient 0.59 Low intensity dry runoff
10	24	2	4.0	3.4	2.4
12	20	2	4.0	3.4	2.4
10	50	5	8.3	7.0	4.9
12	45	5	9.0	7.6	5.3
10	68	10	11.3	9.5	6.7
12	60	10	12.0	10.1	7.1
10	90	25	15.0	12.6	8.9
12	80	25	16.0	13.4	9.4
10	110	50	18.3	15.4	10.8
12	100	50	20.0	16.8	11.8
10	130	100	21.7	18.2	12.8
12	120	100	24.0	20.2	14.2

* The amount of water is calculated as intensity x mins/60, eg 10 x (24/60) = 4

Table 16. Amounts of rainfall calculated from rainfall duration frequency during high rainfall intensity (HI)

Rainfall intensity (mm/hr)	Duration (mins)	Frequency (yr)	Amount*	Wet runoff coefficient 0.91 HI Wet runoff	Dry runoff coefficient 0.79 HI Dry runoff
60	1.5	5	1.5	1.4	1.2
65	1.25	5	1.4	1.2	1.1
60	6	25	6.0	5.5	4.7
65	5	25	5.4	4.9	4.3
60	8	50	8.0	7.3	6.3
65	7	50	7.6	6.9	6.0
70	6	50	7.0	6.4	5.5
60	10	100	10.0	9.1	7.9
65	8.5	100	9.2	8.4	7.3
70	7	100	8.2	7.4	6.5

* The amount of water is calculated as intensity x mins/60, eg 10 x (24/60) = 4

Figure 5 shows a prediction model based on a power curve regression that can give the amounts of runoff for each storm at a given probability. It is difficult to go from this to total runoff in a season unless we know the rainfall intensity classes over the season. However, the graph is useful as it shows it is the low intensity storms of long duration where the greatest potential for runoff collection lies rather than high intensity. That is duration is probably the characteristic that needs to be examined. The thresholds are also interesting, for any low intensity storm we can expect at least 3.17-2.23 mm of runoff, for high intensity only 0.93 mm because of the shorter duration. The graph can be used to predict storm runoff probabilities, or conversely for a given probability, say once in 10 years, we can predict that for high intensity rainfall 2.3 mm, for low intensity 1 in 10 years runoff will be 8.3 mm dry; 5.8 mm wet – this is a very useful procedure for water harvesting assessment.

Figure 5 presents the last two columns of Tables 15 and 16 (runoff amount) versus the frequency, which is presented in Figure 4. The last two columns of Tables 15 and 16 are generated based on the results from the researchers practical experiments in the field.

Significant rainfall tends to occur as short intense thunderstorms of limited areal extent. This leads to a high proportion of runoff that drains via a radial network of wadis to the mudflats near Azraq. Total annual precipitation volume in the Azraq Basin is approximately 1 119 MCM generating a runoff of only 27.85 MCM, occurring mainly during the winter (Ayed 1996). Most of the rain falls from October to April and is characterized by irregularity in intensity and duration. The current researcher feels that the runoff amounts predicted by water balance analyses are not accurate due to a number of reasons. First, there is no accurate data collected about actual runoff amounts in the area due to the lack of devices, personnel and the vandalism of new devices by some local people (Shatnawi 2002). Second, as the rainfall storms are unpredictable, it is difficult to cover a basin of 12 800 km² with gauges to record the rainfall and runoff amounts. And finally, the only available record of the rainfall data is the monthly records. More data are needed especially the daily records or even storm records to be used in calculations and modelling to predict, as far as possible, accurate runoff volumes.

The results of the analysis of rainfall simulation have been applied to monthly rainfall totals to simulate total amounts of runoff generated. These results must be treated with caution because of uncertainties over the structure of monthly rainfalls and the danger of aggregating across large catchments in this manner. These results are inevitably over-estimates of total runoff but in the absence of the data required the current research demonstrates how the results of this investigation may be deployed in the future. The values derived for runoff prediction for different rainfall intensities have also been compared to rainfall-frequency-duration curves for the region. This reveals that duration is by far the most important characteristic of rainfall when predicting total precipitation. Long duration low intensity rainfalls have a higher probability of generating more precipitation than the short-lived, infrequent, high intensity rainfalls. This has significant implications not only for the assessment of water harvesting but also for future modelling of total runoff generation. For both types of analyses to be taken further, however, there is a need for more detailed analysis of precipitation characteristics in the region.

Figure 5. Runoff amounts versus the rainfall intensity frequency

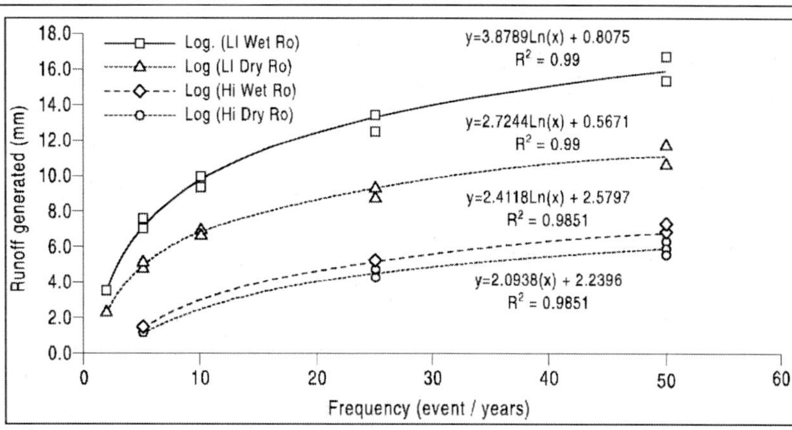

7. Conclusion

Jordan is facing a water crisis. Population growth is expected to continue at over 3.2 % per annum while agricultural intensification is adding to the national water burden. Renewable resources do not meet present water demands and alternative supplies are being sought eg desalination and recycling. Over-abstraction of ground water has resulted in significant falls in the groundwater table in the Azraq basin (Dottridge 1998b) and domestic supply is only provided two to three days a week in Amman. The situation is exacerbated by possible changes in the seasonal pattern of rainfall. Jordan then requires more efficient use of its existing water supplies and this is particularly the case in the drier eastern region, the Badia. One strategy that has been widely practised in the past in the Middle East is water harvesting. This technique aims to make more effective use of precipitation through capturing and directing surface flows or harvesting rainfall directly. But this requires scientific assessment of water harvesting potential.

Assessing runoff potential in a hyper-arid area is the main concern of this study. There are many factors affecting the runoff behaviour such as rainfall intensity, evaporation, infiltration, interception, rock size, drop size, kinetic energy of rainfall, wind, slope and crust. The number and complexity of intervening variables creates demanding experimental work in terms of time, effort and logistics. Some variables are therefore held constant to reduce the permutations and variability. It was also necessary to use a rainfall simulator given the high variability of precipitation in the study region. There are many factors affecting the runoff behaviour such as rainfall intensity, evaporation, infiltration, interception, rock size, drop size, kinetic energy of rainfall, wind, storm duration, volume ratio of the floodwater, slope and crust type. Thus several models have been developed based on the Darcy-Weisbach manning resistance factor. The present investigation has sought to examine the key relationship between rainfall intensities and surface roughness.

A portable drop former mobile rainfall simulator with 2 m height was used. Detailed calibrations were carried out on the simulator board. Attention was paid to the issue of the spatial variability of rainfall, drop size and distribution. Two weeks were invested in calibrations at the beginning of the pilot study and a few days at the beginning of the main field study. A systematic approach was taken to calibration as this design of the simulator is a modified version. It proved to be reliable and gives great confidence to the research in the sense of its stability, reliability, sealing, water saving and easy control over the intensities. The rainfall simulator employed has been widely used in semi-arid areas of the Mediterranean as well as in some areas of northern Europe since the late 1970s as it is easy to setup and operate. It gives excellent control over rainfall intensities, which is an important factor for comparison of results. It is economic in water use and cost of manufacturing. But it is time consuming to use and can become tedious with hundreds of replications.

The work was carried out in three main sites, Tall Hassan, al-Buqi'awiyya and Suway'id. Two main rainfall intensities were used: low at 10 mm/hr, and high is 65 mm/hr (medium intensities were abandoned after the pilot study). Four main surface treatments were considered based upon the pilot study results. These are bare soil, 50 % rock cover, crust removed and loose soil surfaces. The selection was based on the analysis of pilot simulation runs and land cover characteristics. For each plot two simulations took place, dry and wet runs (saturated surface).

720 rainfall simulations were carried out over different sites with different ground cover conditions. The results reveal that the main factor affecting runoff and infiltration rates is the rainfall intensity. Higher rainfall intensity means higher kinetic energy and faster crusting and shorter time to runoff. The surface cover does not have as significant an effect on runoff generation as rainfall intensity. The condition of the surface, wet or

dry, affects the runoff and infiltration rates as the time to runoff for wet surfaces is shorter than the dry surfaces. This is as expected; the wet surface contains amounts of water that minimize the infiltration rates and helps to seal the surface and reduce the infiltration rates so the runoff volume will be greater. In total, 720 runs were carried out over a period of twelve months, including the pilot study. This is believed to be a significant data set and sufficient for the analysis of runoff potential in the Badia region. Some problems occurred during the field study such as the shortage of water supply and the transportation time from the centre to each location. Sometimes high winds and high temperatures delayed the experiments as more precautions were taken. Sand storms were another problem that delayed the transportation and running of the experiments in the field (Abrahams et al 1991, Bowyer-Bower 1993, Boywer-Bower et al 1989, De Roo et al 1989, Emmett 1980, Evans 1980, Higgitt et al 1999, Jetten et al 1998, Patrick 2000, Poesen et al 1996, Takken et al 2000, Wainwright et al 2000)

As stated above, the results show that the most important factor affecting runoff and infiltration rates is the rainfall intensity. It is not surprising to find rainfall intensity and previous amount of water are important, but what is more significant is the greater influence these climatological variables exert on runoff compared to ground cover characteristics. Thus future investigation of water harvesting potential will focus on climatological conditions rather than attempting to map surface rock cover and crusts. This accords with the findings of by Joel et al (2002).

Short intense thunderstorms of limited areal extent lead to a high proportion of runoff that drains via a radial network of wadis to the mudflats near Azraq, as stated above. A much higher density of rain stations is required in order to achieve a degree of accuracy of storm recording that would help locate small dams nearer the places where the rain falls. Over the basin as a whole the water balance computation shows that there is very little runoff available which could lead us to assume that there is little opportunity for runoff harvesting. However, we know water harvesting has been practised in the past in this region and there are a number of dams being constructed. This analysis of annual water balance, however, provides few insights for water harvesting potential. In order to do this and apply the lessons learned from the rainfall simulations we need to examine runoff processes over a shorter time scale and over a smaller area than a basin.

The results of this research revealed that some main outcomes can be predicted from rainfall simulations; while medium to high intensity rainfall generates 79-91 % of precipitation as runoff; low intensity storms still generate 59-84 % of the storm as local runoff. These values are very high and reveal the tremendous potential for water harvesting if the water can be collected locally. Losses through runoff are high, due to seepage and evaporation which increase with catchment size. Runoff needs to be collected within small scale catchments located as close as possible to the slopes generating the runoff, as practised by Bedu by using the *mahafir* technique. Any new developments should take note of the traditional Bedu strategies and serious consideration should be given to deepening the reservoirs of the old *mahafir* to reduce evaporation losses.

There is a major difficulty in using currently available climatological data to apply the results of this research because the simulations are based on storm events whereas the climate data is generally only available for annual or monthly totals. Until daily/storm data become more widely available it is only possible to indicate likely runoff totals by characterizing monthly rainfalls into the two intensity classes and taking account of antecedent conditions. The analysis in the discussion shows that (Table 14) 88 mm per m^2 of slope can be expected which generates 832.4 MCM of runoff, which equals 74.4 % of the annual rainfall received. This is a very promising result which needs to be tested through the deployment of small dams adjacent to slopes in the study region. It also confirms the analysis of the functioning of the *mahafir* by Agnew et al (1995).

For future development of water harvesting systems in the Badia the focus should be on how the water will be used; is it for grazing or for local people's domestic use or ground water recharge? The form of water harvesting may depend upon the answers provided. The next step should be more research on climatological data by installing more modern weather stations especially around the *marab* and *qa'a* catchments. BRDC should setup and manage a new database to collect and analyze the climate data to be used by other researchers and to be available to be used in predicting the size of water harvesting collecting systems. As each catchment could realize a significant runoff it is worth investing a reasonable amount of money for such a project, which may include the study of the possibility of diverting water to nearby catchments. New small-scale dams around the region will be the next step to be considered by decision makers or, better still, to benefit from the old *mahafir* system and redevelop it for water harvesting.

In summary, the results of this research could change the way in which we assess water harvesting potential in this area of Jordan and possibly in other hyper arid areas. Previous work on runoff modelling has placed much emphasis upon the geomorphological controls including roughness, crusting and slope angle. This has led to an emphasis upon classifying different types of land cover and the deployment of image processing and GIS techniques in an attempt to scale up assessment from slope to catchment scales. This research has found conversely that the primary control on local runoff is climatological and that if we are to improve our assessment of water harvesting potential then we need to know much more about storm rainfall characteristics and trends of rainfall in this region. This will require significant investments. A final point of emphasis is that the focus throughout this study has been on local processes and local water harvesting opportunities because it is felt that sustainable solutions are more likely to emerge from this scale of analysis.

Acknowledgements

The research was conducted as part of a PhD thesis at the University of Manchester. The research was funded by the University of Manchester, BRDP, Kareem Rida Said Foundation (British Council), Arab Student International Aid Foundation and Arab-British Commerce Chamber Foundation. Thanks are also due to Michael Clarke and Jone Moore for technical assistance and to Moath Al Zubi and Tareq Al Zubi for help with the experiments.

References

Abrahams AD and Parsons AJ (1991). Relationship between sediment yield and gradient on debris-covered hillslopes, Walnut Gulch, Arizona. Geological Society American Bulletin 103:1109-1113.

Abrahams AD, Parsons AJ and Luk SH (1988). Hydrologic and sediment responses to simulated rainfall on desert hillslopes in southern Arizona. Catena 15:103-117.

Agnew C, Anderson E, Lancaster W and Lancaster F (1995). Mahafir: a water harvesting system in the eastern Jordan (Badia) desert. GeoJournal 37(1):69-80.

AlAyyash S (1993). A development of water harvesting system at Safawi area in northern Jordan Badia. Dissertation (Masters), JUST, Masters dissertation.

Al-Tabini R, al-Khalidi K and al-Soub H (2007). Community based rangeland rehabilitation project: final report. BRDC.

Allison R, Higgitt D, Kirk A, Warburton J, Al-Homoud A, Sunnaa B and White K (1998). Geology, geomorphology, hydrology, groundwater and physical resources. In: Dutton R, Clarke J and Batikhi A (eds), Arid land resources and their management: Jordan's desert margin. London: Kegan Paul International pp. 21-44.

Anderson MG and Burt TP (1990). Process studies in hydrology, New York. Wiley.

Ayed R (1996). Hydrological and hydrogeological study of the Azraq basin-Jordan. PhD Thesis, University of Baghdad.

Bowyer-Bower TAS (1993). Effects of rainfall intensity and antecedent moisture on the steady-state infiltration rate in a semi-arid region. Soil Use and Management 9:69-76.

Boywer-Bower TAS and Burt TP (1989). Rainfall simulators for investigating soil response to rainfall. Soil Technology 2:1-16.

Brown KJ and Dunkerley DL (1996). The influence of hillslope gradient, regolith texture, stone size and stone position on the presence of a vesicular layer and related aspects of hillslope hydrologic processes: a case study from the Australian arid zone. Cantena 26:71-84.

De Lima JLMP (1988). Morphological factors affecting overland flow on slopes. In: Bouma, J and Breget, AK (eds), Land qualities in space and time. Wageningen: Pudoc p. 200.

De Roo APJ, Hazelhoff L and Burrough PA (1989). Soil erosion using "ANSWERS" and a Geographical Information System. Earth surface processes and landforms 14:517-532.

DoM (1998). Department of Metrology files, Jordan.

Dottridge J (1998a). Water resources quality, sustainability and development. In: Dutton R, Clarke J and Battikhi A (eds), Arid land resources and their management, Jordan's Desert Margin. London: Kegan Paul International pp. 21-44.

Dottridge J (1998b). Water resources quality, sustainability and development. In: Dutton RW, Clarke JI and Battikhi A (eds), Arid land resources and their management: Jordan's desert margin. London: Kegan Paul International pp. 67-80.

Emmett WW (1980). Overland flow. In: Kirkby, MJ (ed), Hillslope hydrology. Chichester: Wiley, pp. 145-176.

Evans R (1980). Mechanics of water erosion. In: Kirkby, MJ and Morgan, RPC (eds), Soil erosion. Chichester: Wiley pp. 109-128.

FAO (1994). Water harvesting for agricultural production, Water Reports 3. Rome: FAO.

Fox DM, Bryan RB and Price AG (1997). The influence of slope angle on final infiltration rate for interrill conditions. Geoderma 80:181-194.

Gilley JE and Finkner SC (1991). Hydraulic roughness coefficients as affected by random roughness. Transactions of the ASAE 34:897-903.

Higgitt DL and Allison RJ (1999). Clast exposure on boulder-covered desert slopes. Earth surface processes and landforms 24:111-125.

Imeson AC and Verstraten JM (1986). Erosion and sediment generation in semi-arid and Mediterranean environments: the response of soil to wetting by rainfall. Journal Water Resources 51:388-418.

Jean SJ, Ai KF, Shih K and Hung CC (2000). Stone cover and slope factors influencing hillside surface runoff and infiltration: Laboratory investigation. Hydrological Processes 14:1829-1849.

Jetten V, De Roo APJ and Guerif J (1998). Sensitivity of the model Lisem to variables related to agriculture. In: Boardman, J and Favis-Mortlock, D (eds), Modeling soil erosion by water. NATO ASI Series vol. I 55.

Joel A, Messing I, Seguel O and Casanova M (2002). Measurement of surface water runoff from plots of two different sizes. Hydrological Processes 16:1467-1478.

Lavabre JD, Sempere D and Cernesson F (1993). Changes in hydrological response of a small Mediterranean basin a year after a wildfire. J of Hydrology 142:273-299.

Lawrence DSL (1997). Macroscale surface roughness and frictional resistance in overland flow. Earth surface processes and landforms 22:365-382.

Martinez-Mena M, Albaladejo J and Castillo VM (1998). Factors influencing surface runoff generation in a Mediterranean semi-arid environment: Chicamo watershed, SE Spain. Hydrological Processes 12:741-754.

NES (1991). National Environment Strategy for Jordan, a Resource Book of Information and Guidelines for Action. Ministry of Municipal and Rural Affairs & the Environment, Department of Environment & IUCN, Al Kutba Publishers, Amman.

Patrick E (2000). PhD thesis.

Poesen J (1986). Surface sealing as influenced by slope angle and position of simulated stones in the top layer of lose sediments. Earth surface processes and landforms 11:1-10.

Poesen J and Bunte K (1996). The effects of rock fragments on desertification processes in Mediterranean environments. In: Brandt, CJ and Thornes, JB (eds), Mediterranean desertification and land use. Chichester: Wiley pp. 247-267.

Poeson J (1984). The influence of slope angle on infiltration rate and Hortonian overland flow volume. Zeitschrififur geomorphologie 49:117-131.

Rosser NJ (2002). The geomorphology of coarse clastic surfaces in arid environments. PhD, University of Durham.

Royal Court (1998). Jordan diary, Amman. Jordan: The International Press Office.

Shatnawi RS (2002). Evaluation of surface water resources of Marab Hassan in the north-east Badia of Jordan. PhD thesis, University of Wales.

Shoshan NK (1998). Livestock resources. Jordan Badia Research and Development Programme, Arabic Report.

Takken I and Govers G (2000). Hydraulics of interrill overland flow on rough, bare soil surfaces. Earth surface processes and landforms 25:1387-1402.

UNEP (1983). Rain and stormwater harvesting in rural areas. United Nations Environment Programme, Water Resources Series 5:92-113.

UNESCO (1986). The major regional project on rational utilization and conservation of water resources in the rural areas of the Arab states, with an emphasis on traditional water systems. Paris: UNESCO.

Wainwright J, Parsons AJ and Abrahams AD (2000). Plot-scale studies of vegetation, overland flow and erosion interactions: case studies from Arizona and New Mexico. Hydrological Processes 14(2921-2943).

Yair A and Lavee H, (1976). Runoff generative processes and runoff yield from arid talus mantled slopes. Earth surface processes and landforms 5:205-225.

CHAPTER 7

A brief description and analysis of current waste problems and disposal practices in Jordan

Odeh M. Al-Meshan
Badia Research and Development Centre, P.O. Box 902, Jubeiha 11941, Jordan. E-mail al_meshan@yahoo.com

Ian Foster
Coventry University, UK & Rhodes University, South Africa

Nadhir Al-Ansari
Al-Albayt University, Jordan

A downside of population growth and modernization in Jordan is the production of rapidly growing quantities and varieties of waste. Only in the most recent landfill site, al-Khabawi new site, have modern standards of site selection, preparation and management been applied. Although this is a national issue it is also a Badia issue because, historically, it has been assumed that dumping waste in the wide Badia lands of eastern Jordan will be problem-free. However, it is increasingly being realized that there is a great range of associated problems from soil and groundwater pollution to dust, bad odours and danger to employees and scavengers. Communities near landfill sites are becoming more aware of the problems and less tolerant of them. For all these reasons, establishing criteria and, thence finding locations, for safe waste disposal sites, is urgently required.

Keywords: solid waste, landfill site selection, Jordan Badia

1. Introduction

Jordan is a developing country that has experienced rapid population growth during the last three decades for two main reasons: (i) successive waves of refugees from Palestine since 1948 and from the Gulf countries in the aftermath of the wars in the Gulf in 1991 and 2003; (ii) natural population growth. Recent reports also predict that (based on a population growth, which is ca 3.3 % year-1), about 5.3 million inhabitants are living in Jordan (Maani et al 1998) (excluding the recent migrants from Iraq). The population has jumped dramatically from about one million in 1965 (CVDB 2001, DoS 1998) and is expected to reach 8 million in 2012 (DoS 1998) (Figure 1). This trend in population growth is also coupled with economic development, industrial activities, extensive agricultural activities and the development of infrastructure (Al-Hanbali 2002, CVDB 2001, Dheiafleh 2000, JES 2001, Salameh et al 1993, World Bank 1994). These changes have resulted in a tremendous increase in solid waste generation which has led to serious implications for waste management (Chopra et al 2001, CVDB 2001).

The main aim of this paper is to highlight issues related to the status of solid waste in Jordan and its management, particularly in relation to the need for identification of new landfill sites. The data reported in this chapter derive from a number of sources that include a review of national and international literature and government policy statements, questionnaires conducted as part of a research programme that aimed to develop objective landfill site selection criteria and visits made to most of the major landfill sites in Jordan over the last five years.

Figure 1. The projected population growth (1922-2012) (Source: DoS 1998)

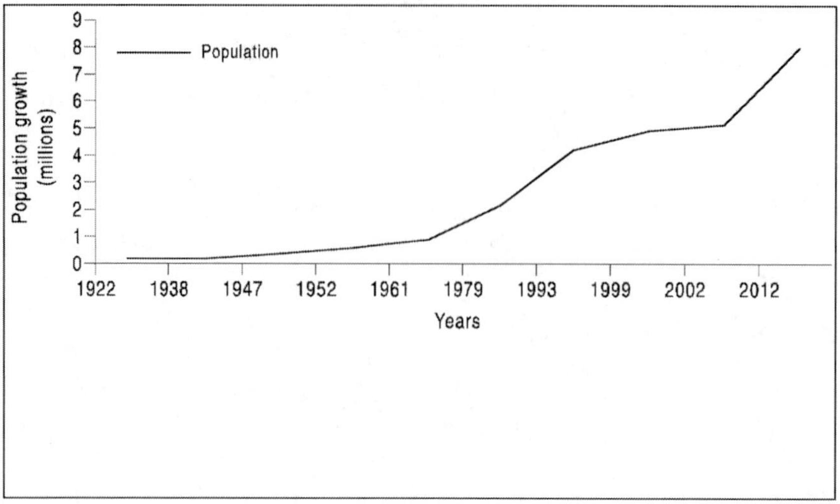

2. Solid waste status and management in Jordan: current practices and strategies

Several environmental studies have discussed solid waste and related issues in Jordan (Abu-Hijleh et al 1998, Al-Hanbali 2002, Chopra et al 2001, CVDB 2001, Kofahi et al 2000, World Bank 1994). These studies indicate that solid waste generation has risen steadily over the last three decades. In fact the amount of waste generated in Jordan is different from one study to another; for example, Chopra et al (2001) and CVDB (2001) reported that the average waste production is about 910 g/person/day. Tables 1 and 2 summarize the estimated solid waste generation in Jordan as reported in several studies and compare Jordanian production rates with other areas of the world. The total generation of solid waste has increased through time due to significant improvements in living standards and conditions and increased industrialization. In 1994, solid waste production was estimated at ca 1.35×10^6 t/yr, and included paper, glass, metals, plastics, food and other wastes (Abu-Hijleh et al 1998). This amount is expected to increase in the future

Table 1. Solid waste production in Jordan in 1994

Governorates	Waste generation (t/yr)
Amman & Zarqa	730 000
Irbid & Jarash	294 000
Balqa	109 500
Karak	54 750
Madaba	40 150
Mafraq	36 500
Aqaba	35 000
Ma'an	29 200
Ajloun	14 600
Tafila	9 125
Total	1 352 825

Source: (Abu-Hijleh et al 1998)

due to economic expansion, population increase and industrialization (CVDB 2001). It appears from Table 2 that waste generation rates in Jordan are relatively high in comparison with some neighbouring countries such as Israel and India, which is attributed to improved living standards and the styles of consumption (Chopra et al 2001).

Solid waste in Jordan derives from six different sources and activities (Table 3) (Chopra et al 2001, GCEP 2001, JES 2001). The domestic sector generates the highest quantity of waste followed by industry. One of the main problems of solid waste in Jordan is the high organic content, which is probably attributable to kitchen waste (Figure 2).

Table 4 compares the typical percentage distribution of solid waste in Jordan and other countries (GCEP 2001).

Table 2. Comparison of solid waste generation rates in Jordan with other countries

Location	Per capita generation rates (g $cap^{-1}day^{-1}$)
Jordan	910
Bangalore, India	400
Israel	400
Manila, Philippines	400
Asuncion, Paraguay	460
Seoul, Korea	2 000
Vienna, Austria	1 180
Mexico City, Mexico	680
Paris, France	1 430
Australia	1 870
Sunnyvale, California, USA	2 000

(Source: Chopra et al 2001)

Table 3. Solid waste types and sources in Jordan (modified from CVDB, 2001)

Type of wastes	Sources
Residential	Single and multiple dwellings
Commercial	Shops of various types, offices, retail stores, entertainment centres, restaurants, hotels, service stations, and green markets
Municipal activities	Demolition and construction, landscaping, parks (tree cuttings), waste treatment residues (sludge), and slaughter houses
Institutional	Universities, schools, hospitals, prisons
Industrial	Chemical and petrochemical, food, medical, trade and crafts, and small scale manufacturing
Agricultural	Animal farm wastes, plant farm wastes (nurseries), and olive mills

Table 4. Comparison of the Jordan solid waste contents with other countries

	Percentage of total weight			
Waste type	Jordan	Asian countries	UK	USA
Organics	50-68	75	3.6	20
Paper	5-10	2	31.2	43
Metals	3-6	0.1	5.3	7
Glass	2-5	0.2	3.8	9
Plastics	4-6	0.1	5.2	5

(Source: GCEP 2001)

Figure 2. Typical waste properties in Jordan (Source: GCEP 2001)

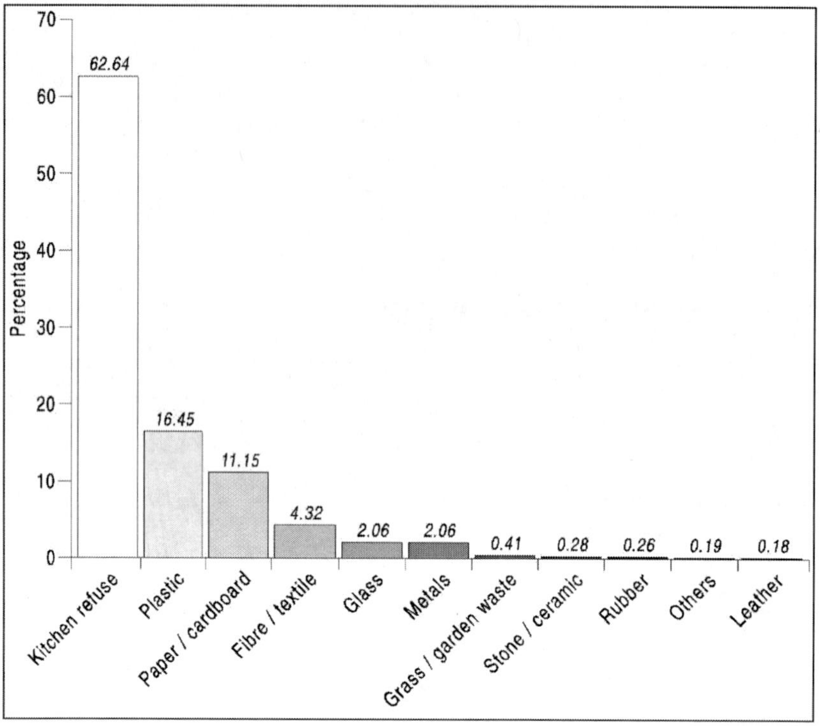

Organic waste contributes the highest proportion of the total in Jordan, followed by paper, metals, glass and plastics. This is attributed to the fact that people in Jordan consume more raw vegetables and less processed 'take away' food. People usually prepare food in their own houses, particularly in the rural areas. For example, meat is sold with bones, and vegetables are bought unprocessed with roots and leaves attached. The inedible parts are discarded but are not recycled. Food waste constitutes about 60 % of the total solid waste which contributes to faster fermentation in landfill sites (Chopra et al 2001, CVDB 2001).

3. Fieldwork survey and outcomes

As part of a questionnaire survey undertaken to assess the suitability of particular areas for siting new domestic landfill sites, the export interviewees were also asked about the waste management and planning process in Jordan. Twelve local community representatives were also interviewed (Table 5).

Lack of landfill planning and siting standards, associated with risk to groundwater, were considered important by more than 36 % of all respondents. Financial constraints, and their influence on the operation, management and construction of landfills were thought to be very important; leading, amongst other things, to insufficient geotechnical investigations. Issues related to the lack of public concern about environmental protection and lack of proper coordination among solid waste management and related agencies were also of major concern for respondents, particularly the planners because these issues caused considerable problems. The decision makers asserted that the real problem of solid waste management in Jordan is that the public does not provide enough

Table 5. Main causes of solid waste management problems in Jordan

No.	Solid waste management problems in Jordan	Opinion survey Group 1* Freq.	%	Group 2* Freq.	%	Group 3* Freq.	%	Overall (%)
1	Lack of management programmes such as sorting, reuse and recycling	14	66.7	12	63.2	7	58	63.5
2	Lack of regulation, legislation and national policies	13	62	15	79	3	25	59.6
3	Financial limitations	17	81	10	53	3	25	57.7
4	Inadequate disposal planning	6	28.6	11	57.9	6	50	44.2
5	Potential threat to groundwater	4	19.0	11	57.9	4	33.3	36.5
6	Lack of landfill standards and criteria	14	66.7	7	36.9	–	–	40.4
7	Lack of public awareness	11	52.3	8	42.1	–	–	36.5
8	Local community opposition	11	52.3	6	31.6	–	–	32.7
9	Lack of community role and opinion	2	9.5	–	–	8	66.7	19.2
10	Lack of geotechnical investigations	5	23.8	5	26.5			19.2
11	Rapid growth of population	5	23.8	4	21.5	–	–	17.3
12	Facilities shortage (trucks, containers)	9	42.9	–	–	–		17.3
13	Lack of trust with decision makers	–	–	–	–	5	41.7	9.6
14	Lack of regional planning and future land use planning	8	38.1	–	–	–		15.4
15	Capacity building	6	28.6	–	–	–		11.5
16	Waste creation and diversity	5	23.8	–	–	–	–	9.6
17	Lack of co-ordination among institutions	–	–	3	15.8	–	–	5.8

Group1*: decision makers; Group 2*: academics and planners;
Group 3*: local community representatives (Source: questionnaire survey results)

support for new landfill sites. Academics and planners have explained this by the lack of awareness programmes focusing on potential risks coming from such facilities. There is also a lack of media attention. All parties agreed that awareness of waste segregation is needed in order to stimulate adequate management and to avoid environmental pollution. The economic situation was seen as the main obstacle to imposing any landfill tax (Al- Fayez 2002).

It was evident that all the problems listed in Table 5 are interrelated and overlap. In fact all parties were aware that increasing population growth made education a high priority. However, respondents were conscious of the fact that Jordan has no legislation, regulations or guidelines. Absence of landfill planning, the lack of a national strategy and a lack of dialogue with local communities regarding the 'pros and cons' of the landfill issue compound the problem of solid waste disposal in Jordan, so there is no clear future plan for land use. All agreed that the current method of disposal must be upgraded to a sanitary landfill method. They also suggested that all environmental parties should meet annually to develop and implement a solid waste management master plan.

4. Hazardous and medical waste in Jordan

With the rapid industrial and economic growth of Jordan, the output of industrial solid waste has increased dramatically. In 1994, the Ministry of Planning commissioned a team from the World Bank to make an inventory of hazardous waste generation in Jordan and develop an action plan for establishing a new central hazardous waste disposal facility. Estimates of current and future hazardous waste were provided. The report was published at the end of 1994. According to the report, untreated hazardous solid wastes totalled about 43 000 t/yr in 1993 (90 % of this figure is liquid hazardous waste). The total was expected to grow at a rate of 7 % per year 1993-2000 and at 3 % per year 2000-15 due to industrialization and also to the increase in the number of institutions. The report also noted that hazardous waste is often disposed of at municipal sites with no control; for example, Suwaqa hazardous landfill site (1.28 km^2; located 120 km south of Amman within the Badia area) was selected in a remote area in order to avoid ownership rights problems and was developed without soil or groundwater investigations.

According to a report by JES (2001) it was estimated that about 32 % of the hazardous waste from hospitals was burned in a Double Chamber incinerator and it was shown that some hospitals have their incinerators dedicated to this purpose. However, most medical waste was still collected and burned with municipal waste with inadequate burning temperatures because there is just a single burning chamber associated with a short chimney. Usually, the burning is not sufficient for complete incineration. Other hospitals throw their waste into the domestic waste stream.

The World Bank team was also commissioned to make on-site geotechnical investigations such as soil and groundwater for the Suwaqa site. The results were:

- There is a possibility of soil contamination from leakage or accidental spillage;
- There is a possibility of groundwater pollution (water depth ca 200 m; limestone formation) that was estimated to be very weak due to the low rainfall in the area (ca 100 mm);
- Leachate may reach groundwater only after 300 years

The above results and assumptions were confirmed by a study carried out by Chopra et al (2001) but the third point above, which suggests delayed pollution of groundwater, may be inaccurate because Chopra et al (2001) used a simulation method for estimating the leachate from al-Akaider waste disposal site (the two sites have similar environments in terms of groundwater depth and weather conditions). The simulation was implemented using default 5-year daily climate records from similar environments. Three barrier factors were used: soil layer, compacted solid waste and soil cover. The daily weather conditions, soil type, thickness and characteristics, type of landfill, topography and geomorphology were calculated. The results of the simulation showed that the volume of leachate generated at the barrier soil layer is very small. According to the research, the two main reasons are low precipitation and high evaporation. This presumably makes the risk to groundwater pollution insignificant so long as the depth to groundwater is greater than 300 m and an adequate lining is established in each new disposal site.

However, improved management has been recommended by the two studies, especially for sites that are used for both domestic and hazardous waste such as Suwaqa, al-Akaider and Mafraq, where these sites are currently receiving a range of wastes from diverse sources, including hazardous waste, which may lead to groundwater pollution.

5. Solid waste management

In common with other developing countries, solid waste disposal in Jordan is becoming more problematic (Kofahi et al 2000, World Bank 1994). Solid waste management (SWM) is the responsibility of local authorities (local Cooperative Services Councils and Municipalities) under the umbrella of the Ministry of Villages & Rural Affairs (GCEP 2001).

Solid waste in Jordan is classified as domestic (non-hazardous) despite the fact that there are considerable amounts of solvents, batteries, medical and industrial waste in addition to agricultural and machinery included. The management of solid waste in Jordan follows the three steps of collection, transfer and disposal. Solid waste collection started in some of the main cities such as Amman, Irbid and Zarqa in the 1950s (Al-Hanbali 2002) on a weekly basis. The current collection process in rural areas is weekly on a house-by-house basis (CVDB 2001), and involves collecting waste directly from individual households. The process is primarily confined to residential areas and shops where the waste is collected in metal containers or plastic bins (1.1 m^3) and is then transferred to special lorries for disposal. The other method of collection, applied to apartments (used in the cities), is where waste is stored in substantial amounts in bulk containers in the basement of the apartments. These containers are transferred on a weekly basis to the bin compound and later are transported to the waste disposal sites. All these methods of collection are implemented by teams with trucks without any transfer stations (Al-Hanbali 2002). Both methods of collection are very labour intensive and time consuming. Separation of solid waste into separate streams at the point of production is not applied.

Until the 1980s, the open dumping and burning of waste was undertaken in the open air without proper management (CVDB 2001). All kinds of solid waste are disposed of in landfills or dump sites in an irregular manner without any treatment (Abu-Hijleh et al 1998). Recycling activities, which are applied in some disposal sites, are not well managed or regulated. Basically it is an ad-hoc arrangement, which is run by the private sector or by individual people living near the sites (Chopra et al 2001). The sorted materials are then removed and collected. According to Chopra et al (2001) about 98 % of these materials, such as soft drinks bottles (particularly), cardboard, school copybooks, and egg containers are sent back to factories/companies for recycling.

The main problem facing the local councils in terms of management is the shortage of a sufficient operational budget. Most of the collection trucks are donated by countries such as Japan (JICA).

Recently, municipal disposal has become a national concern in Jordan. This is attributed to two main reasons: increases in public environmental awareness and improved education; and the government acceptance of the potential negative health and environmental impact of the local disposal system. As a consequence, a modern landfill site (al-Khabawi) was established in 2002 serving more than 50 % of Jordan's population, particularly the Amman and Zarqa regions. An environmental impact assessment was undertaken for the site. According to Al-Tarawneh and Meshagbeh (2003) as the site is about 9 km from the borders of Amman municipality there is no expected adverse impact on the town in terms of smell, dust and noise pollution. The site will be maintained in sanitary condition. There will be a final cover before the landfill is closed and the site will eventually be vegetated.

Generally, the situation of solid waste management in Jordan is not regulated or well managed in contrast with practices in some developed countries such as the USA, Japan and Canada. These countries have their own regulations and laws to manage solid waste. In the USA, for example, the Resource Conservation and Recovery Act, 1976, regulates solid waste management. In Japan it is governed by the Waste Disposal and Public Cleaning Law, 1970 (Petts et al 1994).

The problem of solid waste management in Jordan is not only constrained by technological issues but also has economic, institutional, legal and sociological implications (Chopra et al 2001, CVDB 2001, GCEP 2001, World Bank 1994). These issues have created a difficulty in implementing prevention programmes such as reuse and recycling (Chopra et al 2001). Several studies (eg Chopra et al 2001, JES 2001, World Bank 1994) carried out in Jordan have stressed the need for better management of solid waste, particularly in terms of locating new waste disposal sites.

Jordan is also one of the countries that signed up to the Basel Convention in 1992, which obliges Jordan to take action to ensure the management of hazardous waste and other waste matter in a safe manner. Nevertheless, the country has no existing laws for the treatment and disposal of hazardous waste (World Bank, 1994).

6. Landfill site selection methods in Jordan: current practices

Landfilling is the most common method of solid waste disposal in Jordan. However, this method according to Chopra et al (2001), is not based on feasibility studies that take environmental issues into consideration. According to Al-Fayez, (2001) and Al-Tarawneh and Meshagbeh (2003) there are three disposal methods in Jordan. These methods are:

- Area method: solid waste is spread on flat areas and then compacted. Finally a soil layer is added;
- Excavated cells or trenches: digging holes with different dimensions with a low slope in the bottom. When the hole is full another one is dug;
- Canyon/depression method: used in areas that have natural depressions without the need for digging.

Al-Tarawneh and Meshagbeh (2003) claimed that the trench method is the dominant one because it is the easiest to operate.

Selection of waste disposal sites in Jordan is a controversial and time-consuming issue and it is also attracting increased debate and conflict. This is mainly due to public environmental concerns and local community opposition, particularly in urban areas such as Amman, Zarqa and Irbid, where potential disposal sites are limited (CVDB 2001).

No previous attempts have been made to develop site selection criteria in Jordan except in a study conducted by Al-Hanbali (2002). Al-Hanbali used GIS and remote sensing techniques for identifying new landfill sites in Mafraq governorate and applied the US EPA location criteria. One of the main recommendations of the study was to select new landfill sites based on environmental standards and criteria to avoid polluting the groundwater in Jordan. Another study by Al-Kharaibeh et al (1991) indicated that there was contamination of water resources in Jordan which came from two sources; either from direct contact between the surface water and the dump site materials, resulting in pollution of surface water, or seepage of pollutants down to the groundwater and causing groundwater contamination particularly in the winter season. These two studies have had a major influence on this research because they highlighted a national problem that needed to be addressed.

Decision makers and planners avoid discussing the process of site selection with the local communities due to social perception and attitudes; the communities do not wish to have dump sites located around them (JES 2001). For example, al-Rusaifeh landfill site (Figure 3) is located between Amman and Zarqa and close to the populated area where smoke, smell and bad visibility is especially obvious in the morning. As a consequence, local people complained and a groundwater investigation indicated a serious problem of pollution. Therefore, a decision was made in 2001 by the government to move the site to another location (Al-Tarawneh et al 2003).

7. The local situation of existing landfill sites

There is no consensus among the studies in Jordan on the number of legal and illegal solid waste disposal sites. In fact, the number differs from one study to another. For example, JES (2001) reported 26 sites while Al-Fayez (2002) and Chopra et al (2001) reported 24 sites distributed across the country. Table 6 provides information about these sites, which receive about 1 500 000 t yr^{-1} of waste. This estimated figure does not include the sites that are not listed in Table 6. These sites are located within the research area including those at Safawi and al-Manareh new site.

The Al-Rusaifeh site serves about 2 million people and receives more than half of the waste generated in Jordan (ie 2 200 t day^{-1}). Al-Akaider site is the second largest site located in the northern part of the country and receives about 350 t/day of municipal solid waste. Figure 3 shows the distribution of these sites in addition to al-Khabawi new site and also the location of Suwaqa hazardous site.

In general, these sites are the responsibility of the local authorities (municipalities) for collection and transfer. The management and follow-up of these sites is the responsibility of Co-operative Service Councils (CSCs) and local Municipalities under the umbrella of the Ministry of Villages & Rural Affairs.

Some sites, such as al-Akaider and Safawi dumpsites, use the co-disposal method (sandwich method) of solid and liquid removal.

No data exist on how and when these sites were selected or how they are operated and what criteria were taken into account or applied (GCEP 2001), though apparently they were selected on the basis of ad-hoc methods by the local authorities or were chosen because of the availability of unused government or derelict land JES (2001). More than 70 % of these sites are using open dumps with no control or monitoring (Table 6). They vary in their characteristics and management procedures (Chopra et al 2001, JES 2001).

Several studies from different agencies in Jordan have expressed their concern and worries about future landfill sites. They have identified a problem of finding new ones due to local community opposition, their effect on land value and the potential effects of these sites on groundwater; it is said that sites are selected without taking into consideration environmental or social concerns (Chopra et al 2001, CVDB 2001). Local authorities try to reduce the potential effects of these sites on the environment and the neighbouring populations by using on-site management such as waste placement compacting and then covering with indigenous soils and finally the daily use of chemical sprays against breeding insects, rodents and disease vectors (Chopra et al 2001). In fact, engineering techniques such as lining and groundwater monitoring wells are not applied (except at al-Khabawi new site where new techniques of gas collection and lining were utilized). This is attributed to financial constraints and the lack of local expertise and capacity building. Some of these practices to some extent still use the classical methods of dumping which cannot be defined as a sanitary landfill method. According to Al-Fayez (2001) most of the sites do not comply with safety conditions and are in inappropriate locations. It is apparent from Table 6 that at more than about 70 % of the sites, the classical method of open landfilling is practised. These practices are, to some extent, located between the area and trench methods as practised in the US, and are commonly known as the 'sandwich' method.

8. The potential risk to groundwater

None of the waste disposal sites in Jordan (except al-Khabawi new site) has engineering-management. Recent studies (Jaber et al 2001, Salameh et al 1993) have indicated that all kinds of solid waste pose a potential threat to the quality of both surface and groundwater resources in Jordan; almost all waste disposal sites in Jordan are overlying groundwater aquifer basins. For example, the World Bank (1994) reported that the al-Rusaifeh site was selected and operated without bottom lining and is located above one of the major groundwater basins (Zarqa basin), which it has polluted after only three years of operation (RSS, 1995). One of the main criticisms made by Chopra et al (2001) to the local sites is the proximity to heavily populated areas and areas where extensive pollution to water supplies could be caused by leachate. The Safawi open site was recently used for both solid and liquid waste disposal in a sensitive area (basalt formation) without any lining or monitoring system. 75 % of the sites are located close to residential areas. The al-Rusaifeh site is located adjacent to three large cities; Amman, Zarqa, and al-Rusaifeh. Recent studies have indicated that leachate at this site will be produced for 15 years after closure (RSS 1995, World Bank 1994).

Figure 3. Solid waste disposal sites in Jordan (Source: modified from CVDB 2001)

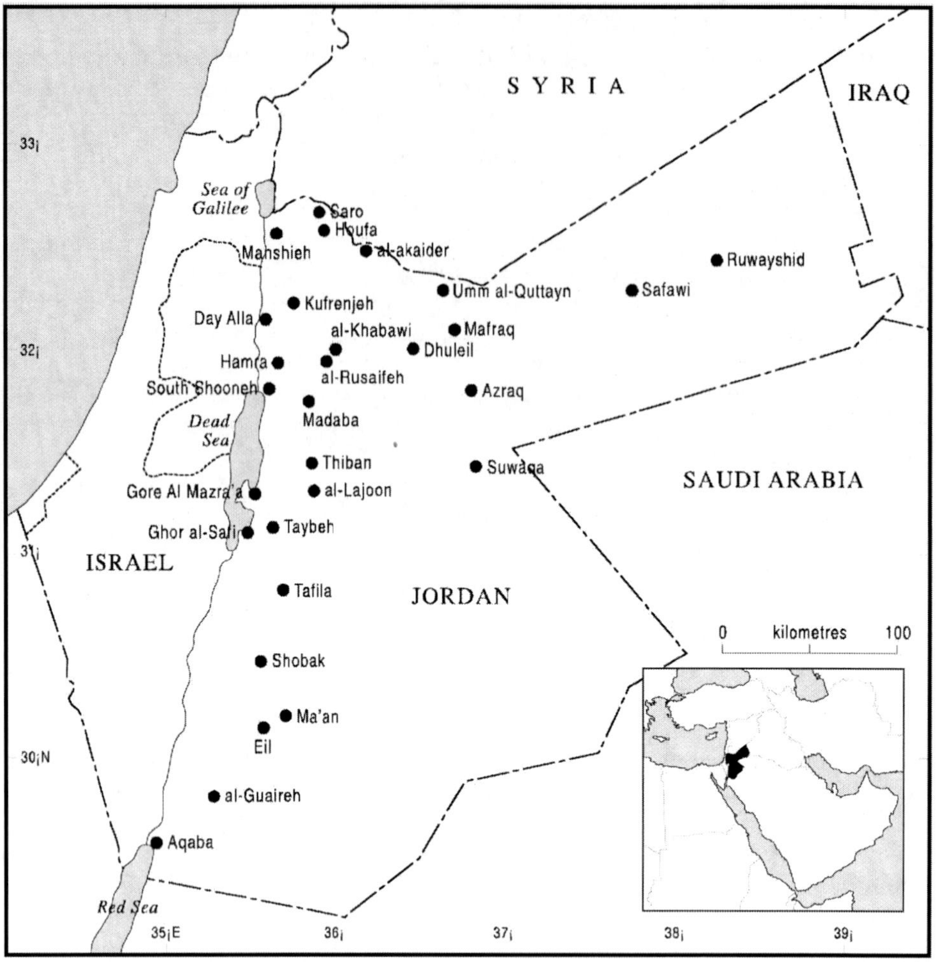

9. Current practices and environmental conditions of solid waste disposal

Based on the experience gained from the fieldwork observations at the existing landfill disposal sites in Jordan, and from interviews with solid waste contractors, a number of management practices, scenarios and implications were noted.

9.1. Waste disposal conditions and the daily practices

Open dumping using burning methods is still practised at some waste disposal sites, creating environmental problems such as air pollution and possibly water pollution. Municipal, untreated industrial and medical wastes such as batteries, fertilizers, metals and aluminium cans, shoes, clothes, agricultural waste, and plastic products were seen disposed of together in all the sites. Carcasses of cows and sheep are buried in trenches dug about 2 m down into the ground at the soil/bedrock interface. Solid waste is stacked to about 2 m high with a limited tipping front. Only some of the main sites, such as al-Khabawi, al-Rusaifeh and Mafraq, were equipped with compaction machinery and bulldozers. Fences were not observed for most of the existing disposal sites and hence there is a potential hazard to the general public. Some of these sites are close to residential areas, valley bottoms and main roads. Daily soil covering was not applied at some sites due to the high daily loading of waste. Where used, the soil cover caused dust problems during the summer and autumn and also made walking and driving on the site difficult during wet seasons.

On-site sorting of refuse material was also seen practised at most of the surveyed sites. Scavenging is the dominant operation in many sites. Waste pickers living in areas adjacent to waste disposal sites were seen searching in these sites during and after regular working hours or during the holidays. Scavengers are either employed by private contractors or are self-employed to collect recyclable materials such as cans, metals and cardboard from these sites and also from the incoming waste trucks and containers immediately after loading. Some individuals used small vehicles to collect recyclable materials directly from individual houses. Workers were seen eating and drinking in close

Table 6. The location and capacity of the available waste disposal sites in Jordan

No.	Site name	Area (ha)	Waste quantity (t/day)	Disposal method	Waste quality
1	al-Akaider	80.6	350	Sandwich*	Municipal industrial and medical wastes
2	Mafraq (Hussiniat)	18.0	100	Area	MSW, industrial and medical
3	Kufrenjeh	7.1	90	Open	MSW
4	North Shuneh	7.8	100	Open	MSW and medical
5	Taybeh	6.0	30	Open	MSW
6	Saro	5.5	80	Open	MSW
7	Umm al-Quttayn	40.0	30	Open	MSW
8	al-Rusaifeh	120.0	2200	Area	MSW, industrial and medical
9	Madaba	8.0	150	Sandwich*	MSW, medical, industrial waste
10	Hamra	27.5	140	Open	MSW and medical
11	al-Dhuleil	7.0	70	Sandwich	MSW
12	Thiban	3.0	20	Open	MSW
13	Deir Alla	20.0	30	Open	MSW
14	Azraq	4.8	10	Open	MSW
15	Aqaba	6.0	80	Open	MSW, medical and industrial
16	Ma'an	50.2	50	Sandwich	Seepage and spent oils
17	al-Lajoon	60.0	85	Sandwich	MSW and seepage
18	Tafila	45.0	50	Open	Medical and seepage
19	Shobak	2.6	20	Open	MSW
20	Eil	28.0	20	Open	MSW
21	al-Guaireh	27.0	20	Open	MSW
22	Huseinyeh	10.0	15	Open	MSW
23	South Shooneh	1.0	40	Open	MSW
24	Ghor al-Safi	15.3	20	Open	MSW
25	Houfa	–	–	Open	MSW
	Total	3800			

*: Used for both liquid and solid waste (Source: modified from Chopra et al 2001, CVDB 2001, GCEP 2001)

proximity to the landfill activities with no attention to safety conditions. None wore masks or hats. Waste pickers therefore pose a safety hazard to themselves and to the landfill employees. Villagers living adjacent to some sites, particularly in rural areas, bring animals such as sheep and goats to the site for grazing. Dogs and donkeys are also commonly seen at waste sites searching for something to eat because there are no barriers to stop them getting into the sites.

The selection of these landfill sites was largely ad-hoc, chosen by local councils and administrators without attention to potential environmental and social problems. Decisions taken to locate these sites were based on the availability of unused governmental land, or any available derelict mining sites.

None of the existing waste disposal sites follow or apply a sanitary landfill system. The locations are often unsuitable with regard to protecting valuable groundwater and surface water resources. Poor practices seem to be encouraged by the absence of government policies regarding sound and proper management programmes. The only exception was al-Khabawi new site. Here there was consideration given to the environmental and social issues when an environmental impact assessment was carried out (Zaki 2002).

At al-Khabawi, good landfill management practices were observed during two fieldwork visits. The site is fenced and is operated to control gas and collect and treat leachate, applying a daily soil cover on waste. Passive ventilation exists for methane and carbon dioxide through pipes installed in the site. Compaction machinery and bulldozers were also available at the site. Plastic sheet layers were used beneath the tip to prevent leaching of contaminants to groundwater. On-site management procedures include inspection and recording of incoming waste, and the collection of dumping fees.

9.2. On-site sorting of materials

Findings based on interviews with private sector contractors who collect materials from the Mafraq site have provided valuable information about the process of on-site material sorting. The contractors claimed that solid waste disposal and management is the same for the whole of Jordan. They also claimed that they were authorized by the local councils in order to sort the reusable materials. Initially, the contractor has to pay JD4000/yr as a tax to the local council (this figure differs from site to site depending on the quantity of waste). Mafraq landfill site is designed to last for about 20 years, with a total potential landfill area of about 38 ha. The site is located about 5 km from the main highway and 0.5 km from the closest settlement. The nearest farm is about 1 km away. The site property belongs to the government and the site has been operating for eight years. Holes 1-3 m deep are dug into the ground to bedrock (depending on the soil depth). Diverse solid wastes are received from 28 local councils, and 12 industrial units in Mafraq district on a daily basis in addition to waste from Al Al-Bayt University. The waste quantity is about 100 t/day (90 t are domestic, 4 t from agriculture, 5 t from industry, and 1 t of medical waste). Daily covering with soil is difficult due to high waste loading. According to Hussein (2002), about 70 t of reusable solid wastes are taken by the factories from the Mafraq disposal site each week. The contractors claimed that they suffered in the winter season when chemical materials were leached by rainfall to pollute surface waters. An individual who was working at the Hussiniat site said that he lost his sense of smell.

On-site workers complain that they are exposed to hospital waste materials such as needles, glass and other materials, which expose them to risks to their own health. They suggest that these materials should be isolated at an early stage and disposed of separately. Workers also claim that they suffer from a 'shame culture' because other people look down on them as waste workers. Some workers tried to attack the researcher when he tried to take pictures at al-Akaider site because they thought he was a journalist. Working in solid waste collection and management is still considered shameful in Jordan, and most of the people who work on solid waste collection and on-site management are from outside Jordan; they are mostly Egyptian. However, more recently the situation has slightly changed since some Jordanians have become involved in this sector as a result of growing awareness by the population of solid waste issues.

10. Summary and conclusions

Solid waste generation is increasing dramatically in Jordan. Programmes of solid waste management, such as recycling, reuse and recovery are not regularly established. Lack of regulations, national policy or guidelines regarding waste disposal complicate the situation of solid waste management. It was felt that most of the landfills in Jordan are operated on the principle of 'out of sight out of mind' with little attention to social and environmental considerations. The principle of site selection is governed by the local economic and technical situation. This method of uncontrolled disposal can lead to the contamination of groundwater resources and to opposition from local communities. Most of the existing disposal sites are located in unsuitable and sensitive areas. Therefore, creating criteria for the designation of safe disposal sites is urgently needed.

References

Abu-Hijleh M, Dwairi R, Kumoos M and Al-Tarazi S (1998). Feasibility study of municipality solid waste incineration plant in Jordan. Energy Converse. Mgmt. Pergman 39 (11):1155-1159.

Al-Fayez K (2002). Head of solid waste division, Ministry of Environment. Personal communication, Jordan, 21 November.

Al-Kharaibeh S, Najeh K and Bani Hani M (1991). Solid waste in Jordan: A case study, HCST. Amman, Jordan.

Al-Fayez K (2001). Solid waste management in Jordan and site selection criteria. Seminar, 23 June, Ajloun, Jordan.

Al-Hanbali SWA (2002). Selection of best possible solid waste disposal sites within Mafraq city using remote sensing and GIS Techniques. MSc Thesis, Al-Albayt University, Al-Mafraq, Jordan.

Al-Tarawneh A and Meshagbeh M (2003). Sanitary landfilling. The second international energy week Seminar, 14 - 17 September, Amman, Jordan.

Chopra M, Reinhart D and Abu-El-Sha'ar W (2001). Report on US-Jordan municipal solid waste management collaborative research. Amman, Jordan.

CVDB (2001). Environmental evaluation for the provision of garbage collection and compactor light trucks for local councils. Report, Cities and Villages Development Bank/Ministry of Municipalities and Rural Affairs, Amman, Jordan.

Dheiafleh RS (2000). Solid waste management in al-Mafraq City. MSc Thesis, Al-Albayt University, Jordan.

DoS (1998). The population and housing census. Internal report, Department of Statistics of Jordan, Amman, Jordan.

GCEP (2001). Report on municipal solid waste composition in Jordan. General Corporation for Environment Protection/Ministry of the Environment, Amman, Jordan.

Jaber JO and Mohsen MS (2001). Evaluation of non-conventional water resources supply in Jordan. Conference on desalination strategies in south Mediterranean countries, Cooperation between Mediterranean countries of Europe and southern rim of the Mediterranean, Tunisia.

JES (2001). Report on Jordan solid waste. Jordan Environment Society, Amman, Jordan.
Kofahi NA, Gharaibeh SH and Oqeili S (2000). Algorithm for selection of domestic solid waste disposal sites in developing countries. Journal of Interdisciplinary Mathematics 3 (2-3):173-190.
Maani M, Hunaiti H and Findlay AM (1998). Demographic change and population projection. In: Dutton, R, Clarke, J and Battikhi, A (eds), Arid land resources and their management, Jordan's desert margin. London: Kegan Paul International pp. 215-246.
Petts J and Eduljee G (1994). Environmental impact assessment for waste treatment and disposal facilities. Chichester: John Wiley.
RSS (1995). Report on developing the Al-Hasa concession phosphate mine. Royal Scientific Society, Amman, Jordan.
Salameh E and Bannayan H (1993). Water resources of Jordan, present status and future potentials. Amman: Friedrich Ebert Stiftung.
World Bank (1994). Industrial pollution control project in Jordan: design of a hazardous waste disposal facility. Final report, Ministry of Planning, Amman, Jordan.
Zaki H (2002). The Greater Amman Municipality Assistant for Health Affairs, Personal communication, 10 September, Amman, Jordan.

CHAPTER 8

Modelling nitrate leaching in the Azraq Basin, Jordan, using GIS

Rida Al-Adamat
Institute of Earth and Environmental Sciences Al Al-Bayt University, Mafraq, Jordan.
E-mail: ridaali@aabu.edu.jo

Serwan M. J. Baban
CLEAR, Department of Surveying and Land Information, University of the West Indies, St. Augustine, Trinidad and Tobago

Ian Foster
Geography, School of Science and the Environment, Coventry University, Coventry CV1 5FB, England

Jordan is a country that faces 'absolute water scarcity' and may not be able to meet its water needs by the year 2025. Groundwater is the major water resource for many areas of the country and the only source of water in some areas. Most of the groundwater basins in Jordan are already exploited beyond their estimated safe yield. Total safe yield for all basins was estimated to be about 418.5x10^6 m^3 yr^{-1} yet the consumed water from these basins was 479x10^6 m^3 yr^{-1}. Groundwater is the second largest contributor to the irrigation sector at 258.4x10^6 m^3 yr^{-1} and it is the largest source for domestic consumption at 182.8x10^6 m^3 yr^{-1}.

A variety of human activities stemming from agricultural, industrial, community and residential sources, as well as the misuse of groundwater resources, have contributed to the deterioration of groundwater quality in Jordan. Since the early 1990s, the Jordanian government has encouraged drilling private wells in the northern part of the Azraq basin to irrigate agricultural land by providing low cost loans to farmers.

Focusing on this area, this paper attempts to produce a GIS-based model to estimate nitrate leaching from cesspools and agricultural land. It was found that nitrate leaching from agricultural sources was much higher than that derived from cesspools. It was estimated that about 0.3-0.7 kg household^{-1} yr^{-1} could leach to groundwater from cesspools in the study area. The estimated nitrate leaching from agricultural sources could reach up to 483 281 kg yr^{-1}.

Several management scenarios were implemented within a GIS environment to minimize nitrate leaching from both cesspools and agricultural fields. It was estimated that emptying the cesspools on a regular basis (about 2 months) could minimize or eliminate nitrate leaching from this source. Farmers could use information on available nutrient concentrations in the soil, irrigation water and organic manures to estimate the optimum fertilizer requirement. This scenario could reduce nitrate leaching by up to 99 %.

Other scenarios were suggested in order to minimize nitrate leaching from cesspools that included better design criteria and the possibility of constructing a local sewage treatment plant in the area.

Keywords: Jordan, groundwater, nitrate, GIS, modelling

1. Introduction

Jordan is a country that faces 'absolute water scarcity' and may not be able to meet its water needs by the year 2025 (Al-Adamat 2002). Groundwater is the major water resource for many areas of the country and the only source of water in some areas. Most of the groundwater basins in Jordan are already exploited beyond their estimated safe yield. Total safe yield for all basins was estimated to be ca 418.5×10^6 m^3 yr^{-1} yet the consumed water from these basins was 479×10^6 m^3 yr^{-1}. Groundwater is the second largest contributor to the irrigation sector at 258.4×10^6 m^3 yr^{-1} and it is the largest source for domestic consumption at 182.8×10^6 m^3 yr^{-1}. The Azraq basin is one of the largest groundwater basins in Jordan with an area of more than 12 000 km^2. It supplies fresh drinking water to the major cities in Jordan and towns within the basin.

A variety of human activities stemming from agricultural, industrial, community and residential sources, as well as the misuse of groundwater resources, have contributed to the deterioration of groundwater quality in Jordan. There are three types of pollution that affect groundwater (i) use and overuse of biocides and fertilizers and irrigation return flows, (ii) cesspools in towns, villages and refugee camps and (iii) use of vehicles with oil spills, lead and corroded particles (Al-Adamat 2002) .

Nitrate is often considered to be a major threat to groundwater quality because nitrate dissolves freely in water and is not held on the soils particles (Addiscott et al 1991) due to its negative charge (Addiscott et al 1991, Stauffer 1998). Therefore, nitrate is most likely to leach into groundwater (Stauffer 1998) by rain or irrigation water passing through the soil down to the groundwater (Addiscott et al 1991).

The major sources of nitrate in groundwater include soil nitrogen, nitrogen–rich geological deposits, and atmospheric deposition (Evans et al 1995). Nitrate could also be sourced in groundwater from fertilizers, septic tank drainage, feedlots, dairy and poultry farming, land disposal of municipal and industrial wastes, dry cultivation of mineralized soils, and the leaching of soil as a result of the application of irrigation water (Evans et al 1995). During the last two decades, many cases of groundwater contamination from extensive application of fertilizer in watersheds developed for agricultural purposes have been reported worldwide (Shamrukh et al 2001). According to Supalla et al (1995), nitrate pollution from agricultural fertilizers is the most serious and widespread threat to groundwater quality. There are several types of inorganic fertilizers on the market these days, which provide farmers with a wide range of choices that suit their crops, soil type and environment. These fertilizers have been developed in order to meet the high nutrient requirement of crops and to supplement available soil nutrients (Finck 2000).

The use of GIS technology covers a wide range of applications such as agricultural, land use planning, municipal applications, global scale applications (Ahn et al 1999), and modelling and management of the natural environment (Burrough 1986). GIS has been used in many aspects of groundwater management and modelling (Merchant 1994). According to Mitasova and Mitas (2002), the use of GIS for environment modelling has increased over the last few years, moving from research to routine applications. The combination of map algebra operations and GIS functions make it relatively easy to implement simple models within GIS. Such an advantage in GIS has made it possible for several research projects to investigate and quantify nitrate leaching to groundwater from various sources in many parts of the world. For example, GIS was used to assess the spatial variability of NO_3 leaching from the root zone from fertilizer applications in Rhode Island, USA (Görres et al 1996), to predict nitrate leaching to groundwater in southern Germany (Rodda et al 1999) and to estimate nitrate leaching from agricultural lands and septic tank systems in Northern Ireland (Jordan 1994). The researchers argued that their research had demonstrated the application of GIS to predict nitrate leaching to groundwater.

This paper is an attempt to develop a methodology for estimating nitrate leaching to groundwater in the northern part of the Azraq basin from both agricultural sources and cesspools using GIS techniques. Also, this paper will be an attempt to suggest sound management scenarios to minimize the possibility of groundwater contamination.

2. The study area

The study area is located in the northern part of the Azraq basin with an area of 867.4 km^2 (Figure 1). In 1993, the total population of the study area was 13 189 living in 33 towns, villages and small settlements (BRDP 1994) (Figure 2). The population is expected to be 21 464 and 25 247 in the years 2008 and 2013 respectively, based on a 3.3 % annual population growth rate (Maani et al 1998).

The climate of the study area is generally characterized by its hot dry summers and cold winters. The area is classified as semi-arid based on its climatic characteristics

Figure 1. The study area within the Azraq basin (after Al-Adamat et al 2004)

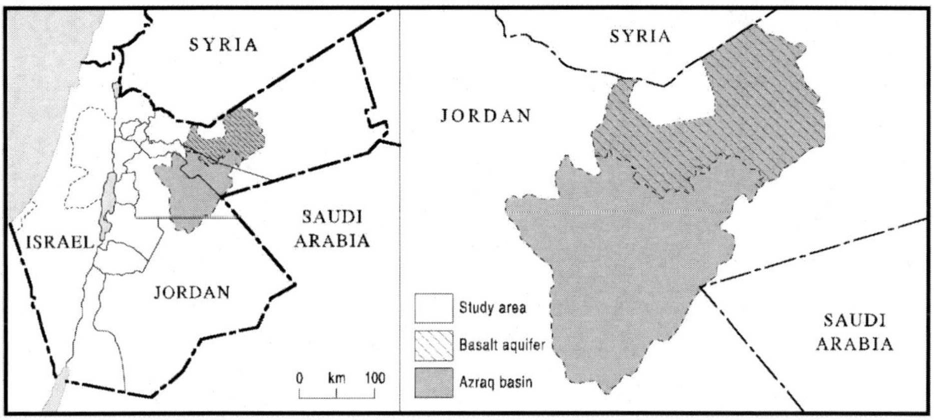

Figure 2. The urban centres in the study area

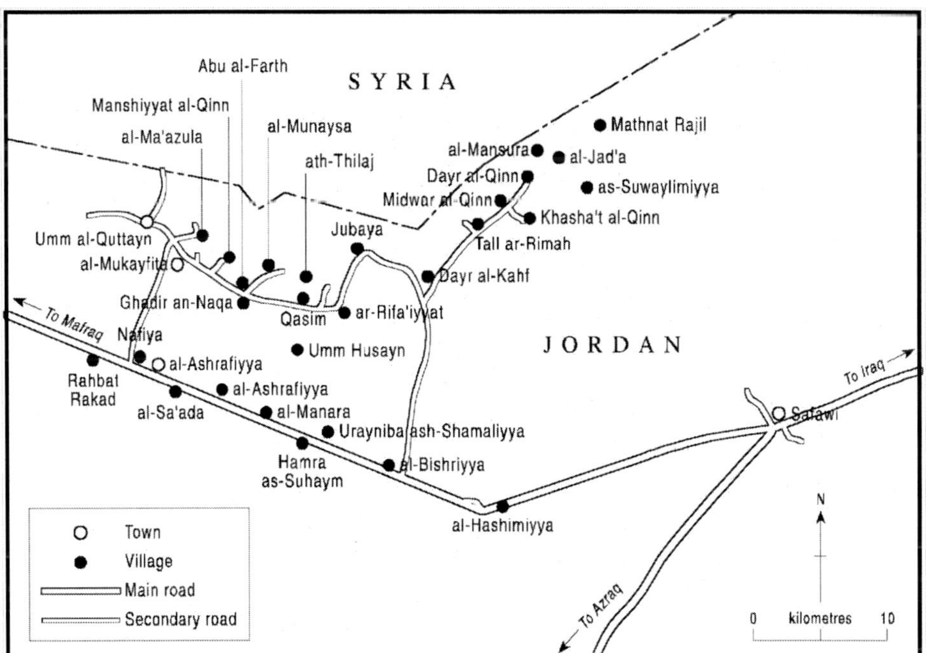

(Salameh et al 1997, Salameh et al 1993). It falls within an area recognized to be a transition zone between the environment of the Jordan valley and the arid interior desert areas of eastern Jordan. Also, it is characterized by low precipitation and high potential evaporation (Allison et al 1998). The duration of rainfall in winter is short with high intensity that mainly generates runoff (Al-Adamat 2002).

The study area is located within the basalt area of the northern Badia of Jordan. The basalt eruptions in the northern Badia are of different ages and appear on the surface in the north and north-east and extend northwards to cover an area of 11×103 km2 (Al-Tarawneh 1996, Allison et al 1998). The basalt eruptions are from what is known as the 'Basalt Plateau', which is related to the North Arabian Volcanic province (Al-Adamat et al 2004).

The soils in the study area (Figure 3) are in general characterized by being shallow and saline with little organic material and are associated with the North Jordan basalt plateau. Soils include well developed xerochrepts on older basalt flows, with gypsitic and calcitic horizons, and weakly developed xerothents on the recent basalt flows and in the wadis (Al-Adamat 2002, Al-Adamat et al 2004). The percentages of clay in all soil units in the study area are 25-26.7 %, silt percentage is 44-50.2 % and sand is 21.3-28.7 %, while the organic matter is 1.21-1.22 % (Al-Adamat 2002, Al-Adamat et al 2004).

Irrigated agriculture in the study area started in the early 1990s after a government decision to allow wells to be dug in order to start cultivating the land in areas close to the Syrian borders (Kirk 1998). All irrigation wells in the study area are drilled in the

Figure 3. The soil units in the study area (after Al-Adamat et al 2004)

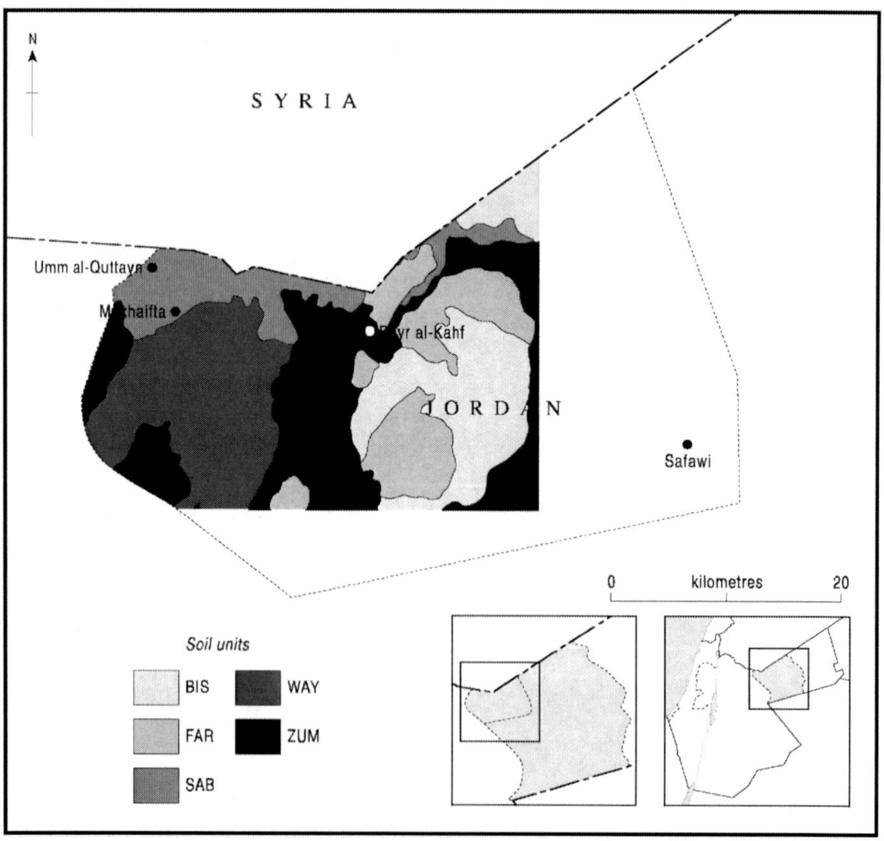

upper aquifer of the Azraq basin and within the basalt area (Al-Hussein 2000, Dottridge 1998). The majority of farmers cultivate vegetables such as tomato and watermelon, as they claim that it generates a rapid profit (Kirk 1998), while other farmers cultivate olive and fruit trees (Al-Hussein 2000, Waddingham 1998). The irrigated agricultural season (vegetables only) starts in April and ends in late November with only two months with no irrigation (December and January) (Al-Hussein 2000, Millington et al 1999, Waddingham 1998). Drip irrigation is the only method used to distribute water to the crops in the study area (Waddingham 1998). In size, vegetable farms are about 10-50 ha (Al-Hussein 2000); while the number of trees on the tree farm is ca 100-40 000 plus (Figure 4).

3. Model design

The model was based on a design described by Gusman and Marino (1999). The model takes into consideration all potential sources supplying nitrate to the soil and all nitrate losses that occur in the environment. These sources include: rainfall; irrigation water; fertilizer applications; existing NO_3 in the soil before cultivation (The existing NO_3 in the soil before cultivation represents the amount of organic-N that had been converted

Figure 4. Agricultural activities in 2000 (after Al-Adamat et al 2004)

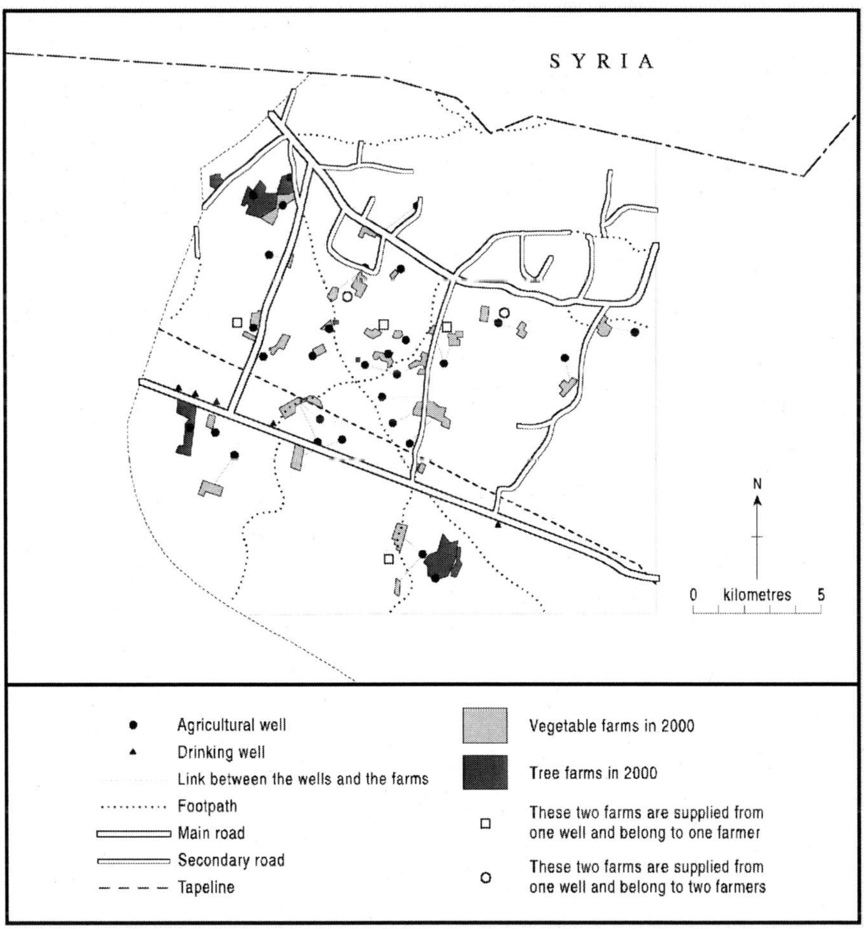

to nitrate through the process of mineralization); and cesspools. Losses of nitrate were defined as: leached; plant uptake; and nitrate decay (denitrification and volatilization).

The model was based on the fact that the nitrate input (sources) should equal the output (losses): equation 1, for agricultural lands, after Al-Adamat, (Al-Adamat 2002).

$$NO_3\,L = [NO_3\,R + NO_3\,IR + NO_3\,SBC + NO_3\,F] - [NO_3\,PU + NO_3\,D] \quad (1)$$

Where:- L: Leached, R: Rainfall, IR: Irrigation water, SBC: Soil before cultivation, F: Fertilizers (organic and inorganic), PU: Plants Uptake, and D: Decay (denitrification and volatilization).

Figure 5. The conceptual model used to estimate nitrate leaching from agricultural land

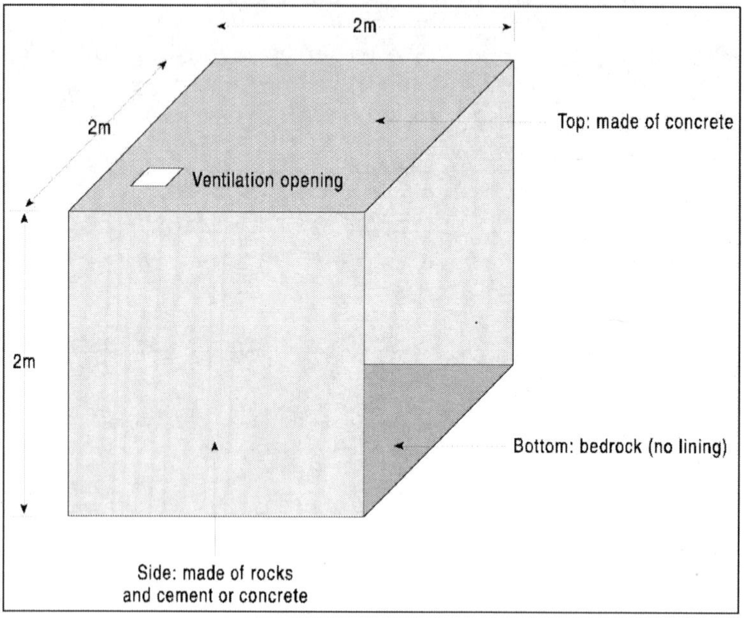

Figure 6. Typical cesspool design in the study area, after Al-Adamat (2002)

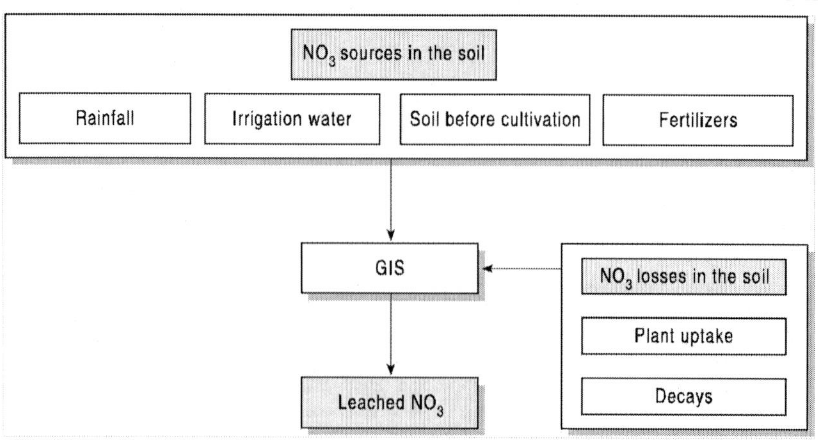

Figure 5 illustrates the conceptual model designed to estimate nitrate leaching to groundwater from agricultural sources.

In the case of cesspools, equation 1 was modified as follows in equation 2, after Al-Adamat, (Al-Adamat 2002):

$$NO_3L = NO_3CP - NO_3\,D \qquad (2)$$

Where CP: cesspools, D: denitrification

This modification results from the fact that cesspools have an estimated depth of 2 m below the surface (Figure 6) and, being below the rooting zone of most crops including grass, nitrogen is not available for plant uptake (Jordan 1994).

4. Data collection

In order to provide quantitative information to estimate the nitrate leaching to groundwater, fieldwork was carried out to collect groundwater samples from all wells used for irrigation purposes to find their nitrate concentration, in April 2000. Also, in May 2001, more fieldwork was conducted to collect soil samples from six different locations in the study area (two in uncultivated lands, two in abandoned farms and two in existing farms in 2001), where nitrate level was measured for three different depths (0-10 cm, 40-50 cm and 90-100 cm). A rainwater sample was collected from a rainfall event that occurred during April, 2001 and it was also subjected to chemical analysis to find the nitrate concentration. Farmers were interviewed to collect information about crops they cultivate and the amounts and types of organic and inorganic fertilizers used in their farms. Chemical fertilizer producers and dealers were interviewed to collect the required information about the chemical composition of their products. Other data such as the agricultural land use, rainfall and soil maps were acquired from several sources (Al-Adamat 2002, Al-Adamat et al 2004, RJGC 1998) (Table 1).

Table 1. The secondary data used in this research

Map type	Year of survey	Format	Original scale	Source
Agricultural land use	2000	GIS	1:25 000	(Al-Adamat 2002), (Al-Adamat et al 2004)
Soil map	1994	GIS	1:50 000	(RJGC 1998)
Rainfall data	2001	Excel		(Al-Adamat 2002)

5. GIS analysis

5.1. Nitrate input in agricultural lands

5.1.1. Rainfall

The nitrate volume (kg ha^{-1}) supplied by rainfall was calculated from equation 3 (modified from Jordan 1994).

$$NO_3(R) = \frac{NO_3(CR) \times R}{100}\ (\text{kg ha}^{-1}) \qquad (3)$$

Where CR: concentration of NO_3 in rainwater (mg l^{-1}), and R: Rainfall amount (mm)

Rainfall data were available from two daily weather stations and three TOTALISER rain gauges (Al-Adamat 2002) (Figure 7). The rainfall data needed to use this model were for the period April 2000-May 2001. Unfortunately, rainfall data were not available from the Dayr al-Kahf station for this period. Data from the Umm al-Quttayn station covered only the first three months of 2000 (January-March). The other three stations at Abu al-Farth, al-Bishriyya and Mathnat Rajil covered the period April 2000-April 2001. Rainfall in the area falls in the months of October-April (Al-Adamat 2002) which means that these three stations covered the required period. An assumption was made to estimate the rainfall at the Dayr al-Kahf station based on the locality of this station in relation to the three other stations. The location of Dayr al-Khaf is between Mathnat Rajil to the north-east (about 17 km) and Abu al-Farth to the west (about 13 km) and al-Bishriyya to the south (about 14 km). The rainfall at this station was assumed to be the average rainfall value of these three stations (about 93 mm). The rainfall amount at the Umm al-Quttayn station for the period April 2000-May 2001 was assumed to be equivalent to the amount of rainfall at Abu al-Farth (about 8 km to the east). This gave Umm al-Quttayn station an estimated rainfall for the whole period of about 82 mm.

The methodology of mapping the rainfall distribution for the study area was based on the Thiessen method. According to Ward and Robinson (2000) the Thiessen polygon method is better than the mean for calculating the areal rainfall because it allows for a non-uniform distribution of rainfall gauges by assigning weights to the measured depths at that gauge. This method involves the connection of the stations on the map by lines. Perpendicular bisectors are drawn to the lines connecting the stations, which generate polygons around each station. Each polygon represents the effective area of each station (ISESCO 1997).

Figure 7. Rainfall stations within the study area (Al-Adamat 2002)

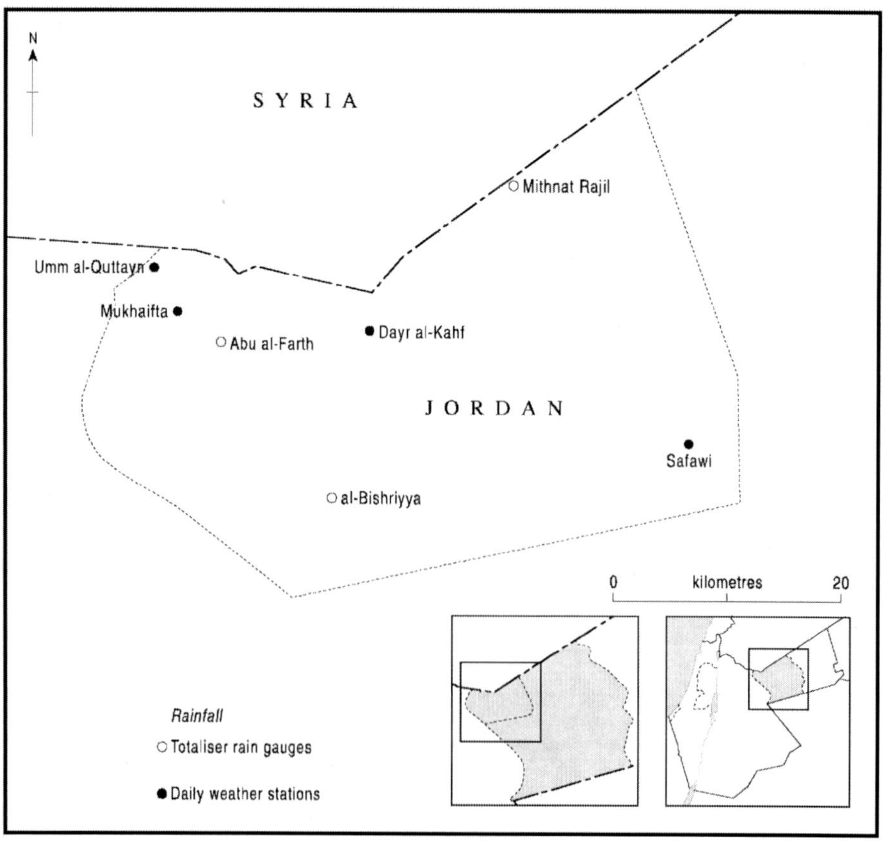

The 'ArealRain Extension' (downloadable at http//:www.esri.com) in ArcView GIS was used to calculate precipitation for the area using the Thiessen Polygon method. Equation 3 was then applied to the attribute file of this map. The outcome of this operation is shown in Figure 8, which shows that the amount of nitrate (NO_3) in the study area from rainfall varied about 5.5-6.75 kg ha^{-1}.

5.1.2. Irrigation water

Since all farming activities in the study area depend on groundwater for irrigation, the nitrate concentration in the water was assumed to be one of the major sources of nitrate in the soil that might be leached to groundwater. The NO_3 concentrations (mg l^{-1}) for irrigation wells were known from the field work of April 2000.

Waddingham (1994) estimated the amount of water abstracted in 1994 for irrigation purposes as ca 6.3×10^6 m^3 used to irrigate 482 ha (ca 13 575 m^3 ha^{-1}). This figure was estimated after interviewing 26 farmers in the study area. It was also estimated that the cultivated area needed only 3.8×10^6 m^3 (ca 7884 m^3 ha^{-1}) if farmers used the optimum amount of water for irrigation. In this research both values were used to estimate the mount of nitrate from irrigation water based on optimum and excessive water usage. The amount of nitrate (kg ha^{-1}) supplied by irrigation water was calculated using equation 4 (Al-Adamat 2002).

Figure 8. The amount of nitrate (NO_3) (kg ha^{-1}) derived from rainfall in the study area (April 2000–May 2001)

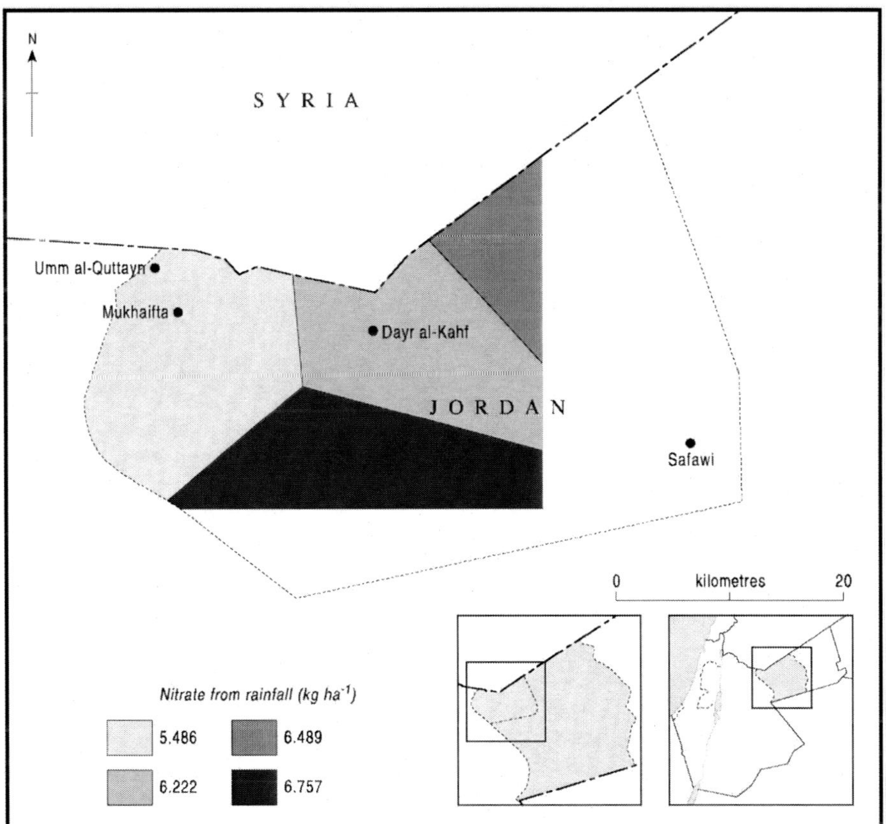

$$NO_3(IR) = NO_3(CIR) \times IR \times 10^{-3} \text{ (kg ha}^{-1}\text{)} \quad (4)$$

Where CIR: concentration of NO_3 in irrigation water (mg l^{-1}), IR: amount of irrigation water (m^3 ha^{-1}).

The agricultural activities map of 2000 (Figure 4) was updated with both estimates of irrigation water (m^3 ha^{-1}) and the nitrate concentrations of the well(s) that supplied each farm with irrigation water in 2000. Equation 4 was then applied to calculate the nitrate volume added to the soil (kg ha^{-1}); Figures 9 and 10 show the estimated nitrate contributed to the soil through irrigation water. Figure 9 represents the minimum nitrate volume added to the soil (kg ha^{-1}) based on the estimated optimum usage of irrigation water, while Figure 10 represents the nitrate (kg ha^{-1}) added if farmers had used excessive irrigation as estimated by Waddingham (1994).

5.1.3. Soil before cultivation

The amount of nitrate in the topsoil before cultivation was calculated as follows (equation 5) (afterAl-Adamat 2002):

$$NO_3(SBC) = NO_3(CSBC) \times B \times 10^{4} \text{ (kg ha}^{-1}\text{)} \quad (5)$$

Figure 9. Estimated nitrate mass (kg ha^{-1}) added to the soil from the optimum use of irrigation water (April-December 2000)

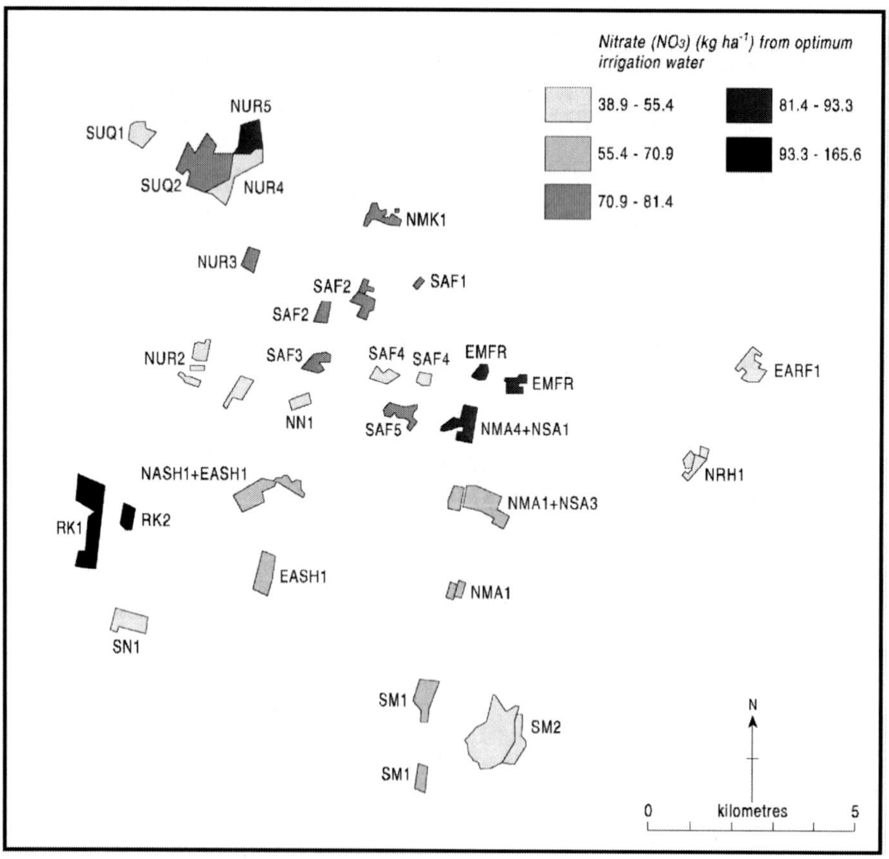

Where:- CSBC: concentration of NO_3 in the soil before cultivation (kg per kg of soil); and B: bulk density of the soil (kg m^{-3}).

The nitrate (NO_3) concentrations from two soil sampling sites (S3 and S6) were used in equation 5. Both samples represent natural soil (uncultivated before) and it has been assumed that the NO_3 concentration found in these soils is what would be found in the farming areas before cultivation. S3 was located in a silt loam soil while S6 was located in a clay loam soil. Nitrate (NO_3) concentrations from both sampling sites were given in mg l^{-1} which meant that this measure had to be transformed to kg (nitrate) per kg soil. In the soil sampling procedure, 10 g of soil were made up to 25 ml (1/40 litre) of soil-water extracts. In order to convert the nitrate concentration (mg l^{-1}) to nitrate per gram of soil, the nitrate concentration was divided by 400 which, when multiplied by 10^{-3}, produced an estimated nitrate concentration in kg per kg of soil.

In ArcView GIS, the attribute file of the soil map (Figure 2) was updated with the bulk density values and the nitrate concentration (kg per kg of soil) based on the assumption that all soil units that have a silt loam texture had 0.4 mg l^{-1} of nitrate if sampled for the topsoil and all soil units that have a clay loam texture had 1.7 mg l^{-1} of NO_3 in the topsoil. The resulting map is shown in Figure 11. This figure shows that the available NO_3 in the topsoil before cultivation was about 1.3-5.7 kg ha^{-1}.

Figure 10. Estimated nitrate mass (kg ha^{-1}) added to the soil from an excessive use of irrigation water (April–December 2000)

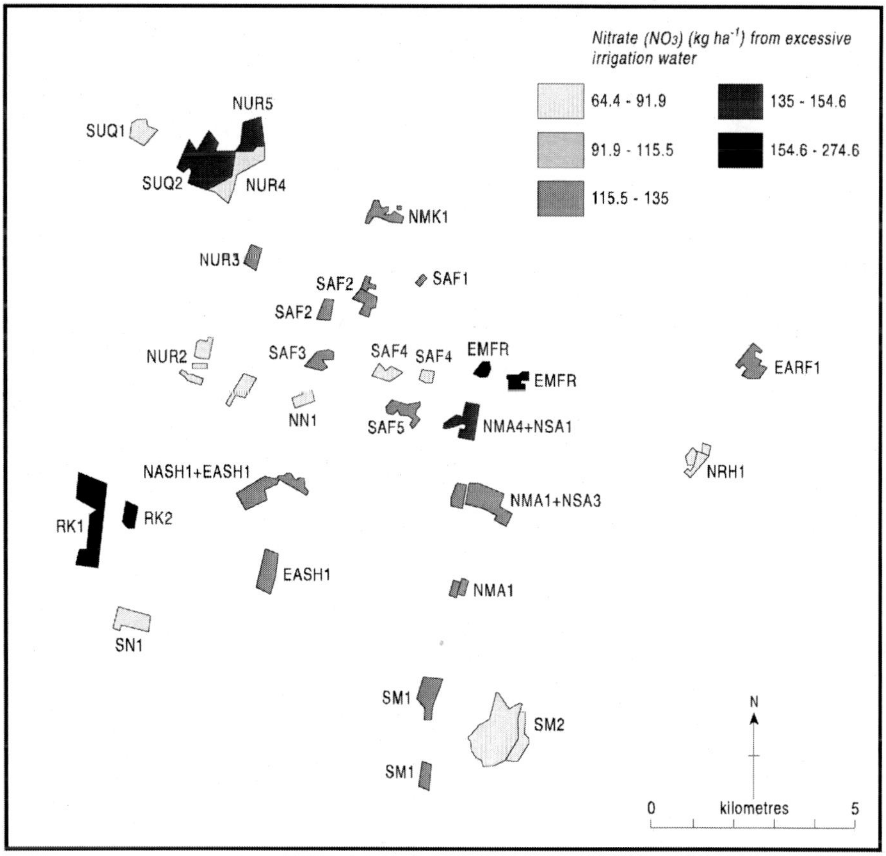

Figure 11. The estimated nitrate mass (kg ha^{-1}) in the soil before cultivation (natural soil)

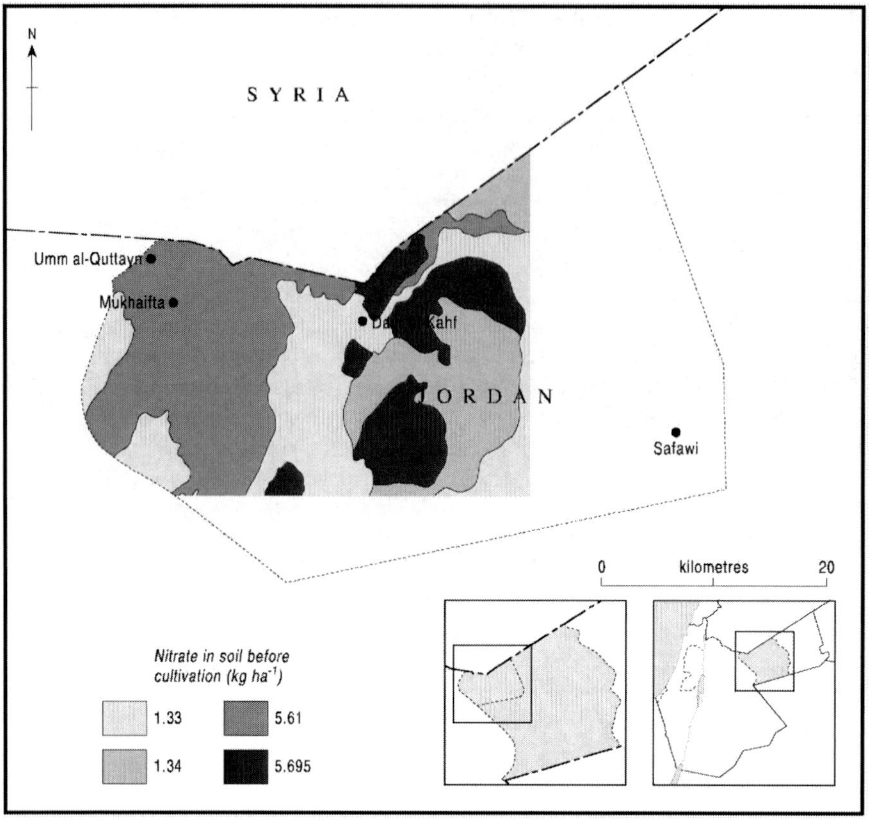

5.1.4. Inorganic fertilizers

The amount of nitrate from the application of fertilizers was known through the field work of April 2000. Each farm was visited where types and amounts of each commercial fertilizer had been documented. Nitrogen N in applied inorganic fertilizers has three forms: N, NO_3 and NH_4. In this model it was assumed that the total amount of nitrogen (Nt) has been transformed to nitrate. Once more, the agriculture activities map of 2000 (Figure 4) was updated with the nitrate estimates in the applied fertilizers as shown in Figure 12.

5.1.5. Organic manures

There are two types of organic manures used by farmers in the study area: (a) chicken manure and (b) sheep manure. Farmers who grow vegetables usually use chicken manure, while farmers who grow fruit and olive trees use sheep manure. The amount of organic fertilizer used was determined from interviews with farmers. Tables 2 and 3 summarize the amount of both chicken and sheep manure used by farmers in the study area.

Various authors have discussed the chemical composition of manure. Table 4 summarizes the chemical composition of organic manure.

Based on Table 4, the percentages given by Cooke (1982) were used as an indication of the chemical composition of organic manures used in the study area. The percentages given by Lekasi (2001) are for chicken and goat manure mixed together which was not applicable to this research. James (1993) gave a high percentage of N, P and K, which

Figure 12. Estimated nitrate mass (kg ha^{-1}) added to the soil from inorganic fertilizers (April–December 2000)

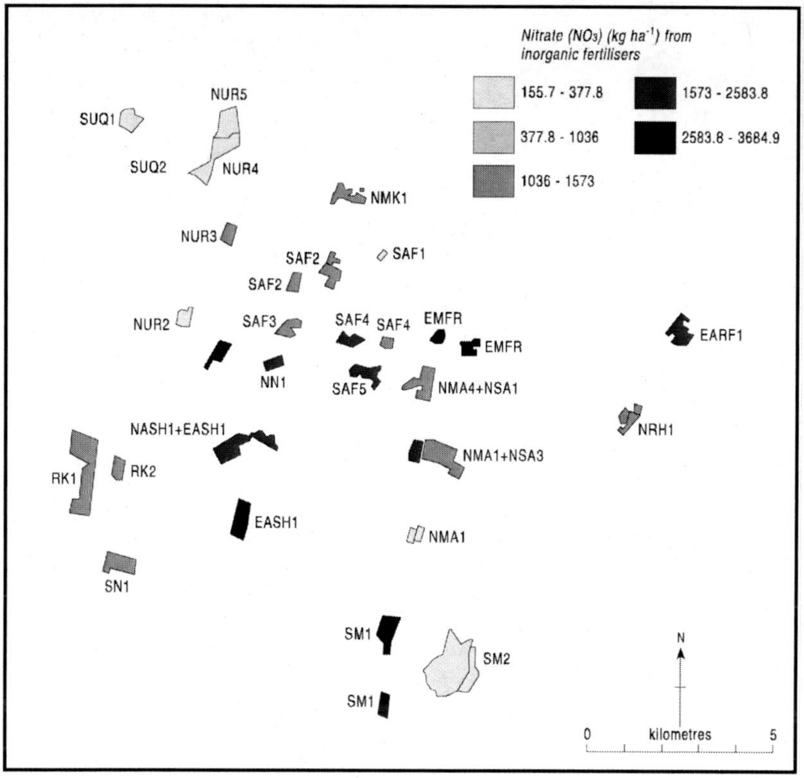

Table 2. The amount of chicken manure used by vegetable farmers in the study area

Farm code	Chicken manure (kg ha^{-1})	Farm code	Chicken manure (kg ha^{-1})
NUR1	5000	NMA1-NMA3	7142.8
NUR2	3980.9	RK1	5000
NUR3	3000	NASH1-EASH1	4909
NUR4	15000	EASH2	8250
NUR5	7000	SM1	9157
SN1	5384.6	EMFR	4000
NN1	5102	NMK1	1700
NSA1-NMA4	8000	EARF1	7142.9
NMA1	5000	SAF3	10000
NMA2	6800	SAF4	5000
SAF1	3000	SAF5	7142.8
SAF2	5000	NRH1	8000

could give an over-estimate of the NPK percentages in chicken and sheep manure in Jordan. This is due to the fact that sheep manure has lower percentages of N, P and K if the animal diet is poor (Cooke 1982) which could be assumed in the study area.

In manures, about one third of the nitrogen is released quite quickly, but much is very resistant and persists for a long time in the soil (Cooke 1982). In this study, it was

Table 3. The amount of sheep manure used by tree farmers in the study area

Farm code	Sheep manure (kg ha^{-1})
NUR4	3448.3
NUR5	9194.3
SUQ1	5970
SUQ2	2395.3
NSA1-NMA4	12000
RK2	10000
SM2	16000

Table 4. The N, P and K percentages in chicken and sheep manures as estimated by various authors

(%)	Chicken	Sheep	Source	Comments
N	0.9	1.5	(Lekasi 2001)	Chicken manure was mixed with goat manure.
P	0.24	0.33		
K	0.45	1.35		
N	4.3	4.0	(James 1993)	Too high: N, P and K composition of manures is highly variable (James 1993).
P	1.6	0.6		
K	1.6	2.9		
N	1.6	1.4	(Cooke 1982)	USA and UK
P	0.4	0.2		
K	0.4	1		
N	3.77	3-5	Parsad and Power, (1997)	P and K for chicken manure given as P205 and K2O respectively. In general the percentages could be considered high for the study area.
P	1.89	0.4-0.8		
K	1.76	2-3		

assumed that only one third of the nitrogen found in organic manures was transformed into nitrate and made available for plant uptake in the study area. Based on that and using the percentages provided by Cooke (1982), it was estimated that the sheep manures had added to the soil in the tree farms, nitrate values of 33.5 kg ha^{-1}, as a minimum value at SUQ2 farm, up to 224 kg ha^{-1} at SM2 farm as shown in Table 5. In the vegetable farms (Table 6), the chicken manure had supplied the soil with nitrate ranging from 27.2 kg ha^{-1} at NMK1 farm to 240 kg ha^{-1} at NUR4 farm.

5.2. Nitrate losses from agricultural land

5.2.1. Plant uptake

Plant uptake was based on values taken from the published literature where the estimated nitrate uptake by plants was found for several crops (eg tomato, watermelon). The plant uptakes were estimated from Finck (2000) as shown in Table 7.

Based on the data collected about crop type in the study area during the farmers' interviews, for each farm the average plant uptake was calculated using equation 6 (Al-Adamat 2002).

$$APU = \frac{\sum_{n=1}^{N} An \times PUn}{A} \text{ (kg ha}^{-1}\text{)} \qquad (6)$$

Where:- APU: Average plant uptake, An: area of crop n (ha), and PUn: plant uptake for crop n (as in Table 8) (kg ha^{-1}).

Table 5. The estimated NO_3 loading contributed by sheep manure in the tree farms

Farm code	NO_3 (kg ha^{-1})
SUQ2	33.5
NUR4	48.3
SUQ1	83.6
NUR5	128.7
RK2	140
NSA1-NMA4	168
SM2	224

Table 6. The estimated NO_3 loadings contributed by chicken manure in the vegetable farms

Farm code	NO_3 (kg ha^{-1})	Farm code	NO_3 (kg ha^{-1})
NMK1	27.2	SAF4	80
NUR3	48	NN1	81.6
SAF1	48	SN1	86.15
NUR2	63.7	NMA2	108.8
EMFR	64	NUR5	112
NASH1-EASH1	78.6	EARF1	114.3
NUR1	80	SAF5	114.3
NMA1	80	NRH1	128
SAF2	80	EASH2	132
RK1	80	SM1	146.5
NSA1-NMA4	128	SAF3	160
NMA1-NMA3	114.3	NUR4	240

Table 7. The estimated plant uptake (kg ha^{-1}) for vegetables and trees in the study area

Vegetables	NO_3 (kg ha^{-1})	Trees	NO_3 (kg ha^{-1})
Tomato	1741	Apple	92
Water Melon	696	Peach	149
Melon	696	Olive	531
Beans	572	Grapes	97
Aubergine	1285	Other fruits	92
Cabbage	536		
Peppers	1772		
Onion	514		
Cucumber	222		
Cauliflower	775		

Table 8. Average plant uptake of NO_3 (kg ha^{-1}) in the tree farms

Farm code	Average plant uptake of NO_3 (kg ha^{-1})
NUR4	101
NUR5	126
RK1	68
SM2	429
SUQ1	116
SUQ2	179

Table 9. Average plant uptake of NO_3 (kg ha^{-1}) in the vegetable farms

Farm code	Average plant uptake of NO_3 (kg ha^{-1})	Farm code	Average plant uptake of NO_3 (kg ha^{-1})
EARF1	584.52	NUR2	468.56
EASH2	456.5	NUR3	498.2
EMFR	331.4	NUR4	456.5
NASH1-EASH1	482.3	NUR5	555.26
NMA1	518.29	RK2	661.76
NMA1-NMA3	546.67	SAF1	605.55
NMA2	513.4	SAF2	750.42
NMA4-NSA1	821.35	SAF3	376.31
NMK1	552.3	SAF4	352.25
NN1	446.9	SAF5	667.34
NRH1	452.24	SM1	551.36
NUR1	599.82	SN1	1320

Table 10. Estimated denitrification rates (based on Rolston 1981)

Denitrification value (%)	Notes
10 - 30	
22	If applied fertilizer is 224 N kg ha^{-1}
45	If applied fertilizer is 300 N kg ha^{-1}
15	If applied fertilizer is 335 N kg ha^{-1}

Table 11. Estimated volatilization losses (based on FAO 2001)

Fertilizer name	Estimated volatilization losses (%)
UREA ($(NH_2)2CO$)	10
MAP (Mono Ammonium Phosphate)	15
NH_4NO_3 (Ammonium Nitrate)	2

Table 12. Estimated nitrate leaching scenarios from agricultural lands

Assumptions	Scenarios SA1	Scenarios SA2
Water usage	Optimum	Excessive
Denitrification rate (inorganic fertilizers only)	20 %	20 %
Volatilization losses	Table 11	
Nitrate in organic manure	33 % of nitrogen in organic manures was transformed into nitrate (NO_3)	
Nitrate in rainfall (kg ha^{-1})	6	
Nitrate in soil before cultivation (kg ha^{-1})	5.61	

Equation 6 was applied to the farm map (Figure 3) after updating its attribute file with crop type, crop area and plant uptake for each crop (Table 7). The results of this operation are shown in Tables 8 and 9.

5.2.2. Nitrate decay (denitrification and volatilization) in inorganic fertilizers

Denitrification and volatilization were assumed to be the only decay processes in inorganic fertilizers in the agricultural lands. Table 10 illustrates the estimated denitrification rates from the applied fertilizers as estimated in a review published by Rolston (1981). Based on this table, it was assumed that the denitrification rate in the study area was about 20 %, which is approximately the average value from these estimates. This percentage was subtracted from the applied inorganic fertilizers within the GIS. The volatilization rate was also estimated from FAO (2001) statistics as shown in Table 11. This table shows the volatilization rate for three types of fertilizers used in the study area. Also, these values were subtracted within GIS for the farms that use such fertilizers.

5.3. Estimated nitrate leaching from the agricultural lands

The estimated nitrate leaching from the agricultural lands was based on having two scenarios as shown in Table 12. These scenarios take into consideration the fact that there are two irrigation water schemes: optimum (SA1) and excessive (SA2). All nitrate inputs and losses were calculated in ArcView GIS to produce the estimated nitrate leaching to groundwater in both the vegetable and tree farms as shown in Tables 13 and 14.

5.4. Nitrate leaching from cesspools

Based on a field survey, Waddingham (1994) estimated the actual per capita water demand for the study area at about 115 l d^{-1}. It was assumed that the average family size

Table 13. Estimated nitrate leaching from the vegetable farms (kg ha^{-1})

Farm code	Scenarios		Farm code	Scenarios	
	SA1 (kg ha^{-1})	SA2 (kg ha^{-1})		SA1 (kg ha^{-1})	SA2 (kg ha^{-1})
EARF1	3116	3162	NUR2[b]	295	341
EASH2	2729	2780	NUR3	765	812
EMFR	4598	4659	NUR4	2608	2651
NASH1+EASH1	1453	1501	NUR5	384	440
NMA1	356	407	RK2	1483	1570
NMA1+NMA3[a]	685	737	SAF1	351	404
NMA2	2046	2095	SAF2	743	793
NMK1	1202	1250	SAF3	1362	1412
NN1	1926	1968	SAF4	1735	1780
NRH1	895	938	SAF5	2126	2178
NSA1+NMA4	277	334	SM1	3309	3357
NUR1	2973	3015	SN1	121	158

[a] S5 soil sample location; [b] S1 soil sample location

Table 14. Estimated nitrate leaching from the tree farms (kg ha^{-1})

Farm code	Scenarios		Farm code	Scenarios	
	SA1 (kg ha^{-1})	SA2 (kg ha^{-1})		SA1 (kg ha^{-1})	SA2 (kg ha^{-1})
NUR4	144	187	SM2	25	61
NUR5	277	333	SUQ1	644	669
RK1	865	974	SUQ2	97	151

in the study area is about six which gives a daily water consumption of about 690 l family^{-1}. It was assumed that only about 10 % of that water would reach the cesspool on a daily basis. This means that the water input to the cesspool is ca 4 m^3 month^{-1}. Based on Figure 6, it was estimated that the cesspool size is about 8 m^3, which means that without drainage or evaporation the cesspool would be full in two months. For this research, evaporation was assumed negligible since the ventilation hole (Figure 5) is too small (ca 10 x 10 cm) to have a major evaporation loss. Equation 2 was modified to estimate nitrate leaching to groundwater based on the period needed to empty the cesspool (equation 7) (after Al-Adamat 2002).

$$NO_3(L) = \frac{NO_3(C)\ x\ NC\ x\ (NM - 2)\ x\ 48\ x\ 10^{-3}}{NM} \qquad (7)$$

Where:- L: leached, C: concentration in sewage (mg l^{-1}) after denitrification, NC: number of cesspools and NM: number of months needed to empty the cesspools. The constant 48 is the estimate of the water consumed that can reach the cesspool in a 12-month period.

This equation is valid only when the period needed to empty the cesspools is 3, 4, 6 or 12 months. It appears from this equation that if the house owner emptied the cesspool every two months, there would be little or no nitrate available for leaching. It would be assumed that the difference between the cesspool size and the water that goes into the cesspool in any period spent after the two months is groundwater recharge (together with its nitrate).

Three scenarios were evaluated in this research project using equation 7. A low risk scenario, where the house owner was assumed to empty the cesspool every three months. In the second scenario, it was assumed that the house owner emptied the cesspool every six months, while in the third scenario, the house owner emptied the cesspool only once every year.

The total nitrogen in sewage before treatment is ca 40 mg l^{-1} of which 15 mg l^{-1} is organic N and the remaining 25 mg l^{-1} is ammonia (NH_4+NH_3) nitrogen and there is no nitrate nitrogen (Schroeder 1981). He explained the absence of nitrate in sewage as a result of the lack of oxygen (O) in the sewer system and the short reaction times. In Jordan, the only available data for nitrate levels in sewage was for the Aqaba Treatment Plant (ATP). According to Salameh and Bannayan (1993) the nitrate concentration at the inlet of the ATP was ca 11.2 mg l^{-1}. The maximum concentration recorded in this plant during the treatment process was ca 34 mg l^{-1}, which might have resulted from the nitrification process. It was also found that the nitrate concentration at the outlet of ATP was ca 18 mg l^{-1} which could result from denitrifying the nitrate in the plant. In this research project, it was assumed that the nitrate concentration in the cesspools after denitrification was ca 18 mg l^{-1}.

In order to apply equation 7, the village map (Figure 2) was updated with the estimated nitrate concentration in sewage and the estimated number of cesspools in each village. In ArcView GIS, equation 7 was applied to calculate the volume of nitrate available in each village based on the above scenarios. The results of this operation are shown in Table 15. The total estimated amount of leached nitrate in the study area supplied from cesspools was estimated at ca 796-1989 kg year^{-1} (ca 0.3-0.7 kg household^{-1} yr^{-1}).

Table 15 shows that the highest estimated nitrate leaching occurred in Abu al-Farth, al-Mukayfita, ar-Rifa'iyyat, Qasim, Dayr al-Kahf, Jubaya and Umm al-Quttayn in the northern part of the study area.

In the southern part also (on the main Road to Baghdad), villages like Rahbat Rakad, Nafiya, al-Sa'ada, al-Manara, Hulaywat al-Masarna, Hamra al-Suhaym and al-Bishriyya have high estimated nitrate leaching to groundwater. The distribution of these villages over a large area (Figure 2) increases the risk of having both chemical and biological groundwater contamination problems.

Table 15a. The estimated leached nitrate (kg year^{-1}) in each village based on three scenarios

Village name	Scenarios			Village name	Scenarios		
	SC1	SC2	SC3		SC1	SC2	SC3
Abu al-Farth	32.3	64.5	80.6	Dayr al-Qinn	7.8	15.6	19.4
al-Ashrafiyya	35.1	70.3	87.8	Ghadir an-Naqa	0.9	1.7	2.2
al-Bishriyya	42.6	85.2	106.6	Hamra as-Suhaym	25.6	51.3	64.1
al-Hashimiyya	2.3	4.6	5.8	Hulaywat al-Masariha	36.0	72.0	90.0
al-Jad'a	5.8	11.5	14.4	Jubaya	32.8	65.7	82.1
al-Ma'azula	5.8	11.5	14.4	Khasha't al-Qinn	7.5	15.0	18.7
al-Manara	43.2	86.4	108.0	Manshiyyat al-Qinn	12.1	24.2	30.2
al-Mansura	2.6	5.2	6.5	Midwar al-Qinn	8.4	16.7	20.9
al-Mukayfita	96.2	192.4	240.5	Mathnat Rajil	8.6	17.3	21.6
al-Munaysa	11.8	23.6	29.5	Nayifa	30.2	60.5	75.6
al-Murayjib	2.3	4.6	5.8	Qasim	26.2	52.4	65.5
as-Sa'ada	25.9	51.8	64.8	Rahbat Rakad	23.3	46.7	58.3
as-Suwaylimiyya	1.2	2.3	2.9	Tall ar-Rimah	16.4	32.8	41.0
ar-Rifa'iyyat	30.8	61.6	77.0	Umm al-Quttayn	150.9	301.8	377.3
ar-Rifa'iyyat ash-Shamaliyya	6.9	13.8	17.3	Umm Hussein	7.8	15.6	19.4
ar-Ranibat an-Na'imat	2.9	5.8	7.2				
Ath-Thilaj	4.3	8.6	10.8				
Dayr al-Kahf	49.2	98.5	123.1				

SC1: cesspool is emptied every three months; SC2: cesspool is emptied every 6 months; and SC3: cesspool is emptied every 12 months

Table 15b. Summary statistics

	SC1	SC2	SC3
Min.	1.2	2.3	2.9
Max.	96.2	192.4	240.5
Mean	22.3	44.6	55.7
SD.	25.0	50.0	62.5

6. Model validation

In this research, the model was validated for the agricultural lands only. In order to validate the model, the NO_3 concentrations (mg 1^{-1}) found in the soil-water extracts from the two cultivated fields in 2000 (S1 and S5) were used to estimate the amount of nitrate lost (leached through the soil).

Equation 5 was used to calculate the amount of NO_3 (kg) in the soil profile. The average NO_3 concentrations in the soil profile for each site were assumed to be the average concentrations from the three different sampling depths. The average NO_3 concentrations at S1 and S5 were 7 mg 1^{-1} and 10.9 mg l^{-1} respectively. S1 was located inside NUR2 farm while S5 was located in NMA1&NMA3 farm. The outcome from this operation showed that the total amount of nitrate in the soil at S1 (depth 0-100 cm) was 320 kg ha^{-1} and 359.7kg ha^{-1} at S5 after cultivating the land for one season. When comparing these values to the estimated nitrate leaching from these farms, it was found that at S1, for the two scenarios, nitrate leaching estimates were similar to what was found in the soil (about 92 % for SA1 and about 107 % for SA2), while at S5, the nitrate leaching estimates were much higher than those found in the soil (ca 190 % for SA1 and about 205 % for SA2). At S1, the estimated nitrate leaching was about 83 % of the total nitrate input, while at S5 it was about 56 % of the total nitrate input.

7. Suggested management scenarios

In this research paper, it was found that inorganic fertilizers pose more threat in the study area. Cesspools also have been identified as potential sites for chemical and biological contamination. Nutrient loading from both contaminant sources (agriculture and cesspools) were estimated in the study area. In order to suggest locally acceptable management options, several scenarios have been examined to overcome the nitrate leaching problem in the study area.

7.1. 'Do nothing' scenario

This scenario is based on taking no management action in order to control nitrate leaching in the study area. This scenario was examined earlier in this paper, where it appeared that nitrate leaching in the study area could cause groundwater contamination. Doing nothing about this situation will mean more nitrate will leach to groundwater from both cesspools and irrigated agriculture. There is also a risk that biological contamination could occur in the study area from the cesspools.

7.2. 'What if scenarios': fertilizer management

In this paper, nitrate leaching was estimated for all farms. It was found that in each farm, farmers had used more fertilizers than required for plant uptake. But what might possibly happen if farmers know:

- The available nitrate in irrigation water (groundwater) (optimum usage). Schepers et al (1983) argued that nitrate in the groundwater is a potential source of nitrogen for plant growth when used for irrigation;
- Available nitrate in the soil before cultivation;
- The nitrate content in organic manure;
- Plant uptake rates.

Such knowledge could be used to calculate how much inorganic fertilizer is needed to compensate the difference, adding about 30 % extra fertilizer as a 'safe margin'. This safe margin was assumed to compensate for any losses that might occur due to denitrification and volatilization processes. This percentage could be modified if the denitrification rate in the area is quantified in the future. This scenario was conducted in a GIS environment where all nitrate data used in estimating the nitrate leaching from agricultural land (inputs and losses) were used again except the amount of nitrate from inorganic fertilizers. The amount of nitrate in inorganic fertilizer was based on the above assumption. Based on this scenario, the estimated nitrate leaching would drop dramatically. The decrease in the estimated nitrate leaching was up to 99 % in some farms (Table 16). This scenario is viable and could be implemented by providing farmers with extension services and basic water and soil tests.

7.3. Cesspool management

There are three possible management scenarios for controlling nitrate leaching from the cesspools. The first scenario is to empty the cesspool on a regular basis as discussed earlier in this paper. It was found that if the house owner emptied the cesspool every two months, nitrate leaching could be minimized. Local government councils could participate by providing such services to the local people as they currently provide the service of collecting garbage for an annual fee (council tax). Currently the cost of taking the sewage from the cesspools in the area could cost up to JD30 in the western part of the study area. For remote villages, this cost could be much higher.

The second management scenario is to have a better design for cesspools. Leaving the bottom of the cesspool without a proper lining overlaying fractured basalt is not an option in any suggested design. Government intervention through the local councils could guarantee that new designs eliminate or minimize nitrate leaching from the cesspool. This scenario could be implemented easily, where local councils in the area could insist on such designs before issuing a building licence. Local people could use the available material in the area such as basalt rocks with cement to line the bottom of the cesspool, similar to what they usually do for the cesspool walls.

The third management scenario is to have a sewerage system to serve the major towns in the area which, as shown before in this paper, are the major contributors of the nitrate originating from cesspools. Villages like Umm al-Quttayn and al-Mukayfita have the potential to have a local sewage treatment plant. The distribution of villages in the area could assist in making this scenario viable. There are two clusters of villages in the study area: on the main road to Baghdad; parallel to the Syrian borders. Any sewage treatment plant in the area could also serve other villages to the west of the study area having the same problem with cesspools. Such sewage treatment plants would have to be assessed environmentally in order to ensure that it does not itself cause groundwater quality deterioration.

Table 16. Managing the amount of inorganic fertilizers used in the vegetable farms

Farm code	Present estimate NO_3 leaching (kg ha^{-1})	Expected leaching after management (kg ha^{-1})	Difference (kg ha^{-1})	Reduction (%)
EARF1	121	100	21	17
EASH2	277	52	225	81
EMFR	351	40	311	89
NASH1+EASH1	295	85	210	71
NMA1	356	94	262	74
NMA1+NMA3	384	86	298	78
NMA2	743	80	663	89
NMK1	685	128	557	81
NN1	765	75	690	90
NRH1	895	56	839	94
NSA1+NMA4	1 202	160	1 042	87
NUR1	1 483	122	1 361	92
NUR2	1 453	88	1 365	94
NUR3	2 126	103	2 023	95
NUR4	1 362	8	1 354	99
NUR5	2 046	88	1 958	96
RK2	1 926	120	1 806	94
SAF1	1 735	133	1 602	92
SAF2	2 973	164	2 809	94
SAF3	3 116	16	3 100	99
SAF4	2 608	47	2 561	98
SAF5	2 729	123	2 606	95
SM1	3 309	76	3 233	98
SN1	4 598	338	4 260	93

8. Summary and conclusion

In this research paper, a GIS-based model was developed to estimate nitrate leaching from the agricultural lands that took into consideration all possible sources of nitrate input and losses from cultivated fields. The nitrate inputs included: rainfall; soil store; irrigation water; organic manure; and inorganic fertilizers. The losses were assumed to be: plant uptake; denitrification and volatilization; and leaching. In the case of cesspools, it was suggested that the only losses of nitrate occurred through either leaching or denitrification.

In the agricultural lands, the estimated amount of leached nitrate from several sources was evaluated. This included rainfall, which contributed to the farmed area ca 5 kg ha^{-1}, and irrigation water with values of ca 39-165.6 kg ha^{-1} if farmers used optimum irrigation water, and ca 64-275 kg ha^{-1} if farmers used excessive amounts of irrigation water. These sources of nitrate also included what existed in the soil before cultivation. This source had contributed to the input of ca 5.6 kg ha^{-1} in the agricultural area. Organic manure and inorganic fertilizers together, with the irrigation water, were found to be the major sources of nitrate in the study area. The nitrate losses in this model included denitrification and volatilization decays and plant uptake. The outcome of this model has shown that the estimated nitrate leaching from the agricultural lands ranges between 121 and 4 598 kg ha^{-1} if farmers use optimum irrigation water and between 158 and 4 659 kg ha^{-1} if farmers use excessive irrigation water.

Nitrate leaching from cesspools was examined on the basis of three scenarios. The results showed that the estimated nitrate loading from cesspools varied from one village to another in conjunction with the population density. Estimates were 1.2-240.5 kg yr^{-1}.

It was also found that nitrate leaching from agricultural lands was much higher than that predicted from the cesspools.

Several management scenarios have been discussed in this research, including having better fertilizer management, which could reduce the estimated nitrate leaching by more than 90 %. This would only be possible, however, if farmers knew the available plant nutrients in water, soil and organic manure and used such knowledge to estimate the required amount of inorganic fertilizers. Nitrate leaching from cesspools could also be reduced or eliminated by having the water extracted on a regular basis, or by having a better design or a local sewerage system.

In conclusion, it seems that both agricultural lands and cesspool pose a threat to future groundwater quality in the study area. Although, in the study area, the estimated nitrate leaching from the agricultural sources is very high when compared to the leaching from the cesspools, the cesspools are also considered a high risk to groundwater quality especially because biological (bacterial/viral) contamination could also result from this source.

Finally, although the groundwater is relatively deep, care should be taken when having developments in these areas due to the importance of the basalt aquifer as a groundwater resource for drinking water supply for the local population and the major cities of Jordan.

References

Addiscott TM, Whitmore AP and Powlson DS (1991). Farming, fertilizers and the nitrate problem. UK: C.A.B International.

Ahn H-I and Chon H-T (1999). Assessment of groundwater contamination using geographical information systems. Environmental Geochemistry and Health 21:273-289.

Al-Adamat R (2002). The use of Geographical Information Systems (GIS) and remote sensing to investigate groundwater quality in the Azraq Basin, Jordan. Ph.D. thesis, Coventry University, UK.

Al-Adamat R, Baban S and Foster I (2004). An examination of land use change due to irrigated agriculture in north-eastern Jordan using geoinformatics. International Journal of Environmental Studies 61(3):337-350.

Al-Hussein S (2000). GIS modelling of land degradation in northern Jordan using satellite imagery. Ph.D. thesis, University of Leicester, UK.

Al-Tarawneh K (1996). The geology of the Al-Hamidiyya (Al-Hababiyya) area. Map Sheet No. 3354II, Bulletin 35, Geology Directorate, Geological Mapping Division, Natural Resources Authority, Amman, Jordan.

Allison RJ, Higgitt D, Kirk AJ, Warburton J, Al-Homoud A, Sunna BF and White K (1998). Geology, geomorphology, hydrology, groundwater and physical resources. In: Dutton R, Clarke J and Battikhi A (eds), Arid land resources and their management, Jordan's Desert Margin. London: Kegan Paul International pp. 21-44.

BRDP (1994). The baseline socio-economic survey of the Badia programme area, internal report. Jordan Badia Research and Development Programme, Amman, Jordan.

Burrough PA (1986). Principles of geographical information systems for land resources assessment. Oxford: Clarendon Press.

Cooke GW (1982). Fertilizing for maximum yield (3rd edition). London: Granada.

Dottridge J (1998). Water resources quality, sustainability and development. In: Dutton R, Clarke J and Battikhi A (eds), Arid land resources and their management, Jordan's Desert Margin. London: Kegan Paul International pp. 21-44.

Evans TA and Maidment DR. A spatial and statistical assessment of the vulnerability of Texas groundwater to nitrate contamination, CRWR online report. http://civil.ce.utexas.edu/centers/crwr/reports/online.html.

FAO (2001). Global estimates of gaseous emissions of NH_3, NO and N_2O from agricultural land, Rome. Italy: Food and Agriculture Organisation.

Finck A (2000). Fertilizers and their effective use: introductory chapter. CD version, In: Wichmann W (ed), World fertilizer use manual. Paris: FAO-IFA.

Görres J and J GA (1996). Incorporating spatial variability into GIS to estimate nitrate leaching at the aquifer scale. Journal of environmental quality 25:491 498.

Gusman AJ and Marino MA (1999). Modelling of nitrogen dynamics in the subsurface environment: impacts of urban growth on surface water and groundwater quality. Proceeding of IUGG 99 Symposium HS5. Birmingham: IAHS Publication No. (259).

ISESCO (1997). Water resources management, Islamic Educational, Scientific and Cultural Organization, Rabat, Morocco.

James DW (1993). Fertilizer composition and reactions in soil. In: W, JD and Topper, KF (eds), Utah fertilizer guide, Utah. USA: Utah State University.

Jordan C (1994). Modelling the nitrate leaching on a regional scale using a GIS. Journal of environmental quality 42:279-298.

Kirk A (1998). The effect of intensive irrigated agricultural upon soil degradation: A case study from Ashrafiyya. In: Dutton R, Clarke J and Battikhi A (eds), Arid land resources and their management, Jordan's Desert Margin. London: Kegan Paul International pp. 127-155.

Lekasi JK (2001). Manure management in the Kenya highlands: practices and potential, (2nd edition). Ryton-on-Dunsmore, UK: HIDRA.

Maani M, Hunaiti H and Findlay AM (1998). Demographic change and population projections 1976-2013. In: Dutton R, Clarke J and Battikhi A (eds), Arid land resources and their management, Jordan's Desert Margin. London: Kegan Paul International pp. 215-246.

Merchant JM (1994). GIS-based groundwater pollution hazard assessment: A critical re-

view of the DRASTIC model. Photogrammetric Engineering & Remote sensing 60(9):1117-1127.
Millington A, Al-Hussein S and Dutton R (1999). Population dynamics, socio-economic change and land colonisation in northern Jordan, with spatial reference to the Badia Research and Development Project area. Applied Geography: 363-386.
Mitasova H and Mitas L (2002). Modelling physical systems. In: Clarke KC, Parks BO and Crane MP (eds), Geographic information systems and environmental modelling. New Jersey: Prentice Hall.
RJGC. GIS maps http://www.rjgc.gov.jo.
Rodda HJ, Demuth S and Shankar U (1999). The application of a GIS-based decision support system to predict nitrate leaching to groundwater in southern Germany. Hydrological Sciences Journal 44(2):221-236.
Rolston DE (1981). Nitrous oxide and nitrogen gas production in fertiliser loss. In: Delwiche CC (ed), Denitrification, nitrification and atmospheric nitrous oxide. New York: John Wiley and Sons.
Salameh E, Al-Ansari N and Al-Nsoor I (1997). Water and environment in the area east of Mafraq and their developmental potentials; report no. (1). Strategic Environment and Water Resources Unit, Al Al-Bayt University, Jordan.
Salameh E and Bannayan H (1993). Water resources of Jordan, present status and future potentials. Amman: Friedrich Ebert Stiftung.
Schepers J, Frank K and Watts D (1983). Influence of irrigation and nitrogen fertilisation on groundwater quality, relation of groundwater quantity and quality. Proceedings of the Hamburg Symposium, August. IAHS publication No. (146).
Schroeder E, D (1981). Denitrification in wastewater management. In: Delwiche CC (ed), Denitrification, nitrification and atmospheric nitrous oxide. New York: John Wiley and Sons.
Shamrukh M, Corapcioglu YM and Hassona FA (2001). Modelling the effect of chemical fertilisers on groundwater quality in the Nile Valley aquifer, Egypt. Journal of Groundwater 39(1):59-67.
Stauffer J (1998). The water crisis, constructing solutions to freshwater pollution. London: Earthscan Publications, Ltd.
Supalla RJ (1995). Adoption of nitrogen and water management practices to improve water quality. Journal of Soil and Water Conservation 50(1):77-82.
Waddingham J (1994). Water demand in the north-east Badia, Jordan. M.Sc. thesis, University of Newcastle Upon Tyne, UK.
Waddingham J (1998). Water demand. In: Dutton R, Clarke J and Battikhi A (eds), Arid land resources and their management, Jordan's Desert Margin. London: Kegan Paul International pp. 127-155.
Ward RC and Robinson M (2000). Principles of hydrology, (4th edition). London: McGraw Hill.

CHAPTER 9

Developing location criteria for domestic and hazardous waste disposal in the Jordanian Badia

Odeh M. Al-Meshan

Badia Research and Development Centre, P.O. Box 902, Jubeiha 11941, Jordan.
E-mail:al_meshan@yahoo.com

Ian Foster

Coventry University, UK & Rhodes University, South Africa

As with all countries, Jordan faces growing problems with waste disposal and the location of landfill sites. This paper combines international experience (obtained from published papers and national and international regulations) with Jordanian expertise and community opinions to arrive at a set of prioritized criteria for landfill site selection. High priority was given by all groups to distance from built-up areas and surface water and groundwater bodies, and to avoiding placing the site upwind of a settlement. The experts showed more awareness, than people from the local communities, of the importance of such technical issues as structural geology (eg avoiding fault lines).

Keywords: Landfill sites, survey techniques

1. Introduction

Jordan is a rapidly developing country with a dramatic population growth. This growth has resulted in the creation of a variety of solid waste types (2001). The process of disposing of these wastes is one of the most serious environmental problems facing the country today and this is likely increase in the future (Chopra et al 2001, GCEP 2001). The problem is compounded by a lack of laws or guidelines dealing with this issue. No regulations or national policies have been enacted in Jordan as yet to dispose of or control these wastes.

The overall aim of this chapter is to develop generic criteria for identifying sites suitable for the disposal of domestic and hazardous waste in the northern Badia region of Jordan. These criteria are designed to help environmental decision makers and planners who work with the public sector on the subject of solid waste disposal.

The criteria were selected on the basis of three principles: meeting environmental and social requirements; minimising the cost of transportation from source to the potential sites; and minimising the effects on human health and on the environment.

According to some authors (Furuseth 1990, Lober 1993, Petts et al 1994, Sloan 1993) there are no regulated criteria or guidance towards identifying waste disposal or landfill sites and that the only solution is to involve interested parties such as decision-makers, planners, academics and the local community in an early planning stage to develop such criteria. The aims of including these parties are to: define site selection criteria; make a judgment on the relative importance of these criteria; improve the quality of decision making; and ensure the success of the site selection process (Baxter et al 1999, Dechapanya et al 1999, Malczewski 1999, McBean et al 1995, Petts et al 1994). However, according to some researchers (Atkinson et al 1995, Baban et al 1998, Dikshit et al 2001, Hussey et al 1996, Kao et al 1996a, Lin et al 1999, Lindquist 1991, Siddiqui et al 1996,

Tchobanoglous et al 1993) there are no universal standards for landfill siting criteria that can be generalized for all countries or even for states within the same country. This is essentially because these criteria are often influenced by a variety of local environmental, economic and socio-economic conditions. These conditions are normally governed by local constraint factors, or 'limitations', such as land use, topography, geology, hydrology, natural resources, cultural resources and soil characteristics (Oweis et al 1990).

Landfill-siting criteria are either exclusionary or restrictive. Exclusionary criteria (synonymous with constraint factors) are those limitations imposed on decision makers to avoid attributes such as residential areas, water bodies, recreational areas, airports, forests and wetlands, or place the facility at a specific distance away, to avoid risks to the environment and human health (Gumbricht 1996, Petts et al 1994). These constraint factors, however, should be attached to weights at an early stage to indicate their relative importance to the siting process before applying further detailed restrictive criteria. A higher weighting is assigned to more important siting factors (Oweis et al 1998, Petts et al 1994).

The term restrictive criteria means that the suitability of the site should be carefully balanced on the basis of specific determined evaluation criteria. Evaluation criteria, according to Malczewski (1999) and Siddiqui et al, (1996), are not always measured on a quantitative scale. For example criteria such as proximity to settlements and airports can be measured using scale units, but criteria such as social attitudes and perceptions towards the impact of establishing a new landfill site cannot be measured on a similar scale. These criteria could be given a subjective description or logical classification. For example, soil characteristics or depth could be given a suitable, possible or unsuitable classification (Bagchi 1990, Oweis et al 1998, Sharma et al 1994). Also, if the soil underlying a potential landfill site has a hydraulic conductivity of less than 10^{-7} cm s^{-1}, it means that the site requires a lining to attenuate the leachate, or cover material to minimize gas emission. Soils with a high Cation Exchange Capacity (CEC) greater than 25 mg/100 g are considered good. If a soil has a pH more than 6.5 it is usually considered suitable for the removal of metal cations (Oweis et al 1998).

Some authors (Lober 1993, Petts et al 1994, Zeiss et al 1992) have claimed that the main concern of landfill site selection is local community acceptance and satisfaction. Therefore, these authors have recommended that consultation with local authorities and statutory agencies must be undertaken in order to rank the acceptability of the selected site(s) and eventually overcome local community resistance to their location, particularly when they know the future benefits that the landfill might bring to them such as parks and revenues (Petts et al 1994). Overall, the ultimate purpose of the site selection process is to build three basic principles; trust, equity and participation (Baxter et al 1999).

2. Landfill siting factors and their importance

Built-up areas, agricultural activities, commercial transportation, conservation areas, mining resources, commercial and public developments, existing airports etc must be avoided when a new landfill site is being selected (Bagchi 1994, Batstone et al 1989). Several authors (Atkinson et al 1995, Bagchi 1994, Basagaoglu et al 1997, Kao et al 1996b, Yegorov et al 2000) claim that locations with these attributes must be avoided in order to alleviate any potential adverse affects on human health and the environment. On the other hand, Batstone et al, (1989) argue that landfill sites could be built and operated in close proximity to both residential and industrial developments, but these facilities must be operated very carefully and should be accepted by the local community. Apparently, these studies try to avoid the not in my backyard 'NIMBY' syndrome. An objective of not placing landfill sites near airports is to avoid attracting birds which could lead to bird-aircraft collision hazards (Tchobanoglous et al 1993).

Archaeological sites and recreational areas are important and need to be considered before establishing any new landfill site for aesthetic and historic reasons. Therefore, a protection zone or buffer area should be designated around the site at a specified distance in order to camouflage the site and limit the visual impact on the surrounding local communities (Bagchi 1994, Batstone et al 1989). Selection of sites near sites that attract tourists could increase the risk of spreading disease through scavenging by animals and birds.

New landfill sites must not be located long distances from transportation routes to avoid the high cost of constructing connecting roads. Availability of soil for covering and lining is also an important issue because, without available indigenous soil, the cost of construction will be high; such extra cost must be taken into account, which complicates the site selection process (Atkinson et al 1995, Bagchi 1994, Basagaoglu et al 1997, Kao et al 1996b).

Topography provides information on the earth's surface and it has a major influence on the suitability of landfills in terms of surface water problems, soil erosion, site stability, visibility and slope angle and length, which eventually reflect drainage patterns and the distribution stream courses (Atkinson et al 1995, Bagchi 1994, Batstone et al 1989, Wentz 1995, Yegorov et al 2000).

Local drainage patterns of the area, where surface runoff may be generated and could run over a dumpsite and seep down inside the landfill subsequently, and could lead to leachate production to groundwater and impact the water quality, are important factors to be considered (Basagaoglu et al 1997). The landfill site must also not be located within a specific distance from water supply wells and surface water bodies in order to ensure that any future potential pollution of these resources is minimized. This is essentially because of the connection with public health and possible future contamination of groundwater (Bagchi 1990, Batstone et al 1989).

Areas of potential floods of about 100-year recurrence interval should be avoided as potential sites for landfill in order to reduce any temporary water storage capacity of the flood plain or allow the wash-out of waste which would pose a hazard to human health and the environment (Bagchi 1994, Sharma et al 1994, Tchobanoglous et al 1993). Landfill sites should also not be located in the flood plain in order to reduce the possibilities of overland drainage polluting surface water resources (EPA 1993, Kao et al 1996b).

Landfill sites ideally should be in climates with low rainfall and high surface evaporation rates to minimize leaching potential (Batstone et al 1989, Wentz 1995). Local weather conditions also have influence on landfill site selection since local conditions may restrict access to the site, especially in the winter season. Wind strength and direction must also be considered to avoid blowing or flying debris. Air quality in the vicinity of the potential sites also must be evaluated as well to minimize objectionable odours (Crawford et al 1985, Tchobanoglous et al 1993).

The ability of subsurface media to transport pollutants from landfills down to the groundwater table through permeable or porous media or through joints and faults, is also important. Fault lines which have had a displacement in the Holocene period may lead to instability in engineering structures, which increases the possibility of site destruction and damage (Kao et al 1996b, Knipe et al 2001). This is primarily because of the potential movement of leachate from landfill down to the groundwater zone leading to groundwater pollution (Tchobanoglous et al 1993). Therefore, on-site subsurface investigations must be carried out to ensure that the potential site will not cause future groundwater pollution problems (Bratley 1983).

Groundwater is a hidden and 'out-of sight' resource (Bachmann 1983, Dorman 1992). It is not always immune from pollution by leachate generated from landfills as long as the waste exists (Gorelick et al 1993, Lee et al 1996). Soil with low permeability has a

Table 1. Major hazardous pollutants in solid waste disposal sites and their regulated limits in drinking water in some selected countries and organizations

Compounds	WHO[5]	EC[6]	Jordan[7]	Japan[8]	Thailand[9]	USA[10]
I. Inorganic	Concentrations (mg/l)					
Arsenic[1,2,3,4]	0.01	0.01		0.01	0.05	0.05
Lead[1,2,3,4]	0.01	0.01	0.01	0.01	0.05	0.015
Cadmium[1,2,3,4]	0.003	0.005	0.003	0.01	0.1	0.005
Beryllium[3]		250				0.004
Cyanide[1]	0.07	0.05	0.07	0.01	0.2	0.2
Asbestos[1]					7 MFL•	
Chromium (hexavalent) /total[1,2,3]	0.05	0.05		0.5	0.05	0.1
Mercury[1,2,3,4]	0.001	0.001	0.002	0.0005	0.002	0.002
Nickel[3,4]	0.02	0.02	0.07	0.01 or less		0.1
II. Volatile organic						
Benzene [1,2,3,4]	10 for 10^{-5} Excess risk	0.001	0.01	0.01		0.005
Phenol[1]		0.0005			0.001	0.001
Dibenz [a,h] enthrance[2]		0.0002 b				0.0003
Benzo(a) pyrene (PAH)[1,2,3]	0.0007	0.00001				0.0002
1,2-Dichloroethane[1,3]	0.03	0.003		0.004		0.005
Trichloroethylene[1,2,3,4]			0.005	0.03		0.005
Chloroform[1,2,3,4]	2for10^{-5} Excess risk	0.1	0.002	0.06		0.6
Barium[3]	2*	1***		1.0		2
PCBs/PCT (spolychloronate biphenyls)[1,2,3]		0.0001		–		0.0005
Carbon tetrachloride[1,3]	0.002	0.001		0.002		0.005
Tetrachloroethylene[1,3,4]			0.005	0.01		0.005
Monochloroethylene[1]						
Dichloromethane (methylenechloride)[1,3]	0.02			0.02		0.005
Methyl ethyl ketone[1]		0.0005				200.0
1,1-Dichloroethane[3,4]		0.9**				0.025 (ppm)
PAHs[3]		0.0001				0.0002
Trans-1,2-Dichloroethene[3]	0.5					0.1
DEHP[2-ethylhexyl] phthalate[3,4]	0.008			0.06		0.006
Vinyl Chloride[3,2,4]	0.005*	0.0005				0.002
Dioxin (2,3,7,8-TCDD)[1,3]	0.00000003					0.00000003
Hexachlorbatadience[2]		1.5				0.5
Benzo [b] Fluoranthene[2]		0.0002				0.0002
DDT (dichloromethane)[2]	0.002	0.0001	0.002			
Endrin[2]		0.0001	0.002	0.01 or less		0.002

[1]: WHO (1990) (Cited in the World Bank, 1994): A priority chemical in hazardous wastes based on estimates of quantity (occurrence in hazardous waste in Europe); [2]: Enger and Smith (1995): Top of hazardous substances; [3]: Johnson and DeRosa (1997): Carcinogen substances based on survey to 1300 disposal sites in the USA; [4]: Friends of the Earth's programme (1999): Hazardous Substances in Municipal Solid Waste Landfills; [5] WHO: Guidelines for Drinking-Water Quality (WHO, 2001); [6] EC: Gary (1994) \\ EC Directive: 98/83/EC: Maximum admissible concentration; [7] Jordan: Maximum Allowable Limits (2001); [8] Japan: Environmental Agency (1997): Maximum standards values allowable in water; [9] Thailand: Ministry of Science (Thailand, 2001): Maximum allowable concentration; [10] USA: Maximum concentration level regulation (US EPA, 2001); • : Million fibres per litre; * : For substances that are considered to be carcinogenic (excess lifetime cancer risk of 10^{-5}); **: Dutch Ministry of Housing, (2000), (The Dutch intervention values); ***: UK (2000). Amounts allowed in drinking water.

major influence on reducing the leachate movement down to groundwater (Öman et al 1993). Clay which has bentonite-embedded zeolites with different clay content (Illite-Montmorillonite and chlorite-Montmorillonite) is preferable for lining and cover materials (Sharma et al 1994). These materials have a fine particle size and a low hydraulic conductivity and they are inexpensive and easy to construct. On the other hand, sands are not recommended as preventative materials but they can be used for trapping leachates for re-pumping in wells (Pereira et al 1997). Lee and Jones-Lee (1994) suggest that monitoring procedures are introduced to ensure that the concentrations of pollutants in potable groundwater should not exceed 'safe limits'.

Four sets (a total of 33) of common hazardous pollutants from landfills have been reported in the literature (ie Enger et al 1995, FoE 1999, Johnson et al 1997, WHO 1990) (in Table 1: WHO[1], Enger and Smith[2], Johnson and DeRosa[3]; FoS[4]). These pollutants are suspected or known to have human health effects due to their carcinogenic properties and acute toxicity (Davis et al 1998, Dorman 1992, Elinder 1991, Enger et al 1995, EPA 1990, Harte et al 1991, Hughes 1996, Krenkel et al 1980, Luxon 1992, Sharma et al 1994, Sitting 1994, WHO 2001).

These pollutants must be within a limited concentration in drinking water to be in line with WHO guidelines, and those of other developed countries, like the USA and the UK, with developing country drinking water regulations and standards such as those for Thailand (Table 1).

3. Research methods and fieldwork preparations

A step-by-step approach was adopted in this study using multi-criteria evaluations (MCE). Information relating to siting criteria was gathered from two domains. First, inputs from regulators academics and the public in Jordan and, secondly, information from the published academic literature and international legislation. Results from the integration of these two sets of information resulted in the development of sound decision-making processes related to landfill siting.

An individual structured questionnaire interview method was adopted for the first domain. The rationale for choosing this method is because it can provide high quality data, assuming accurate and meaningful answers (Bailey 1994, Lober et al 1994, Robson 1997).

Forty key respondents were chosen and interviewed individually. These people were drawn from several disciplines and activities and classified into three groups: 21 decision makers, six planners with 13 academics (considered as one group because some of the academics in Jordan work with the planners). These people were selected because they are either working directly on solid waste management or they are familiar with environmental issues related to waste management in Jordan. All members of the three groups were male. No females were found to be working in the solid waste sector, even within the local community. The respondents' backgrounds ranged from civil engineering, geology, hydrology, EIA specialists, environmentalists and water resource specialists, through to the 12 local community representatives.

Questionnaires were designed using open and closed questions. The questionnaire was initially developed on the basis of an international academic literature review and experiences in dealing with landfill site selection issues. The questionnaire was divided into two parts. Respondent information, such as name, background and job title were recorded.

A ten-point rating scale, ranging from least important (1) to most important (10) was used in the questionnaire. The rating technique, according to Oppenheim (1992), is a numerical value method used to enable the respondents to make some kind of value assessment or judgment. Rating techniques can be applied to everything (Oppenheim 1992). They use ranges in different intervals, such as 1-5, or 1-10, or any other intervals (Oppenheim 1992). The rating technique was used in this study since it does not constrain respondent's responses, as in other fixed point scoring methods, and it is flexible in terms of response measurement (Hajkowicz et al 2000, Malczewski 1999, Oppenheim 1992).

Three pre-defined 'buffer strips', or distances away from specific sites (eg airports, residential areas) were suggested in this study to identify where landfill may or may not be located; the first buffer area is the closest to the site; the second was further away and the third was at the greatest distance. These are respectively called 'unacceptable', 'acceptable' and 'preferable' areas. Both pairs of three alternatives (coded: a,b,c,d,e,f) were attached to each question. The numbers in these questions were selected from other countries' landfill criteria as legislation, or recommended in the published academic literature. Where distances were not cited in the literature, such as political borders, distances were suggested and amended in response to questionnaire outcomes.

4. Data analysis and results

4.1. Identification of constraint factors for landfill site selection

Qualitative assessment and analysis of the interview data resulted in the identification of 27 factors (Table 2) for new landfill site planning in Jordan. Frequency values for each factor were calculated for each of the three groups of interviewees and ordered according to the weighted mean total of each factor. The resultant list gives a priority order of the factors. They may be classified as environmental, social and economic-related issues, rather than engineering based, and they also fall into two domains in terms of measurement and applications:

- Quantitative factors, such as built-up areas, roads and groundwater etc.
- Qualitative factors, such as wind direction and local community acceptance etc.

It was evident that social factors and related issues were the main and most crucial concerns identified by all groups. All respondents agreed that factors related to the local population must be given a high priority. For the local community, health and protection of water resources were the prime issues. For each group built-up areas were given the highest priority to overcome local community opposition) followed by groundwater and then surface water protection. Water resources were given priorities because they are considered a precious and valuable resource in Jordan since they are very limited. Wind direction was also considered an important factor because air pollution contributes to local community opposition to landfill sites. The decision makers highlighted issues related to economic considerations such as haulage distance and transportation cost. Academics and planners concentrated more on socio-economic and geological issues. Other technical issues, like soil type and slope, were considered important by the decision makers and academics because they are aware of the importance of these factors in site selection, construction and operation. Local community representatives raised the role of local community opinion and acceptance before any landfill could be established.

Population and groundwater were assigned high scores (Table 2). This is attributed to the fact that most of the respondents considered groundwater contamination to result from the weakness of solid waste disposal management which, subsequently, could lead to human health problems.

4.2. The relative importance of the constraint factors

Two groups answered the question about the weight they placed on constraint factors. The first, as before, comprised the decision makers while the second included the aca-

Table 2. Landfill constraint factors and their percentage identified by respondents

No.	Planning constraint factors identified by the respondents	Responses analysis *1		*2		*3		Overall
		Freq.	%	Freq.	%	Freq.	%	(%)
1	Built-up areas	21	100	16	84.2	12	100	94.2
2	Groundwater	18	85.7	15	78.9	10	83.3	82.7
3	Surface water bodies (ponds and dams)	15	71.4	12	63.2	10	83.3	71.2
4	Dominant wind direction	15	71.4	9	47.4	9	75	63.5
5	Wadis	15	71.4	7	36.8	6	50.0	53.8
6	Topography	16	76.1	12	63.2	–		53.8
7	Soil availability	12	57.1	7	36.8	2	16.7	40.4
8	Structure geology	11	52.4	10	52.6	–	–	40.4
9	Geotechnical investigations	12	57.1	6	31.6	3	25.0	40.4
10	100-year flood plains	10	47.6	6	31.6	4	33.3	38.5
11	Access roads and infrastructure	11	52.4	9	47.4	–	–	38.5
12	Soil type	9	42.9	7	36.8	2	16.7	34.6
13	Land value and ownership	–	–	10	52.6	7	58.3	32.7
14	Local community acceptance	1	4.7	6	31.6	9	75	30.8
15	Transportation cost	10	47.6	6	31.6	–	–	30.8
16	Recreational sites	6	28.6	3	15.8	6	50.0	28.8
17	Site capacity	6	28.6	2	10.5	4	33.3	23.0
18	Present land use	10	47.6	4	21.0	–	–	26.9
19	Future land use planning	3	14.3	7	36.8	–		19.2
20	Visibility to the public	2	9.5	2	10.5	4	33.3	15.4
21	Population growth	4	19.1	4	21.0	–		15.4
22	Airports	2	9.5	4	21.0			11.5
23	Availability of financial resources	2	9.5		–	3	25.0	7.4
24	Waste type and composition	2	9.5	3	15.8	–	–	7.4
25	Political borders	3	14.3	2	10.5	–	–	7.4
26	Climate conditions	–	–	3	15.8	–	–	5.8
27	Rangeland and agricultural areas	–	–	–	–	1	8.3	1.9

*1: Decision makers group; *2: Academics and planners group; *3: Local community representatives group. (Source: Survey analysis).

demics and planners (Table 3). However not all of these people answered the question completely. Of them 65-68 % assigned different weights to these factors to express their relative importance. The remaining 32-35 % were hesitant since they claimed that they had no experience of the weighting process; higher scores imply higher importance and smaller scores imply lower importance.

Frequency analyses were carried out to calculate the majority opinion for each factor for all experts. Tables 3 and 4 list the range, mean, median and standard deviations for the two groups combined. The median values were adopted in this study to measure central tendency because (i) the distribution of responses was not normal in some cases because there are many outlying values which might have distorted the mean, (ii) avoiding decimal numbers, which might have caused some confusion and (iii) the median is considered more meaningful when presenting the central tendency of a group's response (Bryman et al 1999, Tashakkori et al 1998).

Despite some minor differences in the respondents' answers, there was a consensus among them about some factors to be given a greater weight than others due to safety and socio-economic reasons. All of them agreed that geology, hydrology, built-up areas and socio-economic factors should be given high weights. Constraint factors, such as

Table 3. Weight calculations assigned to the constraint factors by the two groups combined

No.	Constraint factors	Weight Range Min.	Max.	Mean	Median	SD
1.	Faults	5	10	8.4	9	3.730
2.	Slope	1	10	6.3	6	4.401
3.	Flood plains	5	10	8	8	4.053
4.	Surface water bodies	5	10	8.2	9	3.941
5.	Unconfined aquifer	4	10	8.5	9	3.897
6.	Confined aquifer	1	8	3	3	4.942
7.	Oil and gas wells	1	8	4	4	4.655
8.	Oil and gas pipeline	1	9	5.1	5	4.868
9.	Mining activities	1	8	4.5	5	4.595
10.	Soil availability	3	10	8	8	3.730
11.	Built-up areas	5	10	8.6	9	3.747
12.	Wind direct	6	10	8.3	8.5	3.799
13.	Road network	5	10	7.9	8	3.815
14.	Railway	2	7	4.7	5	4.549
15.	Airport	3	10	7	7	2.710
16.	Agricultural areas	5	9	7	7	3.908
17.	Political borders	3	10	7.1	7	3.949
18.	(CEC)	2	10	6.4	7.5	4.003
19.	pH	2	10	6.2	7	4.133
20.	Clay and silt contents	3	10	7.7	8	3.981
21.	Sand contents	1	8	3.9	3.5	4.846

Table. 4. Calculation of relative importance of median weights

No.	Constraint factors	Median 1*	Median 2*	Overall Median
1.	Faults	8	9	9
2.	Slope	6	6	6
3.	Flood plains	9	7	8
4.	Surface water bodies	9	8	9
5.	Unconfined aquifer	9	9	9
6.	Confined aquifer	3	3	3
7.	Oil and gas wells	5	3.5	4
8.	Oil and gas pipelines	5	5	5
9.	Mining activities	4	6	5
10.	Soil availability	9	8	8
11.	Built-up areas	9	9	9
12.	Wind direction	9	8	8.5
13.	Road network	8	8	8
14.	Railway	5	5	5
15.	Airport	8	7	7
16.	Agricultural areas	8	7	7
17.	Political borders	7.5	6	7
18.	(CEC)	6	8	7.5
19.	pH	7	6	7
20.	Clay and silt contents	8	7	8
21.	Sand contents	5	4	3.5

*1: decision maker, *2: academic and planners

water resources and their protection, in particular unconfined aquifers and wells, and settlements were considered important because of their influence on human health and the environment. Road networks and soil availability were also considered important because of the need to minimize the cost of transportation and construction. Political borders were considered important because there are five countries surrounding Jordan.

As shown in Tables 3 and 4 there are some differences within and between the two groups regarding some factors. For example built-up areas, faults and water resources were considered extremely important, which is attributed to the awareness of the interviewees to the social and water pollution issues and their future consequences for Jordan. In most cases the median values of the two groups (Table 4) was within a single point of each other.

4.3. Identification of landfill site selection criteria

Responses were analysed in relation to three buffer areas, each representing a specific criterion. Criterion one is an exclusion area which cannot be used as a potential area for landfill. Criterion two forms an area that could be 'acceptable' as a potential landfill area, and criterion three was considered 'preferable' for landfill locations.

Box plot tools in SPSS were used to analyse and illustrate the distribution of the questionnaire answers. According to Pallanat (2001) boxplots are useful for three reasons:

- They show the median, minimum, maximum and extreme response values;
- They provide information about the shape and dispersion of the distribution;
- They are useful when the distributions of variable scores from different groups need to be compared.

4.4 Results of analysis

Criteria were re-classified into two main groups based on measurement type. The first group comprise quantitative criteria (Table 5), while the second comprise the qualitative criteria group (Table 6). The first group are measured in numerical units, such as distances to main roads and built-up areas, or the angle of slope in degrees. The second group, as shown in Table 6, are descriptive. These criteria are measured on the basis of site evaluation and assessment. A summary of the analyses of classes is presented in Table 5.

There are some outliers in terms of the responses while other respondents answers were considered extreme (Pallanat 2001). These outliers and extremes values may probably be attributed to lack of experience or bias towards the importance of particular attributes to be protected, such as surface water, wells and groundwater.

For economic reasons, and based on the published academic literature (Baban et al 1998, Siddiqui et al 1996), it was suggested in this study that landfill sites should be located within 10 km of a built up area or source of waste. It was also suggested for economic reasons that landfill sites should not lie more than 2 km from main roads. This complies with some case studies used worldwide. For example, in Illinois County, USA, Lindquist (1991) selected landfill sites within 1 230 m of the main roads. In other case studies (eg Dikshit et al 2001, Kao et al 1996b, Lin et al 1999) from India and Taiwan landfill selection models were developed by applying criteria of less than 1 km from main roads. Based on the international academic literature and legislation, a 2 km maximum distance has been suggested in this study.

The site capacity lifetime, recommended by some respondents during the fieldwork and also from available documents in Jordan, should be at least 15 years. This recommendation is based on the local condition of site selection as a criterion to allow establishment of a site for a reasonable period of time.

Criteria concerned with proximity to airports are recommended in this study based on US EPA (1993) standards and regulations. The criteria require that, for safety considerations and complying with limitations contained in Subtitle D of the Resource Conservation and Recovery Act, landfill sites must not be located within 10 000 ft (ca 3 000 m) of airport runways used by turbojet aircraft and 5 000 ft (ca 1 500 m) of airport runways used by piston-type aircraft (Daniel 1993, Lindquist 1991, Sharma et al 1994).

Other criteria pertaining to the required soil characteristics are adopted from the international literature which recommends that soils with high silt and clay content and containing clay minerals such as kaolinite, montmorillonite, illite and chlorite are most acceptable. According to Bagchi (1990) and Sharma and Lewis (1994) the typical soil hydraulic conductivity required for landfill lining and covering should not exceed 1×10^{-7} cm s^{-1}. Chapius (1990) (cited in Kayabli (1996) suggests that the ideal hydraulic

Table 5. General quantitative criteria and constraint factor weights defined by the respondents

Criterion type	Constraint factors	Buffer distances and specifications Buffer 1*: No landfill sites should be located:	
	Built-up areas	< 3 km and more than 10 km	
	Farms	< 300 m	
	Wells	< 500 m	
	Ponds	< 400 m	
	Wadis	< 200 m	
(1) Exclusive	Flood plains	Not within 100-year flood frequency event on flood plains	
	Faults	< 90 m	
	Roads	< 500 m and > 2 km	
	Slope	> 20 %	
	Political borders	< 1 km	
		Buffer 2**: Acceptable site locations:	Buffer 3†: Preferred site locations:
	Built-up area	3 km to 5 km	5 km to 10 km
	Farms	300 m to 400 m	> 400 m
	Wells	–	> 500 m
	Ponds	400 m to 500m	> 500 m
	Wadis	200 m to 275m	> 275 m
(2) Restrictive	Flood plains	<100 m	> 100 m
	Faults	90 m to 200m	> 200 m
	Roads	1 km to 2 km	500 m to 1 km
	Slope	20 % to 5 %	< 5 %
	Political borders	1 km to3 km	> 3 km
	Oil pipeline	–	> 125 m
	Site capacity	Sites of less than 20 ha should be excluded	
		Qualitative criteria	
	Local community	Selected sites must be accepted by local community	
	Wind direction	Sites with upwind dominant direction must be avoided	
	Visibility	Sites must not be visible to the public	

*: buffer weight is 0; **: buffer weight is 5; †: buffer weight is 10.

Table 6. A set of qualitative criteria adopted from the literature

No.	Criteria	Explanation
1.	Soil availability	Available on-site cover materials to meet the need for immediate lining and final cover (silt and clay are preferable).
2.	Cation exchange capacity (CEC)	Areas with low CEC must be avoided and areas with a high CEC are preferable (The typical soil hydrologic conductivity required for landfill lining and covering is less than or equal to 1×10^{-7}cm s^{-1}).
3.	Clay and silt content	Areas with high percentage of silt and clay are preferable. (Kaolinite, Montmorillonite, Illite and Chlorite soils are preferable.)
4.	Sand and gravel content	Sites must not be located in sandy soils or where there are high percentages of sand and gravel
5.	Wind direction	Sites upwind of urban areas must be avoided
6.	Community acceptance	Sites must be accepted by host communities
7.	Visibility	No visibility of the proposed site from any built-up and recreational areas

conductivity should be 10^{-9} to 10^{-6} cm s^{-1} and should be able to retard the transport of pollutants (Pereira et al 1997). Table 6 shows the CEC of these minerals.

Once the criteria (Tables 5 and 6) had been finally identified, they were validated by six academics and decision makers in Jordan to assess the reaction of these people to the results of the analysis. All of them agreed that these criteria would be suitable for Jordan. The only reservation was the proximity to water resources for protection purposes. They suggested that the landfill should be located in areas at least 1 000 m from valleys and water bodies. This suggestion was not accepted because of the high density of the existing drainage system in the study area. The distances deemed preferable by this study (ie > 275 m from wadis and > 500 m from water bodies) (Table 5) were retained. The experts' suggestions could be attributed to their fears about the current operational practices at current disposal sites and their poor management. However, with good management, these problems can be overcome.

5. Summary and conclusions

Discussing the methodological issues underpinning this research indicated that it would be most suitable to use quantitative research methods, relying more on survey, to arrive at major findings. Initially, a review of the research methods and techniques was highlighted. Subsequently, questionnaire interviews were prepared and designed on the basis of the international academic literature reviews, local experts and international legislation. The questionnaire development and design procedures were then discussed and explained. Visits were made to several organizations and parties in Jordan to find key players who could provide answers to the research questions.

The gathering of expert opinions using interviews and questionnaires provided an objective assessment of criteria and factor weightings that should be used to identify sites potentially useful for landfill. Experts from various fields and levels, from representatives of governmental organizations to the private sector, from universities and other specialists from scientists to landfill site managers, provided opinions from a broad spectrum of parties, including local communities, involved in landfill site selection, which could be used in the development of a subsequent GIS-based landfill site selection package.

Eventually, 27 constraint factors and 25 quantitative and qualitative siting criteria were identified as essential for deciding on the location of a new landfill site in Jordan.

References

Atkinson SF, Schoolmaster FA, Lyons DI and Coffey JM (1995). A Geographic Information Systems approach to sanitary landfill siting procedures: a case study, professional report. Journal of the Environmental Professional 17:20-26.

Baban MJ and Flannagan J (1998). Developing and implementing GIS-assisted constraints criteria for planning landfill sites in the UK. Journal of Planning & Research 13 (2):139-151.

Bachmann G (1983). Health and environmental aspects of solid waste management in developing countries. Symposium on solid waste management in developing countries, October p1-17. Geneva, Switzerland.

Bagchi A (1990). Design, construction, and monitoring of landfills (1st edition). Wisconsin, USA: John Wiley & Sons, Inc.

Bagchi A (1994). Design, construction, and monitoring of landfills (2nd edition). Wisconsin, USA: John Wiley & Sons, Inc.

Bailey K (1994). Methods of social research (4th edition). Canada: Maxwell Macmillan Inc.

Basagaoglu H et al (1997). Selection of waste disposal sites using GIS. Journal of the American Water Resources Association 33 (2):455-464.

Batstone R, Smith JEJ and Wilson D (1989). The safe disposal of hazardous waste: The special needs and problems of developing countries. Vol, I, II and III, Washington, D.C: World Bank Technical Papers.

Baxter JW, Eyles JD and Elliott A (1999). From siting principles to siting practices: a case study of discard among trust, equity and community participation. Journal of Environmental Planning and Management 42 (4):501-525.

Bratley KJ (1983). Techniques of sanitary landfill. In: R.J, H (ed), Practical waste management. UK: Wiley & Sons Ltd p.565.

Bryman A and Cramer D (1999). Quantitative data analysis with SPSS release 8 for windows: a guide for social scientists. London: Routledge.

Chopra M, Reinhart D and Abu-El-Sha'ar W (2001). Report on US-Jordan municipal solid waste management collaborative research, Amman, Jordan.

Crawford JF and Smith PG (1985). Landfill technology. London: Butterworth.

Daniel DE (1993). Landfills and impoundments. In: Daniel, DE (ed), Geotecnical practice for waste disposal. London: Chapman & Hall.

Davis ML and Cornwell DA (1998). Introduction to environmental engineering (3rd edition). New York: WCB/McGraw-Hill Series.

Dechapanya W, Rogers TN, Baker JR and Radecki PP (1999). Application of the analytical hierarchy process for integrating environmental considerations into process design decision: Proceeding of the NATO advanced research workshop on tools and methods for pollution prevention. Kluwer Academic publisher, Netherlands pp. 367-383.

Dikshit AK, Padmavathi T and Das RK (2001). Locating potential landfill sites using Geographic Information System. Journal of Environmental Systems 28(1):43-54.

Dorman D. Water quality problems: health and household, http://www.ces.uga.edu/pubcd/c819-8w.html.

Elinder CG (1991). Biological monitoring of cadmium, London: John Wiley & Sons.

Enger ED and Smith BF (1995). Environmental science, a study of interrelationships (6th edition). Washington: WCB, McGraw-Hill.

EPA (1990). Handbook groundwater: volume I: groundwater and contamination. EPA/625/6-90/016a, Cincinnati, USA: United State Environmental Protection Agency.

EPA (1993). Draft technical manual for solid waste disposal facility criteria. 40 CFR, part 258. Washington, DC: United State Environmental Protection Agency.

FoE, Friends of the Earth's Programme to Assist Gross Roots Activities; landfill manual introduction, http: // www.foe.org/ptp/chapter1.html <http://www.foe.org/ptp/chapter1.html>.

Furuseth OJ (1990). Impact of a sanitary landfill: spatial and non-special effects on the surrounding community. Journal of Environmental Management 31 (3):269-277.

GCEP (2001). Report on municipal solid waste composition in Jordan. General Corporation for Environment Protection/Ministry of the Environment, Amman, Jordan.

Gorelick SM, Freeze RA, Donohue O and Keeley JF (1993). Groundwater contamination: optimal capture and contaminant. London: Lewis Publisher.

Gumbricht T (1996). Application of GIS in training for environmental management. Journal of Environmental Management 46:17-30.

Hajkowicz SA, McDonald GT and Smith PN (2000). An evaluation of multiple objective decision support weighting techniques in natural resource management. Journal of Environmental Planning and Management 43(4):505-518.

Harte J, Holdren C, Schneider R and Shirley C (1991). Toxics A to Z: A guideline to everyday pollution hazards. Berkeley: California University Press.

Hughes WW (1996). Essentials of environmental toxicology: the effect of environmentally hazardous substances on human health. Washington: Taylor & Francis.

Hussey V, Dodd VA and Dennison GJ (1996). Locating a landfill site for Dublin using Geographic Information Systems. Proceeding of the Institution of Civil Engineers-Municipal Engineer 115 (3):125-133.

Johnson BL and DeRosa CT (1997). The toxicological hazard of superfund hazardous waste sites. Agency for Toxic Substances and Disease Registry, Public Health Services, Reviews on Environmental Health, Atlanta, Georgia USA: Freund Publishing House Ltd.

Kao JJ, Chen WY, Chen W, Lin H and Guo S (1996a). Network expert Geographic Information System for landfill siting. Journal of Computing and Civil Engineering 10 (4):307-317.

Kao JJ and Lin HY (1996b). Multi factors spatial analysis for landfill siting. Journal of Environmental Engineering 122(10):902-907.

Kayabli K (1996). Engineering geological aspects of replacing a solid waste disposal site with a sanitary landfill. Journal of Engineering Geology 44:103-212.

Knipe RJ, Rutter EH and Holdsorth RE (2001). The nature and tectonic significance of fault zone weakening. UK: Geological Society special publication, N.186.

Krenkel P and Novotny V (1980). Water quality management. London: Academic Press, Inc.

Lee GF and Jones-Lee A (1996). Dry tomb landfills. Available at: http://www.gfredlee.com/hwlf-pap.htm, MSW Management 6 (1):82-89.

Lee GF and Jones-Lee A (1994). Landfilling of solid waste & hazardous waste: facing long-term liability. Proceeding of the second conference on federal environmental restoration iii and waste minimisation, Hazardous Materials Control Resources Institute, Rockville MD pp. 1610-1618.

Lin HY and Kao JJ (1999). Enhanced spatial model for landfill siting analysis. Journal of Environmental Engineering 125 (9):845-852.

Lindquist RC (1991). Illinois cleans up: using GIS for landfill siting. Geo Info Systems 1(2):30-35.

Lober DD (1993). Beyond self-interest: a model of public attitudes towards waste facility siting. Journal of Environmental Planning and Management 36 (3):345-359.

Lober DJ and Green DP (1994). NIMBY or NIABY: a Logit model of opposition to solid-waste-disposal facility siting. Journal of Environmental Management 40:33-50.
Luxon SG (1992). Hazards in the chemical laboratory. Cambridge: Royal Society of Chemistry.
Malczewski J (1999). GIS and multicriteria decision analysis. New York: John Wiley & Sons Inc.
McBean EA, Rovers GJ and Farquhar GJ (1995). Solid waste landfill engineering and design. New Jersey, USA: Prentice-Hall, Englewood Cliffs.
Öman C and Hynning PA (1993). Identification of organic compounds in municipal landfill leachate. Journal of Environmental pollution 80:265-271.
Oppenheim AN (1992). Questionnaire design, interviewing and attitude measurement (new edition). London: Pinter Publishers.
Oweis IS and Khera RP (1990). Geo-technology of waste management. Boston MA: PWS and Company Publisher Ltd.
Oweis IS and Khera RP (1998). Geotechnology of waste management (2nd edition). Boston: PWS Publishing Company.
Pallanat J (2001). A step by step guide to data analysis using SPSS for windows; version 10 and 11. Buckingham: Open University Press.
Pereira JM, Didier G, Winiarski T and Coelho AG (1997). Multivariable statistical analysis: A useful tool to evaluate some geotechnical properties of fine-grained soils for their use as geochemical barriers. Proceedings of the sixth international conference on landfill, Sardinia, Italy (13-17 October) pp. 9-25.
Petts J and Eduljee G (1994). Environmental impact assessment for waste treatment and disposal facilities. Chichester: John Wiley.
Robson C (1997). Real world research: a resource for scientists and practitioner researchers. UK, USA: Blackwell Publishers Inc.
Sharma H and S L (1994). Waste containment systems, waste stabilisation, and landfills: design and evaluation. New York, USA: John Wiley & Sons, Inc.
Siddiqui MZ, JW E and Vieux BE (1996). Landfill siting using Geographic Information System: a demonstration. Journal of Environmental Engineering 122 (6):515-523.
Sitting M (1994). Worldwide limits for toxic and hazardous chemical in air, water and soil. New Jersey, USA: Noyes Publications.
Sloan WM (1993). Site selection for new hazardous waste management facilities.
Tashakkori A and Teddlie C (1998). Mixed methodology: combining qualitative and quantitative approaches; vol.46. London: Sage Publications, Inc.
Tchobanoglous G, Theisen H and Vigil SA (1993). Integrated solid waste management. New York: McGraw-Hill, Inc.
Wentz CA (1995). Hazardous waste management (2nd edition). New York, USA: McGraw-Hill.
WHO (1990). Environmental health criteria, 118, 134, 194: aluminium, inorganic mercury and cadmium. Geneva: WHO UNEP Library Cataloguing publications, International Programme on Chemical Safety (1990, 1991, 1992, 1997).
WHO (2001). Guidelines for drinking water quality (3rd edition), July. Available at: http://www.who.nt/m/topics./drinking water quality/en [Access date: 1-1-02], Geneva: World Health Organization.
Yegorov YK and Zerkal OV (2000). Geological data management for urban planning using GIS. Proceeding of the 8th international congress for engineering geology and the environment, Vancouver, Canada IAEG Congress.
Zeiss C and Paddon B (1992). Management principles for negotiating waste facility siting agreement. Journal of Air and Waste Manage 42 (10):1296-1304.

CHAPTER 10

Biodiversity of reptiles and mammals in the Jordan Badia

Zuhair S. Amr
Department of Biology, Jordan University of Science & Technology, P. O. Box 3030, Irbid, Jordan.
E-mail: amrz@just.edu.jo

The paper describes the habitats in the Badia, with particular reference to those factors that condition the presence of reptiles and mammals and their survival in what would normally be considered a hostile environment. A full list of amphibians, reptiles and mammals reported from the Badia is given and the unique or special features of some of them are recorded.

Keywords: Biodiversity, mammals, reptiles

1. Environment and habitat

The Badia encompasses a wide and significant part of Jordan. It covers an area of approximately 72 600 km², which constitutes 81 % of the total area (89 400 km²) of the country. The region is sub-divided into three geographical areas as follows (Allison et al 1998):

- Northern Badia which comprises 35 % (25 930 km²) of the Badia total area;
- Middle Badia which comprises 13 % (9 634 km²) of the Badia total area;
- Southern Badia which comprises 51 % (37 096 km²) of the Badia total area.

The Badia falls in an arid climatic zone. Rainfall is erratic both spatially and temporally, with a maximum of 200 mm annually. Air temperature fluctuates widely from a daily mean minimum of 10 °C with a mean maximum of 24.5 °C and a mean daily temperature of 17.5 °C. Occasionally absolute minimum and maximum temperatures might reach -5 °C and 46 °C respectively.

The original Jordan Badia Research and Development Programme (BRDP), comprised some 11 210 km² or about 14 % if Jordan's total land area (Figure 1). The area is bounded by Syria to the north and Saudi Arabia in the south. The east and west margins approximately follow the perimeter of extensive basalt outcrops, which cover a large part of the ground surface between the towns of Azraq and Ruwayshid (Figures 1 and 2).

Tertiary-Quaternary continental basalt flows and tuffs cover almost the entire BRDP area. They form distinct strata dating from 23 to less than 0.5 million years ago for some exposed rocks that originated in lava flows centred on the Jabal al-Arab. Significant mineral resources have been identified including zeolite, basalt, tuff, porcellanite, bentonite and others (Allison et al 1998). Their economic value remains to be confirmed but some of them could become a significant focus for future development. Some small gasfields and oilfields have also been identified.

Topographically the area is not complex. The highest ground, 1 100-1 200 m, occurs to the north-west, south of the Jabal al-Arab in Syria. The other relatively high ground is formed by the north-south chain of volcanic cones towards the eastern edge of the study area. Between the two, from the north, the weakly defined wadi Rajil descends towards the Azraq basin which is the low point of the area. Much of the land surface is composed of gently undulating low hills with elevation differences of 25-30 m within short distances of each other, and gentle concavo-convex slopes.

In the basalt fields the dominant feature is large basalt boulders overlying poorly

Figure 1. BRDP area within the Jordan Badia

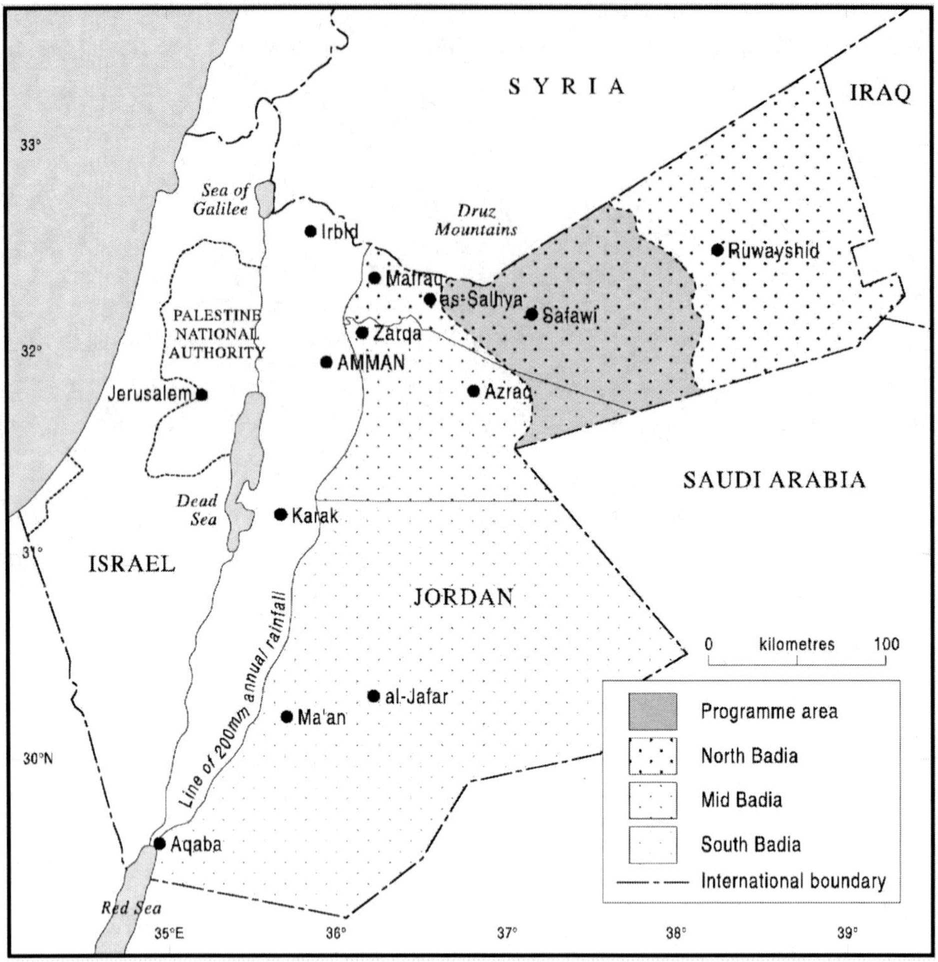

weathered basaltic soils, but towards the south-east small, angular chert fragments are more common. Another significant geomorphological zone in the BRDP area is characterized by wind-blown sediment deposits, which are encroaching on the southern boundary of the Badia (Figures 3a-b on p158). The aeolian deposits represent the margin of an extensive sand-sea in Saudi Arabia.

A combination of wind and water erosion has washed some of the soils into the low points which show up on satellite imagery as distinct off-white patches between the black boulder-covered higher land. Some of the low points are basins, or near basins, locally known as *qa'a*, which collect water after the rains and tend to be covered with fine deposits of very limited permeability and which may be somewhat saline. Other low points have a small but distinct gradient so that water flows through them leaving coarser grained non-saline deposits that are more permeable, and are locally known as *marab*.

A better understanding of the characteristics of the basalt boulder surfaces (Higgitt et al 1998), of surface water infiltration (Warburton 1998) and of soil capping characteristics (Kirk 1998b) under the impact of rainfall and irrigation, all studied in an earlier phase of the Badia Programme, will materially assist with hydrological modelling and water management.

Figure 2. The black lava desert with black stone covered by lichens. These lichens provide several small invertebrates and other animals with food

On this high steppe land (mean elevation 800 m) far from marine influence the climate is continental with big diurnal and seasonal temperature fluctuations. In winter, mean minimal temperatures may drop to 2-9 °C (Meteorological Department 1988). Snowfall and sub-zero temperatures are known, and the chill factor can be quite severe when cold air sweeps down from the north-west. In the summer, absolute maxima may exceed 46 °C, and mean maxima 35-38 °C. Summer air has low humidity, and the potential evapotranspiration is estimated to be over thirty times the mean annual rainfall. Occasional very strong winds create dust-storms. Though generally pleasant in the spring and autumn, in winter and summer it is a harsh climatic regime, and one that creates conditions that can become very unpleasant for all forms of life (Dutton 1998).

The mean rainfall is very low (Kirk 1998a), and there is a strong sub-regional variation from up to 250 mm in the north-west on the high land along the boundary with Syria, to less than 50 mm in the south. The variation from west to east is less pronounced, meaning that most of the precipitation is in the northern half of the study area. Most of the rain falls between December and March but otherwise the temporal and spatial variation is highly variable, as is the intensity of individual showers and storms. Some catchment areas are quite large and the vegetation sparse, so runoff and spate wadi flow are more frequently generated than might otherwise be expected.

Runoff is sufficient, in places, to have justified the construction of a dam in ancient times at Burqu, and for it to be rebuilt in the twentieth century (Figure 4 on page 159) and to retain enough water for sheep and goats, also to support a large number of raptors and other birds in the migration seasons.

Broadly speaking, on the older basalt flows the landscape is more rounded, with well developed drainage patterns and a fine colluvial rill network. The more recent flows, on the other hand, show highly irregular topography with many silt filled depressions and a poorly connected drainage network.

Most of the ground surface is boulder-strewn. The size of individual clasts varies. The size distribution is principally controlled by the ground surface geology, with some characteristics of sorting, ground cover and relative exposure being influenced by slope position. In some of the basalt fields large boulders are predominant, while in other places small, angular clasts are more common. Towards the north-west the basalt boulders become increasingly covered by lichen, an indication of the increased moisture

Figure 3a. Sand dunes along the margins of Marab Salam

Figure 3b. Sand dune habitats with wild palms, *Phoenix dactylifera* and *Nitraria retusa*

availability (Figure 2). The basalt clasts generally form a mantle of stones overlying fine-grained, brown to orange-coloured sediment. The orange deposit is unlithified, laterally extensive and of variable depth. Exposures in basalt quarries suggest that the sediment is a product of basalt weathering, although there may be an input of fines from aeolian processes (Allison et al 1998).

The wadi systems which drain the BRDP area are extensive. The general flow direction is from the north to the south and south-west, with much of the water entering the Azraq basin. Some of the wadi systems extend for many kilometres. Wadi Rajil, for example, stretches from north of the border with Syria into the Azraq basin and Saudi Arabia. Peak flows during wet seasons can be considerable although gauging records are limited; information from the local population suggests that during parts of the year lack of water is not a problem.

The soils of the BRDP area comprise three distinct types. The major groups broadly follow geological variations, in particular the spatial distribution of basalt lava flows and outcrops of other more recent sediments.

There are soils associated with the East Jordan Limestone Plateau. Soil parent materials in the Plateau area include limestone, chalk, calcareous sands, gypsum, marls and Tertiary cherts. The soils are highly variable, as a result of repeated cycles of erosion and deposition and include shallow to deep entisols and aridisols. Down-slope, bedrock surfaces merge into sequences of gravel fan and terrace deposits, with entisols frequently occurring in wadi beds and depressional mudflats. Second, there are soils associated with the north Jordan basalt plateau. The main soil divisions are related to topographic and weathering differences. A common feature of the soils is the presence of a thick, secondary caliche deposit on the basalt bedrock. Third, there are the soils of the north-east Jordan Basalt Plateau, which are classified on the age and physical characteristics of lava flows and eruptive centres, yielding a wide range of soil types.

Erosion and redistribution of sediment and soils has a number of components: environmental controls on sediment production and detachment; aeolian mobilization

Figure 4. Burqu area with a permanent water body in the middle of the eastern desert

processes; runoff mobilization processes and factors promoting contemporary sedimentation. Key variable controls on erosion systems include vegetation and topography. Erosion by water is a consequence of high magnitude, low frequency storms, the absence of a protective vegetation cover across much of the ground surface and the low stability of soil aggregates. A surface crust frequently develops which seals the ground, reduces infiltration, promotes runoff and aggravates erosion (Allison et al 1998).

Erosion by wind is most frequently aggravated when ground surface disturbance, for example by livestock or vehicles, provides locations where aeolian erosion processes can be initiated. There is extensive evidence of wind erosion, particularly during long, dry summers, when soil surfaces dry out, crack, lose cohesion and begin to fragment. Extensive tracts of uninterrupted land with little or no topographic variation promote rapid wind speeds. It is all too frequently the case that the top, most productive soil layers, are lost at times when little or no vegetation is present and wind speeds are at their most severe. The same is the case where basalt boulders have been removed from the ground surface, exposing the underlying sediments to deflation.

All these factors have strongly influenced the types and densities of the vegetation and cover and, thence, animal biodiversity including the occurrence of both mammals and reptiles.

In recent years the reptiles and mammals of the eastern desert have been studied extensively; this was facilitated and financially supported by BRDP and, in the years 1995-9, by funding from the UK's Darwin Initiative (Dutton et al 1996). More than 20 publications investigated the taxonomy, distribution, ecology and threats associated with these animals. This study of the biodiversity of the reptiles and mammals of the BRDP area comments also on the species richness and endemic species and makes comparison with the species composition of the fauna of other parts of Jordan.

Two biogeographical elements characterize the eastern desert of Jordan. The Irano-Turanian region that forms a narrow strip that extends along the Syrian borders, and is characterized by higher altitude (400-1200 m). The Saharo-Arabian region is wide and constitutes more than 70 % of the total study area, including basaltic fields and boulders (lava desert), locally known as *harra,* where it is covered with the bare black basaltic boulders described above which originated from ancient volcanic activities. Vegetation of the area is poor, lichens cover most rocks in the area turning them white. Plant cover is concentrated in the *marab* and small wadis where soil moisture is greatest. Examples of this habitat include Azraq ad-Druz, al-Buqi'awiyya, Hashad, Jawa, Qasr Burqu, Safawi and Wadi Salma.

In the *marab* vegetation cover is highly dependent on the season. Dominant plants include: *Artemisia* spp, *Origanum* spp and *Achillea fragrantissima.*

In the sandy and sand dune areas vegetation is dominated by wild palms, *Phoenix dactylifera, Nitraria retusa* and *Tamarix* spp. Other plants include; *Peganum harmala, Retama raetam, Alhaji maurorum, Astragalus dactylocorpus,* and *Atriplix* spp. Hazim, Faydat ad-Dahikiya and Azraq ash-Shishan are examples of places in which this habitat and plant assemblage are found.

Red sand dune areas were seen as either patches in the middle of the *hammada* vegetation south of Ruwayshid, or near the Saudi border in vegetated low wadis surrounded by rocky cliffs and basaltic hills in the al-Wisad area. The sand dunes are mainly dominated by *Hammada salicornica.*

Some Aeolian sand deposits are found in the gravel deserts. Sand stabilizes in low areas and may give rise to a healthy vegetation cover of *Seidlitzia rosmarinus.*

In the *hammada,* vegetation is dominated by low shrubs of *Seidlitzia rosmarinus* and some annual shrubs and succulent plants such as *Spergularia diandra, Herniaria hirsute* and *Anthemis deserti.*

2. Amphibians and reptiles

The reptiles and amphibians of the Badia have been studied extensively (Parker 1935, Schmidt 1930, Schmidt 1939, Werner 1971, Werner 1991, 1992, Wettstein 1951). Since 1995 several herpetological expeditions to the Badia have yielded new records to Jordan (Abu Baker et al 2005, Al-Disi 1997, Al-Melhim et al 1997), as well as additional localities for 42 species previously identified in the country (Disi et al 1999). For examples, see Figures 5-16 on pp 162, 163. For a full listing see Appendix 1 on page 170. A total of 49 species of amphibians and reptiles have been reported in the Badia (Table 1).

The diversity of amphibians and reptiles in the Badia is higher than in other regions of Jordan and represents about half of the known species. They represent a heterogeneous mixture of species of different origins and zoogeographic affinities. In part the diversity is due to the great variety of habitats ranging from sand dunes, to *harra*, to *hammada*, to wetlands. The Badia acts as a filter for some Mediterranean species that are restricted to the western mountains of Jordan. Some Mediterranean species expanded their range eastwards to the lava deserts with relatively mild temperatures around the southern ends of Jabal al-Arab within Jordan (eg *Ophiomorus latastii, Eirenis rothi, Ptyodactylus puiseuxi* and *Testudo graeca*). Snakes of the family Colubridae constituted the highest number of species (24.4 %), followed by lizards of the family Lacertidae (18.4 %), while Bufonidae, Ranidae, Testudine, Chamaeleonidae, Varanidae, Leptotyphlopidae, Boidae, and Elapidae are represented by a single species for each family (Table 1).

Several relicts were identified within the Badia; Schmidt's Whip Snake (*Coluber schmidti*) occurs sympatrically with *Coluber ravergieri* and both are known from Jawa (Figures 15, 16). It seems that Jawa is the southernmost outpost for both these species of Palaearctic origin. Here, certain desert microhabitats apparently provide the minimum requirements that allow these species to persist and coexist in this very delicate and fragile habitat (Abu Baker et al 2002, Disi 1993).

Hernidactylus turcicus lavadeserticus (Figure 7) is an endemic subspecies confined to the black lava desert in Syria (Moravec et al 1997) and was recently collected from Jordan. It was found to inhabit the narrow strip of the black lava desert that extends across northern Saudi Arabia, Jordan and southern Syria. *Laudakia stellio picea* (Figure 11) is consid-

Table 1. Amphibian and reptilian families known from the Badia

Family	No. of species	%
Bufonidae	1	2.1
Ranidae	1	2.1
Testudinidae	1	2.1
Gekkonidae	6	12.3
Chamaeleonidae	1	2.1
Lacertidae	9	18.4
Agamidae	7	13.3
Scincidae	5	10.2
Varanidae	1	2.1
Leptotyphlopidae	1	2.1
Boidae	1	2.1
Colubridae	12	24.4
Elapidae	1	2.1
Viperidae	2	4.2
Total	49	100

Diversity of reptiles in the Badia

Figure 5. The Striped legless Skink, *Ophiomorus latastii*, represents a Mediterranean element in the Badia

Figure 6. The Sandfish, *Scincus scincus meccensis*, is a sand dune inhabitant found in the Hazim area

Figure 7. *Hemidactylus turcicus lavadeserticus* is an endemic species of the black lava desert

Figure 8. The Persian Agama, *Trapelus persicus fieldi*, is an Arabian species that reaches its most north-western distribution in the Jordanian Badia

Figure 9. The Dabb, *Uromastyx aegyptia microlepis*, is an endangered species dwelling in selected habitats in the Badia

Figure 10. The Pale Agama, *Trapelus pallidus haasi*, a common species in gravel areas in the Badia

Diversity of reptiles in the Badia

Figure 11. The Starred Agama, *Laudakia stellio picea,* is an endemic species to the black lava desert

Figure 12. The Horned Viper, *Cerastes gasperetti,* is a sand inhabitant that is found in sand dune belts on the eastern borders of Jordan with Saudi Arabia

Figure 13. The False-horned Viper, *Pseudocerastes persicus fieldi,* dwells among black lava rocks

Figure 14. The Tree Snake, *Psanunophis schokari,* is a diurnal species found in most of the Badia region

Figure 15. Ravergier's Whip Snake, *Coluber ravergieri,* has relict populations in the slopes of Jabal al-Arab. This is a very rare species that requires conservation

Figure 16. Schmidt's Whip Snake, *Coluber schmidti,* is another relict species situated on the southern slopes of Jabal al-Arab

ered to be an endemic subspecies distributed along the black lava desert. Evidence of the Saharo-Arabian influence is well exemplified by the presence of two agamid lizards (*Trapelus persicus fieldi* (Figure 8) and *Phrynocephalus maculatus*), and one gecko (*Stenodactylus doriae*). Jordan represents their most north-eastern range of distribution.

The oasis of Azraq is the only permanent wetland in the heart of the Badia. This oasis is fed by wadis (notably Wadi Rajil) and used to be fed by springs from the basalt aquifers until they stopped flowing the early 1990s. Several otherwise widespread amphibian and reptile species are restricted to the Azraq oasis, evidently representing relict species, which survived aridification of the surrounding region (eg *Rana bedriagae, Chamaeleo chamaeleon, Mabuya vittata, Malpolon monspessulanus insignitus* and *Natrix tessellata*). The nearest populations of the above species are found more than one 100 km west of Azraq, separated by an arid area bare of suitable habitats (Disi et al 1999).

Most of the remaining species of reptiles in the Badia are a confluence of several origins of a wide distribution extending from North Africa or the Arabian Peninsula into Iraq and parts of Turkey. Reptiles of oriental affinities with a wide range of distribution includes: *Bunopus tuberculatus, Trapelus ruderatus, Acanthodactylus grandis*. African forms are represented by *Pseudotrapelus sinaitus, Acanthodactylus boskianus, Mesalina guttulata, Mesalina oliviera*, while Arabian forms include *Acanthodaclylus opheodurus, Acanthodactylus schmidti* and *Phrynocephalus maculates*. For a full listing see Appendix 1 on p 170.

3. Mammals (see pp166-169 for Figures)

Mammals of the Badia have only recently been studied extensively, facilitated by BRDP. Several publications have reported on different mammalian groups, mainly the rodents (Abu Baker et al 2003a, Abu Baker et al 2005, Rifai et al 2000, Scott et al 2000) and the carnivores (Amr 2000a, Amr et al 2004, Bunaian et al 2001, Bunaian et al 1998, Searight 1987). However, the bats of the eastern desert are still not well studied (Amr et al in press). The mammalian fauna of the Badia is quite rich and includes about 42 % of the known mammals of Jordan. This group is represented by 6 orders, 14 families, 23 genera and 33 species (Table 2); see also Appendix 2 on pp 170, 171.

The mammalian diversity of Jordan is remarkable considering its location at the meeting point of three different faunal assemblages; African, Oriental and Palaearctic. Mammals originating in these locations are found in addition to a few endemics. Jordan's location has resulted in a huge faunal diversity (compared to the surrounding countries) allowing heterogeneity and range expansion of different species. It shelters a huge assembly of mammals of different zoogeographical affinities (Amr 2000b). Most remarkably, Jordan represents biogeographical boundaries for the extreme distribution limits of several African mammals (eg *Procavia capensis* and *Rousettus aegyptiacus*) and Palaearctic mammals (eg *Erinaceus concolor, Sciurus anomalus, Apodemus mystacinus, Lutra lutra* and *Meles meles*).

Table 2. Mammalian orders, genera and species known from the Badia

Order	No. of genera	No. of species
Insectivora	4	4
Chiroptera	3	3
Carnivora	7	11
Artiodactyla	1	1
Lagomorpha	1	1
Rodentia	7	13
Total	23	33

To date, a total of 78 mammals has been recorded from Jordan (Amr 2000b). Mammals of the eastern desert are represented by 33 species that constitutes 42.3 % of Jordan's total mammalian fauna. Endemism is very low and is represented by only one species (*A. lewisi*). The distribution pattern for the other mammals has no significant zoogeographical importance, since the eastern desert acts as a filter for some Mediterranean or Palaearctic species common in the western mountains. The southern slopes of Jabal al-Arab in Syria may represent the most eastern distribution range of some Mediterranean carnivores such as the Eurasian badger, *Meles meles*.

The insectivores of the Badia are represented by four species; two belong to the hedgehog family (Erinaceidae: *Paraechinus aethiopicus* and *Hemiechinus auritus*), and two shrews from the family Soricidae (*Crocidura suaveolens* and *Suncus etruscus*). Both shrews are confined to the Azraq oasis with ample humidity and thick vegetation cover. The Ethiopian hedgehog, *P. aethiopicus* is a true desert species in that it is found in extreme dry habitats in the Badia, while the long-eared hedgehog, *H. auritus* is associated with the Irano-Turanian ecozone.

The Order Chiroptera is poorly represented in the Badia, with only three species recorded: *Pipistrellus kuhli, Otonycteris hemprichi* and *Tadarida teniotis*. This is due the limited studies on this group which has been well-studied in other parts of Jordan (Qumsiyeh et al 1998).

The carnivores of the Badia are well diversified, including 11 species, whereas canids are represented by four species (the wolf, *Canis lupus*, the Asiatic jackal, *Canis aureus*, the Sand fox, *Vulpes rueppellii* and the Red fox, *Vulpes vulpes*). The Sand fox remains the only specialist in terms of habitat selection, where it is associated with the sand desert. The felines include three species (*Caracal caracal, Felis silvestris* and *Felis margarita*; Figure 23), of which the Sand Cat (*F. margarita*) is considered globally endangered. Family Mustelidae is represented by three species (*Meles meles, Mellivora capensis* and *Vormela peregusna*). All these species are associated with the mild Irano-Turanian ecozone and tend to be distributed towards the north-west, and are usually close to farms and human settlements. Wolves, Asiatic jackals and hyenas are the carnivores most frequently attacked by human beings. They are killed instantly when encountered in many parts of the country. The Striped Hyena, *Hyaena hyaena*, is the only species of the family Hyaenidae. This species is in severe decline, attacked due to many superstitions and misunderstanding (Amr 2000a, Bunaian et al 2001, Qarqaz et al 2004). During a previous study, three freshly killed hyenas (one adult and two young individuals) in the al-Hashad area were encountered. Also, in Dayr al-Kahf village, three cubs and their mother were mutilated in a cave overlooking the village.

The rodents of the eastern desert have been subjected to intensive studies. They include 13 species within five families. Family Dipodidae includes two jerboas, *Allactaga euphratica* and *Jaculus jaculus*. Family Gerbillidae includes the highest number of species (*Meriones libycus, Meriones crassus, Meriones tristrami, Psammomys obesus, Gerbillus dasyurus, Gerbillus cheesmani, Gerbillus nanus* and *Gerbillus henleyi*). Some are habitat specialists, for instance *Gerbillus cheesmani* (Figure 19) is a sand dune inhabitant. In general, all gerbils are desert-adapted species with tremendous tolerance to extreme arid habitats. Each of the remaining three families (Gliridae, Muridae and Cricetidae) is represented by a single species, *Eliomys melanurus* (Figure 20), *Acomys lewisi* and *Cricetulus migratorius* respectively. The Spiny Mouse, *Acomys lewisi*, is an endemic species to the black lava desert that extends across northern Arabia, Jordan and southern Syria (Figure 21).

Both orders Artiodactyla and Leporidae are represented by a single species, *Gazella subgutturosa* and *Lepus capensis* respectively. The former species is considered extinct. A skull was collected from H-5 (Safawi) in 1950.

Rodents show some remarkable habitat selections in the Badia. *G. cheesmani*, for example, is an inhabitant of sand dunes and its distribution range extends over Arabia to Iran (Abu Baker et al 2003a, 2003b). This species utilizes sand dune islands that are scattered across deserts. Other rodent species are confined to rocky habitats of various types. *Acomys russatus lewisi* (Figure 21) is associated with black lava desert habitats. Relicts of earlier geological periods are exemplified by the presence of *E. melanurus* (Figure 20) in extreme desert habitats, despite the arboreal activity of this species.

Almost all the carnivores of Jordan are considered threatened or endangered, due to accelerating population growth and habitat changes as well as prevailing human attitudes to wild animals (Amr et al 2002, Amr 2000a). Wolves, jackals, hyenas and other small mustelids are declining in their natural habitats. Several studies have addressed the status of the carnivores in Jordan and suggested further conservation efforts (Abu Baker et al 2002, Bunaian et al 2001, Qumsiyeh et al 1993, Rifai et al 1999).

Fortunately, two operating nature reserves are located within the Badia. Shaumari Reserve (22 km^2), established in 1975, lies within the Saharo-Arabian zone and is represented by the *hammada* vegetation type. Shaumari Reserve is a flat gravely area penetrated by a medium-sized wadi that floods in wintertime. The main objective of Shaumari Reserve is the reintroduction and breeding of the Arabian oryx, the Syrian ostrich and the Syrian wild ass (Amr et al 2004).

The second reserve is the Azraq Wetland Reserve, a silt dune area with some open water and marshes. It lies within the Saharo-Arabian biogeographical zone and is represented by the *hammada* vegetation type in most of the area and a small part of the *qa'a* vegetation type in the eastern part of the reserve. Azraq Wetland Reserve was established in 1977, with a total area of 12 km^2. The main objective of this wetland reserve is to protect migratory birds and conserve wetland habitats. A large number of migrant

Figure 17. The Hemprich Bat, *Otonycteris hemprichi*, is desert species that can survive extreme desert conditions

Figure 18. The Ethiopian Hedgehog, *Paraechinus aethiopicus*, is a true desert adapted species

Figure 19. Cheesman Gerbil, *Gerbillus cheesmani*, is an inhabitant of sand dunes in the Badia

Figure 20. The Garden Mouse, *Eliomys melanurus*, represents a relict species in the Badia

Figure 21. The Black Spiny Mouse, *Acomys russatus lewisi*, is an endemic species to the black lava desert

Figure 22. The Euphrates Jerboa, *Allactaga euphratica*, inhabits open Hamada with scattered vegetation. This species is endangered with a considerable decline in its populations

Figure 23. The Sand Cat, *Felis margarita*, is one of the most graceful cats inhabiting the Badia. This species is threatened and requires protection

and wintering wildfowl and waders use the oasis (Amr et al 2004). Additionally, two proposed nature reserves lie within the Badia, at Wadi Rajil and Burqu.

Such nature reserves will help to maintain and conserve the remaining Badia animals (reptiles, birds and mammals) as well as other invertebrates, though Azraq reserve is threatened by the pumping of water for agriculture and for Amman.

Appendix 1. List of amphibians and reptiles reported from the eastern desert

Family Bufonidae
Bufo viridis
Family Ranidae
Rana bedriagae
Family Testudinidae
Testudo graeca
Family Gekkonidae
Bunopus tuberculatus
Cyrtopodion scabrum
Hernidactylus turcicus lavadeserticus
Hernidactylus turcicus turcicus
Ptyodacty!us puiseuxi
Stenodactylus doriae
Stenodactylus grandiceps
Family Chamaeleonidae
Chamaeleo chamaeleon recticrista
Family Agamidae
Laudakia stellio picea
Pseudotrapelus sinaitus
Trapelus pallidus haasi
Trapelus persicus fieldi
Trapelus ruderatus
Uromastyx aegyptia microlepis
Family Lacertidae
Acanthodactylus boskianus
Acanthodactylus grandis
Acanthodaclylus opheodurus
Acanthodactylus robustus
Acanthodactylus schmidti
Mesalina brevirostris
Mesalina guttulata
Mesalina oliviera
Ophisops elegans blanfordi
Family Scincidae
Chalcides ocellatus ocellatus
Eumeces schneideri
Mabuya vittata
Ophiomorus latasti
Scincus scincus meccensis
Family Varanidae
Varanus griseus
Family Leptotypholpidae
Leptotyphlops macrorhynchus
Family Boidae
Eryx jaculus
Family Colubridae
Coluber ravergieri
Coluber rogersi
Coluber schmidti
Eirenis coronella
Eirenis rothi
Lytorhynchus diadema
Maipolon moilensis
Malpolon monspessulanus insignitus
Natrix tessellata
Psanunophis schokari
Rhynchocalamus melanocephalus
Spalerosophis diadema cliffordii
Family Elapidae
Walterinnesia aegyptia
Family Viperidae
Cerastes gasperetti
Pseudocerastes persicus fieldi

Appendix 2. List of mammals reported from the Badia

Order Insectivora
Family Erinaceidae
Paraechinus aethiopicus
Hemiechinus auritus
Family Soricidae
Crocidura suaveolens
Suncus etruscus
Order Chiroptera
Pipistrellus kuhli
Otonycteris hemprichi
Tadarida teniotis
Order Carnivora
Family Canidae
Canis lupus
Canis aureus
Vulpes rueppelli
Vulpes vulpes

Appendix 2 continued. List of mammals reported from the Badia

Family Felidae
- *Caracal caracal*
- *Felis silvestris*
- *Felis margarita*

Family Mustelidae
- *Meles meles*
- *Mellivora capensis*
- *Vormela peregusna*

Family Hyaenidae
- *Hyaena hyaena*

Order Artiodactyla

Family Bovidae
- *Gazella subgutturosa*

Order Lagomorpha

Family Leporidae
- *Lepus capensis*

Order Rodentia

Family Dipodidae
- *Allactaga euphratica*
- *Jaculus jaculus*

Family Gliridae
- *Eliomys melanurus*

Family Muridae
- *Acomys russatus lewisi*

Family Cricetidae
- *Cricetulus migratorius*

Family Gerbillidae
- *Gerbillus dasyurus*
- *Gerbillus cheesmani*
- *Gerbillus nanus*
- *Gerbillus henleyi*
- *Meriones libycus*
- *Meriones crassus*
- *Meriones tristrami*
- *Psammomys obesus*

References

Abu Baker M and Amr Z (2003a). A morphometric and taxonomic revision of the genus Gerbillus in Jordan with notes on its current distribution. Zoologische Abhandlungen (Dresden) 53:177-204.

Abu Baker M and Amr Z (2003b). Rodent diversity in the Northeastern Desert of Jordan, with special reference on the ecology of Gerbillus cheesmani. (Mammalia: Rodentia), Casopis Národního Muzea, Rada prirodovedná 172:141-152.

Abu Baker M, Qarqaz M and Amr Z (2005). Small mammal remains recovered from pellets of the Barn Owl (Tyto alba) at Shawmari Wildlife Reserve, eastern Jordan. Casopis Národního Muzea, Rada prirodovedná 174:127-129.

Abu Baker M, Rifai L, Joger U, Nagy Z, Wink M and Amr Z (2002). Occurrence of Coluber (Hierophis) schmidti Nikolsky, 1909 in Jordan (Squamata: Colubridae). Herpetozoa, Wien 15(1/2):29-36.

Al-Disi AM, (1997). Reptiles survey, October to December 1996. Amman: CORD/BRDP, RSCN p. 70 plus annexes.

Al-Melhim WN, Amr ZS, Disi AM and Katbeh-Bader A (1997). On the diet of the Little Owl, Athene noctua, from Safawi area, eastern Jordan. Zoology in the Middle East 15:19-28.

Allison RJ, Higgitt DL, Kirk AJ, Warburton J, Al-Homoud A, Sunna BF and White K (1998). Geology, geomorphology, hydrology, groundwater and physical resources. In: Dutton, RW, Clarke, JI and Battikhi, A (eds), Arid land resources and their management: Jordan's desert margin. London: Kegan Paul International pp. 21-46.

Amr Z, Abu Baker M and Qumsiyeh M. Bat diversity and conservation in Jordan. Turkish Journal of Zoology.

Amr Z and Quatrameez M (2002). Wildlife conservation in Jordan: a cultural and Islamic perspective. In: M, SV (ed), Heaven and earth and I: ethics of nature conservation in Asia: Menon. Penguin Enterprise pp. 172-184.

Amr ZS (2000a). Conservation of the carnivores of the northeastern Badia. World Conservation 2:24.

Amr ZS (2000b). Mammals of Jordan. Amman: United Nations Environment Programme.
Amr ZS, Hamidan N and Quatrameez M (2004). Nature conservation in Jordan. Denisia 14:467-477.
Bunaian F, Hatough A, Ababaneh D, Mashaqbeh S, Yuosef M and Amr ZS (2001). The carnivores of the northeastern Badia, Jordan. Turkish Journal of Zoology 25:19-25.
Bunaian F, Mashaqbeh S, Yousef M, Budairi A and Amr ZS (1998). A new record of the Sand cat, Felis margarita Loche, 1858 from Jordan. Zoology in the Middle East 16:5-7.
Disi AM (1993). A contribution to the knowledge of the herpetofauna of Jordan. V. New records of three colubrid snakes from Jordan: The Snake 25:109-113.
Disi AM, Modry D, Bunian F, Al-Oran R and Amr ZS (1999). Amphibians and reptiles of the Badia region of Jordan. Herpetozoa, Wien 12 (3-4):135-146.
Dutton RW and Brown K (1996). Darwin initiative for the protection of species project: Management of biodiversity in the Badia region, north-east Jordan. Annual report year one: CORD, University of Durham, BRDP report p. 12.
Higgitt DL and Allison RJ (1998). Characteristics of basalt boulder surfaces. In: Dutton RW, Clarke JI and Battikhi A (eds), Arid land resources and their management: Jordan's desert margin. London: Kegan Paul International pp. 171-182.
Kirk AJ (1998a). The effect of intensive irrigated agriculture upon soil degradation: a case study from Ashrafiyya. In: Dutton RW, Clarke JI and Battikhi A (eds), Arid land resources and their management: Jordan's desert margin. London: Kegan Paul International pp. 127-156.
Kirk AJ (1998b). A synthesis of climatic data with specific interest in the precipitation record. In: Dutton RW, Clarke JI and Battikhi A (eds), Arid land resources and their management: Jordan's desert margin, London: Kegan Paul International pp. 47-66.
Moravec J and Böhme W (1997). A new subspecies of the Mediterranean gecko, Hemidactylus turcicus from the Syrian lava desert (Squamata: Sauria: Gekkonidae). Herpetozoa, Wien 10 (3/4):121-128.
Parker HW (1935). A new melanic lizard from Transjordania, and some speculations concerning melanism. Proceedings of the Zoological Society of London 1935:137-142.
Qarqaz M, Abu Baker M and Amr ZS (2004). Status and ecology of the Striped Hyaena, Hyaena hyaena, in Jordan. Zoology in the Middle East 33:87-92.
Qumsiyeh MB, Amr ZS and Al-Oran RM (1998). Further records of bats from Jordan and a synopsis. Turkish Journal of Zoology 22:277-284.
Qumsiyeh MB, Amr ZS and Shafei D (1993). The status and conservation of carnivores in Jordan. Mammalia 57(1):55-62.
Rifai L, Al-Shafee D, Al-Melhim W and Amr ZS (1999). Status of the Marbled polecat, Vormela peregusna (Güldenstaedt, 1770), in Jordan. Zoology in the Middle East 17:5-8.
Rifai LB, Al-Melhim WN, Gharaibeh BM and Amr ZS (2000). The diet of the Desert Eagle Owl, Bubo bubo ascalaphus, in the eastern desert of Jordan. Journal of Arid environments 44(3):369-372.
Schmidt KP (1930). Reptiles of Marshall Field north Arabian desert expeditions 1927-1928. Field Mus. Nat. Hist., Zool. Ser 17:223-230.
Schmidt KP (1939). Reptiles and amphibians from Southwestern Asia. Field Mus. Nat. Hist., Zool. Ser 24:49-92.
Scott DM and Dunstone N (2000). Environmental determinants of the composition of desert-living rodent communities in the north-east Badia region of Jordan. Journal of Zoology 251:481-494.
Searight A (1987). Some records of mammals from north eastern Jordan. Proceedings of the symposium on the fauna and zoogeography of the Middle East pp. 311-317.
Warburton J (1998). A reconnaissance survey of surface water infiltration characteristics. In: Dutton RW, Clarke JI and Battikhi A (eds), Arid land resources and their man-

agement: Jordan's desert margin. London: Kegan Paul International pp. 157-170.
Werner YL (1971). Lizards and snakes from Transjordan, recently acquired by the British Museum (Natural History). Bulletin of the British Museum (Natural History), Zoology, London 21(6):213-256.
Werner YL (1991). Notable herpetofaunal records from Transjordan. Zoology in the Middle East 5:37-42.
Werner YL (1992). Identity and distribution of Agama stellio picea Parker (Sauria: Agamidae), endemic to the volcanic desert of Jordan. Zoology in the Middle East 6:41-44.
Wettstein O (1951). Ergebnisac dat Oster reichischen Iran Expedition 1994/1950, Sitzber. Oster reichischen Akad. Wiss., Abt. I Wien 160 (5):427-448.

CHAPTER 11

Parasitic fungi and seed plants associated with the natural vegetation in the Jordan Badia

K. M. Hameed
Department of Plant Production, Faculty of Agriculture, Jordan University of Science and Technology (JUST), P. O. Box 3030 Irbid 22110, Jordan. E-mail: hameed@just.edu.jo, kmhameed@hotmail.com

H. Z. Ghosheh
Department of Plant Production, Faculty of Agriculture, JUST

A. F.Al-Jamali,
Department of Plant Production, Faculty of Agriculture, JUST

H. K. Shannag
Department of Plant Production, Faculty of Agriculture, JUST

Z. Al-Shyab,
Department of Plant Production, Faculty of Agriculture, JUST

Broomrape (Halouk in Arabic) (*Orobanche* spp) and dodder (Hamoul in Arabic) (*Cuscuta* spp), in addition to the fungi associated with the root systems of the natural vegetation in the Jordan Badia, were investigated. Different locations of the Badia field study area, ie Safawi, Marab Suway'id, Marab Shubaika, Marab Salma-1, Marab Salma-2 and Tall Hassan were visited several times during the period 1999-2000, and 2000-2. Representative soil samples from each location were carefully collected from the root zones of the natural vegetation cover and bare ground. Soils of the rhizosphere, rhizoplane and bare ground were studied for their fungal population by serial dilution and plating on acidified potato dextrose agar (PDA) and malt agar (MA). Segments of plant roots were plated for recovery of intimately associated fungi. Those segments were also cleared and stained and examined for the presence of endophyte fungi and VA mycorrhizae. Soil samples were also processed for their VA fungal populations by extracting VA spores.

Visited locations were searched for the presence of Orobanche and Cuscuta. Studies related to the biology and control of Orobanche were also conducted. Fungal populations of the following values; 51×10^3, 91.4×10^3, 18.2×10^3, 1.9×10^3 and 206×10^3, 500×10^3, 25.6×10^3, 4.3×10^3 were recorded for the locations Marab Salma, Marab Suway'id, Marab Shubaika, and Tall Hassan, on acidified PDA and MA respectively. Several genera of soil-borne fungi were observed, but four *Penicillium* spp, *Puluria* spp, *Dactylospors* spp, two *Asperigillus* spp, and yeasts were the most abundant. Several *Actinomycetes* were also observed. It was revealed that the Badia soil is very rich in mycorrhizae fungi and that *Glomus* spp and *Aculospora* spp were the most abundant. Under the current circumstances, our field visits did not lead to any observations of *Orobanche*. However, this parasitic flowering plant is widely present in association with wild and cultivated vegetation towards Mafraq. Dodder was encountered parasitizing the above ground growth of *Prosopis glandulosa* at the south-eastern side of the valley leading to Marab Salma. Extended work on *Orobanche* substantiated the validity of olive pulp (the solid by-product from olive mills, or *jift*) as a bioherbicide against *Orobanche* and a wide range of weed plants. *Jift* application *in vitro* and *in vivo* proved to have a potential herbicidal effect against *Orobanche* seeds, broad leaf and grass weeds. Several soil-borne fungi were investigated as potential agents for the biological control of *Orobanche*.

Keywords: Jordan Badia, fungi, mycorrhizae, actinomycetes, Orobanche, endophytes

1. Introduction

This chapter summarizes results of the sub-project; 'Parasitic seed plants and fungi associated with natural vegetation in the Jordan Badia region' (Abd-El-Nasser et al 1973, Ali et al 1998, Hameed et al 1991), which fitted under the umbrella of the macro project, 'Biodiversity and resources in the Jordanian Badia' that was funded by the Jordanian Badia Research and Development Programme (BRDP). Results of continued exploration along aspects of parasitic seed plants plus pathogenic and beneficial fungi that are associated with root systems of wild plants are presented. This presentation is a synthesis of all findings and conclusions of the 2–3 year investigations. Severe drought characterized the first year (1999-2000) of the study in the Badia, but good rainfall encouraged the presence of better vegetation cover in the following seasons.

2. Work plan

During the years 1999-2002 many field trips were made to Safawi, Marab Suway'id, Marab Shubaika, Marab Salma and Tall Hassan. Soil samples from both bare land and around the plant root regions were collected, and taken to the laboratory at JUST. All of the collected soil samples were examined to verify the existing fungal population, Vesicular-Arbuscular (VA) Mycorrhizae fungi and *Orobanche* seeds. Extensive isolation from underground plant parts was conducted in search of pathogenic and non-pathogenic fungi intimately associated with the root systems of the existing vegetation cover within those locations. Scouting for the presence of the root parasitic flowering seed plant, *Orobanche*, and the stem parasitic flowering seed plant, *Cuscuta*, was performed. Further investigation on the biology and control of *Orobanche* under *in vivo* and *in vitro* conditions was conducted (Fayadh et al 1990).

3. Materials and methods

3.1. Field trip and soil sampling

Field trips conducted during the period 1999-2002 included several visits to the locations of Safawi, Marab Salma-1, Marab Salma-2, Marab Suway'id, Marab Shubaika and Tall Hassan, all of which are within the BRDP study area. The final field trip was made 14th-16th March, 2002. During this trip, the following three locations, Marab Salma, Marab Sway'id and Marab Shubaika, were visited to collect soil samples with and without plant root systems. Within each location soil sample composites were taken from root zones of the vegetation cover and bare land.

Soil samples from 8-10 spots per stop and 3-5 stops per location were taken from each site in order to ensure good coverage and complete representation of each location. From each spot 300-500 g of soil were collected from the sub-soil, after removing the top soil (3-5 cm deep), and placed inside properly labelled zip-locked polyethylene bags (Goodchild et al 1987). Vegetation cover and the plant root systems were dug out along with associated soil to ensure proper sampling of the roots and soil. All bags were kept in the shade and stored inside the cold room for further analyses.

3.2 Soil analyses for pH, EC, and total organic matter (TO)

Each soil sample was thoroughly mixed and then sub-sampled for pH, EC, and TO analyses. Each sample was analyzed in three replicates. Soil pH and electrical conductivity (EC) for each sample were determined by mixing equal amounts of soil and distilled water, 1:1, (V/V) ratio. The pH was measured with the aid of a pH meter by immersing the electrode inside the mixture with continuous stirring and the EC was measured using a conductivity meter.

Total organic matter in each soil sample was measured by combustion inside a muffle furnace. Exactly weighed soil samples inside porcelain crucibles were first de-moisturized and then combusted inside the muffle furnace to burn all organic matter in those samples. A total organic matter content of the samples was calculated as the weight lost by each sample after combustion.

3.3 Isolation of associated fungi

3.3.1. Isolation from soil samples

Isolation of fungi from soil and bathing solutions of roots was made by plating the soil solutions in a series of dilutions ranging from 1/10 to 1/10 000 (Abd-El-Nasser et al 1973). Two different microbial media, Acidified Potato Dextrose Agar (PDA) and Malt Agar (MA), were used for plating, except for the 2001-2 samples which were plated on acidified PDA only. Each soil sample was run in triplicate and repeated twice. Inoculated plates were incubated at 25 °C for three days and examined for growth of fungal colonies. Total fungal colonies were counted in each dilution. Fungal populations in the original 10 g of soil used were calculated on the basis of the average population counted at all dilutions, and recorded as fungal population per gram of soil. Culture plates were further incubated for a longer time to allow fungi to develop their characteristic reproductive, non-sexual and sexual structures. Identification of those fungi to genus and species was performed according to mycological manuals and keys available in the laboratory. The frequency of each fungus was calculated according to the following equation:

$$\text{Frequency} = \text{No. of colonies of the fungus} / \text{Total No. of Fungal colonies} \times 100 \qquad (1)$$

Stock cultures of the isolated fungi were stored on PDA slant culture tubes under cold conditions (4 °C), inside a refrigerator.

3.3.2. Isolation from plant roots

Isolation of fungi from roots involved plating of surface sterilized root segments 1 cm long on agar plates. Fungi were also isolated from the bathing solution of those roots, plated on fungal culture media and included within those considered as rhizosphere organisms (Assaf 2001).

3.3.3. Isolation and counting of mycorrhizae spores

Soil samples were also processed for the presence of the Vascular Arbuscular mycorrhiza (VA) spores. The modified Flotation-Adhesion method (Ianson et al 1986) was adopted for this purpose. According to this method, 10 g soil samples were mixed with 150 ml of water and vigorously shaken in order to break loose the spores from soil particles. The soil solution was placed inside a 500 ml separatory funnel and allowed to drip at a rate of 70-90 drops per minute. Floating mycorrhizae spores gradually adhered to the internal surface of the funnel. Adhered spores were washed down into a regular funnel with a filter paper. Mycorrhizae spores captured on the filter paper were counted under magnification. The VA spore population was determined as spores per gram of soil. The different VA fungal genera were identified according to the spore morphology. Short segments of root systems were fixed inside formalin-acetic acid (FAA) solution, cleared by boiling inside 10 % KOH and then stained with cotton-blue. Those segments were examined for the presence of mycorrhizae and other endophyte fungi (Assaf 2001).

Table 1. Measurements of soil pH, EC and OM of soil samples collected from the Jordan Badia during March 2002

Sample	Organic matter mg/g	EC mmohs/cm	pH
Location: Marab Suway'id			
S1	33	272	6.43
S2	57	268	5.95
S3	80	188	6.61
S4	31	192	5.91
S5	33	563	6.05
S6	35	208	6.04
S7	40	200	6.25
S8	57	180	6.06
S9	210	235	6.06
S10	52	215	6.15
S11	41	258	5.87
S12	65	356	5.78
S13	61	139	6.35
S14	48	541	6.34
S15	71	258	6.54
S16	64	317	6.15
S17	100	271	6.47
S18	55	241	6.28
S19	52	316	6.13
S20	43	228	6.19
S21	65	209	6.57
S22	129	292	6.23
S23	74	197	6.35
S24	23	282	6.25
S25	31	322	6.38
S26	53	340	6.05
S27	44	228	6.38
Average	61	271	6.22
Location: Marab Salma			
SL1	69	227	5.95
SL2	85	270	6.42
SL3	92	260	6.38
SL4	165	254	6.08
SL5	44	286	6.14
SL6	31	248	6.37
SL7	79	308	6.14
SL8	105	352	6.16
SL9	58	289	6.24
SL10	102	271	6.38
SL11	103	382	6.23
Average	85	275	6.3
Location: Marab Shubaika			
SH1	76	243	6.25
SH2	82	656	5.98
SH3	79	439	6.29
SH4	74	291	6.05
SH5	84	447	5.98
SH6	57	565	6.31
SH7	20	325	6.09
SH8	63	337	6.40
SH9	86	252	6.00
SH10	110	281	6.55
SH11	55	318	6.32
SH12	77	200	6.53
SH13	57	263	6.30
SH14	25	402	6.42
SH15	59	365	6.34
SH16	23	288	6.36
SH17	60	319	6.00
SH18	67	200	6.02
SH19	122	227	6.38
SH20	100	322	6.42
SH21	112	878	6.14
Average	71	363	6.2

Table 2. Isolation of fungi from soils using Acidified Potato Dextrose Agar (PDA)

Sample no.	Fungal population on acidified PDA x 1000 per g soil			
	Peni-cillium spp1	*Peni-cillium* spp2	*Asper-gillus* spp1	Total
Location: Marab Suway'id				
S1	1000	0	0	1000
S2	400	0	0	400
S3	9500	0	3015	12515
S4	1850	500	0	2350
S5	2035	0	0	2035
S6	1021	2000	0	3021
S7	2022	0	0	2022
S8	30000	0	0	30000
S9	1215	535	0	1750
S10	415	450	0	865
S11	511	311	0	822
S12	0	0	0	0
S13	223	255	0	478
S14	0	0	0	0
S15	715	330	145	1190
S16	751	321	0	1072
S17	572	410	0	982
S18	6950	11365	0	18315
S19	750	635	0	1385
S20	0	0	0	0
S21	300	0	0	300
S22	435000	20000	0	455000
S23	0	10	0	10
S24	250	1234	521	2005
S25	642	754	0	1396
S26	244	128	0	372
S27	270	1070	0	1340
Location: Marab Salma				
SL1	9300	265	0	9565
SL2	0	10535	0	10535
SL3	850	210	0	1060
SL4	0	0	0	0
SL5	2650	0	0	2650
SL6	556	874	0	1430
SL7	2215	450	0	2665
SL8	3765	1050	0	4815
SL9	865	3080	0	3945
SL10	1800	235	0	2035
SL11	0	0	0	0
Location: Marab Shubaika				
SH1	10380	0	0	10380
SH2	254	0	0	254
SH3	331500	165000	0	496500
SH4	165	0	0	165
SH5	380	0	0	380
SH6	765	1015	0	1780
SH7	0	0	0	0
SH8	2415	1050	0	3465
SH9	585	0	0	585
SH10	585	0	0	585
SH11	131500	195000	0	326500
SH12	435000	80000	0	515000
SH13	0	0	0	0
SH14	737	0	0	737
SH15	0	535	90	625
SH16	0	0	0	0
SH17	684	1000	0	1684
SH18	0	0	0	0
SH19	395	0	215	610
SH20	0	0	0	0
SH21	30	0	0	30

Fungal population is recorded as number of propagules per gram of soil, for the period 2001-2002

4. Results and discussion

4.1. Soil pH and electrical conductivity (EC) and total organic matter (OM)

Measurements of pH, EC and OM for the soil samples collected from different locations revealed variations among locations (Table 1). Those results indicated that collected soil samples were slightly acidic (pH 6) with low to moderate salt content (271-363 mmohs/cm). The relatively high organic matter content (61-84 mg/g) could be attributed to the presence of good vegetative cover on the date and at the place from where the samples were collected.

There was a dense vegetation cover of broad leaf plants and grasses under which root systems were not easy to separate or identify for each plant species. Therefore isolation from plant roots was made on a random sample of root conglomerates. Roots were first surface-sterilized by immersing them inside 10 % commercial bleach solution containing 5.25 Na-hypochlorate for three minutes and then plated on mycological media (Abd-El-Nasser et al 1973).

4.2. Isolation of fungi: common soil inhabitant fungi

Results of the isolation of fungi from soil revealed that the Badia soils become very rich in fungal populations following winter and spring rainfalls (Table 2) compared to previous years where rainfall was below average (Tables 3 and 4).

The most prevalent fungi are those belonging to the class Ascomycetes, within the order Eurotiales (Arshad et al 1998). This order includes the fungi that are known for their imperfect stage or the non-sexual phase; the so-called conidial phase. These fungi are represented by genera such as Penicillium and Aspergillus. These and other similar fungi are commonly grouped within the form-class Fungi Imperfecti, as they are classified on the basis of their non-sexual phase. Some other imperfect fungi were also recovered, (Tables 2, 3, 4) (Abd-El-Nasser et al 1973). Fungal biodiversity seemed to be less than in previous years. This may be attributed to the prevalence of highly sporulating

Table 3. Isolation of fungi from soils using acidified potato dextrose agar (PDA)

Soil sample	Fungal population on acidified PDA, X 1000 per g soil											
From marab	P1	P2	P3	P4	PU	DA	ACR	A1	A2	Y	Total	
Salma	1	11.2	30.1	0	0	0	0	0	0.08	0	110	151
Suwyid	2	122	0	0	0	0	0	0	0	0	0	122
	3	58.4	0.85	0	0	0.09	0	0	0.08	0	1.3	60.7
Salma	4	0.09	1.63	0	0	0	0	0	0	0	0	1.71
	5	13.5	0	0	1.11	0	2.7	0	0.93	2.02	0	20.3
	6	3.78	3.83	0	0	0	0	0.18	0	0	3.7	11.5
	7	3.35	0.8	0	0	0.07	0	0	4.76	0	42	50.9
Shubaika	8	0	0	0	1.9	0	0	0	0	0	0	1.9
Total		212	37.2	0	3.01	0.16	2.7	0.18	5.85	2.02	157	

Fungal population is recorded as number of propagules per gram of soil (1999-2000). Total values are presented to closest decimal point. P1, *Penicillium sp1;* P2, *Penicillium sp2;* P3, *Penicillium sp3;* P4, *Penicillium sp4;* PU, *Pullularia sp;* DA, *Dactylospora sp;* ACR, *Acremonia sp;* A1, *Aspergillus sp1;* A2, *Aspergillus sp2;* Y, *Yeast.*

fungi on the one hand and the scarcity of other fungi isolated in previous years (Assaf 2001). Also it could be recognized here that specific media could have revealed the presence of those fungi.

4.3. Vascular arbuscular mycorrhiza

4.3.1. Isolation of spores from soil

The results of soil washing and application of the flotation-adhesion technique revealed the presence of high populations of mycorrhizal fungi (Al-Karaki et al 1999, Cui et al 1992). This is a second documentation of VA mycorrhizae, by our team, in the Jordan Badia for the project. These fungi flourished as the vegetation cover increased following the rainy season of winter 2001-2 which came after two years of severe drought. The mycorrhizae were recovered from soil adjacent to roots of plants such as al-Sheeh (*Artimizia herba alba* L.), Harmala (*Piganum harmala* L), and many other small broad leaf plants and grasses of the natural vegetation (Collier et al 2003). The mycorrhizal fungi increased the depletion zone, around the root, of water and mineral nutrients mainly phosphorus; they extract water and minerals from a wider and deeper area around the plant root system (Al-Karaki et al 1999, Bell et al 2003, Cui et al 1992, Jansa et al 2003). These fungi could be selected and reared on their proper host under controlled conditions to be produced in large quantities (Al-Karaki et al 1999, Munro et al 1999). Such VA mycorrhizal fungi could be used as inoculum in supporting newly introduced plants under Badia conditions in future re-vegetation plans. These fungi are a natural resource of the endogenous genes of this old habitat (Ianson et al 1986). Tables 5 and 6 show the population of the VAM spores in soil samples collected during the 1999-2000, 2000-1 and 2002 seasons. Notice that the genus *Glomus* was abundant and may be represented by several of its species. This genus needs to be further cultured under controlled condition on proper host plants, and possibly utilized as a useful inoculum for establishing mycorrhizae. The second genus recovered was *Aculospora*. This fungus appears to be an equally significant mycorrhizal fungus that needs to be considered for further investigation (Assaf 2001).

Table 4. Isolation of fungi from soils using Malt Agar (MA)

Soil sample		Fungal population on acidified MA, X 1000 per g soil										
From marab		P1	P2	P3	P4	PU	DA	ACR	A1	A2	y	Total
Salma	1	79	14.5	0	0	0	0	0	0	0	112	206
Suwyid	2	0	114	0	0	0	0	0	0	0	0	114
	3	33.7	0.5	0	0	0.08	0	0.9	850	0	1.3	886
Salma	4	2.43	0	0	0	0	0	0	0	0	0	2.43
	5	28.9	0	0	0.86	0	0	0	0.77	2.1	0	32.6
	6	0	10	0	0	0	0	1	0	0	3	14
	7	2.92	0	0	0	0	0	0.06	1	0	49.6	53.6
Shubaika	8	1	1	2.16	0.09	0	0.09	0	0	0	0	4.34
Total		148	140	2.16	0.95	0.08	0.09	1.14	852	2.1	166	

Fungal population is recorded as number of propagules per gram of soil (1999-2000). Total values are presented to closest decimal point. P1, *Penicillium spp1;* P2, *Penicillium spp2;* P3, *Penicillium spp3;* P4, *Penicillium spp4;* PU, *Pullularia spp;* DA, *Dactylospora spp;* ACR, *Acremonia spp;* A1, *Aspergillus spp1;* A2, *Aspergillus spp2;* Y, *Yeast.*

4.4. Direct observations on the endophyte fungi

Endophytes are organisms that grow within living plants but not necessarily in the living cells, deriving nutrients without been noticed and not detrimental to the host plant (Assaf 2001). Direct examination of cleared plant roots revealed the presence of fungal hyphae that permeate plant-root tissues. Fungi of the mycorrhizae and other fungi are capable of establishing this biological-physical relation with plant roots (Ali et al 1998, Arnold et al 2003). This is expected under stress conditions as a means of self-defense and accommodation to conditions. Present results revealed this phenomenon in the Jordan Badia. In Tables 7-9 the isolation of fungi from those roots which was done by direct plating on Acidified Potato Dextrose Agar (PDA) revealed this phenomenon. All fungal genera and species were identified and recorded on the bases of their appearance on those root segments.

Fungal hyphae were present inside the root tissues of 20 out of the 35 plant samples where roots were included (Table 7) indicating the presence of endophyte fungi and supporting the idea mentioned above.

4.5. Other parasitic flowering plants

4.5.1. Dodder (Cuscuta spp) in the Badia

Dodder is an important parasitic flowering plant that attaches to the stems and foliage of the green plant host (Fayadh et al 1990). This parasitic weed belongs to the family Cuscutidacae. It grows from seeds germinating in soil, putting out a slender shoot which grows upright and is characterized by an oscillating movement along its longitudinal axis. In this way the shoot is searching for a substratum, which should be a stem or any other vegetative part of its future host plant. Once it touches that plant, it starts winding around it and piercing and forcing into its tissue tiny pegs known as haustoria. Each haustorium is an outgrowth of the parasite into the cortex of the stem of the host reaching to the vascular bundle of that host. Those haustoria uptake water, minerals and organic compounds (carbohydrates, proteins, and other growth nutrients) produced by the host plant and needed by the parasite. Once the parasite is in successful contact with its host it no longer retains contact with the soil. Dodder depends on its host for nutrients because it is devoid of chlorophyll and therefore cannot photosynthesize. Dodder has a slender and very tender yellow stem which winds around its host plant in all directions. The advanced stage of infection makes the host plant look like a tumbled mass of winding yellow threads similar to yarn. In the final stage the host plant becomes exhausted and it collapses. Dodder (*Cuscuta campestris*) has been previously reported only one time in the Badia, and a specimen was deposited in the herbarium at the Badia Centre - Safawi. During the season 2000-2, dodder was discovered along the side of the valley leading from Marab Suway'id to Marab Salma. The parasite invaded *Prosopis glandulosa* over a wide area in which this plant is predominant; it is considered to be the main staple food of camels. The parasite reduced the population and the vigor of *P. glandulosa*. The seeds of dodder may also get disseminated by camels once they consume plants infected by the parasite; the seeds can survive passing through the digestive tract of camels and are able to germinate at new sites wherever the camel's dung is deposited. Hence, the parasite may infest new areas and more plants. This discovery opened avenues into basic and applied research in relation to the significance and importance of this parasitic weed in the Jordan Badia.

4.5.2. Broomrape (Orobanche spp) in the Badia

Work on basic and applied research relevant to the biology and control of *Orobanche* was undertaken. *Orobanche* was not encountered during the present investigation, but it was previously discovered and documented by a herbarium specimen of *Orobanche ramosa* in the herbarium of the Badia Centre at Safawi. This parasitic seed plant, however, is widely distributed in territories stretching from Safawi towards Mafraq. Based upon the potential significance of *Orobanche*, the biology and control of this problem was addressed and it was established that olive pulp (*jift*) can act as a bioherbicide against Orobanche, and several other broad leaf and grass weeds. *Jift* was shown to be a very effective bioherbicide in fields of tolerant crops such as faba beans (*Vicia faba*) (Bedi 1994, Garcia Torres et al 1994a, Garcia Torres et al 1994b, Hameed et al 2001). The potential biological control of *Orobanche* by fungi isolated from soil in Jordan was also investigated (Hameed et al 1999, 2000, Hameed et al 2001, Kroschel et al 1995).

Table 5. Isolation of VAM spores from soil samples collected from the Jordan Badia, 2001-2

no	Spores of VAM in 1 gm soil		Total	Frequency %	
	GL	AC		GL.	AC
Location: Marab Suway'id					
S1	32	1	33	97	3
S2	51	0	51	100	0
S3	56	6	62	90	10
S4	52	0	52	100	0
S5	15	0	15	100	0
S6	12	2	14	12	2
S7	30	0	30	100	0
S8	121	0	121	100	0
S9	29	0	29	100	0
S10	1	0	1	100	0
S11	32	0	32	100	0
S12	21	0	21	100	0
S13	150	0	150	100	0
S14	15	2	17	88	12
S15	41	0	41	100	0
S16	31	0	31	100	0
S17	14	0	14	100	0
S18	76	2	78	96	4
S19	20	4	24	83	17
S20	4	0	4	100	0
S21	36	3	39	92	8
S22	56	2	58	97	3
S23	32	0	32	100	0
S24	6	4	10	60	40
S25	33	2	35	94	6
S26	210	7	217	97	3
S27	16	1	17	94	6
Location: Marab Salma					
SL1	120	0	120	100	0
SL2	100	32	132	76	24
SL3	96	0	96	100	0
SL4	92	0	92	10	0
SL5	120	0	120	100	0
SL6	32	0	32	10	0
SL7	46	0	46	100	0
SL8	24	0	24	100	0
SL9	4	0	4	100	0
SL10	58	0	58	100	0
SL11	8	0	8	100	0
Location: Marab Shubaika					
SH1	16	0	16	100	0
SH2	27	0	27	100	0
SH3	4	0	4	100	0
SH4	40	0	40	100	0
SH5	28	0	28	100	0
SH6	57	0	57	100	0
SH7	116	7	123	94	6
SH8	9	0	9	100	0
SH9	25	0	25	100	0
SH10	24	3	27	89	11
SH11	12	0	12	100	0
SH12	16	0	16	100	0
SH13	38	15	53	88	12
SH14	101	0	101	10	0
SH15	122	0	122	100	0
SH16	12	0	12	100	0
SH17	22	2	24	92	8
SH18	44	2	46	96	4
SH19	10	0	10	100	0
SH20	10	0	10	100	0

GL = *Glomus* spp AC = *Acaulspora* spp

Table 6. Isolation of VAM spores from soil samples collected from the Badia, 1999-2000

No.	Spores of VAM in 1 gm soil			Frequency %	
	GL	*AC*	Total	*GL*	*AC*
		Marab Suway'id			
S1	74	2	76	97.4	2.6
S2	118	0	118	100	0
S3	170	1	171	99.4	0.6
S4	80	0	80	100	0
S5	23	0	23	100	0
S6	14	0	14	100	0
S7	20	0	20	100	0
S8	49	2	51	96	4
S9	106	0	106	100	0
S10	5	0	5	100	0
S11	1	0	1	100	0
S12	45	0	45	100	0
S13	16	0	16	100	0
S14	31	0	31	100	0
S15	9	0	9	100	0
S16	28	1	29	96.5	3.5
		Salma, beside road			
SL1	68	0	68	100	0
SL2	28	0	28	100	0
SL3	272	3	275	99	1
SL4	36	2	38	94.7	5.3
SL5	9	0	9	100	0
SL6	14	0	14	100	0
SL7	6	1	7	85.7	14.3
SL8	39	0	39	100	0
		Marab Salma			
SM1	37	0	37	100	0
SM2	152	0	152	100	0
SM3	27	0	27	100	0
SM4	39	1	40	97.5	2.5
SM5	180	0	180	100	0
SM6	134	0	134	100	0
SM7	15	0	15	100	0
SM8	26	0	26	100	0
		Marab Shubaika			
SH1	159	4	163	97.5	2.5
SH2	16	0	16	100	0
SH3	3	0	3	100	0
SH4	36	2	38	94.7	4.3
SH5	66	1	67	98.5	1.5

GL = *Glomus* spp AC = *Acaulspora* spp

Table 7. The presence of fungal hyphae within cleared root samples for the season of 2000-2

Plant root sample no.	Fungal hyphae present (+) or absent (-)
Marab Suway'id	
S1	+
S2	+
S3	+
S4	-
S5	+
S6	+
S7	-
S8	+
S9	+
S10	-
S11	NO ROOT
S12	NO ROOT
S13	NO ROOT
S14	-
S15	+
S16	-
S17	+
S18	NO ROOT
S19	NO ROOT
S20	-
S21	+
S22	NO ROOT
S23	NO ROOT
S24	NO ROOT
S25	NO ROOT
S26	-
S27	NO ROOT
Marab Salma	
SL1	NO ROOT
SL2	NO ROOT
SL3	+
SL4	+
SL5	NO ROOT
SL6	NO ROOT
SL7	NO ROOT
SL8	+
SL9	NO ROOT
SL10	NO ROOT
SL11	-
Marab Shubaika	
SH1	+
SH2	-
SH3	NO ROOT
SH4	+
SH5	+
SH6	+
SH7	NO ROOT
SH8	-
SH9	-
SH10	NO ROOT
SH11	-
SH12	NO ROOT
SH13	-
SH14	-
SH15	NO ROOT
SH16	NO ROOT
SH17	+
SH18	-
SH19	+
SH20	+

Table 8. Fungi isolated from roots of plants collected in the designated locations in the Jordan Badia for the season of 2000-2

Plant root sample no.	Fungal species
	Marab Suway'id
S1	*Fusarium oxysporum.*
S2	*Fusarium oxysporum.* .
S3	*Spicuria* spp, *Fusarium oxysporum.*
S4	*Fusarium oxysporum.*
S5	*Fusarium oxysporum.*
S6	*Spicuria* spp, *Fusarium oxysporum.*
S7	*Fusarium oxysporum., Dactylosporium* spp, *Alternaria solani.*
S8	*Fusarium oxysporum.*
S9	*Fusarium oxysporum., Dactylosporium* spp, *Alternaria solani.*
S11	*Fusarium oxysporum., Dactylosporium* spp, *Spicuria* spp
S12	*Fusarium oxysporum.*
S13	*Fusarium oxysporum.*
S14	*Spicuria* spp, *Fusarium* spp
S15	*Alternaria soalni., Fusarium oxysporum.*
S16	*Cephalosporium* spp
S17	*Fusarium oxysporum.*
S18	*Spicuria* spp, *Fusarium* spp
S19	*Fusarium oxysporum., Dactylosporium* spp
S20	*Trichoderma sp., Dactylosporium* spp, *Fusarium* spp
S21	*Alternaria solani., Fusarium oxysporum.*
S27	*Helminthosporium* spp, *Fusarium oxysporum.*
	Marab Salma
SL4	*Penecillum* spp
SL8	*Dactylosporium* spp, *Alternaria* spp, *Helminthosporium* spp
	Marab Shubaika
SH1	*Helminthosporium* spp, *Fusarium* sp., *Rhizoctonia*
SH4	*Alternaria* spp, *Fusarium oxysporum., Sicuria* spp
SH5	*Helminthosporium* spp, *Fusarium* spp
SH6	*Fusarium oxysporum., Rhizoctonia* spp
SH8	*Fusarium oxysporum.*
SH11	*Spicoria* spp, *Fusarium* spp
SH13	*Alternaria solani, Fusarium oxysporum., Sicuria* spp
SH14	*Fusarium oxysporum.*
SH16	*Dactylosporium* spp
SH17	*Fusarium oxysporum., Rhizoctonia* spp
SH19	*Dactylosporium* spp, *Fusarium oxysporum*

Table 9. Fungi isolated from roots of plants collected in the designated locations in the Jordan Badia for the season of 1999-2000

Plant root sample no.	Fungal species
Marab Suway'id	
S1	*Helminthosporium* spp
S2	*Hormiscium* spp
S3	*Botyritis* spp, *Dactylosporium* spp
S4	-
S5	*Basidomycets* spp, *Rhizoctonia* spp
S6	*Botyritis* sp
S7	*Trichoderma* spp, *Basidomycets* spp
S8	-
S9	*Botyritis* spp, *Basidomycets* spp
S10	*Dactylosporium* spp, *Fusarium* spp 1
S11	*Dactylosporium* spp, *Fusarium* spp 1
S12	*Fusarium* spp 1
S13	*Fusarium* spp 1
S14	-
S15	*Helminthosporium* spp
S16	*Fusarium* spp 2
Salma beside road	
SL1	*Rhizoctonia* spp, *Helminthosporium* spp, *Cephalosporium* spp
SL2	-
SL3	*Trichoderma* spp, *Dactylosporium* spp, *Plerothecium* spp
SL4	-
SL5	*Rhizoctonia* spp
SL6	*Rhizoctonia* spp, *Fusarium* spp 1
SL7	*Botyritis* spp, *Helminthosporium* spp, *Fusarium* spp 2
SL8	*Dactylosporium* spp, *Fusarium* spp 1
Marab Salma	
SM1	-
SM2	-
SM3	*Fusarium* spp 2
SM4	*Trichoderma* spp, *Helminthosporium* spp
SM5	*Dactylosporium* spp
SM6	*Botyritis* spp, *Dactylosporium* spp, *Penicillium* spp
SM7	-
SM8	-
Marab Shubaika	
SH1	*Fusarium* spp 1, *Dactylosporium* spp
SH2	-
SH3	-
SH4	*Fusarium* spp 2
SH5	-

Acknowledgements

Members of this research group wish to express their gratitude to BRDP for funding this project, and to the Deanship of research at JUST for handling the finances.

References

Abd-El-Nasser M and Moawad H (1973). Occurrence of certain autotrophic and heterotrophic soil microorganisms in the rhizosphere and rhizosplane of different plants. Zbl. Bakt.II 128:397-404.

Al-Karaki G and Clark R (1999). Varied rates of mycorrhizal inoculum on growth and nutrient acquisition by barley grown with drought stress. J. Plant Nutr 22(11):1775-1784.

Ali MA, El-Saeedy MT, Moawad H and Abou-Shanab RAI (1998). Host range diversity and impact of vegetation on the desert truffles of Egypt. J. Union Arab Biol 5(B): 43-57.

Arnold A, Mejia L, Kyllo D, Rojas E, Maynard Z, Robbins N and Herre E (2003). Fungal endophytes limit pathogen damage in a tropical tree. Proceedings Of the National Academy Of Sciences of the USA 100(26):15649-15654.

Arshad M and Frankenberger W (1998). Plant growth-regulating substances in the rhizosphere: microbial production and functions. Advances in Agronomy 62(62):45-151.

Assaf TA (2001). Effect of soil amendment with olive pomace (jift) under various soil treatments on productivity of three legumes and associated mycorrhiza. MSc thesis, Faculty of Agriculture, Jordan University of Science and technology 131

Bedi JS (1994). Further studies on control of sunflower broomrape with Fusarium oxysporum f. sp. orthoceras – a potential mycoherbicide. In: Peterse, AH, Verkleig, JAC and ter Borg, SJ (eds), Proceedings of the 3rd International Workshop on Orobanche and Related Striga Research, Amsterdam, The Netherlands: Royal Tropical Institute pp. 539-544.

Bell J, Wells S, Jasper D and Abbott L (2003). Field inoculation with arbuscular mycorrhizal fungi in rehabilitation of mine sites with native vegetation, including Acacia spp. Australian Systematic Botany 16(1):131-138.

Collier SC, Yarnes CT and Herman RP (2003). Mycorrhizal dependency of Chihuahuan Desert plants is influenced by life history strategy and root morphology. J. of Arid Environments 55(2):223-229.

Cui M and Noble PL (1992). Nutrient status, water-uptake and gas exchange for 3 desert succulents infected with mycorrhizal fungi. New Phytologist 122(4):463-469.

Fayadh AH, Hameed KM and Al-ani HA (1990). Dodder (Cuscuta campestris Yunck) blight caused by Alternaria alternata (Fr.) Keissler and Geotricum candidum Link ex. Pres. Arab Journal of Plant Protection 8:55-59.

Garcia Torres L, Castejon-Munoz M and Lopez-Granados F (1994a). The problem of Orobanche and its management in Spain. In: Peterse, AH, Verkleig, JAC and ter Borg, SJ (eds), Proceedings of the 3rd International Workshop on Orobanche and Related Striga Research, Amsterdam, The Netherlands: Royal Tropical Institute pp. 623-627.

Garcia Torres L, Lopez-Granados F and Castejon-Munoz M (1994b). Pre-emergence herbicides for the control of broomrape (Orobanche cernua Loefl.) in sunflower (Helianthus annus L). Weed Research 34:395-402.

Goodchild DJ and Myers LF (1987). Rhizosheaths – a neglected phenomenon in Australian agriculture. Australian Journal of Agricultural Research 38(3):559-563.

Hameed KM and Foy CL (1991). Observations on the primary haustorium formation by germinated Orbanche ramosa seeds in relation to suscept and non-suspect plant. In: Ransom, JK, Musselman, LT, Worsham, AD and Parker, C (eds), Proceeding of the Fifth Internation Symposium on Parasitic Weeds, Nairobi, Kenya pp. 36-42.

Hameed KM and Foy CL (1999). Inhibition of Orobanche seed germination and radicle growth by aqueous extract of olive pomace in vitro. 53rd Annual Meeting of the Southern Weed Science Society of America, Jan. 24-26, Tulsa, Oklahoma, USA.

Hameed KM and Foy CL (2000). Potential utilization of Olive (Olea euoropaea) pomace (Jift) from oil mills as a bioherbicide. In: Amerecanos, PG, Abu-Irmaileh, BE and Saghir, AR (eds), Improved weed management in the Near East Cairo, Egypt.

Hameed KM, Saadoun IM and Al-Shyab Z (2001). Potential biological control of Orobanche by fungi islated from diseased specimens in Jordan. Plant Pathology Journal 17 (4):257-263.

Ianson DC and Allen MF (1986). The effects of soil texture on extraction of vasicular-arbuscular mycorrhizal fungal spores from arid sites. Mycologia 78(2):164-168.

Jansa J, Mozafar A and Frossard E (2003). Long-distance transport of P and Zn through the hyphae of an arbuscular mycorrhizal fungus in symbiosis with maize. AGRONOMIE 23(5-6):481-488.

Kroschel J and Klien O (1995). Biological control of Orobanche spp with Phytomyza orobanchia Kalt., a review. In: Kroschel, J, Abderabihi, J and Betz, H (eds), Advances in parasitic weed control at onfarm level. Vol. II. Joint action to control Orobanche in the WANA region, Weikersheim, Germany: Margrat Verlag pp. 135-159.

Munro RC, Wilson J, Jefwa J and Mbuthia KW (1999). A low-cost method of mycorrhizal inoculation improves growth of Acacia tortilis seedlings in the nursery. For. Ecol. Manage 113(1):51-56.

CHAPTER 12

Diversity of soil bacteria in the Badia region: studies of the genera Bacillus, Pseudomonas *and* Streptomyces

Ismail Saadoun
Department of Applied Biological Sciences, Jordan University of Science and Technology (JUST), Irbid-22110, Jordan. E-mail: isaadoun@just.edu.jo

Fouad Al-Momani
Department of Applied Biological Sciences, JUST

Ahmad Al-Kofahi
Department of Medicinal Chemistry & Pharmacognosy, JUST

In this study, enumeration of bacteria and assessment of microbial diversity have been conducted for different soils collected from the Badia region in north-east Jordan. Aerobic bacterial populations of cultivated and uncultivated soils showed total viable counts that ranged between 2.84 x 10^7 and 38.9 x 10^7 CFU/g and with average counts of 25.52 x 10^7 and 20.60 x 10^7 CFU/g of dry soil, respectively. Uncultivated soils showed counts that ranged between 13.44 x 10^7 and 35.83 x 10^7 CFU/g with an average count of 20.60 x 10^7 CFU/g. However, counts in the control soils ranged between 0.85 x 10^7 and 22.92 x 10^7 CFU/g and with average counts of 8.17 x 10^7 CFU/g of dry soil. Two to five different colony types of bacterial strains were recovered on the agar plates and their phenotypic identification revealed the dominance of members of the genus *Bacillus*, represented by the following species: *B. cereus, B. subtilis, B. megaterium, B. lentimorbus* and *B. azotoformans*. Other bacterial genera were also identified, namely: *Streptomyces, Corynebacterium, Micrococcus, Erysiplothrix, Listeria, Acinetobacter, Bordetella* and *Proteus*. Data indicated that the bacterial count of the control soils is 68 % less than that of cultivated soils and 60 % less than that of uncultivated soils. Prevalence of members of the genus *Bacillus* and their wide distribution in soils of the Badia could indicate their potential to withstand unfavourable or harsh conditions as they harbour the endospores that are resistant to inactivation by heat and desiccation.

Sixty different *Streptomyces* isolates were recovered from 12 soil samples which were then characterized by conventional methods and assessed for their antagonistic activity against three pathogenic and antibiotic-resistant bacteria: *Escherichia coli, Klebsiella pneumoniae* and *Staphylococcus aureus*. Results indicated that gray and white colour series were the most abundant with 30 % of all isolates active against, at least, one of the test pathogenic organisms. Screening results revealed the characterization of four *Streptomyces* isolates (I_7, AC_{32}, Bb_{36} and Z_{11}) which inhibited the growth of all tested pathogens. The unusual antibiotic profile of these isolates stressed their potential as a source of novel antibiotics. These active antibiotic-producing *Streptomyces* isolates were screened for their plasmid content and data revealed no observation of low molecular weight plasmids suggesting that antibiotic production in these strains is likely to be chromosomally-encoded DNA.

Potential hydrocarbon degrader bacteria were isolated from soil samples that had been exposed to crude petroleum oil spills. Aerobic bacterial populations of these polluted soils showed total viable counts ranging between 9.5 x 10^5 and 237.5 x 10^5 CFU/g soil with two different colony types of bacterial strains which had been recovered on the agar plates. Results indicated that longer-aged contamination exhibited a greater number of microorganisms. Phenotypic examination of the recovered bacteria revealed that they belong mainly to the genus *Pseudomonas*. Physical appearance was used as an indication of the ability of these bacteria to grow on diesel.

The low nutrient levels and harsh environmental conditions in the Badia region explain the low soil bacterial count and diversity and could trigger or favour unusual metabolite biosynthesis by the isolated bacteria.

Keywords: Soil bacteria, *Bacillus, Pseudomonas, Streptomyces*

1. Introduction – general background

Jordan like the other countries that are concerned with the problem of land aridization decided, through the Badia Research and Development Programme (BRDP), to study the diversity of bacteria in Badia soils. The Badia falls almost exclusively in the arid climate zone in east Jordan, and covers about 80 % of the country, extending from north to south in eastern Jordan, and totalling 72 660 km^2 of Jordan's total land area of 89 840 km^2. The Badia is part of the Syrian desert and lies on the north-western edge of the Arabian desert.

The natural vegetation cover, though typically sparse, plays an important conservation role. Most farm animals in Jordan (61 %) are located in the Badia and livestock production includes sheep (74 % of the national herd), goats, cattle (where they can be protected from the climatic extremes) and camels. The limited resources of moisture and soil nutrients play a big role in determining the concentration of soil microorganisms and the survival of plants in that environment. The interaction between plants, animals and soil microorganisms is also important. Vast numbers of organisms populate soil. By far the greater proportion of these is composed of bacteria, fungi, actinomycetes and algae, which form the microflora. Innumerable genera and species of bacteria and actinomycetes can be found in almost any soil sample and the relative proportion of these microorganisms is influenced to some extent by the environment. The microorganisms play an important role in organic matter decomposition and mineral recycling, making them available to plants and, as a result, to animals.

The microbial diversity of Jordan's Badia soils has not been studied before. Adaptation of these living organisms to the harsh conditions of the desert environment of the Badia could potentially lead them to produce novel secondary metabolites. Thus, expansion of screening of these valuable organisms may lead to important discoveries.

Several studies have been conducted on the soil microflora of different parts of Jordan (Saadoun et al 1999a, Saadoun et al 1996, 1997, Saadoun et al 1998), however, the Badia is the least studied area. This justifies the present study which focuses on soil bacterial flora (nb the genus *Bacillus*) in the Badia. Their distribution and morphological and cultural characters were investigated.

A second focus was on bacteria (nb *Pseudomonas* spp) that can obtain nutrition from petroleum fuel spills resulting from pipeline ruptures, tank failures and various production storage and transportation accidents. They are considered to be the most frequent organic pollutants of soil and ground water (Bossert et al 1984, Margesin et al 1997) and are classified as hazardous waste (Bartha et al 1984).

The ability of different bacterial isolates to degrade diesel fuel was also investigated. Diesel was tested since it is one of the fuel oils used in terrestrial situations and is a complex hydrocarbon product (Wang et al 1990)

The third focus was on *Streptomyces* species, which are widely distributed in both terrestrial and aquatic environments. Their distribution and predominance depend mainly on several factors such as nutrient availability, temperature, pH, moisture, soil type, season and climate (Goodfellow et al 1987). Several studies have already been conducted on the soil *Streptomyces* of Jordan (Saadoun et al 1999a, Saadoun et al 1996, 1997, Saadoun et al 1998).

Actinomycetes and particularly *Streptomyces* synthesize about two-thirds of currently known antibiotically active substances belonging to a great variety of chemical groups (Crandall et al 1986). It is therefore understandable that their economic importance has

Figure 1. The BRDC field area and the six collection sites

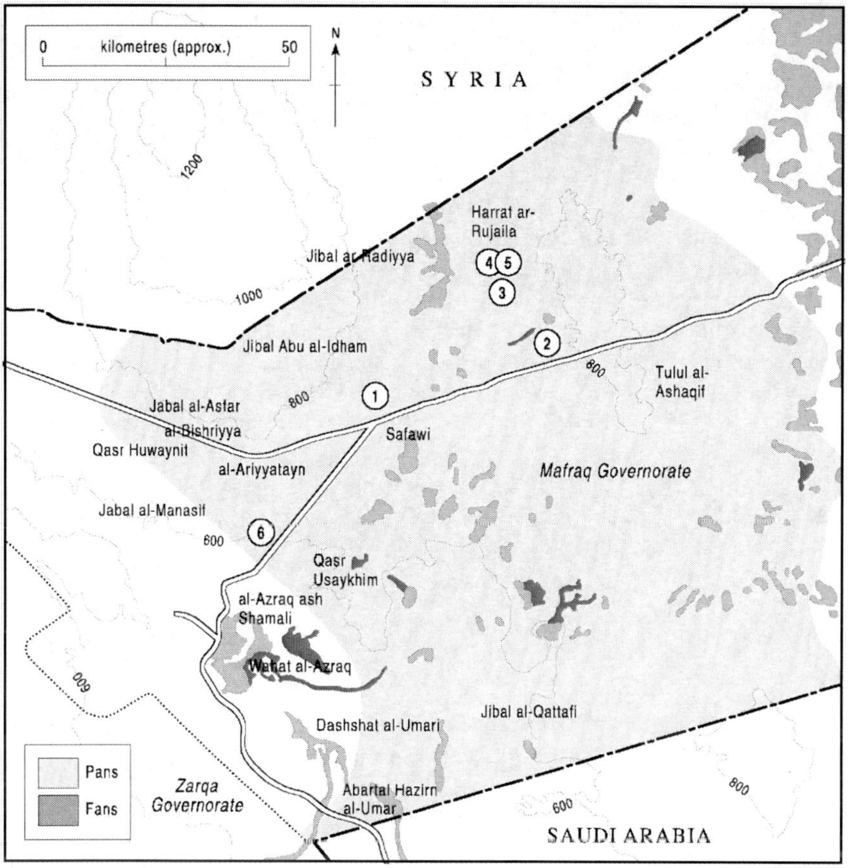

1-Safawi; 2-Marab Suway'id; 3-Shubaika; 4-Marab Salma; 5-Wadi Salma; 6-Tall Hassan

led to greatly increased interest in the genetic aspects of antibiotic biosynthesis in these organisms (Fayerman 1986, Gramajo et al 1993, Hopwood et al 1985, Jensen et al 1993, Stutzmzn-Engwall et al 1992).

Many of the *Streptomyces* isolated from nature carry detectable extra-chromosomal elements (plasmids) and, in most cases, plasmids were detected in the form of covalently closed circular (CCC)-DNA, but, occasionally, linear elements were also found (Chardon-Loriax et al 1986, Haruyasu et al 1987, Hayakawa et al 1979, Hirochika et al 1982, 1984, Ishibashi 1992, Keen et al 1988, Saadoun et al 1997). So far, a number of different *Streptomyces* have been investigated for plasmids which may be involved in the genetic control of the production of many antibiotics in *Streptomyces* species (Chater et al 1985, Hayakawa et al 1979, Hirochika et al 1982, Hopwood 1983, Hopwood et al 1985, Kirby et al 1977). However, more detailed analyses have shown that antibiotic biosynthetic structural genes lie mostly on the chromosome (Hutter et al 1988).

In this investigation, 60 pure isolates of *Streptomyces* were studied for their distribution, morphological and cultural characters, and their inhibitory effect against pathogenic bacteria. Also studied were genetic aspects of antibiotic synthesis by the most active isolates.

2. Materials and methods

2.1. Location, sampling and sample processing

For each of the three bacterial studies, soil samples were collected from up to six locations (Figure 1 and Table 1). For the *Bacillus* study 17 soil samples were collected from 4 different locations representing different agricultural and grazing fields (*marab*), and the steppe (control sites) which is covered with boulders of Tertiary and Quaternary basalt origin. It was assumed that the boulder-strewn areas are subject to much less impact from sheep and goats than the *marab*. For each of the bacterial studies 1-4 samples were collected from each location. After removing approximately 3 cm of the soil surface, samples were taken at depths down to 10 cm. The samples were placed in polyethylene bags, closed tightly and stored in a refrigerator. Each soil sample was crushed, thoroughly mixed and sieved through a 2 mm pore size sieve (Retsch, Germany), to get rid of large debris, and the sieved soil was used in the bacterial isolation process.

Streptomyces were recovered from 12 soil samples collected from five different locations in the north-east Badia.

For the Pseudomonas study, samples were collected from three different locations contaminated with crude petroleum oil spills resulting mainly from oil tanker accidents along the highway connecting Iraq and Jordan. One sample of 1 kg from each location was collected in summer 2001.

2.2. Soil analysis

The soil samples collected from each site, of a similar plant residue status, were combined and thoroughly mixed. The soil pH was measured in a 1:1 soil:water suspension and organic matter was measured by oxidation. One gram of each sieved soil sample was placed in a crucible dish then heated in an oven (Supertek, India) at 45 °C for 12 h (Williams et al 1972). After drying, soil samples were mixed with 0.1 g of $CaCO_3$ then incubated at 26 °C for seven days in a water bath (GFL, Germany) (El-Nakeeb et al 1963).

2.3. Isolation

Sub samples of 1 g were suspended in 100 ml sterile distilled water then incubated in an orbital shaker incubator (TEQ, Portugal) at 28 °C with shaking at 140 rpm for 30 min. Mixtures were allowed to settle, then serial dilutions up to 10^{-6} (10^{-4} in the case of *Streptomyces*) were prepared. From each dilution, 0.1 ml was taken and distributed evenly over the surface of starch casein nitrate agar (SCNA) (El-Nakeeb et al 1963, Kuster et al 1964) plates (in triplicate), using a sterile L-shaped glass rod, then incubated at 28 °C for 10 days.

Table 1. The six soil collection sites

Locality	Latitude	Longitude
1-Safawi	31° 10′ N	37° 07′ E
2-Marab Suway'id	32° 18′ N	37° 27′ E
3-Shubaika	32° 24′ N	37° 14′ E
4-Marab Salma	31° 47′ N	35° 45′ E
5-Wadi Salma	32° 25′ N	37° 15′ E
6-Tall Hassan	31° 57′ N	36° 55′ E

In the case of *Bacillus*, plated dilutions that gave 20-200 colonies were chosen for further isolation. Repeated streaking on starch casein nitrate agar (SCNA) plates (El-Nakeeb et al 1963, Kuster et al 1964) purified bacterial colonies that showed characteristics of the different genera under examination.

2.4. Characterization of the isolates

The morphological characterization of each isolate was first performed, including colour, size, and colony characteristics (form, margin, and elevation). A Gram stain test was performed for each isolate and the ability of certain isolates to grow at 42 °C was also investigated.

Streptomyces colonies were characterized morphologically and physiologically following the directions given for the International *Streptomyces* project (ISP) (Shirling et al 1966). General morphology was determined on oatmeal agar plates, incubated in the dark at 28 °C for 21 days, by direct light microscopy examination of the surface of cross-hatched cultures. Colours were determined according to the scale adopted by Prauser (1964) and isolates were grouped into separate colour series according to the system proposed by Nonomura (1974). Melanin reactions were detected by growing the isolates on at least one of the ISP media (no. 6 and no. 7) (Shirling et al 1966).

The following biochemical tests were used in the bacterial isolates' identification process: gelatin liquefaction; citrate utilization; oxidase; catalase; growth at 6.5 % sodium chloride; fluorescent pigment production; indole formation; glucose fermentation and nitrate reduction (Cappuccino et al 1996). The isolates were placed in phenol red glucose broth to determine glucose fermentation as well as gas production.

2.5. Spore morphology and staining

Each formed *Bacillus* colony was screened for the presence of endospores by staining bacterial smears with carbol fuchsin (Stahly et al 1992) stain, and then examined under light microscope (Nikon, Germany) to observe the morphology of the spores.

2.6. Pseudonomas and hydrocarbons

Bacterial strains were identified based on Bergey's Manual of Systematic Bacteriology (Palleroni 1984). *R. erythropolis* was used as a control for its ability to degrade different hydrocarbons (Saadoun et al 1999b).

Growth on diesel was determined by the 'hole-plate diffusion method' that is reported here for the first time. 20 mm of mineral salts (Leadbetter et al 1958) agar medium were poured into agar plates. Plates were then inoculated with the test organisms using a sterile swab. Plugs of 6 mm diameter were removed from the agar and 50 µl of filter-sterilized diesel were placed in the holes. Sterile distilled water was used as the control. Agar plates with the bacterial isolates were incubated overnight at 28 °C. The results were recorded after 48 h by the physical appearance of the bacterial growth surrounding the holes.

2.7. *Streptomyces* and antimicrobial activity

Antimicrobial activity was tested by the Bauer-Kirby method (Bauer et al 1966) against antibiotic-resistant Gram-negative bacteria; *Escherichia coli* and *Klebsiella pneumoniae*, and Gram-positive bacteria; *Staphylococcus aureus*. *Escherichia coli* was resistant to ofloxacin (Novecin), augmantin, teicoplanin, tetracycline, cefamandole, chloramphenicol, tobramycin, ciprofloxacin, sulphamethoxazole and impecillin, but sensitive to streptomycin, cefotaxime and cefuroxime sodium. *Klebsiella pneumoniae* was resistant to gentamicin, amoxicillin, streptomycin, ofloxacin (Novecin) and carbenicillin, but sensitive to cefotaxime. However, *Staphylococcus aureus* was resistant to gentamicin, but sensitive to streptomycin, cotimoxazol, erythromycin, oxaxillin, penicillin, vancomycin and cephaloyhin. These pathogens were isolated from patients with urinary tract (UT) and respiratory tract (UP) infections and identified by the clinical laboratory of Princess Basma hospital, Irbid, Jordan. Isolates of *Streptomyces* were grown on oatmeal agar for 14 days; then three discs (5 mm in diameter) were transferred to Mueller Hinton agar previously seeded with a lawn of the test pathogens and incubated at 28 C. The isolates that showed an inhibition zone with diameter more than or equal to 18 mm were considered as active isolates. Inhibition zones were visually detected after 24 h.

2.8. Genetic aspects of antibiotic biosynthesis

Three microbial cultures were used in this study: (a) *Streptomyces lividans* ATCC 35287, containing pIJ702 plasmid (5.65 kb in size) (Katz et al 1983), and used as a positive control; (b) *Escherichia coli* strain X2556 standards (same as VA 517 according to Macrina et al, (1978), containing eight plasmids ranging in size from 1.4 megadalton (Mdal) to 35.6 Mdal [1 Mdal corresponds to 1500 base-pairs (bp), ie 1 kb equals 0.66 Mdal], used as a source of molecular weight marker; (c) antibiotic-producing *Streptomyces* strains (I_7, AC_{32}, Bb_{36} and Z_{11}) isolated from soils in the Badia (Saadoun et al 2002).

Streptomyces were cultured on a yeast extract-malt extract (YEME) broth (Hopwood et al 1985) [per litre: 3 g YE, 5 g Bacto-peptone, 3 g malt extract, 10 g glucose, 340 g sucrose, 0.5 % glycine and 2 ml of 2.5 M $MgCl_2.6H_2O$; the latter was added after autoclaving]. *Streptomyces* were grown in broth cultures at 28 °C with shaking at 100 rpm for three days. *E. coli* was cultured on yeast-extract dextrose (YD) broth [per litre: 10 g dextrose, 10 g yeast extract (YE), 0.5 % (v/v) glycerin, pH 7.5], and grown at 37 °C with shaking at 100 rpm overnight.

The alkaline lysis method of Kieser (1984) was followed for detection of small molecular weight plasmid DNA (covalently closed circular or CCC). Gel electrophoresis was usually conducted with 0.7 % agarose gels using a Tris-borate (0.045 M)-EDTA (0.001M) buffer (0.5X TBE). Electrophoresis was carried out at a constant current of 35 mA. All gels were stained with 0.5 μg/ml ethidium bromide solution for 30-60 minutes at room temperature followed by 30-60 minutes destaining in distilled water. Gels were viewed using an ultraviolet (300 nm) transilluminator model TM-20 (Ultra Violet Products Inc., Upland, CA), then photographed using a Polaroid DS-34 direct screen instant camera (Polaroid Gel Cam, U.K.) fitted with a Tiffen 40.5 mm deep yellow 15 filter (Tiffen Manufacturing Co., Hauppauge, NY).

3. Results and discussion

3.1. Bacillus

The bacterial population of cultivated soils showed counts ranged between 2.84 x 10^7 and 38.9 x 10^7 CFU/g with an average count of 25.52 x 10^7 CFU/g dry soil. Uncultivated soils showed counts between 13.84 x 10^7 and 35.83 x 10^7 CFU/g with an average count of 20.60 x 10^7 CFU/g (Table 2). However, counts in the control soils ranged between 0.85 x 10^7 and 22.92 x 10^7 CFU/g with an average count of 8.17 x 10^7 CFU/g dry soil (Table 2).

The bacterial count of the control soils is 68 % and 60 % less than that of cultivated and uncultivated soils, respectively (Table 2).

Table 2. Bacterial population count and some soil characteristics of the sampling sites

Site*	Plant residues	Bacteria x 10^7 CFU/gm	Number of different colonies	pH	Moisture (%)	Organic matter (%)
Aa$_1$	Undecayed/tilled	2.84	4	7.60	3.10	6.70
Aa$_2$	Undecayed/tilled	10.16	3	8.50	3.70	9.05
Aa$_4$	Undecayed/untilled	14.80	5	7.87	6.89	14.50
Aa$_5$	Undecayed/untilled	30.56	4	8.10	5.00	13.60
Aa$_6$	Control	0.85	3	7.93	2.89	5.93
Ba$_1$	Undecayed/untilled	25.84	3	8.04	6.80	11.90
Ba$_2$	Decayed/untilled	35.83	3	7.90	3.79	8.65
Ba$_3$	Undecayed/untilled	23.68	4	8.18	5.50	8.22
Bb$_5$	Cultivated/barley	22.40	4	8.35	5.30	10.60
Bb$_6$	Control	22.92	3	8.25	4.89	11.10
Bb$_8$	Cultivated/wheat	38.90	3	8.50	9.20	8.55
C$_2$	Undecayed/untilled	13.44	3	7.88	4.69	13.30
C$_3$	Control	1.08	2	8.22	5.19	9.49
D$_1$	Cultivated/barley	16.48	3	8.10	2.80	9.34
D$_2$	Undecayed/barley	24.32	4	8.45	2.19	9.30
D$_3$	Undecayed	28.34	3	8.50	2.19	8.63
D$_4$	Control	7.84	3	8.29	2.10	9.09
		Average number of aerobic viable bacterial population, diversity and soil characteristics of the different fields.				
	Cultivated or tilled	25.52	4	8.35	4.87	9.45
	Uncultivated	20.60	4	8.06	4.63	12.48
	Control (desert plain)	8.17	3	8.17	3.77	8.90

*A: Marab Suway'id; B: Shubaika; C: Marab Salma; D: Tall Hassan.

Two to five different colony types of bacterial strains were recovered on the agar plates and their phenotypic identification revealed the dominance of members of the genus *Bacillus*, represented by the following species: *B. cereus, B. subtilis, B. megaterium, B. lentimorbus* and *B. azotoformans*. (Table 3).

When these different *Bacillus* species were characterized culturally and microscopically, the data showed that all of them exhibited smooth and opaque colonies and that the cells have central and oval spores except *B. azotoformans* which has terminal spores (Table 4).

Table 3. Characteristics of different Bacillus isolates recovered from different sites

No. of isolates	Gram stain	Shape	Spore former	Motility	Indole	Oxidase	Catalase	Starch hydrolysis	Identified isolates
16	+		+	+	-	-	+	+	*Bacillus cereus*
18	+		+	+	-	+	+	+	*Bacillus subtilis*
2	+		+	-	-	+	+	+	*Bacillus megaterium*
1	+		+	-	-		-	-	*Bacillus lentimorbus*
1	+		+	-	-	+	-	-	*Bacillus azotoformans*

Table 4. Microscopic and macroscopic properties of the different *Bacillus* isolates

Bacillus sp.	Micro-morphology Sporangium	Spore shape	Spore position	Smooth	Macro-morphology Rhizoid	Opaque	Shiny
Bacillus cereus	Not swollen	Oval	Central	+	+	+	-
Bacillus subtilis	Not swollen	Oval	Central	+	-	+	-
Bacillus megaterium	Not swollen	Oval	Central	+	-	+	+
Bacillus lentimorbus	Swollen	Oval	Central	+	-	+	-
Bacillus azotoformans	Swollen	Oval	Terminal	+	-	+	-

Table 5. Characteristics of different bacterial isolates recovered from different sites

No. of isolates	Gram stain	Shape	Spore former	Motility	Indole	Oxidase	Catalase	O/F	Identified isolates
7	+	Rod chains	-	-	-	-	+		*Corynebacterium*
1	+	Curved	-	-	-	-	-		*Erysiplothrix*
2	+	Rod	-	+	-	-	+		*Listeria*
1	+	Cocci	-	-	-	-	+	-	*Micrococcus*
6	-	Rod	-	-	-	-	+	-	*Acinetobacter*
3	-	Rod	-	-	-	+	+		*Bordetella*
1	-	Rod	-	+	-	-	+		*Proteus*

Dominanace of members of the genus *Bacillus* and their wide distribution in Badia soils could explain their potential to withstand unfavourable or harsh conditions as they harbour the endospores that are resistant to inactivation by heat, desiccation, and organic solvents (Stahly et al 1992).

Other bacterial genera were also identified, namely: *Streptomyces, Corynebacterium, Micrococcus, Erysiplothrix, Listeria, Acinetobacter, Bordetella* and *Proteus*. (Table 5). Saadoun and Gharaibeh (2002) examined members of the genus *Streptomyces* (not listed in Table 5) for their distribution, morphological, cultural characters and their inhibitory effect against pathogenic bacteria in another study.

3.2. Pseudonomas

The bacterial population of the crude petroleum-oil polluted soil showed counts between 9.5×10^5 and 237.5×10^5 CFU/g soil with two different colony types of bacterial strains recovered on the agar plates (Table 6).

Although microbial enumeration is not a direct measure of their activity in soils, it provides an indication of microbial vitality and/or biodegradative potential, in a crude petroleum-oil contaminated soil, by measuring how well the soil supports microbial growth (Bossert et al 1995). In addition, the distribution of the type and the number of microorganisms at a site may help to characterize the site with respect to the concentration and age of the contaminant. Fresh spills and/or high levels of contaminants often kill or inhibit a large sector of the soil microbiota, whereas soils with lower levels of aged contamination show a greater number and diversity of microorganisms (Bossert et al 1984, Dean-Ross 1989, Leahy et al 1990, Walker et al 1976). The current study also indicated that longer-aged contamination showed a greater number of microorganisms, whereas fresh spills reduced the bacterial number in the crude petroleum oil-polluted soil.

Table 6. Total bacterial count in soils (at 10 cm depth) polluted with crude petroleum oil

Locality and soil colour*	Latitude/Longitude	CFU x 10^5/gm
1-Safawi (Centre of the north-east Badia) (light black)	31° 10′ N/37° 07′ E	152
2-Suway'id (dark brown)	32° 18′ N/37° 27′ E	9.5
3-Tall Hassan (light black)	31° 57′ N/36° 55′ E	237.5

* Time of exposure to petroleum oil spill at locality 1, 2 and 3 was 2, 1 and 5 years, respectively. Colony types recovered from these localities were 2, 2 and 1, respectively.

Table 7. Morphological and physiological properties of the different *Pseudomonas* isolates[a]

Species	Biochemical and cultural criteria[b]		
	Citrateutilization	Nitrate reduction	Growth at 42 °C
Sp. 1 (B1) *Pseudomonas maltophilia*	-	+	Heavy growth
Sp. 2 (B3) *Pseudomonas putida*	+	-	No growth
Sp. 3 (B5) *Pseudomonas mallei*	-	+	Heavy growth

[a]All of the isolates gave negative results for the indole, gelatin and Mr/VP tests;
[b]All of the isolates were motile, oxidase negative and gave K/A result for TSI test.

Three different *Pseudomonas* species (*P. maltophilia, P. putida* and *P. mallei*) were recovered from different soils contaminated with fuel spills (Table 7). Their ability to degrade diesel fuel was studied. Other bacterial genera were also identified namely: *Enterobacter cloacae* and *Acinetobacter lowffi* (data not shown). Radwan et al (1995) reported the predominance of the members of the genus *Pseudomonas,* in addition to *Bacillus, Streptomyces* and *Rhodococcus,* in the various oil-polluted Kuwaiti desert soil samples subjected to various types of management.

The prevalence of members of the genus *Pseudomonas* in all soils tested confirms previous reports about the widespread distribution of such bacteria in hydrocarbon-polluted soils and reflects their potential in utilizing these hydrocarbon contaminants for growth and thus to clean these polluted sites (Cork et al 1991). The growth response of the different *Pseudomonas* cultures on diesel fuel indicates that the growth of *P. putida* and *P. mallei* were more prominent than that of the other bacterial cultures (data not shown).

3.3. Streptomyces: genetic aspects of antibiotic biosynthesis

By employing the enrichment methods, a total of 60 different *Streptomyces* isolates were recovered from 12 soil samples that were collected from different locations in the Badia. All of these isolates fitted the genus description as reported by several investigators (Nonomura 1974, Shirling et al 1966, Williams et al 1983). *Streptomyces* isolates were categorized into six colour series according to the colour of their mature sporulated aerial mycelium, with gray and white colour series being the most abundant (Table 8). Data indicated that 75 %, 35 % and 30 % of the isolates exhibited reverse-side pigmentation and were able to produce soluble and melanin pigments, respectively (Table 8). Microscopic examination of the sporophore morphology revealed that most of the isolates bore rectiflexible sporophores (RF) and spirales (S) that are represented by 52 % and 42 %, respectively. Retinaculiaperti (RA) bearers were less observed (6 %) and were limited to only gray colour series (Table 8).

All of the 60 isolates were tested for their ability to produce inhibitory substances against three test bacteria. As shown in Table 8, only 18 isolates representing 30 % of all isolates showed antibacterial activity against at least one of the test organisms. However, the percentage of active isolates varies within each colour series, with higher production of such compounds recorded in the gray series. *Staphylococcus aureus* was inhibited by almost all of the white and gray active isolates; while Gram-negative test bacteria were less inhibited.

Table 9 shows the characteristics of four *Streptomyces* isolates (I_7, AC_{32}, Bb_{36} and Z_{11}) with the ability to inhibit all of the antibiotic-resistant bacteria.

The extraction method of Kieser (1984) revealed at least one plasmid band (~ 5.4 kbp) from *Streptomyces lividans* ATCC 35287 (data not shown), a plasmid-containing control *Streptomyces*. No low molecular weight plasmids DNA were observed in the antibiotic-producing *Streptomyces* strain No. Bb_{36}, *Streptomyces* strain No. Z_{11}; *Streptomyces* strain No. I_7 and *Streptomyces* strain No. AC_{32}.

Table 8. Activity of the different *Streptomyces* isolates on antibiotic-resistant bacteria and their morphological and cultural characteristics

	Colour Series						
	White	Grey	Yellow	Red	Blue	Variable (violet; orange; pink)	Total
No. of isolates	20	20	13	3	2	2	60
Active isolates	8	9	1	0	0	0	18 (30 %)
Activity against							
S. aureus	8	9	1	0	0	0	18 (100 %)
E. coli	4	3	0	0	0	0	7 (39 %)
K. pneumoniae	2	2	0	0	0	0	4 (22 %)
Pigment production:							
melanin	6	5	5	0	1	1	18 (30 %)
Reverse side	16	17	5	3	2	2	45 (75 %)
Soluble	9	7	4	1	0	1	21 (35 %)
	Sporophore morphology						
Rectiflexible (RF)	13	5	12	1	0	0	31 (52 %)
Spirales (S)	6	12	1	2	2	2	25 (42 %)
Retinaculiaperti (RA)	1	3	0	0	0	0	4 (6 %)

Numbers between parentheses represent the percentage out of the total isolates.

Table 9. Characteristics of the active *Streptomyces* strains against all of the antibiotic-resistant bacteria

Isolate no.	Aerial mycelium	Reverse side colour	Soluble pigment colour	Melanin pigment	Sporophore morphology	Antimicrobial activity *S. aureus*	*E. coli*	*K. pneumoniae*
I_7	Gray	Gray-black	-	-	Rectiflexibiles	+	+	+
AC_{32}	Gray	Green	-	-	Rectiflexibiles	+	+	+
Bb_{36}	White	Yellow	Yellow	-	Spirales	+	+	+
Z_{11}	White	Yellow	Yellow	-	Rectiflexibiles	+	+	+

Similar results were reported for Jordanian *Streptomyces* isolates by Saadoun and Al-Momani (1997) and Saadoun et al (1999b) and for Iraqi *Streptomyces* isolates by Hamdi et al (1981) when they showed that the gray and the white isolates had the highest occurrence in soils of both countries. Furthermore, the cultural and morphological properties of both countries' isolates were also almost the same.

The unusual activity of I_7, AC_{32}, Bb_{36} and Z_{11} isolates against antibiotic-resistant bacteria was tested by their ability to inhibit the growth of all the tested bacterial pathogens. These isolates were recovered from a region with low nutrient levels and harsh environmental conditions which may support the idea of producing novel antibiotics under such conditions. The unusual antibiotic profile of these active strains stressed their potential as a source of novel antibiotics.

Several studies reported that antibiotic production in *Streptomyces* spp is encoded by either small or giant linear plasmids (Hershberger et al 1989, Hirochika et al 1984, Keen et al 1988, Kinashi et al 1987). However, the possible presence of a low copy number plasmid cannot be excluded. Thus far, separation of large linear DNA plasmids from chromosomal DNA by conventional techniques in *Streptomyces* has been unsuccessful (Kinashi et al 1987). Large linear DNAs of more than 30 kb have been shown to comigrate in conventional agarose gel electrophoresis and are not separable. Thus pulsed field gel electrophoresis (PFGE) has been developed to solve this problem. This study has indicated that antibiotic production in these isolates is likely to be chromosomally encoded. Because the role of giant linear plasmids in *Streptomyces*, which code for antibiotic production, is possible, further study is recommended to isolate such plasmid(s) from these organisms.

References

Bartha R and Bossert ID (1984). The treatment and disposal of petroleum refinery wastes. In: Atlas, RM (ed), Petroleum microbiology. New York: Macmillan Publishing Co pp. 1-61.

Bauer AW, Kirby WM, Sherris JC and Turk M (1966). Antibiotic susceptibility testing by a standardized single disk method. Am. J. Clinic. Pathol. 45, 493-496.

Bossert ID and Bartha R (1984). In: Atlas, RM (ed), The fate of petroleum in soil ecosystems. New York: Macmillan Publishing Co, pp. 435-474.

Bossert ID and Compeau GC (1995). Cleanup of petroleum hydrocarbon contamination in soil. In: Young, LY and Cerniglia, CE (eds), Microbial transformation and degradation of toxic organic chemicals. New York: Wiley-Liss, Inc pp. 77-126.

Cappuccino JG and Sherman N (1996). Microbiology: A laboratory manual. New York: The benjamin/Cummings Publishing Company, Inc.

Chardon-Loriax I, Charpentier M and Percherson F (1986). Isolation and characterization of a linear plasmid from Streptomyces rimosus, FEMS Microbiol. Lett 35:151-155.

Chater KF and Bruton CJ (1985). Resistance, regulatory and production genes for the antibiotic methylenomycin are clustered. EMBO J 4:1893-1898.

Cork DJ and Krueger JP (1991). Microbial transformation of herbicides and pesticides. Adv. Appl. Microbiol 36:1-66.

Crandall LW, Crandall N and Hamill RL (1986). Antibiotics produced by Streptomyces: Major structural classes. In: Queener, SW and Day, LE (eds), Antibiotic-producing Streptomyces, the bacteria, volume IX. New York: Academic Press pp. 355-401.

Dean-Ross D (1989). Bacterial abundance and activity in hazardous waste-contaminated soil. Bull. Environ. Contam. Toxicol 43:511-517.

El-Nakeeb MA and Lechevalier HA (1963). Selective isolation of aerobic actinomycetes. Appl. Microbiol 11:75-77.

Fayerman JT (1986). New developments in gene cloning in antibiotic-producing microorganisms. Biotechnology 4:786-789.

Goodfellow M and Simpson KE (1987). Ecology of streptomycetes. Front. Appl. Microbiol 2:97-125.

Gramajo HC, Takano E and Bibb MJ (1993). Stationary-phase production of the antibiotic actinorhodin in Streptomyces coelicolor A3(2) is transcriptionally regulated. Mol. Microbiol 7(6):837-843.

Hamdi YA, Ahmed D and Al-tai A (1981). Genera and species of actinomycetes isolated from Iraqi soils. Egypt: J. Microbiol 15:7-22.

Haruyasu K, Shimaji M and Sakai A (1987). Giant linear plasmids in Streptomyces which code for antibiotic biosynthesis genes. Nature (328).

Hayakawa T, Kanaka T, Sakaguchi N, Otake N and Yonchara H (1979). A linear plasmid-like DNA in Streptomyces sp. producing lankacidin group antibiotics. J. Gen. Appl Microbiol 25:255-260.

Hershberger CL, Arnold JGB, Larson J, Skatrud P, Reynolds P, Szoke P, Rosteck PR, Swartling J and Mcgilvray D (1989). Role of giant linear plasmids in the biosynthesis of macrolide and polyketide antibiotics. In: Hershberger CL, Queener SW and Hegeman G (eds), Genetics and molecular biology of industrial microorganisms. Washington, D.C: American Society for Microbiology pp. 147-155.

Hirochika H, Nakamura K and Sakaguchi K (1982). Analysis of linear plasmids isolated from Streptomyces: association of protein with the ends of the plasmid DNA. Plasmid 7:59-65.

Hirochika H, Nakamura K and Sakaguchi K (1984). A linear DNA plasmid from Streptomyces rochei with an inverted terminal repetition of 614 base pairs. J. EMBO 3:761-766.

Hopwood DA (1983). Actinomycete genetics and antibiotic production. In: Vining, LC (ed), Biochemistry and genetic regulation for commercially important antibiotics. Reading, Mass: Adison -Wesley pp. 1-23.

Hopwood DA, Bibb MJ, Chater KF, Kieser T, Bruton CJ, Kieser H, Lydiate DJ, Smith CP, Ward JM and Schrempf H (1985). Genetic manipulation of Streptomyces. A laboratory manual. Norwich: The John Innes Foundation.

Hutter R and Eckhardt T (1988). Genetic manipulation. In: Goodfellow, M, William, ST and Mordarski, M (eds), Actinomycetes in biotechnology. London: Academic Press, Inc. (London), Ltd pp. 89-184.

Ishibashi Y (1992). Genetic studies into musty odor production by actinomycetes. Water Sci. Tech 25:171-176.

Jensen SE, Alexander DC, Paradkar AS and Aidoo KA (1993). Extending the B-Lactam biosynthetic gene cluster in Streptomyces clavuligerus. Industrial Microorganisms: Basic and Appl. Mol. Genet:169-176.

Katz E, Thompson CJ and Hopwood DA (1983). Cloning and expression of the tyrosinase gene from Streptomyces antibioticus in Streptomyces lividans. J. Gen. Microbiol 129:2703-2714.

Keen CLS, Mendeloritz G, Cohen Y, Haronowitz A and Roy KL (1988). Isolation and characterization of a linear DNA plasmid from Streptomyces clavuligereus. Mol. Gen. Genet 212:172-176.

Kieser T (1984). Factors affecting the isolation of DNA from Streptomyces lividans and Escherichia coli. Plasmid 12:19-36.

Kinashi H and Shimaji M (1987). Detection of giant linear plasmid in antibiotic producing strains of Streptomyces by the OFAGE technique. J. Antibiotics 40:913-916.

Kirby R and Hopwood DA (1977). Genetic determination of methylenomycin biosynthesis by the SCP1 plasmid of Streptomyces coelicolor A3(2). J. Gen. Microbiol 98:239-252.

Kuster E and Williams ST (1964). Selection of media for isolation of streptomycetes. Nature 202:928-929.

Leadbetter ER and Foster JW (1958). Studies of some methane utilizing bacteria. Arch. Microbiol 30:91-118.

Leahy JG and Colwell RR (1990). Microbial degradation of hydrocarbons in the environment. Microbiol. Rev 54:305-315.

Marcina FL, Kopeck DJ, Jones-Lee AR, Ayers DJ and Mccowen SM (1978). A multiple plasmid-containing Escherichia coli strain: convenient source of size reference plasmid molecules. Plasmid 1:417-420.

Margesin R and Schinnur F (1997). Efficiency of indogenous and inoculated cold-adapted soil microorganism for biodegradation of diesel oil in Alpine soils. Appl. Environ. Microbiol 63:2660-2664.

Nonomura H (1974). Key for classification and identification of 485 species of the streptomycetes included in the ISP. J. Ferm. Technol 52:78-92.

Palleroni NJ (1984). In: Holt, JG and Krieg, NR (eds), Bergey's Manual of Systematic Bacteriology, vol 1, Baltimore. Md: Williams and Wilkins pp. 141-199.

Prauser H (1964). Aptness and application of colour for exact description of colours of Streptomyces. Zeitschrift Fur Allgemeine Mikrobiologie 4:95-99.

Radwan SS, Sorkhoh NA, Fardoun F and Al-Hasan H (1995). Soil management enhancing hydrocarbon biodegradation of the polluted Kuwaiti desert. Appl. Microbiol. Biotechnol 44:265-270.

Saadoun I, Al-Akhras MA and Abu-Ashour J (1999a). Bacterial degradation of hydrocarbons as evidenced by respirometric analysis. Microbios 100:19-25.

Saadoun I and Al-Momani F (1996). Bacterial and Streptomyces flora of some Jordan Valley soils. Actinomycetes 7(3):95-99.

Saadoun I and Al-Momani F (1997). Studies on soil Streptomycetes from Jordan. Actinomycetes 8(3):42-48.

Saadoun I, Al-Momani F, Malkawi HI and Mohammad MJ (1999b). Isolation, identification and analysis of antibacterial activity of soil streptomycetes isolates from north Jordan. Microbios 100:41-46.

Saadoun I and Gharaibeh R (2002). The Streptomyces flora of the Badia region of Jordan and its potential as a source of antibiotics against resistant bacteria against antibiotic-resistant bacteria. J. Arid Environments (53):365-371.

Saadoun I, Mohammad MJ, Malkawi HI, Al-Momani F and Meqdam M (1998). Diversity of soil streptomycetes in Northern Jordan. Actinomycetes 9 (3):52-60.

Shirling EB and Gottlieb D (1966). Methods for characterization of Streptomyces species. Inter. J. Syst. Bacteriol 16:313-340.

Stahly D, Andrews R and Yousten A (1992). The genus Bacillus insect pathogens. In: Balows A, Truper H, Dworkin M, Harder W and Schleifer K (eds), The Prokaryotes. New York: Springer-Verlage pp. 1697-1744.

Stutzmzn-Engwall KJ, Otten S and Hutchinson CR (1992). Regulation of secondary metabolism in Streptomyces spp., and overproduction of daunorubicin in Streptomyces peucetius. J. Bacteriol 174:144-154.

Walker JD and Colwell RR (1976). Enumeration of petroleum-degrading microorganisms. App. Environ. Microbiol 31:198-207.

Wang X, Yu X and Bartha R (1990). Effect of Bioremediation on polycyclic aromatic hydrocarbon residues in soil. Environ. Sci. Technol 24:1086-1089.

Williams ST, Goodfellow M, Alderson G, Wellington EM, Sneath PH and Sackin MJ (1983). Numerical classification of Streptomyces and related genera. J. Gen. Microbiol 129:1743-1813.

CHAPTER 13

Induction of oestrus in Awassi ewes in the Jordanian Badia using intravaginal progestagen sponges and eCG

Rami T. Kridli

Department of Animal Production, Faculty of Agriculture, Jordan University of Science and Technology (JUST), P.O. Box 3030, Irbid 22110, Jordan. E-mail: rkridli@just.edu.jo

Mustafa Q. Husein

Department of Animal Production, Faculty of Agriculture, JUST

Abdullah Y. Abdullah

Department of Animal Production, Faculty of Agriculture, JUST

This study was conducted to evaluate the effect of eCG (equine chorionic gonadotropin) dosage on reproductive parameters of Awassi ewes synchronized using intravaginal progestagen sponges. One hundred and forty Awassi ewes in three different flocks (50, 70 and 20 ewes in flocks 1, 2 and 3, respectively) were fitted with flourogestone acetate sponges for a period of fourteen days. At the time of sponge removal, ewes in each flock were divided into two groups to receive either 400 IU eCG (n=72) or 600 IU eCG (n=68) per head (i.m.). Ewes in flocks 1 and 3 were divided equally between the two treatments; in flock 2, 37 ewes received 400 IU eCG and 33 ewes received 600 IU eCG. Rams were allowed with the ewes immediately following eCG administration and oestrus was monitored for five days. There was no eCG dose by flock interaction in any of the tested parameters. Oestrus expression was similar between the groups receiving 400 and 600 IU eCG. The interval from eCG administration to onset of oestrus was shorter ($P < 0.05$) in the group receiving 600 IU eCG (51.1±3.2 h and 42.1±3.3 h in ewes receiving 400 and 600 IU eCG, respectively). Pregnancy rate was similar between treatments while the number of multiple births tended to be greater ($P < 0.10$) in ewes receiving 600 IU eCG. The number of lambs born per served ewe and the number of lambs born per lambed ewe were greater ($P < 0.05$) in ewes receiving 600 IU eCG. As for the effect of flock, oestrus expression, the interval from eCG administration to onset of oestrus, and pregnancy rate were greater ($P < 0.05$) in flock 1 than flock 3 while flock 2 was intermediate. The percentage of ewes pregnant with multiple births and the number of lambs born per served ewe were greater ($P < 0.01$) in flock 1 than flocks 2 and 3, while the number of lambs born per lambed ewe was greater in flock 1 than in flock 2. Results of the present study indicate differences in reproductive performance among different flocks of sheep following an oestrus induction programme. Additionally, the administration of 600 IU eCG at the time of sponge removal under the Badia conditions resulted in better reproductive performance than the administration of 400 IU eCG.

Keywords: Oestrus induction, eCG, Awassi sheep, reproduction

1. Introduction

The sheep industry in Jordan is important from economic and social points of view. Economically, sheep provide around 50 % of the red meat consumed in Jordan. Despite this fact, the self-sufficiency of red meat remains below 30 % (MoA 2003). Socially, and because of the arid nature of the Jordanian Badia, the majority of the people have traditionally depended on sheep production as the main source of income. In recent years the sheep industry has been in decline for a variety of reasons which has led to many people departing for the cities, often with no real expectations of finding work, thus increasing the level of unemployment. This fact, together with the rising cost of sheep and meat imports, makes it essential to provide the sheep owners with the means to improve their production efficiency.

The fat-tailed Awassi is the local breed of sheep in Jordan (Epstein 1982, 1985) and is the most important breed in the semi-arid regions of the near east countries (Epstein 1985, Zarkawi et al 1999). It has several desirable traits such as the popularity of its meat and milk, and its high adaptability to different ecosystems. Resistance to diseases, tolerance to extreme temperatures, ability to walk long distances for grazing, strong flocking instinct and endurance of adverse management and feeding conditions have encouraged breeders in Arab countries to rear this breed (Epstein 1985, Thomson et al 2003).

Awassi sheep are seasonal breeders. Under natural conditions Awassi ewes breed during the summer and autumn, thus lambing only once a year (Epstein 1982, 1985, Rottensten et al 1971) to produce 1.05 lambs per ewe (Epstein 1985). They mate between late June through early September to lamb between late November and early February (Thomson et al 2003, Zarkawi 1997).

Awassi sheep reproductive performance tends to be low under traditional management conditions (Abdullah et al 2002). Such low performance may be attributed to several factors including genetics, nutrition, health and other environmental conditions. Hormone administration has been successfully used to improve reproductive performance (Robinson 1967). Progesterone (or progestagen) combined with equine chorionic gonadotropin (eCG) was one of the earliest methods of inducing oestrus in anoestrous ewes during the non-breeding season (Hafez 1993). During the normal breeding season, using eCG in addition to intravaginal progesterone treatment results in more precise synchronization and enables pre-set insemination without oestrus detection (Laliotis et al 1998).

1.1 Rationale and objective

Fertility and prolificacy in the Awassi are lower than in most other sheep breeds. Improving reproductive performance increases lamb production during the lifetime of a female, thus providing higher returns for the producer. Hormonal treatments can be used to induce and synchronize the breeding season. Different commercial preparations of eCG are available on the local market ranging in concentration 400-600 IU eCG per dose. In a previous study (Abdullah et al 2002), we attempted to identify the most effective eCG dose to improve reproductive parameters. However, due to some factors beyond our control, the overall pregnancy rates in that study were extremely low. For this reason, this study was designed to re-investigate the effects of eCG dosage, following the administration of intravaginal progestagen sponges, on oestrus induction and the reproductive performance of Awassi ewes.

2. Materials and methods

2.1. Location and animals

The study was conducted in July in the Dayr al-Kahf area in the northern Badia of Jordan. One hundred and forty ewes in three different flocks (50, 70 and 20 ewes in flocks 1, 2 and 3, respectively) were used. Rams and ewes were run together in all flocks before the start of the experiment. Ewes in all flocks were ear tagged at the beginning of the experiment and were treated with anti-parasitic drugs. Age, date of last lambing and lactation status were recorded for each ewe. About 20 % of the ewes in all flocks were lactating at the time of sponge insertion and removal. Rams were isolated from the flocks 10 days prior to sponge insertion. None of the ewes was cycling by that time, as declared by the sheep owners.

All experimental animals were in good body condition at the start of the experiment. Animals were raised in a semi-extensive management system. They were allowed to graze during the day and received 0.3 kg of barley plus 1 kg of wheat straw·head^{-1}·day^{-1} in the afternoon. Animals were flushed with 0.5 kg of concentrate mixture plus 1 kg of wheat straw·head^{-1}·day^{-1} during the period in which sponges were in place and continued for four weeks following sponge removal. Vitamin supplements were administered in drinking water when sponges were in place. Additionally, the amount of feed offered was increased by 25 % during the last month of gestation and into the lambing season. Enterotoxemia vaccine was administered to all ewes (1 cc/head; s.c.) around six weeks before the anticipated lambing date.

2.2. Treatments and data collection

All ewes were fitted with flourogestone acetate sponges (FGA, 40 mg, Ceva Sante Animale, France). 0.5 ml of antibiotic (oxytetracycline, Tenaline®, Ceva) was injected into each sponge to prevent microbial contamination. Sponges were maintained for 14 days. No sponges were lost during this period. At the time of sponge removal, ewes in each flock were divided into two groups to receive either 400 or 600 IU eCG per head (i.m.). Flocks 1 and 3 were divided equally between the two treatments; in flock 2, 37 ewes received 400 IU eCG and 33 ewes received 600 IU eCG. Such doses of eCG were selected because they were the most effective in our previous study (Abdullah et al 2002).

Fertile Awassi rams were allowed with the ewes at the time of sponge removal. Oestrus expression and the interval to onset of oestrus were monitored for the three flocks by the flock owners. Lambing data (lambing date, the number of offspring, and lamb sex) were recorded five months later. Definitions of some of the variables tested in this study were similar to those reported by Zeleke et al (2005) and were as follows:

- Pregnancy rate – the number of ewes lambing as a percentage of ewes mated;
- Lambing rate – the number of lambs born per ewe mated;
- Fecundity – the number of lambs born per ewe lambing.

2.3. Statistical analyses

Data were analyzed to a completely randomized design, using a general linear model procedure of the SAS Institute (SAS 1997). The effect of lactation status on reproductive responses was included in the original model. However, because it had no significant effects on any of the tested parameters, it was omitted from further analyses. The effect of eCG treatment by flock and the flock/treatment interaction on the interval to onset of oestrus, lambing rate and fecundity were evaluated by using the pairwise t test. Oestrus expression, pregnancy rate and the number of multiple births were analyzed by Chi Square (SAS 1997).

3. Results

No eCG treatment-by-flock interactions were detected for any of the studied parameters. For this reason, data will be presented as main effect means.

3.1. The eCG effect

Table 1 shows the effect of eCG dose on oestrus and lambing of Awassi ewes. There was no difference in oestrus expression between the groups receiving 400 IU and 600 IU eCG, while the interval to onset of oestrus was shorter in ewes receiving 600 IU eCG. Pregnancy rate was similar between the two eCG treatments while the percentage of multiple births tended to be greater (P < 0.10) in ewes receiving 600 IU eCG. The difference in multiple births resulted in higher lambing rate and fecundity (P < 0.05) in ewes receiving 600 IU eCG.

3.2. The flock effect

Table 2 shows oestrus and lambing data in the three experimental flocks. The overall oestrus expression was 87 %. More (P < 0.05) ewes in flock 1 expressed oestrus compared with flock 3 while flock 2 was intermediate. Similarly, oestrus occurred earlier (P < 0.05) in flock 1 than flock 3. Interval to oestrus onset was 43.9±3.5, 47.2±3.3 and 52.9±4.1 h in flocks 1, 2 and 3, respectively.

The overall pregnancy rate was around 78 % in the three flocks. Pregnancy rate was greater (P < 0.05) in flock 1 (86 %) than flock 3 (60 %) with flock 2 being intermediate (77 %). The percentage of ewes with multiple births and the number of lambs born per served ewe were greater (P < 0.01) in flock 1 than flocks 2 and 3, while the number of lambs born per lambed ewe was greater in flock 1 than in flock 2 (Table 2).

Table 1. Oestrus and lambing data of Awassi ewes induced with FGA sponges and different doses of eCG

Variable	eCG dose 400 IU (n=72)	 600 IU (n=68)
Oestrus expression	62 (86%)	60 (88%)
Interval to onset of oestrus (mean±SE)	51.1^{a}±3.2	42.1^{b}±3.3
Number of lambed ewes	54	55
Pregnancy rate (%)	75	81
No. of multiple births	$13/54^{c}$	$22/55^{d}$
% of ewes with multiple births	24.1^{c}	40.0^{d}
Lambs/ served ewe (mean±SE)	0.94^{a}±0.1	1.19^{b}±0.1
Lambs/ lambed ewe (mean±SE)	1.24^{a}±0.1	1.47^{b}±0.1

^{a,b} Means within a row with different superscripts differ (P < 0.05); ^{c,d}Means within a row with different superscripts tend to differ (P < 0.1)

Table 2. Oestrus and lambing data in three flocks of Awassi ewes induced with FGA intravaginal sponges and eCG

Variable	Flock 1 (n=50)	Flock Flock 2 (n=70)	Flock 3 (n=20)
Oestrus expression	48 (96.0^{a})	59 (84.2^{ab})	15 (75.0^{b})
Interval to onset of oestrus (mean±SE)	43.9^{a}±3.5	47.2^{ab}±3.3	52.9^{b}±4.1
Number of lambed ewes	43	54	12
Pregnancy rate (%)	86^{a}	77^{ab}	60^{b}
No. of multiple births	22/43	10/54	3/12
% of ewes with multiple births	51.2^{c}	18.5^{d}	25.0^{d}
Lambs/served ewe (mean±SE)	1.38^{c}±0.1	1.01^{d}±0.1	0.81^{d}±0.1
Lambs/lambed ewe (mean±SE)	1.53^{a}±0.1	1.22^{b}±0.1	1.33^{ab}±0.2

^{a,b}Means within a row with different superscripts differ (P < 0.05); ^{c,d}Means within a row with different superscripts differ (P < 0.01)

4. Discussion

Even though the breeding season of Awassi ewes begins in July (Epstein 1982), none of the ewes was cycling at the beginning of the experiment in July which was similar to what we reported in a previous study (Abdullah et al 2002). The northern Badia of Jordan is arid to semi arid. The availability of feed is an important factor affecting reproductive performance of animals possibly through the effect of energy on luteinizing hormone release (Sen et al 1979). The fact that feed availability was limited during the period preceding the start of this experiment may have resulted in the delay in cyclicity.

4.1. The eCG effect

Oestrus expression during the five days following sponge removal was similar between both groups. The overall oestrus expression was 87 % which was similar to previous studies (Zarkawi et al 1999) while being better than our previous experiment conducted under similar Badia conditions (Abdullah et al 2002). Treatment with eCG reduces the interval from progestagen removal to onset of oestrus (Cline et al 2001). Kridli and Al-Khetib (2006) reported a shorter interval to onset of oestrus in ewes treated with intravaginal sponges and eCG compared to ewes receiving intravaginal sponges alone. The interval from sponge removal to onset of oestrus in the present study was shorter in ewes treated with 600 IU eCG. Such reduction may be related to the action of the higher eCG dose on improving follicular development (Husein et al 1998, Oliveira et al 2001).

No differences in pregnancy rates were observed between the two treatments. The overall pregnancy rate in the present study (78 %) was similar to that previously reported in ewes synchronized with progestagen and eCG (Boscos et al 2002, Kridli et al 2006, Zarkawi et al 1999). Zarkawi (1999) reported no differences in pregnancy rate in Syrian Awassi ewes synchronized with progestagen with or without eCG administration, while Zeleke et al (2005) reported a higher pregnancy rate in ewes receiving progestagen and eCG compared with those receiving progestagen alone.

The number of multiple births tended to be greater in ewes receiving 600 IU than those receiving 400 IU eCG. Lambing rate and fecundity were also greater in ewes receiving 600 IU eCG. These results are in agreement with our previous study (Abdullah et al 2002) in which 600 IU eCG resulted in a higher lambing rate and fecundity than other eCG doses. Several studies have reported improvements in the incidence of multiple births in sheep following treatment with eCG (Abdullah et al 2002, Kridli et al 2006, Zarkawi 1999). The use of eCG following progestagen treatment has been shown to improve fecundity due to having more follicles ovulated (Cunningham et al 1980). Even though both treatments in the present study improved fecundity in Awassi ewes [compared with 1.05 lambs per ewe under natural conditions (Epstein 1985)] the higher eCG dosage seems to be more effective in inducing follicular development and multiple ovulations.

4.2. The flock effect

There were significant differences in oestrus response, pregnancy rate, lambing rate, multiple births and fecundity among the experimental flocks. Ewes in flock 1 responded better to hormonal treatment. Variation in reproductive responses among flocks, despite the similarity in treatments, may be attributed primarily to differences in flock management and genetics. Similar results were reported in our previous studies in the northern Badia (Abdullah et al 2002) where smaller flocks had a greater percentage of multiple births while the medium-sized flocks were intermediate. Smaller flocks allow the farmers to provide better management per animal especially during the summer season when feed availability is minimal in the Badia. Additionally, despite the fact that they were all Awassi ewes, there may be some differences in genetics among the three flocks resulting in more fertile animals in flock 1.

5. Conclusion

Results of the present study indicate that hormonal treatment can effectively induce oestrus and improve reproductive performance of Awassi ewes. The administration of 600 IU eCG at the time of FGA sponge withdrawal is more effective than administering 400 IU eCG in improving lambing rate and fecundity. These results re-confirm our previous findings under the Badia conditions. Variations in ewe responses to hormonal treatments in the three experimental flocks may be attributed to differences in flock management and genetics.

Acknowledgements

The authors wish to acknowledge The Jordan Badia Research and Development Programme (BRDP; now a Centre) and The Higher Council for Science and Technology (HCST) for funding this project. Great appreciation is expressed to all the sheep owners in the Jordanian Badia who offered their flocks for conducting the project and to Mr. A. N. Al-Yacoub for assistance in data collection. Thanks are also due to the personnel at the Safawi Centre for their cooperation in locating the sheep owners.

References

Abdullah AY, Husein MQ and Kridli RT (2002). Protocols for estrus synchronization in Awassi ewes under arid environmental conditions. Asian-Aust. J. Anim. Sci 15:957-962.

Boscos CM, Samartzi FC, Dellis S, Rogge A, Stefanakis A and Krambovitis E (2002). Use of progestagen-gonadotropin treatments in estrus synchronization of sheep. Theriogenology 58:1261-1272.

Cline MA, Ralston JN, Seals RC and Lewist GS (2001). Intervals from norgestoment withdrawal and injection of equine chorionic gonadotropin or PG 600 to estrus and ovulation in ewes. J. Anim. Sci 79:589-594.

Cunningham NF, Saba N, Boarer CDH and Hattersley JJP (1980). Plasma hormone levels and reproductive behavior in anestrous ewes after treatment with progesterone and PMSG. J. Reprod. Fert 60:177-185.

Epstein H (1982). Awassi sheep. World Anim. Rev 44:9-18.

Epstein H (1985). Biology of reproduction, suckling regimes, growth, and development, The Awassi sheep with special reference to the improved dairy type, Rome. Italy: Food and Agriculture Organization of the United Nations pp. 81-140.

Hafez ESE (1993). Reproductive cycles, sheep and goats. Reproduction in farm animals, 6th edn Philadelphia: Lea & Febiger pp. 94-113.

Husein MQ, Bailey MT, Ababneh MM, Romano JE, Crabo BG and Wheaton JE (1998). Effect of eCG on the pregnancy rate of ewes transcervically inseminated with frozen-thawed semen outside the breeding season. Theriogenology 49:997-1005.

Kridli RT and Al-Khetib SS (2006). Reproductive responses in ewes treated with eCG or increasing doses of royal jelly. Anim. Reprod. Sci 92:75-85.

Laliotis V, Vosniakou A, Zafrakas A, Lymberopoulos A and Alifakiotis T (1998). The effect of melatonin on lambing and litter size in milking ewes after advancing the breeding season with progestagen and PMSG followed by artificial insemination. Small Rumin. Res 31:79-81.

MoA (2003). Annual Report of the Livestock Production Department, Amman, Jordan: Ministry of Agriculture.
Oliveira MAL, Guido SI and Lima PF (2001). Comparison of different protocols used to induce and synchronize estrus cycle of Saanen goats. Small Rum. Res 40:149-153.
Robinson TJ (1967). The control of the ovarian cycle in sheep. Sydney Sydney Univ. Press.
Rottensten K and Ampy F (1971). Studies on Awassi sheep in Lebanon: I-production traits of a flock. J. Agric. Sci 77:371-373.
SAS (1997). SAS/STAT user's guide, Version 6, Carry, NC: SAS Institute Inc.
Sen KK, Azhar S and Menon KJM (1979). Evidence for the involvement of an energy-dependent process in gonadotropin-releasing hormone-stimulated luteinzing hormone release by rat anterior pituitary. Endocrinology 105:1158-1161.
Thomson EF, Martini MA and Tutwiler RN (2003). Sheep management practices in Iraq, Jordan and Syria: the case of reproduction and fertility. Integrated natural resource management research report series no. 2, Aleppo, Syria: International Center for Agricultural Research in the Dry Areas (ICARDA).
Zarkawi M (1997). Monitoring the reproductive performance in Awassi ewes using croaker one radioimmunoassay. Small. Rumin. Res 26:291-294.
Zarkawi M (1999). Oestrous synchronisation and twinning rate of Syrian Awassi ewes treated with progestagen and PMSG during the breeding season. New Zealand. J. Agric. Res 44:159-163.
Zarkawi M, Al-Merestani R and Wardeh MF (1999). Induction of synchronized oestrous and early pregnancy diagnosis in Syrian Awassi ewes outside the breeding season. Small Rumin. Res 33:99-102.
Zelke M, Greyling JPC, Schwalbach LMJ, Muller T and Erasmus JA (2005). Effect of progestagen and PMSG on oestrous synchronization and fertility in Dorper ewes during the transition period. Small Rumin. Res, 56:47-53.

CHAPTER 14

Pathological studies on parasitic infestations of Arabian camels in Jordan

Falah K. Al-Ani
P.O. Box 2970, Code 111 Muscat, Oman. E-mail: alani99@hotmail.com

Ali M. Shotar
Department of Legal Medicine, Toxicological and Forensic Medicine, Faculty of Medicine, Jordan University of Science and Technology, Jordan

Labib A. Sharif
Department of Animal Health, Jordan University of Science and Technology, Jordan

To study the pathological changes of parasitic infestation, 178 camels slaughtered at Ramtha abattoir 1999-2005 were used. Before slaughtering, a complete examination of the animals was undertaken. Blood smears were prepared and stained. Faecal samples were also collected to check for gastrointestinal parasites. Skin scrapings from camels with skin lesions were prepared for detection of mites. Following slaughtering, complete examination of all organs was undertaken. Lesions were carefully examined using the naked eye and histopathological methods.

Results indicated the presence of the following parasites: *Sarcocystis cameli* cysts were found in 25 (14.0 %) of the hearts, oesophaguses, tongues and diaphragms examined. Oocysts of *Cryptosporidium parvum* were seen in one (0.56 %) faecal stained samples. The cysts appeared red in acid-fast stain and of 5-6 µm. Schizonts of *Eimeria* spp were seen in one (0.56 %) intestinal section of a young calf. *Trypanosoma evansi* were seen in 14 (7.8 %) of the camel blood samples examined.

Cephalopina titillator (camel nasal fly) were seen in 69 (38.8 %) of the camel heads examined. Most of the first-instar larvae were found in the labyrinth of the ethmoid bone while second and third-instar larvae were mostly seen in the pharyngeal cavity. Onchocercaisis due to *Onchocerca fasciata* was found in 5 (2.8 %) of the cases examined. The nodules were mainly located in the subcutaneous tissues of the chest and belly areas. 98 % of the camels were found carrying gastrointestinal helminthes; *Trichostrongylus probolurus, Camelostronglus mentulatus* and *Monezia expansa* were the most common. *Echinococcus granulosus* (Hydatid) cysts in the lungs and liver were found in 28 (15.7 %) of the samples examined. *Cysticercus tenuicollis* was seen in 1 (0.56 %) case in which the cyst was found hanging in the mesentery with no pathological changes. Tick and mite infestations were the most common ectoparasites seen. Tick problems due to *Hyalomma dromedarii* and *Hyalomma anatolicum* were seen in 174 (97.7 %) and mange due to *Sarcoptes scabie var cameli* was found in 31 (17.4 %) of the skin scrapings examined.

Keywords: Parasites, camels

1. Introduction

The Arabian camel is widely distributed in the Arabian Peninsula, East Asia, and North Africa. In Jordan, the population of Arabian camels is around 18 000. Camels play a vital role in the subsistence economy of pastoral peoples in the Jordan Badia (Al-Ani 2003). Bedouin use camels for production of milk, meat and hair, for tourism and for transportation in the desert. Camels were formerly considered resistant to most of the diseases commonly affecting livestock, but as more research has been conducted, camels have been shown to be susceptible to a large number of pathogenic agents (Al-Yaman 1985, Saliba 1977, Sherkov et al 1976, Sherkov et al 1978). Parasites are important pathogens of Arabian camels.

The literature on parasitic diseases of the camels has been reviewed extensively in different parts of the world (Al-Ani 2004, Al-Ani et al 1991, El-Bihari 1980, Sharrif et al 1996). Prevalence of infection at various times of the year has been studied in several countries of Asia and Africa. The present paper provides a critical pathological account on camel parasitic infestations in Jordan.

2. Materials and methods

To study the pathological changes of parasitic infestation, 178 camels slaughtered at Ramtha abattoir 1999-2005 were examined. Before slaughtering, a complete examination of the animals was undertaken. Blood smears were prepared and stained. Faecal samples were also collected to check for gastrointestinal parasites. Skin scrapings from camels with skin lesions were prepared for the detection of mites. Following slaughtering, examination of all organs was undertaken. Lesions were carefully examined using the naked eye and histopathological methods. The gross damage to the local tissue was described and photographed. Selectively tissue samples were collected, preserved in buffered 10 % formalin solution and processed by standard histopathological techniques and stained with haematoxylin and eosin. Eggs, larvae and adult parasites were identified in the parasitology laboratory.

3. Results

The following parasites were discovered:

Sarcocystis cameli cysts were found in 25 (14.0 %) of the hearts, oesophaguses, tongues and diaphragms examined (Figure 1). The cysts were seen between muscle fibres as elongated cylindrical bodies measuring 4-9 μm in breadth and 12-16 μm in length. In most cases, no pathological changes were seen in the examined tissues.

Trypanosoma evansi was seen in 14 (7.8 %) of the camel blood samples examined. Camels with chronic infection showed signs of emaciation, oedema of the belly and hind legs, and weakness (Figure 2).

Figure 1 Sarcocystis cyst between the fibres of heart muscle (H & E stain)

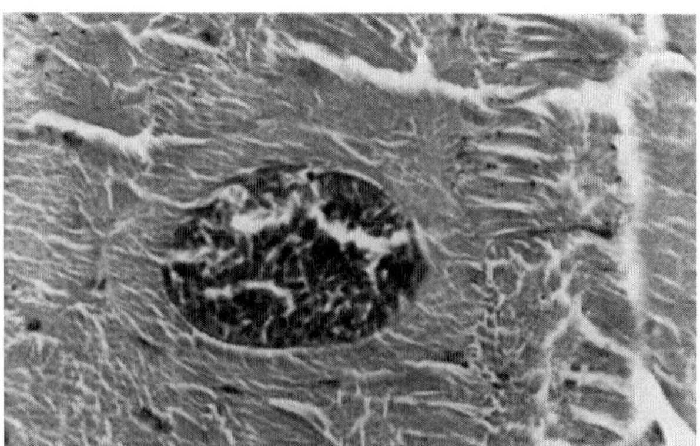

Figure 2. Oedema of the ventral parts of the abdomen due to chronic trypanosomiasis

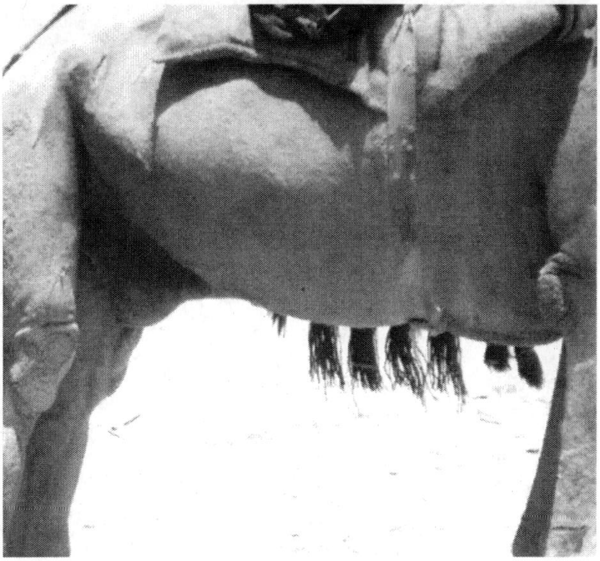

Figure 3. Gross lesion of the camel nasal fly larvae in the pharyngeal area

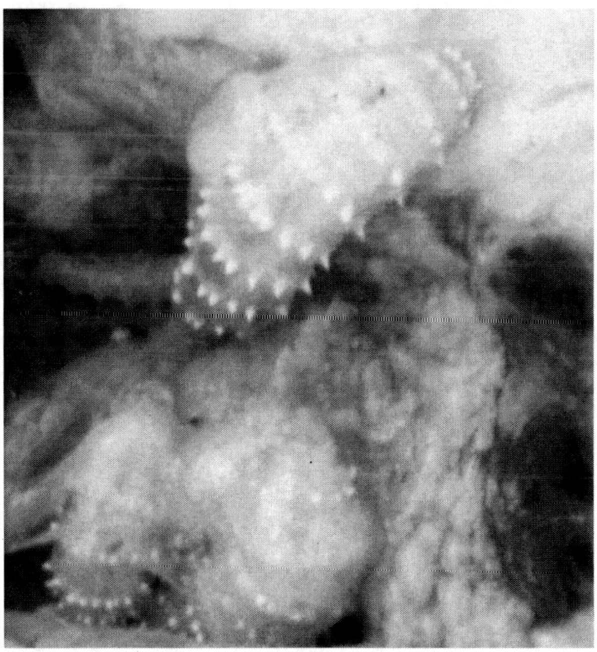

Cephalopina titillator (camel nasal fly) was seen in 69 (38.8 %) of the camel heads examined. Most of the first-instar larvae were found in the labyrinth of the ethmoid bone while second and third-instar larvae were mostly seen in the pharyngeal cavity (Figure

Figure 4. Histopathological section of the parasite nodule in the subcutaneous tissues

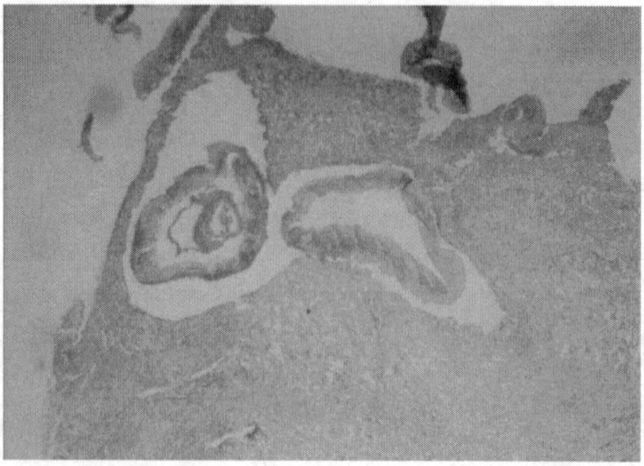

3). The nasal cavities and sinuses were congested where some of larvae were entrapped; red nodules in the nasopharyngeal mucosa indicate the site of larval attachment.

Onchocerca fasciata was found in 13 (23.14 %) of the samples. The nodules were mainly located in the subcutaneous tissues of the chest and belly areas. The nodules contain live and degenerate worms. Most of these worms were fertile females with microfilariae in the uterus. Histologically, nodules contained the parasites and were surrounded by a connective tissue capsule infiltrated by mononuclear cells. Figure 4 shows cross sections of the parasites inside the nodule (H & E stain).

Examinations of the faecal samples revealed the infestation of camels by different types of parasites. Some 98 % of the camels were found to carry gastrointestinal helminthes (Figure 5); *Trichostrongylus probolurus, Haemonchus* and *Monezia expansa* were the most common. Other parasites of less frequent occurrence were *Ascaris spp, Cooperia onchophora* and *Trichuris species*. Protozoan parasitic species seen in the camel faeces were *Eimeria* and *Cryptosporidium*. Schizonts of *Eimeria* spp were seen in one (0.56 % of the total) intestinal section of a 6-month-old camel calf. The histological section showed a large number of coccidian stages

Figure 5. Histopathological section of camel intestine

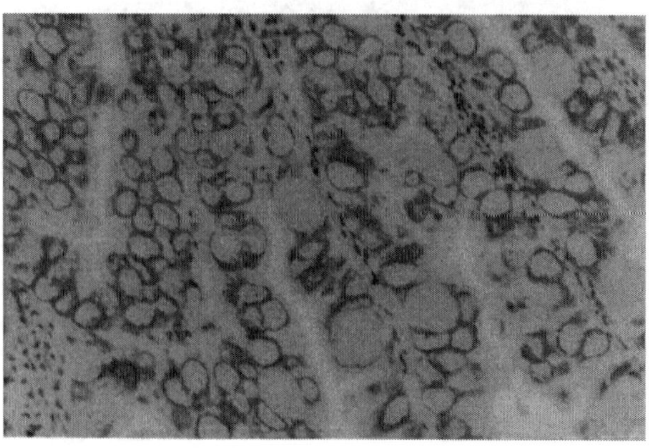

Figure 6. Histopathological section of hydatid cyst in the lung of a camel (H & E stain)

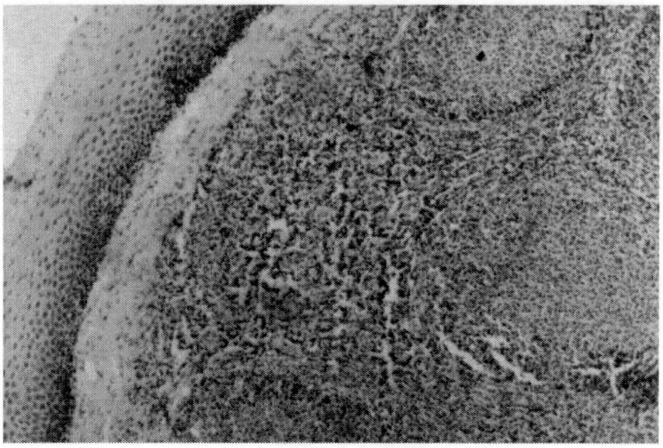

Figure 7. Hair losses due to *Sarcoptes scabie* var *cameli mange*

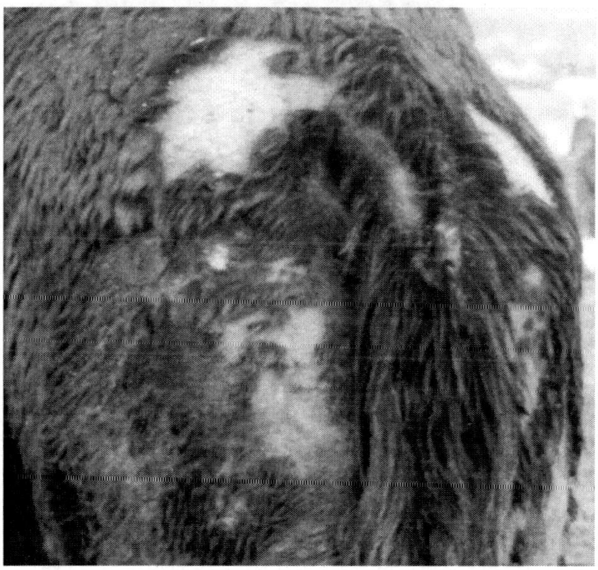

in the intestinal villi. Oocysts of *Cryptosporidium parvum* were seen in one (0.56 %) of the faecal stained samples. The cysts appeared red in acid-fast stain and of 5-6 μm in size. See the cross section of the parasite in the intestine wall (H & E stain) in Figure 5.

Echinococcus granulosus (Hydatid) cysts in the lungs and liver were found in 28 (15.7 %) of the samples examined. The cysts were of variable sizes and contained fluid, and some were calcified. Histologically, the cysts were surrounded by a fibrous connective tissue and infiltrated by mononuclear cells, eosinophils, and multinucleated giant cells (Figure 6).

Sarcoptes scabie var *cameli* was found in 31 (17.4 %) of the skin scrapings examined. Affected camels lack proper rest because of the intense pruritus and they spend much time biting, scratching and rubbing the infected area. They usually have low feed intake which often leading to weakness, emaciation and a marked reduction in milk production

Figure 8. Ticks engorged with blood on the base of the tail of a camel heavily infested by *Hyalomma dromedarii* ticks

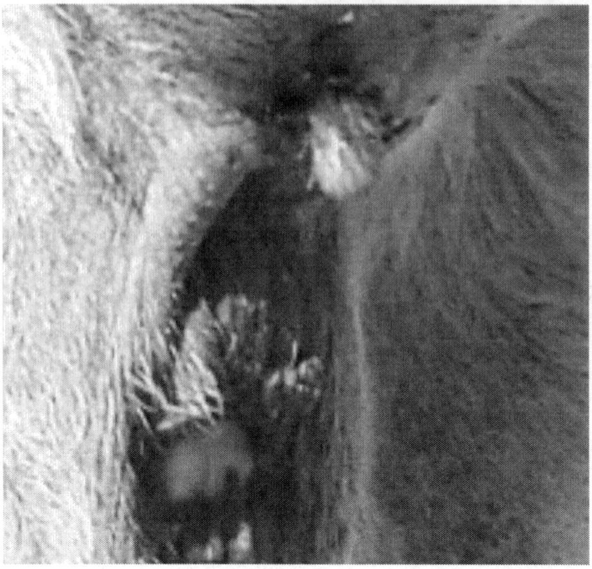

and working capacity. The skin on the infected areas was markedly thickening and alopecic (Figure 7). Microscopically, the epidermis was variably acanthotic, hyperkeratotic, showing oedema and congestion in the dermis. Sarcoptic mites were also observed in the epidermal layers. The dermal lesions consisted of superficial vasodilatation, oedema, and extensive infiltration by lymphocytes, plasma cells and histiocytes.

In 174 camels (97.7 % of the total; an extremely high rate) several types of ticks were identified; *Hyalomma dromedarii* and *Hyalomma anatolicum* were the most common (Figure 8). Skin damage, decreased productivity, vectors for diseases and tick paralysis have been observed.

4. Discussion

Parasitic infestations are a big problem for camels in different parts of the world (Al-Ani et al 1995, Al-Ani et al 1991, Al-Ani et al 1998, El-Bihari 1980). According to our knowledge this is the first report of *Sarcocystis cameli* infection in Arabian camels in Jordan. Sarcocystosis due to *Sarcocystis* species is a common parasitic infestation of the skeletal and cardiac muscles of mammals, birds and reptiles. Wild and domestic ruminants can be infected by different species of *Sarcocystis*. Mason (1910) gave the first description of *Sarcocystis cameli* in an Arabian camel in Egypt. Since then the parasite has been reported in camels in Afghanistan, Egypt, Iran, Sudan, Saudi Arabia, Iraq and Russia (Dubey et al 1989, Fatani et al 1996).

Studies on diseases of the Arabian camels have suggested that trypanosomiasis and helminthiasis infections are the major cause of impaired milk and meat production and decline in calving rates (Richard 1979). Trypanosomiasis, caused by *Trypanosoma evansi*, is transmitted by biting flies. The parasite commonly leads to trypanosomiasis in camels in the Jordan Valley but is of low prevalence in the other areas of Jordan (Al-Rawashdeh et al 1999). The disease usually takes two different clinical forms: the acute form and the chronic form (Al-Ani 2004). Also, variations in the percentage of infection were seen in Jordan from month to month with a peak from April to October (Al-Rawashdeh et al 1999).

In our study only one camel calf was found infected with coccidiosis. There were numerous reports about coccidiosis in camels in different parts of the world. Levine (1985) reported the occurrence of six species of intestinal coccidian parasites in Arabian camels, Asiatic camels and South American camelids. Camel coccidiosis has been diagnosed in newly born camel calves in the United Arab Emirates (Kinne et al 2001). Massive numbers of *Isospora orlovi* stages were detected in the lamina properia of the intestine by histopathological studies. Severe coccidiosis causing enteritis and a mortality rate up to 10 % in young camels has been reported (Kinne et al 2001, Levine 1985). The invasion with destruction of intestinal mucosa by the coccidian is responsible for watery diarrhoea, anorexia, dehydration and even death of infected animals. These symptoms may be mild and transient lasting for few days or may become severe enough to cause death of the affected animals (Al-Ani 2004).

Our study showed that camels in Jordan are highly infested with gastrointestinal parasites; 94.4 % of the camels examined. Moreover, mixed infestations comprized of two or even three parasites present in the same camels were diagnosed. These findings were in agreement with the finding of Selim and Rahman (1972) in Egypt and Altaif (1974) in Iraq.

The adult tapeworm, *Echinococcus granulosus*, lives in the small intestine of dogs and other *Canidae*. Camels, along with other animals, become infected by ingesting the parasite's eggs containing an embryo. Hydatid cysts of various sizes develop in various camel organs, especially lungs and liver (Azlaf et al 2006).

The present study is the first to report the prevalence of *Onchocerca fasciata* in Arabian camels in Jordan. The histopathological findings of *O. fasciata* nodules of live, degenerate and dead worms in the present study is in accordance with the results of Cheema et al (1984) and Al-Yousif et al (1994).

In most camel-raising areas, ectoparasites play a major role in spreading many camel diseases (Higgins 1984). Camels and their tick parasites harbour several viruses and rickettsia of biomedical importance (Sherkov et al 1976). Camels are believed to play an important epidemiological role in the maintenance and distribution of Rift Valley fever virus and possibly in the epidemiology of Congo Crimean Hemorrhagic fever (Hoogstraal 1979, Sellers et al 1982). *Sarcoptes scabiei var cameli* is the most important species affecting camels. It is reported in camels in Iraq, Jordan, Syria, India, Egypt and Saudi Arabia.

5. Conclusions

The cooperation of the staff of the Jordan Badia Research and Development Centre (BRDC) to the researchers helped us to study the socio-economic importance of camels in the Jordan Badia. In this study, we found that parasitic infestations occur quite frequently in camels in Jordan. According to the disease occurrence, we classify the incidence of camel parasites in Jordan into 3 different categories: common, less common and rare. Among the common parasites were gastrointestinal parasites, nasal myiasis and tick infestations. Less commonly occurring parasites were associated with were hydated cysts, trypanosomiasis, and sarcosporidiosis. Among rare parasites were *Eimeria* spp, *Cysticercus tenuicollis* and *Cryptosporidium parvum*, causing infections.

Acknowledgements

The authors thank Mr. Mohammed Shahbaz and his staff members at BRDC for their cooperation, for their help in gaining access to the camel herders and visiting different locations, and in the collection of samples from camels in the Badia. The financial support from BRDC is appreciated.

References

Al-Ani FK (2003). Camel encyclopedia; 2nd edition. Amman, Jordan: Dar Ammar Book Publication.
Al-Ani FK (2004). Camel management and diseases; 1st edition. Amman Jordan: Dar Ammar Book Publication.
Al-Ani FK and Al-Shareefi MR (1995). Observation on medical leech (Limnatis nilotica) in a camel in Iraq. J. Camel Pract. Res 2:145.
Al-Ani FK, Khamas WA and Zenad KH (1991). Camel nasal myiasis: clinical, epidemiological and pathological studies in Iraq. Indian J. Anim. Sci (61):576-578.
Al-Ani FK, Sharrif LA, Al-Rawashdeh OF and Al-Hammi Y (1998). Proceedings of the third annual meeting for animal production under arid conditions pp. 77-92.
Al-Rawashdeh OF, Sharif LA, Al-Qudah K and Al-Ani FK (1999). Trypanosoma evansi infection in camels in Jordan. Revue Elev. Med. Vet. Pays. Trop 52:233-237.
Al-Yaman FM (1985). Prevalence of hydatidosis in slaughtered animals from North Jordan. Ann. Trop. Med. Parasito 79:501-506.
Al-Yousif M and H.S H (1994). Onchoceriasis in the Arabian camel (Camelus dromedarius) in the central region (Riyadh and Qasim Provinces) of Saudi Arabia, Arab Gulf. J. Sci. Res 12:361-383.
Altaif KI (1974). Helminthes in camels in Iraq. Trop. Anim. Health. Prod 6:55-57.
Azlaf R and Dakkak A (2006). Epidemiological study of the cystic echinococcosis in Morocco. Vet. Parasitol 137:83-93.
Cheema AH, Al-Bihari S, Ashour NA and Ali AA (1984). Onchcerciasis in camels (Camelus dromedaries) in Saudi Arabia. J. Helminthol 58:279-285.
Dubey JP, Speer CA and Fayer R (1989). Sarcocystosis in animals and man. Boca Raton, Florida: CRC Press.
El-Bihari S (1980). Helminthes of the camel: a review. Br. Vet. J 141:315-326.
Fatani A, Hilali M, S, A-A and S, A-S (1996). Prevalence of Sarcocystis in camels (Camelus dromedaries) from Al-Ahsa, Saudi Arabia. Vet. Parasitol 62:241-245.
Higgins N (1984). The camel in health and disease. London: Bailliere Tindall.
Hoogstraal H (1979). The epidemiology of tick-borne Crimean Congo Hemorrhagic Fever in Asia, Europe and Africa. J. Med. Entomol 15:307-417.
Kinne J, Ali M and Wernery U (2001). Camel coccidiosis caused by Isospora orlovi in the United Arab Emirates. Emir. J. Agri. Sci 13:62-65.
Levine ND (1985). Veterinary protozoology. 1st ed., Ames, Iowa. USA: Iowa State University Press.
Mason FP (1910). Sarcocystis in the camel in Egypt. J. Comp. Pathol. Ther 23:168-176.
Richard D (1979). Dromedary pathology and productions, Workshop on Camels, 18-20 Dec, 1979, at Khartoum, Sudan. USA: International Foundation for Science pp. 409-429.
Saliba EK (1977). Records of Tabanidae from Azraq Oasis Jordan. J. Med. Ent 14:468.
Selim MK and Rahman MS (1972). Enteric nematodes of camels in Egypt. Eygyptian J. Vet. Sci 9:75-80.
Sellers RF, D.E, P and Tucker MR (1982). Rift Valley Fever, Egypt 1977: disease spread by windborne insect vectors? Vet. Rec, 110:73-77.
Sharrif LA, Al-Qudah K and Al-Ani FK (1996). Prevalence of gastrointestinal parasites in camels in Jordan. J. Camel Pract.Res, 5:1-4.
Sherkov S, Leitch B and El-Rabie Y (1976). A survey of sarcosporidia in domestic animals in Jordan. Egyptian J. Vet. Sci 13:45-51.
Sherkov SN and El-Rabie Y (1978). A survey of Linguatula serrata in domestic animals in Jordan. Egyptian J. Vet. Sci 15:89-97.

CHAPTER 15

Livestock marketing in the Jordan Badia before the removal of the feed subsidy in 1996

Salem Al-Oun

Department of Business Administration, Faculty of Economics and Administrative Sciences, Al Al-Bayt University, PO Box 13004 Mafraq 25113, Jordan. E-mail: al_oun@aabu.edu.jo

Little work has been undertaken on livestock marketing systems in the Badia in spite of the traditional importance of livestock to the formal and informal economy of the Bedu. The paper compares the major sheep production statistics from the 1990s and today but focuses on field work undertaken in the mid 1990s before the removal of a major food subsidy in 1996-7. It thus not only provides an understanding of the many factors that governed sheep marketing at that time but also acts as a baseline for any similar study today a decade after the removal of the subsidy. A large number of farmers responded to questionnaires and traders were interviewed. The relationship between producers and traders is explored and it is clear that sales were made for a complex mix of commercial and non-commercial reasons. The paper appeals to the government to better understand Bedu society and requirements before embarking on future interventions that affect the livestock industry.

Keywords: Jordan Badia, sheep, goats, marketing, development

1. Introduction

The marketing system is an integral part of the society within which it operates; it responds to society's needs. Marketing in theory is regarded as a tool for economic and social development (Drucker 1958, Mitchell 1985). Markets potentially contribute to development by providing ways of allocating resources to ensure the highest value of production. By improving the skills and capacities of farmers, marketing contributes to an increase in productivity, speeding up economic development and improving the farmers' welfare. The absence of markets or the inadequacy of existing markets can be an obstacle to both agricultural and economic development. Markets may stimulate growth by promoting technology, innovation and increased supply and demand. Improvement in marketing services, especially for small farmers, has a positive effect (Abbott 1987, Mitchell 1985).

The term 'Bedu' is used instead of 'Bedouin' and refers to mobile, settled and semi-settled pastoralists. The term 'producer' refers to the farmer who is considered in this study as the head of the household and who is responsible for trade and business decisions. The term 'outlet market' is applied to livestock markets usually found in major cities and towns in Jordan. These daily markets open early in the morning and close before midday.

2. Definition of the problem

The Badia of Jordan is the main area of livestock production, and it plays an important role in the supply of meat to local and regional markets where demand has grown in an unprecedented manner, partly as a result of urbanization (BRDP 1994). The rapidly developing growth in animal production in recent decades led to extremely degraded grazing areas, some of which have been irreversibly damaged and desertified. The increase in production costs was counterbalanced by government intervention in the form of feed subsidies for the two decades up to the mid-1990s. However, this intervention encouraged Bedu to expand their flocks, thereby worsening the problem of overgrazing (Randhawa 1990). Unfortunately, agricultural marketing policies and institutions in Jordan have been formulated mainly for the benefit of arable farming and are less relevant to livestock farmers. Indeed, these policies have contributed to the inefficiency of livestock marketing. In the 1990s, the Jordanian government implemented a trade liberalization programme that reduced direct control over parts of agricultural production and subsidies. Tariff exemptions and quantitative restrictions were also reduced. This policy was motivated by the fact that Jordan planned to join the World Trade Organization, which it did in December 1999.

In spite of the importance of sheep in the household economy of the Badia, little emphasis has been given to livestock marketing research, with most research concentrating on crop production. There is little understanding of the socio-economic changes in Bedu life, and in the likely responses to these changes and the workings of their economy (Livingstone 1984). According to Randhawa (1990), research into the animal sub-sector has received the lowest priority. The few data available on the livestock system are mainly government reports and statistics. The Badia Research and Development Centre (BRDC[1]), whose mandate is to promote agricultural development at the regional level, has, so far, received very little reliable or relevant agricultural marketing information. Consequently, present knowledge about livestock marketing in the area is incomplete.

3. Objectives of the study

The research on which this paper is based is part of a comprehensive socio-economic set of studies investigating the extent to which changes in the agricultural policy of the Jordanian government have affected the people of the Badia region. The overall objective of this paper is to provide a better understanding of the processes of the Bedu livestock marketing system in Jordan, particularly farmers' marketing decisions and how they are linked to the market environment. The specific objectives of this research are:

- To explore and analyze the marketing of livestock in order to determine factors that affect producers' marketing decisions
- To provide an analysis of farmers' marketing behaviour
- To analyze the livestock marketing channels and processes in terms of the organization and role of major market participants
- To suggest policy recommendations for livestock marketing in the Badia of Jordan in the longer term.

4. Livestock in the Jordanian Badia as an economy and its present conditions

Before the 1920s, Bedu neither owned nor cultivated land, although most tribes held claim to a certain grazing area. Droughts, limitation of movement, and the expansion of state control were all factors that made many young Bedu give up their nomadic lifestyle and join the military forces in Jordan and Syria (Abu Jaber 1987, Abu Jaber et al 1981). Settlement was also accelerated in the early 1970s when large numbers of Bedu were drawn to the Gulf States for work. The Bedu, both nomads and settled, who now remain

[1] The Jordan Badia Research and Development Programme was launched as a joint three-year programme between the Royal Geographical Society (RGS), The Higher Council for Science and Technology (HCST) in Jordan and The Centre for Overseas Research and Development (CORD), University of Durham. The main aim of the programme has been research and development in the northern region of the Jordan Badia. BRDP became a national centre – BRDC – in 2005)

in the Badia, constitute only a small percentage of the Jordanian population. Their numbers in 1989 were estimated to be146 105, about 4.7 % of the total population of 4.15 million, with only 5 % of Bedu still living as nomads (MoP 1989). A typical Bedu can no longer be defined as having a nomadic lifestyle. The Bedu can be nomadic, semi-nomadic, or settled. More than the way he lives, it is the strength of the Bedu mentality that is important for the classification of a Bedu. The three types share similarities in social values and customs, and they only differ in the physical terms such as movement and size of flock. Tribal membership is still important for all three types as it gives them certain rights over the strategic resources found in their territory, such as land and employment opportunities (Jabbur et al 1995, Nesheiwat 1991).

Because settled Jordanian Bedu outnumber nomadic Bedu, the Bedu who have persisted with pastoralism appear to be more affluent now than before. With new-found wealth in the region, Bedu have been able, albeit with government subsidies, to balance comfortably a semi-nomadic lifestyle with a dwelling near the town. The increased numbers of settled Bedu have created a larger market which has benefited the nomadic Bedu. Bedu society in the villages is quite different from that of the nomadic kind. Less than 5 km from the border with Syria, a different form of society has developed. These northern Badia settlements have evolved into efficiently organized societies based on the active participation of the inhabitants. The settlements are dependent on government aid, but have made steps towards self-sufficiency. The life of the animal owner has had to adapt to the new conditions; some flocks graze in the eastern Badia, but the majority are grazed locally around the northern villages. Movement with flocks has become a factor distinguishing nomads from sedentary people (Livingstone 1984). The semi-nomads cultivate crops in the Badia, primarily barley and wheat alongside tomatoes and watermelons, and then graze their flocks of sheep and goats. With population pressures and extensive farming, many rangelands have become overused. Limited pastures and growing numbers of sheep have added to overgrazing and deterioration of the Bedu areas (Livingstone 1984, Marx 1984).

In the past, when the Bedu were less marginalized than they are today, their rhythm of life was more influenced by the seasonal movements of their animals, following rainfalls, to different areas within the Badia. Raising animals in this way caused no danger of environmental degradation or scarcity of animal feed. One of the main constraints facing them in recent times is that of obtaining sufficient feed for their animals (Chatty 1986). Bedu settlement was on a tribal basis with all members settling on land given to the tribe by the government as a way of encouraging their self-settlement. As these settlements are far away from urban centres, the Bedu were able to keep their cultural values and traditions apart from other populations, with their economy still dependent on livestock and its products (Abu Jaber et al 1981).

For some years, few Bedu have relied on livestock as sole source of income. Instead, livestock farming has become only one of several options which include salaried employment in the army and the civil service (Abu Jaber et al 1981, Galaty et al 1991). Lacking the qualifications to enter the workforce, many Bedu became manual workers and so received low wages (Kay 1978, Nesheiwat 1991). The interdependence between the Bedu and the urban population presumably represents a balance of power between these two. The urban people use up enormous quantities of meat, milk, and milk products coming from the Bedu. In Jordan, in the 1990s the Bedu were estimated to own about 74 % of sheep and 70 % of the country's animal products. The remaining 30 % were largely owned by farmers, merchants, and civil servants (BRDP 1994). Recent figures of sheep numbers indicate that about 40 % of the sheep and goat production in Jordan comes from the north-east Badia alone; Mafraq governorate (Department of Statistics 2007). In 1995, the Mafraq governorate's contribution to national output was about 27 %

for lambs and 13 % for ewes/new-born lambs; some 444 000 lambs, ewes and new born lambs, valued at over JD18 million (Oakeley 1996a). However, by 2005 Mafraq governorate's contribution to national output was about 47 % for lambs, 35 % for ewes and 50 % new-born lambs; a total of 896 330 lambs, ewes and new born lambs (Department of Statistics 2007).

In the past, the nomads' reason for keeping livestock was mainly to provide themselves with milk and milk products, meat and wool. In more recent times it became a source of income and employment, being both more accessible and more reliable for Bedu than other means of income. This latter point is emphasized by the 1988-9 economic crash in Jordan which resulted in the average income per head falling from the 1985 sum of $1 570 to $980 in 1989 (EIU 1993). Furthermore, the 1992 figures show that for a typical family of 6.8 persons, the abject poverty line was JD61 and the absolute poverty line was JD97 for a family not paying rent and JD119 for a family paying rent. The percentage of families below the abject poverty line increased from 1.5 % in 1987 to 6.6 % in 1992, while the percentage of families below the poverty line increased from 18.7 % in 1987 to 21.3 % in 1992 (MoP 1993), and to 30 % in 2001 (CIA 2006, Wikipedia Encyclopedia 2006). Today, their cash needs have risen sharply with the introduction of technology and increased interaction with the town populations. The Jordanian Bedu economy has also changed in response to economic growth and the population increase at the regional level.

In the past, the nomads' mobility allowed them to adjust to many constraints and to profit from widely dispersed resources whose availability varied from one year to another. The Bedu's production management system for herding their animals suited their environment. Animals were herded eastward and westward, allowing natural vegetation to recover (Nesheiwat 1991). At the same time, most Bedu found themselves able to shift from herding camels to small livestock production and act as intermediary traders between different countries. The sheep grazing system fell into two dominant categories: the semi-settled and the nomadic. According to Nesheiwat (1991) the nomadic grazing system only existed in the Badia. The livestock economy advances and becomes a success among the Bedu only when they become both flock owners and operators, not relying on hired shepherds. Livestock raising was considered a worthwhile activity by the Bedu, because in addition to providing them with food, wool, and goat-hair, it enabled them to fulfil obligations on social and religious occasions, and strengthened their feelings of identity.

For some decades, the increased fragility of the Badia due to soil deterioration appears to be one of the most disturbing effects on livestock development. The large extent of rangelands in the Badia is found where rainfall is low, variable and unreliable, and where long-range annual precipitation averages below 75 mm. Pastures in the area deteriorated and became incapable of maintaining the number of animals found in the country. Government policy had both major positive and negative effects on livestock production (Kay 1978, Kerven 1992, Livingstone 1984, Nesheiwat 1991, Oakeley 1996b, Salzman et al 1990). Many government programmes failed or harmed pastoralists because they were based on urban judgment and the estimates of government officials (Briggs 1985). They failed to understand social values. The Bedu have not only tribalism, but also a form of political and economic organization, with social activities within and outside tribal affairs (Abu Jaber et al 1981, Leybourne 1993).

In the past, the Jordanian government maintained a stable policy toward land ownership in the Badia in that the land belonged to the state, and the tribes held additional usufruct rights (Abu Jaber 1987). In more recent years, the government introduced a new policy by transferring the ownership of land to individual tribal members, which in turn encouraged the sale of land in the Badia. Government attempts to bring tribal lands under private title have ended up removing large areas of land from grazing because of the growing interest in using pastures as farms, although this type of farming has proved to be both uneconomical and harmful to the environment, particularly water resources. According to 1990s estimates, forage production from rangelands had dropped 50 % during the previous 15 years and would continue to drop by 33 % in the coming 15 years (Oakeley 1997).

All the above factors contributed to high costs of livestock production. Estimation of supplementary feed accounted for 68 % of production costs (Roe 1996). Before the lifting of feed subsidies in 1996, farmers were able to make profits from livestock farming, an estimated JD13.5 average profit per head. This figure then fell to JD1.62 when subsidies were lifted, which shows the dependence of the livestock production system on such subsidies (Oakeley 1996b, Roe 1997). Such conditions have both a short and medium term effect on Bedu marketing. In the short term, as livestock are threatened and the price of feed rises, the market is flooded with animals being off-loaded and more expensive feed being bought. In the medium term, Bedu producers had to rethink and some were expected to give up production (Oakeley 1996b).

4.1. Livestock economy, 1996-07

Livestock has remained one of the main constituents of the agricultural sector, which has contributed about 60 % of the agricultural output and provided a major source of income to 250 000 people. The contribution of the different sectors of livestock to agricultural produce is variable. About 48 % of the Badia rangeland communities depend on this sector for their livelihood, and women assume a great role in the production activities through assisting in livestock raising, animal feeding, flock milking and produce processing. The number of people living directly or indirectly on pastoral activities is now about 185 000 persons living in 170 communities. Approximately 48 % of the families in these communities own livestock (MoA 2007).

Available statistics indicate that self-sufficiency in livestock produce in the 1990s was about 30 % for red meat and 53 % for milk and dairy products. The value of imported live animals and livestock products was JD114 million in 1999, out of which JD73 million was for live or slaughtered animals and frozen meat, and JD41 million for milk and dairy products. In 2000, the livestock sector included about 1 484 090 head of sheep and 472 460 head of goats. The livestock output in 2000 was estimated at 15 000 tons of red meat. Very recently, Bedu farmers have significantly decreased their sheep flocks but have also somewhat increased their numbers of camels and goats. There were 2 671 317 sheep, 452 953 goats, and 10 867 camels in Jordan in 1990 (Directorate of Statistics 1991), and 1 901 160 sheep, 516 130 goats, and 13 000 camels in 2005 (Department of Statistics 2007).

Since 1997 many factors have caused fluctuation of livestock numbers in Jordan especially in the Badia (Table 1). For example, in 1998 numbers dropped to about 1.6 million sheep, and dropped again in 2000, but since 2001 the combined sheep and goat numbers have grown (MoA 2006).

Factors that have influenced the increase since 2001 are:

- The recent oil boom and comprehensive economic reforms in 2001 (AFCR 2005, Ignatius 2005), which increased demand for red meat in the Gulf countries including Saudi Arabia, leading to a growth in exports from Jordan to these countries;
- Sanctions imposed in September 2000 and 2001 by Saudi Arabia on livestock exported from Africa because it was infected with Rift Valley Fever[2]. This again increased demand for red meat from Jordan;

---

[2] Rift Valley Fever is a viral zoonosis that may cause severe disease in both animals and humans leading to high morbidity and mortality (MoA 2002), Nature Management and Fisheries 2002; www.foodcontamination.ca/animalnet 2007).

Table 1. Sheep and goat numbers in Jordan and Mafraq, 1997-2005 (Source: MoA 2007)

Year and location		Number of livestock by type Sheep	Goats
1997	Jordan	1 935 100	649 590
	Mafraq	786 600	101 310
1998	Jordan	1 581 130	631 410
	Mafraq	507 960	66 540
1999	Jordan	1 933 990	461 390
	Mafraq	759 770	72 540
2000	Jordan	1 484 090	472 460
	Mafraq	477 490	40 710
2001	Jordan	1 457 920	425 910
	Mafraq	604 970	54 450
2002	Jordan	1 433 320	557 290
	Mafraq	538 280	93 550
2003	Jordan	1 476 450	547 470
	Mafraq	584 780	60 540
2004	Jordan	1 529 100	501 130
	Irbid	99 280	57 410
2005	Jordan	1 890 450	516 130
	Mafraq	896 330	69 560

- The government continued to subsidize animal feed for socio-economic reasons, as government policies were aimed at ensuring continued macroeconomic stability and supporting economic growth, despite the conflict Jordan had with the International Monetary Fund (IMF 2007) about eliminating feed subsidies. The role of feed subsidies, especially for feed imports, in promoting livestock numbers is indicated from the fact that production of animal feed in Jordan is limited to barley and other grains, which cover only 23 % of total needs, and to dry forage, which covers only 52 % of the forage needs. Jordan produces 268 000 tonnes of all kinds of animals feed which cover about 21 % of the livestock sector needs. Jordan still lags behind other countries in using agricultural by-products, which have a potential of providing an estimated 300 000 tons of animal feed. The most important agricultural by-products include those coming from field crop harvesting, slaughterhouses, poultry farms, olive presses, tomato factories and others (MoA 2006). The value of livestock feed imports in 1999 was JD162 million, which included JD56 million for barley, JD57 million for corn and JD49 million for feed concentrates. The total quantity of imported feed products amounted to 1.7 million tons in 2000;
- The spread of bird flu (pathogenic avian influenza H5N1) in poultry in Jordan in March 2006 (USDHHS 2006), which increased demand for red meat at both the national and regional levels;
- Liberalization of trade agricultural commodities:
    - In 1998, Jordan joined the Great Arab Free Trade Zone, which stipulated phasing out custom duties by 10 % annually and completely cancelling them by 2007;
    - In 2000, Jordan signed and implemented the Jordan-EU Association Agreement, which provided for facilities and for the establishment of a free trade area, as well as exemption of custom duties on Jordanian exports of agricultural products;
    - In December 1999, Jordan became a full member of WTO, which made Jordan's local and foreign trade subject to the provisions of WTO agreements;

- In 2001, Jordan signed a Free Trade Agreement (FTA) with the USA, which provided for a wide range of Jordanian agricultural products to enter the American market, exempted from custom duties or at reduced rates;
- Jordan has also signed several bilateral trade agreements to establish free trade zones with some Arab countries.

These changes have increased livestock export activities and demand, thereby creating more interest in livestock investment. However, livestock numbers may keep fluctuating in response to any major events that may occur in the future.

5. Agricultural marketing theoretical framework

The analysis of livestock production and marketing under traditional systems cannot be confined to a particular approach or concept. Applying one approach can provide some, but not all, of the answers to research questions. While a precisely defined concept of the market and the marketing system is essential for the investigation of current livestock marketing in the Badia, with its strong connection to production behaviour of producers and market conditions, a narrowed down concept is not very helpful. Any analysis has to examine the impact of the varying conditions of production and marketing on the Bedu. People seek to satisfy their needs as they perceive them, whether these needs be biological or psychological. But the means used to attain them differ from one society to another, in the light of the technological, historical, social and psychological factors that are at play. Thus, economic motivation may not be the only, or even the dominant, explanatory variable of the behaviour of many small-scale farmers (Alcantara 1992).

As Ekins and Max-Neef (1992) argued, economists have insisted on employing very narrow approaches when focusing on rational economic behaviour. For these economists, this implies the input to output ratio of production and dismisses all other aspects of a farmers' life that should be taken into account, such as household production and satisfaction. Farmers do not commit themselves lightly to a definition of optimal efficiency, when the economy does not provide the resources that would allow production to reach an ideal Western standard of efficiency (Ekins 1986, Ekins et al 1992, Sachs 1992, Schultz 1974, Shaffer 1980). In many cases, the economist decides what that standard is, taking an industrial society as his or her model (Sachs 1992). However, there seems to be a growing consensus among agricultural economists that aligns with the broader, more dynamic view of marketing as a major element in the development of the agricultural sector and in coordinating agriculture with growth and development in other sectors (Alcantara 1992, Kohls et al 1985).

In this study the Environment-Behavioural-Performance (E-B-P) model was chosen because it allows more room to consider the environmental and behavioural factors that affect Bedu marketing behaviour and decision making. It provides a strict number of criteria when it comes to determining market performance, and focuses mainly on economic efficiency, marketing margins, profit margins, price efficiency, etc. Desirable market performance is directly related to the competitiveness of an industry because distortions thereof tend to impede price efficiency (Abbott 1993, Drucker 1958, Fialor 1994, Kinsey 1982, Shaffer 1980).

6. Method of the study

This part of the paper deals with livestock and livestock marketing in the mid-1990s before the removal of the main livestock concentrate feed subsidy in August 1996. This is a time, therefore, when it was possible to make a profit of JD13.5 per head instead of the much more meagre JD1.62 per head after the subsidy had been removed; a notional profit margin which, in practice, left many farmers making a loss.

The Bedu livestock system comprises interacting environmental, biological, cultural, and economic factors. Attention therefore firstly centres upon the household characteristics and the household member, usually the head of household, who is the key decision-maker with regard to livestock production and marketing practices. Secondly, attention is focused on the livestock trader, who plays the principal marketing role at the first stage. Consideration is given to environmental conditions, socio-economic circumstances, and national policies.

6.1. Hypotheses of the study

It is hypothesized that livestock marketing in the Badia and specifically the BRDP area was, up to the mid-1990s, performing well because:

- Under prevailing livestock marketing circumstances, the sale of livestock from Bedu households was a reflection of demand for live animals from local and regional markets;
- Neither producers nor traders had excess bargaining strength within the marketing system;
- Large-scale producers had more alternatives in their marketing decisions than small-scale producers;
- The majority of marketing decisions were made on production considerations; and
- The importance which farmers place on marketing varied according to the number of animals they had and on various socio-economic factors.

6.2. Choice of the study area

The majority of Jordan is arid and semi-arid land. Over 81.3 % is defined as rangelands, receiving an annual rainfall of less than 200 mm. This area is known as the Badia. The north-east Badia is located in Mafraq governorate and extends north and eastwards to the Syrian and Iraqi borders and comprises 25 930 km^2. The middle Badia extends into central and southern Jordan and comprises 9 634 km^2. The southern Badia is located in the Ma'an governorate, extends to the Saudi border and comprises 37 096 km^2 (Figure 1) (Abu Jaber 1987, BRDP 1994). The Jordan Badia receives low priority in national and international policy initiatives, despite its economic and structural vulnerability. Livestock production accounts for approximately 38-45 % of the gross value of agriculture production; 74 % of livestock is found in the Badia.

The study area, the north-east Badia covered by the original Badia Programme[3], extends for some 11 210 km^2 and constitutes over 15 % of the total Jordanian Badia area (Figure 1). The BRDP area represented an important part of livestock production in Jordan, including about 107 000 sheep and goats (MoA 1992). The total population of the BRDP area was estimated to be 15 318 in 1993, divided among 34 villages with additional 152 bayt as-sha'r (tents) (BRDP 1994). In this area, the main tribes are Beni Khaled, Al-Sirhan, Al-Isa, Ahl Al-Jabal, and Al-Sardiyyeh (Abu Jaber et al 1987). Whilst the population of the Badia could not simply be described as pastoralist, most of the people in the area depended, at least in part, on livestock for their livelihood. In 22 % of house-

[3] In May 2005 the Badia Programme (BRDP) became the Badia Research and Development Centre (BRDC) and its remit was widened to cover all the dry lands in Jordan. A second field station has since been opened at Ma'an in southern Jordan.

Figure 1. Map of the study area and the Badia zones (Source: BRDP 1994)

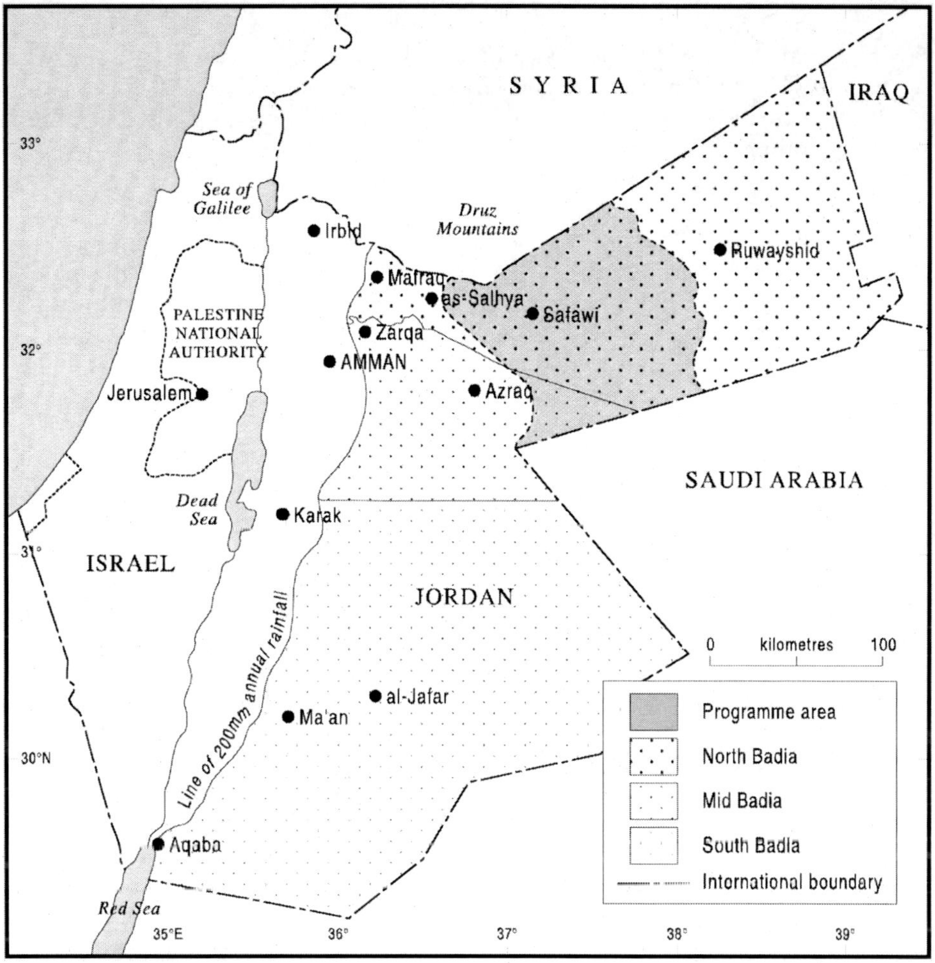

holds, livestock formed the main source of income, while 35 % of the people listed livestock production as their major activity (BRDP 1994).

6.3. Method of data collection

There has been little work done on livestock and livestock products or on red meat markets in Jordan. The majority of the literature published focuses on nomadic, socio-economic and anthropological studies. Consequently, the following data collection methods were used in this research:

- Official publications and statistics from different institutions. The presentation of agricultural statistics is reasonably comprehensive on the statistical side, but research suffers from the disadvantage of being limited to a few government animal experimental stations (MoA 1990, 1991, 1992, 1993, 1994, 1995).
- Informal field observations and discussion with farmers, traders, exporters, butchers and experts in livestock production. Livestock markets such as Amman, Mafraq and Ruwayshid, were visited on more than one occasion. Visits were also

paid to different government institutions such as the Ministry of Supply, Ministry of Agriculture, and some customs departments, especially in Mafraq governorate.
- As neither of these two sources provided enough detailed information on livestock marketing, a questionnaire for farmers was designed and interviews with traders were undertaken. A basic structured questionnaire was chosen for farmers because they greatly outnumbered traders making it not feasible to interview a large number of farmers.

6.3.1. The questionnaire

As producers were widely scattered over the BRDP area, a personal face-to-face questionnaire was therefore proposed. This allowed more time to travel along the marketing chain and to meet more participants in the markets. The study required information from producers on price, reasons for sale, their relationships with traders and access to market information. The questionnaires were directed towards collecting detailed information on producers' marketing behaviour and decisions and on examining household production, as well as livestock marketing patterns.

According to the Ministry of Agriculture, Mafraq governorate office (MoA 1995), the number of livestock holders holding 1-100 head represents 50.0 %, 101-200 head 20.5 %, and 201-300 head 9.5 %. For the purposes of this study, the decision maker in the household for management of flocks was classified, according to the Directorate of Statistics 1991-1992 Census' definition, as the owner of livestock. The owner of a flock is also seen as the head of the household who is responsible for household affairs. The household is defined in this study as including all members who are the direct kin (wife/wives), son(s), daughter(s), brother(s), sister(s) father and mother) of the head, and family members who work away from the household but nevertheless invest in the household. There was a choice of three sources of information concerning livestock owners in the BRDP area from which to select the sample: The demographic and socio-economic survey for BRDP (1994), Campbell and Roe (Campbell et al 1995b), and the Directorate of Statistics (Directorate of Statistics 1991). These sources gave no information on the actual number of animals in any household.

The target population was the livestock holders given in the 1991-2 census conducted by the Directorate of Statistics. The census collected information on the sex, age, size and type of flocks and farmers' reasons for keeping them, as well as providing the names and addresses of animal holders. On the basis of statistical records obtained from the Ministry of Agriculture's branch in Mafraq (MoA 1995), an estimate of livestock farmers was made using the 1992 figures. Based on flock size and number of sheep per household, a stratified random sampling procedure was used to represent the different classes of farmers. The number of sheep and goats in flocks reflects not only the farmer's preference for either species, but also the suitability of these species to socio-economic conditions in the different parts of the Badia.

The stratified random sample covered 23.4 % of the total population of farmers, and was carried out in July-November 1995. In this study the sample size was 200 households, of a total of 800 in the study area, distributed among 34 villages. The stratified sampling technique divided the total population of livestock farmers into seven groups according to the size of flocks they had, and so included different stockholders, especially small-holders of livestock as follows: 1-100 represented by 101 households (50.5 %), 101-200 represented by 41 households (20.5 %), 201-300 by 19 households (9.5 %), 301-400 by 11 households (5.5 %), 401-500 by 9 households (4.5 %), 501-600 by 6 households (3 %), and 601+ by 13 households (6.5 %).

6.3.2. The interviews

It was decided to adopt an unstructured interview to collect information from traders since they were many fewer in number than farmers and could therefore be held in discussion over a longer time period. This type of interview also allowed for group discussion about important marketing issues. The topics and approaches were carefully focused and refined in response to the traders' contributions to the pilot study carried out in the summer of 1994. The intention was to include all types of traders in the interviews, including exporters, and to include different types of participants at the farm-gate level. In most cases, selection of research sites, subjects, and dominant issues for the study were made with the intention of giving the broadest sampling possible. The selection of traders was also based on a careful consideration of the characteristics of the marketing environment, such as geographical and residential factors, economic scale, especially capital, and destination of sale.

The interview plan was to follow the marketing chains from farmers to the outlet markets. Categorizing traders according to their role and financial ability in the markets was considered, as some carried out their activities within the Badia and others in the outlet markets and others acted as exporters. The group discussions were centred on the following areas:

- Identification of the major participants in the marketing chain;
- Identification of major outlet markets;
- Price information, source of supply, transport costs, sales destinations;
- Importance of regional markets;
- Marketing arrangements; and
- Cross-border trade.

Personal information about and economic activities of traders were recorded in advance by relying on the data collected on households by BRDP in 1993, and from relatives to traders.

The interviews were carried out in July-November 1995. Five unstructured interviews (discussion groups totalling 19 traders) were conducted; one at Mafraq market (seven traders) and the other four (three traders each) at the home of one of the traders. Information from traders was backed up by a variety of visits to outlet markets, such as Ruwayshid, Mafraq, and Amman markets and some governmental and private institutions such as Mafraq Customs Department and some government offices in the governorate.

7. Results and discussion

7.1. Producers' sales pattern

7.1.1. Household marketing management, weaning, and sales strategies and preferences

Household marketing strategies: Changes in the structure of livestock organizations had contributed to the shrinking number of animals per household and to an increase in other households. Inputs in the production of livestock, especially feed, had become one of the main constraints forcing this kind of reduction. Wealthier farmers tended to keep animals for longer and then sell them at 3-4 months of age. A steady flow of young lambs seemed to be the case among some of the small livestock farmers. Table 2 indicates the sale of animals by type. Sales of male lambs dominated, followed by female lambs. The sale of livestock in the study area appeared to take place throughout the year. Lambs came on to the market as early as November or December and continued until September. The sale of sheep and new-born lambs generally occurred in the early part of the year. The sale of lambs reached its peak from late February to early May.

Under Bedu management, the sale of animals takes place every year and sale strategies vary. In the past, young females were retained to replace older ewes in order to

Table 2. Type of animals sold by flock size (Source: Field study conducted by the researcher 1995)

Flock size	Male lambs No. of flocks	Ewes with male lambs No. of flocks	Ewes & new-born No. of flocks	Total (%)
1-100	102	3	1	106 (54.9)
101-200	30	2	0	32 (16.6)
201-400	30	1	0	31 (16.2)
401-800	16	0	1	17 (8.8)
801+	7	0	0	7 (3.6)
Total (%)	185 (95.9 %)	6 (3.1 %)	2 (1.0 %)	193 (100.0 %)

build up breeding stock, and only a few males were kept for reproduction whilst the rest were sold. Some livestock farmers held animals as a store of wealth, and so only sold animals to meet cash needs rather than to maximize income. In other words, this was the case when farmers focused on production rather than commerce. By the mid-1990s, some changes in their attitudes to sale could be identified, but they are still far from being commercial producers. Since livestock production served a multitude of purposes, households made production/marketing decisions under circumstances which were different from those made by farmers involved in commercial production. Concerning livestock marketing, a household had to evaluate the entire household welfare function before making a decision to sell an animal or not. That is to say, factors external to the livestock enterprise were constantly incorporated into the livestock management decision process and could have a significant impact on marketing decisions of all enterprises.

Weaning strategies: In the study, producers were not directly involved in fattening, as they tended to sell their animals directly in the market. However, large flock owners had developed a new strategy in weaning practices to get better prices by extending the weaning age[4] for as long as possible. Most producers stated that they only weaned their lambs at the end of the spring, with the exception of early born lambs, which were sold directly to cover some of the expenses. In many cases, it was found that some producers practised very late weaning to get a better price for their lambs (Table 3). Other producers were aware of the good prices they would receive if they let their lambs grow to a weight of 30 kg or more. In reality, lambs were sold at a young age with an average weight of 20 kg (Blench 1995). The age of sold lambs ranged from 2-6 months depending on the type of household. Weaning styles varied among livestock holders in the area. Farmers with large flocks let lambs suckle up to 4-5 months until the date of sale, whereas small stock farmers tended to separate lambs during the day and freed them at night. This can be explained by the fact that small farmers relied on milk for household and commercial purposes and large stock farmers did not.

Male lambs might have been partially weaned at 6-10 weeks of age if the fodder situation allowed. They were usually completely weaned at approximately 2-3 months of age. Complete weaning of both male and female lambs at the same time was followed by 1-3 months of milking time. Partial weaning could take place one month before complete weaning if the forage supply was adequate. This system was practised by almost all small farmers. Sometimes, all lambs born in the same period were completely weaned at the same time and herded by young people. Ewe-lambs were returned to the breeding flock after one month in this weaner flock.

Sale strategies and preferences: The majority of statements made by the household sam-

[4] The majority of Bedu farmers believe 75 days is the minimum age to wean lambs.

Table 3. Weaning patterns by flock size (%) (Source: Field study by the researcher 1995)

Flock size	Less than 3 months No. of flocks	Over 3 months No. of flocks	Did not wean at all No. of flocks	No. of farmers Total
1-100	68 (64.2)	19 (17.9)	19 (17.9)	106
101-200	17 (53.1)	6 (18.8)	9 (28.1)	32
201-400	5 (16.1)	8 (25.8)	18 (58.1)	31
401-800	4 (23.5)	5 (29.4)	8 (47.1)	17
801+	1 (14.3)	2 (28.6)	4 (57.1)	7
Total	95	40	58	193

ple favoured selling animals on a per head basis, rather than by weight. Of the total 193 answers, 191, nearly 99 %, favoured this method of sale. Two factors emerged as being of particular importance in influencing decisions favouring selling on this basis. These were lack of trust (47.2 %) and the habit of selling on a per capita basis (23.8 %). Whilst no single farmer in the sample sold his animals on the basis of weight, farmers who sold in the Amman market sometimes did so. More than 26 % of the farmers in the sample replied that they sold because of disease and the cost of feed. In terms of wealth, the extent to which household finances were flexible is extremely important as they could affect freedom of choice, particularly with regard to decisions about age and date of sale. Beside the need for cash, other factors may have forced farmers to market their animals against their wishes. Weather conditions, especially high temperatures, were among the many factors which forced farmers to sell early.

The variable 'date of sale' was altered here to facilitate some of the analysis. Livestock sales were categorized into three periods: winter (November-March), spring (March-May) and summer (May-November). Figure 2 shows the seasonal variations in livestock sales by number in the sample. The sale of livestock in the study area appears to have taken place throughout the year. Early lambs came on to the market as early as November or December and continued until September. The sale of sheep and new-born lambs generally occurred in the early part of the year. The sale of lambs reached its peak from late February until early May. With the intention of keeping their flocks healthy, some farmers sold new-born lambs together with the old and weak sheep in order to secure the sale. The timing of livestock sales was usually a response to different constraints such as the health of the animals, the cost of feed, disease, risk, the cash needs of the producers, and the demand for animals. Purchase of animal feed and payment of debts were the main reasons for sale. The early flow of animals in January and February to the market came from both small (less than 200 head) and large (more than 200 head) flock producers. There was a significant difference in the timing of sales between small and large holders. Moreover, well-off households benefited from their wealth and were able to withhold animals for longer in order to wait for better prices; lambs are in their best physical condition after spring (early May) when they have finished suckling.

When respondents were asked about their satisfaction with the prices they received that season compared to the previous season, 52.8 % were not satisfied, compared with 3.6 % who were satisfied. Forty-three per cent said prices were the same as those received in 1994. However, about 64 % said low export potential was the reason affecting prices, indicating their dissatisfaction with the price of the current year compared with the previous year. Asked about the best place to sell animals, all respondents preferred to sell at the location of the flock. Farmers stated many reasons for this. Taking animals to market always involved expense, especially transport. Another reason stated was that taking animals to market limited their choice of sale as, they claimed, it was an indica-

Figure 2. Time of livestock sale within the sample[5] (Source: Field study conducted by the researcher, 1995)

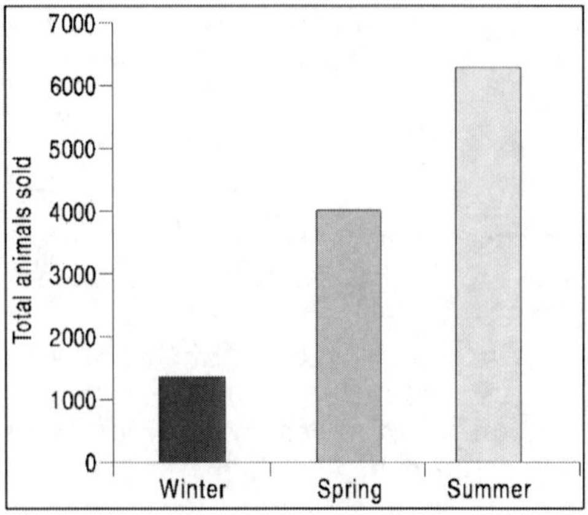

tion of an emergency sale which in turn might incur the cost of bringing unsold animals back home again, making them more vulnerable while bargaining. Traders usually roamed the Badia searching for animals to purchase. It was observed that owners of large flocks were targeted by these traders. In the sample, about 39 % of respondents were visited more than once that season by buyers in search of livestock.

Sale on short-term credit was observed in the study area. According to both farmers and merchants, this pattern of sale used to be common. However, lack of trust among producers had made this style of sale rare as it required high credibility of traders. This is because many producers had experienced no repayment for their animals. Sixty-six per cent of producers claimed they did not sell on credit at all, whereas 32 % still practised this type of sale if they knew the traders personally. The marketing strategy of small-scale producers differed markedly from that of larger holders of animals. It was observed that small-scale producers deliberately held back some male stock of marketable age. The idea behind this was to reduce economic risks by ensuring that there was always some saleable stock available to generate income when circumstances demanded. Clearly, this was not only the poor households' insurance strategy, but it also served as a banking strategy to accumulate wealth for larger owners. For productive females, the situation was more complex. Any sale of such animals would have implied reduction of a herder's capital and thus less security. Bedu livestock is a risky form of capital; herds are exposed at all times to the hazards of diseases and predators. In difficult periods, producers were forced to sell all classes of animals, even the breeding stock.

7.1.2. Marketing decisions and present environmental conditions

Diversity in livestock production in traditional systems was a result of climatic, economic, social, and political forces. These factors affected livestock mortality, reproduction and sales as well as the movement and distribution of livestock. Consequently, producers' behaviour was a result of the complex interaction of these forces. In this study, Bedu marketing practices were found to some extent to be similar to pastoralists' marketing behaviour in other developing countries. Even though sale decisions had more to do with individual motivation, other environmental factors such as harsh climatic con-

[5] Farmers were asked to state their total number of sales per annum.

Figure 3. Reasons for sale by flock size (less than and over 200 head) (Source: Field study conducted by the researcher, 1995)

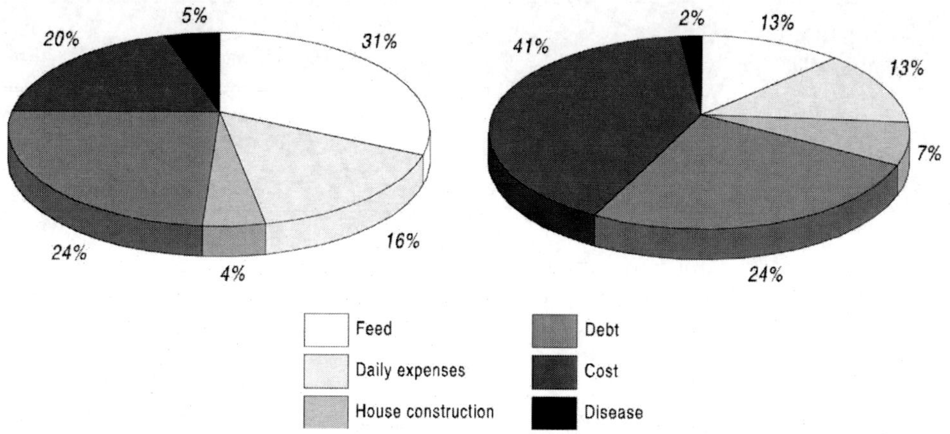

ditions, government policy and economic and social changes, as well as household conditions, were also to be taken into account (Cossins 1983, Kerven 1992).

Many farmers had, however, a reasonably good understanding of the most profitable times to market their animals, albeit they were often compelled to sell in response to farm input needs. In the study, farmers were asked why they sold at a particular time and what they intended to do with sales revenue. Their responses suggested purchase of feed to be the most important factor. Some households had to sell their lambs as repayment of debts. From the households studied, the factors which appeared to determine and motivate distress sales of animals were mainly the need for money due to: livestock purchase, condition of livestock, disease, shortage of labour, shepherds' salaries, cash needs and size of flock.

Purchase of livestock feed: In the study, farmers suggested that the purchase of feed was the most important factor. Most of these households were heavily dependent on feed purchases. Other farmers indicated that they often sold livestock to pay off debts incurred through the purchase of feed on credit or cash borrowed for the purpose of purchasing feed earlier in the year. Some households had to sell their lambs to the same lenders as repayment of their debts (Figure 3). Feed and debt, the two main reasons for sale, were interrelated. Some farmers believed that keeping animals over the summer period would put them at risk because of weather conditions and diseases possibly affecting lambs, which would increase feed costs. Fattening strategies were absent from their management policy. There was a clear indication that most of them were unaware of when lambs would stop putting on more weight.

One of the areas of greatest importance to this study was the relationship between interventions aimed at supplying subsidized feed to farmers and their impact on livestock sale. Feed was one of the major inputs in livestock production in the area, and although it was available for purchase by farmers, the withdrawal of government subsidy in late 1996 made it expensive and less affordable to small-scale farmers. Supplementary feed costs accounted for 68 % of production costs for producers in the Badia prior to price rises in August 1996 (Roe 1996). They then account for 74 % of production costs.

The Bedu livestock production system was (and remains) under pressure. They were no longer allowed to adjust naturally to phenomena such as overgrazing, but were in-

Table 4. Lamb losses by flock size (Source: Field study by the researcher, 1995)

Flock size	Mean lamb losses	Total farmers
1-100	2.31	106
101-200	9.2	32
201-400	11.4	31
401-800	23.4	17
801+	29.0	7
Entire population	7.7	193

creasingly being brought into the mainstream of development by indirect forces beyond their control, and into situations where traditional pastoral strategies might become less efficient. Traditional strategies, therefore, became less useful and so Bedu became vulnerable and were left with no room to manoeuvre. Movement patterns had changed considerably over the years, and livestock production in the Badia remained largely determined by feed, and access to it. Right from the 1920s, with the establishment of national boundaries and extensive agricultural and urban development, livestock owners have had increasingly limited access to grazing pastures (Galaty 1981). However, despite the high cost of stock rearing and the low prices of fresh or frozen imported meat, the number of local sheep and goats was still increasing, to about 3.5 million, according to Masri (1986).

Condition of livestock: Climatic variability has important implications for livestock marketing, and can greatly influence the behaviour of the producers and traders involved in livestock trade and export. For example, Bedu livestock gain weight and are in the best physical condition in late spring and early summer. In order to take best advantage of this and to avoid cost and risk in late summer and winter, farmers preferred to sell their animals during this period.

Disease: The study revealed that disease was the main cause of distress sales among producers and traders. Animals sold because of disease fetched the lowest prices, with some farmers selling animals for a quarter of their expected price. Furthermore, farmers and traders changed their selling or trading strategies by practicing early sales as a result of the increased risk of diseases. Most farmers stated that they had lost lambs, with an average loss of about 7.7 per household during the study season (see Table 4). Disease was becoming one of the drivers of early sales in the Badia. During the fieldwork, some farmers had been forced to sell their animals at very low prices because of disease, with prices falling from JD65 to JD30 and even JD20. Animals were often at greater risk of catching diseases during dry, hot summers; lambs born later in the summer were sold at a young age in order to avoid disease. One of the main reasons for sale stated by farmers was to avoid risk, and this accounted for 25.9 % of all reasons given for sale.

Livestock diseases and other problems such as pneumonia, parasitic diseases, enteritis, ectoparasites and viral diseases, which include foot and mouth, blue tongue and sheep pox, have been commonly reported in the Badia. Viral diseases are endemic in Jordan, and there is evidence that the incidence of blue-tongue was near 80 % and PPR was at 60 % and rising. These and other diseases were commonly found in village-based flocks (Oakeley 1997). Low production was identified by experts from BRDC to result from animal diseases. Campbell, Jones and Roe (Campbell et al 1995a) indicated that the veterinary system had not coped well with this situation. A further constraint concerning animal health arose as a result of plastic waste which was increasingly being used for industrial and agricultural purposes[6]. Farmers wanted to reduce the cost of production by grazing on rangelands but they faced a great risk of losing flocks due to the toxic effects of this plastic (Blench 1995).

Shortage of labour: Another important factor behind early sales was the shortage of labour. Whilst family labour was an important input in raising livestock, some families had to make a choice between sending their children to school and limiting the number of animals they kept. The scarcity of labour was obvious for large livestock holders. Many households suffered from a shortage of male and female labour, unappreciative of Bedu traditional life. While a flock needed to be large enough to meet the subsistence needs of a family, family labour had equally to be sufficient to maintain this herd and to take full advantage of its resources. The majority asserted that the younger generation are more oriented toward salaried jobs than livestock production, leaving the older generation with the responsibility of managing small numbers of animals. Following the settlement of the Bedu, children started to go to school and a larger number of educated people were attracted to an urban lifestyle, securing jobs away from livestock production. Older people had started to recognize the importance of education for their children and this pushed some of them either to reduce or give up livestock production altogether. The younger generation seemed to prefer a settled lifestyle and expressed their bitterness towards the traditional nomadic lifestyle. Upon entering marriage, many young women laid down conditions for a more settled lifestyle, refusing to participate in livestock production. In addition, some women began to secure jobs in teaching and nursing (Abdullahi 1990, Ensminger 1992, Kay 1978, Marx 1984, Salzman et al 1990).

Shepherds' salaries: During the period of field work the average monthly payment for a shepherd was about JD105. These shepherds came from neighbouring countries and generally worked for a year. Paid employment attracted a large number of younger Bedu men to work outside the livestock organization and this caused the Bedu to rely on hired labour. Shepherding became one of the main vacancies to be filled by outside labour. Shepherding activities varied according to the flock size and location. Farmers within villages tended to have one or two family members, usually children, who wandered with the animals during the day. Large stock owners maintained a full employment strategy by hiring a shepherd on a monthly basis. Generally speaking, Bedu who owned more than 70 head of sheep hired a shepherd, with payment every three months.

Cash needs: Empirical data gathered indicated that small flock holders in the study area generally received lower prices for their animals than larger flock farmers. Abdul lahi (1990) supports this and reported that large-scale farmers were in a better position when they made their sale. Owners of larger flocks rarely took their animals to the outlet markets and most of their sales occurred on location, leaving them with the choice to sell or not to sell and enabling them to escape transport costs. They also benefited from sale on credit, and sometimes sold without a down payment if they trusted the trader. This sale pattern usually attracted traders who were willing to offer higher prices. Furthermore, their animals were in better condition, especially when they let lambs suckle for a longer period, and when their wealth enabled them to keep their animals until an opportunity for a good sale appeared.

Liquidity was a further constraint which limited their sale choices, and sometimes high transport costs forced farmers to dispose of their animals in outlet markets and receive low prices for their animals. The sale of sheep, especially ewes, was strongly associated with pressing cash needs for the purchase of feed. Household expenses viewed as optional expenditures could be timed so as to maximize the terms of trade between daily expenditure and the income of households. There were certain occasions when household expenses reached a peak: school expenses, weddings, the Holy Month of Ramadhan and the two major Moslem Eid celebrations. School starts in early September

[6] Black plastic is used widely in agricultural areas, bordering rangelands, to reduce evaporation in fields under drip irrigation.

and represented a financial constraint to most households, especially if a household had many children, as was (and remains) often the case in the Badia. Construction was another reason behind the sale of animals, especially sheep. The building of new houses and extension of existing ones were witnessed during the field work. Machinery expenses were also one of the reasons mentioned by some farmers; vehicles such as tankers, trucks and pickups, are necessary when flocks travel. Maintenance of these machines sometimes forced farmers to sell sheep in order to raise cash needed for that purpose.

Cash income derived from animal sales was one of the most important reasons for farm households raising livestock. The decision to sell could also be framed as an evaluation of the relationship between a household's financial needs and the use of a household's livestock holdings to meet these needs. Cash spending showed that animal sale income was used for feed, food, goods, school supplies, transport, services, rent, and emergency needs. Investments in increasing flock size were only found in wealthy households financed by lamb sales. Additional cash expenses included hired labour, shepherds, and other inputs. These production costs came at a set time of year and could be postponed or foregone only at the cost of reducing or endangering production. Some minimal cash needs could be described as fixed; it might be claimed that some amount of food and feed was a fixed minimum for survival but beyond this the need for cash was flexible over time. From the information obtained in the field work, it can be said that families generally hesitated to sell sheep unless they were forced to do so. Investment opportunities apart from livestock production were rare in the Badia, so sheep had become an important factor in the process of wealth accumulation.

7.2 Markets and marketing systems and functioning

This section examines trade organization by focusing on the identification and characterization of the livestock marketing intermediaries, and describing the marketing channels. Cross-border trade was the most sensitive issue for the traders. For both producers and traders, cross-border trade was a constraint due to its influence over the demand for Jordanian animals. Traders also needed to be aware of the origins of the animals they were dealing with. Exporters benefited from cross-border trade when they were able to export female animals which were not allowed to be slaughtered or exported from Jordan. They were able to sell Jordanian female animals and export them to Saudi Arabia where small traders had no such access.

People involved in the cross-border trade from Syria to Jordan could make good money, but small traders were left out since they did not have the capital to act as exporters and there was no market for Syrian animals, which were not favoured for sale within the country. The same can be said about animals brought from Iraq, with the exception of male lambs. Imports from Iraq might have caused producers to face unfair deals in the Saudi market. Importing animals from Iraq was risky, and the few Jordanians who had access to such business tended to have access to the Saudi market as well. The majority of transactions of livestock were carried out by Bedu traders. There were, however, a few non Bedu who had developed a strong relationship with Bedu producers over the years.

7.2.1. Marketing price information, grading, evaluation, and analysis

Price information: While some producers were up-to-date on detailed and immediate changes in market conditions at regional markets or the export marketing chain, they all seemed to have at least some information and awareness of general changes in market conditions that ultimately affected them. Most farmers claimed an awareness of price variations. Almost 70 % of the sample indicated that they collected price information before the sale of their animals, generally by word of mouth. The farmers' sources of information included other farmers, traders, friends and relatives, and visits to main markets. Farmers claimed an awareness of price information even when they did not sell. Figure 4 gives the difference between the higher prices received by informed sellers and the lower prices received by those without market information.

Grading: Bedu producers favoured sale per head rather than weight for different reasons. Producers stated that they did not trust buyers who bought by weight. This can be explained by the fact that illiteracy among producers in the Badia was high, around 90 %, and many producers were unable to read the scales. Producers thought that traders could move elsewhere with the scales and did not want to take such a risk. Producers also mixed weak or unhealthy animals with the good ones in order to sell them. Ideally, traders wanted farmers to stop mixing the odd sick animals with the rest of the purchased animals. It is probably the case that farmers did not understand or they underestimated the gains to be made from selling on a grading basis. Grading and sorting out animals for sale according to their condition could have contributed to both operational and pricing efficiency. In addition to gaining traders' trust, grading could have generated more accurate market information and price discovery. Grading could have reduced the transport costs when good animals could have been sold further afield and animals in a poorer condition sold at local markets. Farmers might also have benefited from the competition and price efficiency resulting from a good grading system.

Price evaluation: The evaluation of livestock prices is a difficult task due to the complexity of the livestock business. Prices for livestock are reached through bargaining between seller(s) and buyer(s). Animals can be sold either as a group or individually, or sometimes in mixed groups; in most cases, prices are determined on the same basis. Buyers at the time of the fieldwork valued the animal for the immediate return of meat production, and manually checked the quantity and quality of meat that each animal would yield. Traders also valued the condition and health of the animal. Livestock prices were expected to vary from one area or market to another according to transport costs. The assumption was that prices varied with the distance from outlet markets or points of export embarkation. However, variations in prices between markets were difficult to evaluate in practice because of the lack of information on prices and animal weight and condition in different markets. Since prices were given per head and not in units of weight, it was difficult to make comparisons.

Illegal trade and cross-border trade of animals from Syria and Iraq could have affected prices, especially if large numbers entered the market at one time. It was impos-

Figure 4. Average price received by price information (Source: Field study conducted by the researcher (1995)

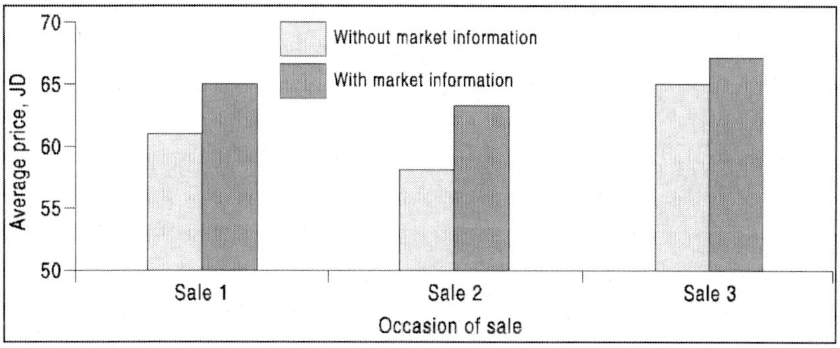

sible to estimate the volume and value of cross-border trade. Oakeley (1996a) found prices in a market close to the production area higher than those in a more distant market for the same animal[7]. A likely explanation for this could have been that animals were smuggled across the border to Saudi Arabia. Another problem involving prices was that, as mentioned above, farmers tended to mix weak animals with good animals and this in turn reduced the price most of the time. It was then difficult to estimate the price per head. Traders were sometimes able to pick animals from a flock and this required a higher price to be paid. At other times traders bought all animals, save some, and so required a different price arrangement. The age of animals was an important selling factor. For example, two-month-old lambs attracted better prices than older lambs. Social values based on the ownership of livestock were also still important, and many Bedu chose to increase their stock rather than sell. As a result, if the price was high, fewer animals were likely to be sold because farmers would have been able to meet their costs more easily. Although the Bedu preferred to build their stock number and hold animals, other factors encouraged them to market more animals:

- Increased demand for meat: The increased demand for meat also motivated Bedu to trade, and for many this had become an important source of income. As Bedu crossed national borders[8], or camped close to regional borders, it was almost impossible to work out exactly how many of the animals sold in Jordan were actually produced in the country (Oakeley 1996a). The rapid increase in livestock exports from the late 1970s into the 1990s shows there was a market response to growing marketing opportunities (Randhawa 1990).
- Religious festivals: The sale of livestock was heavily influenced by religious festivals in the region. This usually happened about two months before these festivals took place, when sellers of animals held their sale in expectation of higher prices. In Ramadhan and the Eid festival, price rises were significant because of the greater consumption of red meat during these periods. The Eid Al-Adha also is a very important time for sales since most households in the region slaughter at least one animal, in accordance with the Islamic creed. Furthermore, each Moslem practising pilgrimage (Haj) has to slaughter a lamb or sheep while in Mecca. As a result, the number of exported animals increased dramatically. Live exports of animals in 1995 reached a peak in April as a result of the Eid Al-Adha, and a peak in January 1996 after Ramadhan, as shown in Figure 5. The timing of these determining factors of supply and demand is not fixed, but varies yearly according to the lunar month (Oakeley 1996a). Some of the farmers interviewed stated that they took such occasions into consideration when they wanted to sell animals, by holding their sale till these times.
- Consumer taste: Consumers of red meat in the Middle East are known for their loyalty to Awassi meat. Nevertheless, Awassi meat is still classified according to age, sex and weight, with lambs preferred over sheep, male lambs favoured over females, and younger males over older males. Consumer taste usually attracted producers into selling animals under the age of 75 days. Typically, prices for two-month-old or younger male lambs were high and their meat was considered to be the most expensive. The consumer preferences for meat of young lambs has continued, it as assumed, to encourage the tendency of buyers toward the purchase of young animals as well as the early sale of lambs, despite the economic argument that higher sale weights would increase producers' profitability as Oakeley (1996a)

---

[7] Ruwayshid market is located in the centre of the livestock production area and therefore prices in this market are expected to be less than those in the Mafraq market.

[8] Jordanian Bedu were allowed to cross into Iraq and Saudi Arabia until the mid 1980s.

stated. This management practised by farmers could be seen as a response to consumer taste in Jordan and in neighbouring countries where the slaughter of young lambs is favoured. This might explain why farmers preferred a per head basis for sales rather than weight. Nearly 99 % of statements made by the household sample favoured selling animals on a per capita basis.

Figure 5. Live lamb and ewes exported to the Gulf States January 1995 to March 1996 (Source: Oakeley, 1996a)

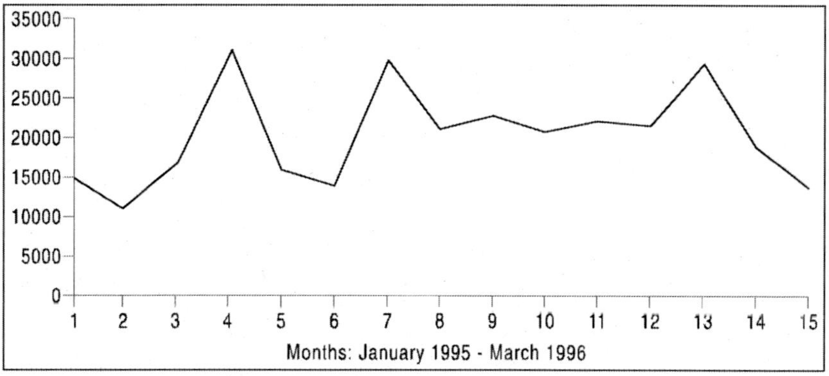

Price analysis: In this study, analysis of price variations is only an attempt to estimate and test for effects on price of several important variables using analysis of variance and applying different variables where possible. Livestock price analysis requires knowledge of animal characteristics, general supply and demand conditions, and conditions of the transaction. Livestock price per head and live-weight can be analysed according to animal characteristics such as breed, sex, age, season, volume of supply and the number of buyers in the market (Bekure et al 1983). Analysis of variance, along with Pearson correlation coefficients, were seen as appropriate statistical techniques. Besides sex, breed, and age, the variables of time and place of sale, price information, and buyers were included in the analysis. Some households sold animals at different times of the year, according to the characteristics of their household condition and number of animals they owned. The analysis was only undertaken for male and female lambs. Data on sale of mature sheep were omitted to facilitate the analysis of variance. Price analysis included the following variables: farmers' price information, type of animals and age of animals sold[9], date of sale, place of sale, reason for sale[10], type of payment and buyers' destination.

7.2.2. Functioning of the livestock marketing

This discussion is presented under the following headings: the marketing channels, market information, and market conditions:

Marketing channels: The livestock marketing system can be divided into local and export. The main participants in the former were mobile middlemen, butchers and some producers, while the latter involved exporters and agents who maintained trade links in regional markets, mainly in the Gulf. Local trade was dominated by a large number of participants, but the majority of these were mobile traders who roamed the area search-

[9] Due the absence of weight information, information on age was collected to give an indication of the weight of animals. It is expected the older the animal the greater its weight.

[10] Reasons for sale were generated by farmers

ing for animals. In contrast, the export trade seemed to have fewer participants because there were many restrictions such as bureaucracy, capital, and regional trade links in other markets. The middlemen traders were relatively active in their procurement and distribution activities. They were also able to maintain an informal but effective communication link with the markets and livestock production areas in which they operated. Due to the mobility of production and the lack of affordable credit, the traders tended to purchase at flock location and where farmers were willing to sell on credit. The degree of market power exhibited by export traders was determined by their financial ability. They were small in number and had more access to domestic and regional traders.

Functionally, two classes of markets existed in the area: the location of animals (farm-gate) and major outlet markets. The field work indicated that the total recorded number of animals sold through the market channel was 39 247 during 1995. The location/village markets were generally affected by climatic conditions and location of producers. About 83.2 % of farmers' sales took place in such markets in the Badia, whereas only 16.8 % of sales were in outlet markets. Most transactions took place where the animals were pastured, since the long distance to any market dissuaded farmers from selling there. Family ties were important in trading channels, either easing entry for some or making obstacles for others, especially when such trade required the availability of capital for traders. Co-operative arrangements among mobile traders were found to be common and could cause a lack of competitiveness in the livestock trade because market information was shared through their personal relationships rather than because of purely business interests.

Market information: Marketing might also have been affected by lack of information about the best time to sell or the ability to sell at a more opportune time (in terms of price and demand) due to the isolation of the Badia or high transportation costs. Most producers indicated their knowledge of market information, with almost 70 % stating that they collected price information before making a sale. Sources of information were based on verbal communication from other farmers, traders, friends and relatives, and visits to main markets. Further, agreement on transactions between buyers and sellers was usually based on the last transaction that took place in the area on which they had information. Farmers who were informed about current prices before they made their sale obtained significantly higher prices than uninformed households. Traders were more able to maintain contact with other markets through their visits and mobility between markets, and so gained an advantage over farmers by weakening their bargaining ability. Even though traders were considered one of the main sources for market information, they did not necessarily share all their information with farmers. The lack of market information was seen as one of the constraints inhibiting the smooth performance of livestock marketing, since it exposed farmers to abuse from other bodies and agents.

To ensure the effectiveness and efficiency of marketing, it is of paramount importance that all parties should have access to relevant market information. The absence of information can lead to high transaction costs. For farmers, this can put them at a disadvantage when bargaining with traders, and some farmers are tied to credit relations which might prevent them from achieving a reasonable price. Bedu farmers and traders tried to extract information on all aspects of a transaction by asking about the type of animals sold, their age and condition, the method of payment and the buyers and sellers involved. The Bedu were always interested in such information, regardless of whether or not they themselves had animals for sale. Producers were expected to share information with each other about transactions, as they needed to maintain good relationships with other producers in order to gain information from them for future sales.

Market conditions: The results show that the vast majority of animal sales occurred at the location of the flock (Figure 6). Most farmers did not favour sales in the outlet markets, as selling animals away from outlet markets saved time and money, and left farm-

Figure 6. Percentage of animal sales, by location (Source: Field study conducted by the researcher, 1995)

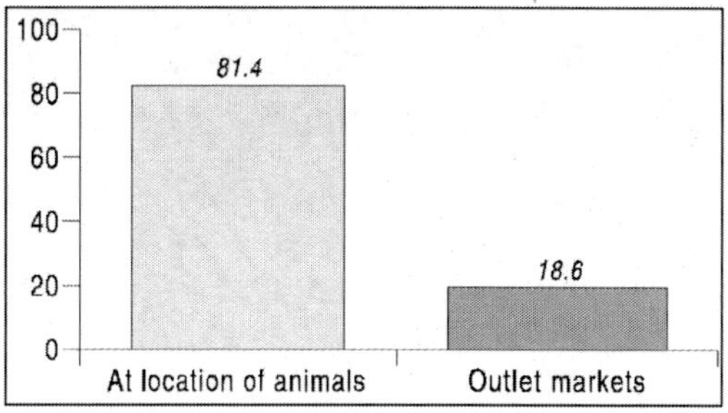

ers with more choice of and control over sales. The farmers indicated that outlet market sales occurred when there was an urgent need for cash.

Infrastructure: The existence of marketing facilities not only affected marketing costs, but also influenced the market opportunities open to livestock producers and traders. Movement of livestock was mainly carried out through the road network. However, accessibility to producing areas depended on the quality and mileage of roads and the ability to reduce marketing costs. The cost of transportation could have been an obstacle for some farmers and traders. Other farmers did not transport their stock, maintaining the view that selling away from the flock location put them in a disadvantageous bargaining position. This attitude seemed to be rooted in traditional beliefs which still heavily influenced marketing strategies.

There was a positive relationship between infrastructure and livestock marketing. Some remote parts still suffered from poor infrastructure, and the lack of facilities in outlet markets had some negative impact on the sale of livestock. Any future improvement, such as fences, paved surfaces, watering and feed facilities, and public conveniences might encourage more producers to sell at outlet markets. Whilst inadequate marketing facilities and services were partly responsible for high marketing costs, communication facilities and market information were thought likely to reduce the effects of such constraints if introduced or improved (Blench 1995, Oakeley 1996b, Randhawa 1990).

Sales arrangements and credits: in the Badia, there were many sales arrangements that governed relationships between sellers and buyers, such as cash sales, sales on short or long term credit and exchange of animals. There was therefore no guarantee that the seller or buyer would not cancel the sale, if he suspected that he might be cheated or if he was offered an unreasonable price for his commodity. Transactions and contractual obligations were governed by a system developed over decades of tradition, personal ethics, reputation, tribal links and social standards. The sale of animals was mostly carried out by the farmer himself, but there were occasions when a farmer would ask for help from his town, especially from traders, if the sale was going to take place in outlet markets. The majority of sales taking place in outlet markets were for cash. Agreement on price and the transaction of ownership of animals was carried out without any paper work. In outlet markets, bargaining involved many people who circled round a producer until one of them offered the final price.

Investment capital constituted another major input that was difficult to acquire. Borrowed finances from institutional sources for agriculture were difficult to obtain due to high lending rates, the collateral security required by the banks and religious beliefs. This shifted attention to private money-lenders, who were mainly traders and wealthy farmers. The lack of choice in borrowing sources and the domination of the agricultural credit market by traders resulted in credit transactions that tended to link credit to other obligations. For farmers, this took the form of linking livestock inputs to future livestock output, and for some farmers this resulted in a limit to whom they could sell stock. In balancing cash needs against the desire for household security, the farmers also considered the possibility of borrowing money if their livestock and income from other sources were insufficient. The interviews revealed no interest on borrowed money. Interest in practice took different forms, such as selling or purchasing animals at a higher price than their market value. According to farmers, repayment for their animals was not always met. Others indicated that traders did pay, but not the total cash, despite the farmer being given a written cheque as a guarantee by traders. When those farmers who sold their animals on credit were asked how they made the decision to sell to one trader rather than another, they claimed it was usually because the trader was known to them and they had sold to him previously.

Sales of animals outside Jordan sometimes meant loss of money, and repayment rarely occurred, especially from Saudi Arabia. Traders interviewed in this study did not deny the producers' claims. On the contrary, they agreed that this did happen and had affected them when some producers insisted on getting cash for their stock. They added that a new generation of traders had started to appear and approach producers, and would then just disappear to avoid payment. They also claimed that lax laws in the country enabled such people to abuse producers by giving them false cheques and then receiving no punishment for this kind of act. Whilst it is true that the introduction of modern methods into the market facilitated transactions, some negative effects accompanied these changes. The problems with the use of cheques resulted in a lack of trust, a fundamental element of the traditional market. The use of cheque books had both a positive and a negative effect on the organization of trade in the Badia. On the one hand, they made exchange easier, but, on the other, misuse by some traders and a deficiency in the financial laws in the country resulted in a lack of trust among market participants, as some traders did not fulfil their contractual obligations. It is interesting to note that when farmers and traders compared past and present conditions surrounding the enforcement of contractual obligations, they stated that previously failure to meet such obligations had been rare.

8. Conclusions and recommendations

Cultural beliefs remained one of the factors that attracted households to livestock production, which was also a source of wealth, subsistence and occupation. Domestic animals were regarded as giving a family nobility and prestige, and the Bedu identified themselves with their animals in an inextricable manner though today some people, particularly some women and young adults, have less interest in livestock keeping (Oughton 2008). Animal slaughter was an example of when the Bedu relied on their self-production to meet social customs confined to community occasions such as acts of hospitality, marriage and death. Since the Bedu had a basic knowledge of rearing animals, this sub-sector of the Badia economy remained crucial for their employment at the time of the survey. Animals were also seen as an important capital stock, largely secure, but a convertible asset. At a time when basic structural traits of Bedu society appeared to be unchanged, constant small modifications within it had considerable impact on it as well as on its physical and social environment.

Household production under traditional systems was found to have a series of characteristics that differentiated it from commercial production. With the high cost of production in this sub-sector, labour could be costly when hired. Consequently, family labour was the most important input in livestock raising, involving mainly guarding or herding animals, although this may have forced some families into making a choice between sending their children to school or limiting the number of animals they kept. The issue of schooling was the main motive behind settlement and the reduction in animal numbers in many households. One aspect of the relationship between labour and household livestock marketing is the residual time allocation of labour to herd lambs when they had to be separated from flocks. Limited overall family resources to meet financial obligations and the few alternative income opportunities available, may explain this time-intensive use of family labour.

While the Bedu subsisted mainly on their income from livestock production, there were other sources of income such as wage labour, arable farming, trade and smuggling that cannot be neglected even though some were relatively insignificant. It was clear that in those circumstances livestock production was neither a commercial nor a traditional system, but one that could best be described as transitional. Basically, the marketing system did not fit into the society of small livestock holders. Small-holders lacked information about and access to the different marketing channels. Furthermore, through their activities, traders in the different marketing channels sometimes did not share their information, preferring to utilize it to their own advantage. The result was that the different marketing channels were incompatible with the needs of farmers in areas of production.

Traders with little capital dealt with small farmers who tended to stay in or around villages. Those with larger flocks made their own deals with exporters and sold on credit. Some arrangements were fixed and so larger flock producers were seen to sell to the same traders every year. Small traders bought from small flock holders and big traders from larger flock holders. Small traders with little capital usually had no access to credit and, if they did, for not more than a week.

The marketing system as a whole could be considered a stable one. The system was capable of responding to changes in supply and demand patterns and had adapted to the use of modern transportation and labour opportunities. Furthermore, the system showed low-cost of trade and low and narrow trade margins. An assessment of the efficiency of a marketing system was normally based on the measurement of the extent of total marketing margins in final consumer prices. Mobile traders' profits were low, but there were indications that high differences existed between mobile traders and exporters or butchers. The system had not been tested against real export embargo, competition, and shortage of feed supply. There were signs that the traditional system was under strain following the government decision to lift the subsidies and to ease policy on imports of animals in late 1996. The market was oversupplied by local and foreign animals and farmers either delayed their sales or disposed of their animals at low prices[11].

8.1. Market structure

The existence of outlet markets is strongly related to population size, and so livestock markets in Jordan are only found in cities. Their operations start early in the morning and last for three to four hours. The four main livestock outlet markets in Jordan which were of importance to Bedu producers at the time of the study are Mafraq, Ruwayshid, Irbid and Amman. Alongside sales made at the flock location, outlet markets also served as principal sales points for live animals, but they tended to attract principal buyers, mainly speculators, dominating the markets. Outlet markets were lacking in many areas, and they did not reflect, in any way, the needs and expectations of the farmers. Farmers

---

[11] Information on such issues was obtained from informal discussions with farmers (different to the ones interviewed in 1995) during a visit to the Badia in March, 1997. This information was confirmed by Alan Roe, a researcher in livestock production in the Badia, in October 1997 during a meeting at Durham University.

clearly wanted more freedom of choice in the place of sale.

The analysis of the marketing channels verified that farmers had many alternatives for selling their animals. However, in reality only two marketing channels, middlemen and exporters, were dominant in the Badia. Most farmers did not favour sale in outlet markets unless they were forced to do so. Important reasons perceived to be responsible for farmers' dissatisfaction with outlet markets were low prices and the limited number of buyers. Farmers also experienced some constraints including lack of transportation and higher transportation costs as well as the possibility that the sale might not take place. Producers who sold in outlet markets usually did so because of a financial crisis. Traders indicated that producers, who had no immediate need for cash, kept their animals and announced them for sale. Traders stated that most of the time weak and unhealthy animals, rather than good animals, found their way to outlet markets.

There was an absence of formal channels for market information, which could have led to high transaction costs. Some small livestock holders might have been affected by a lack of knowledge about the best time to sell or the lack of opportunity to sell at such times. This put farmers at a disadvantage when bargaining with traders, as some farmers were tied to credit relations which might have prevented them from achieving a reasonable market price. The preference for selling on location also made the need for information through formal channels important. Traders had better and quicker access to market information than the farmers with whom they traded, and so were able to bargain on the basis of up-to-date market information.

Middlemen performed the different post-farm marketing functions in various ways that might not necessarily have matched conventional standards. Being the dominant marketing group, they were able to establish their own standard operating procedures that linked the production and the marketing sectors. The presence of a great number of middlemen linking production areas to consumers was a vital element in the Bedu livestock production. They provided producers with more marketing options, making the raising of cash accessible with less effort and lower transport costs. The large group of middlemen with little capital enabled mainly small-scale producers to maintain their marketing choices.

Despite the high level of competition in the red meat market in the Middle East and despite the traditional marketing methods and regulations that govern market arrangements among market participants, livestock marketing in the Badia was showing a certain stability conditioned by the stability of the market. Livestock production in the Badia was strongly connected to the Gulf markets, where high demand was reflected in high prices. Moreover, higher prices offered for *baladi* red meat in the local and regional markets allowed the continuation of Bedu livestock production. However, there were many pointers towards a decline in demand for Jordanian animals in such markets. The Saudi market had the greatest influence on the Jordanian market. In other words, if the price was high in Saudi Arabia then this was reflected in Jordan; the price was usually about JD20 higher in Saudi Arabia. Exporters from Mafraq and Zarqa usually employed Bedu from the area to buy animals for them, and this was clearly the case in cross-border trade from Syria. The peak of demand for lambs is centred around the Eid festivals, and both traders and producers were aware of such demand. Sale of animals usually reached its peak two weeks before the month of Ramadhan. Eid Al-Adha is the most important occasion, for sale and prices usually rise before the Eid. The consumption and the export demand within the Ramadhan-Haj season have important implications for domestic prices. As these festivals have no fixed date, producers cannot plan production for a fixed time of the year.

Export traders maintained control over the export of animals to the Gulf market, providing mainly large-scale producers with marketing choices. Export traders maintained strong relationships with producers through their long presence in the trade industry.

Tribal links shaped the organization of the trade, especially cross-border trade from Syria. The livestock market provided labour opportunities mainly for mobile Bedu traders who occupied an important position in livestock marketing. They bought animals directly from farmers who could then determine the price of their stock rather than being price takers.

The marketing system as a whole suffered from the lack of a standard system, standard grades, trust, formal regulations, outlet market facilities and a relatively small number of exporters. The private credit system, though appearing to be more unfavourable to farmers than any other category, remained a reliable source of capital for the production of livestock. The low price of some animals seemed to reflect the importance of other factors in explaining the sale behaviour of households. The presence of bankrupt traders, the loss of faith in finance systems, and the arrival of newcomers, especially the younger generation, in the trade business, resulted in the traders' lack of credibility and trust, and this in turn affected the marketing arrangements and became a new challenge facing the organization of trade in the Badia.

Trends since 1997 indicate that marketing channels and players have stayed the same. However, in the context of recent changes although middlemen have remained players in the market channel, their numbers have decreased; wholesaler activity also increased as exporting market activity grew from 2001 to 2005.

8.2. Farmers' and traders' behaviour

With regard to farmers' behaviour in relation to the timing and place of sales, the study showed that they were clearly aware of the benefit of a late sale, but due to production circumstances and household conditions they fell short of doing that. In addition to variation according to seasonality and forage production, the organization of livestock trade was affected by cost and availability of feed, animal health, and demand from neighbouring countries. According to traders, grazing was a very important factor in determining the health of animals, especially in the lambing seasons. Grazing is important for ewes and lambs during weaning and this was confirmed by mobile Bedu receiving higher prices for these animals.

Prices for livestock were agreed following bargaining between seller(s) and buyer(s). Animals could be sold either as a group or individually, or sometimes in mixed groups; in most cases, prices were determined on the same basis. Price was found to be strongly correlated with age, information, payment, place, reasons for sale and type of animals sold, whereas date of sale, place of sale and type of buyer showed no significant correlation with price. The Badia farmers suffered from lack of financing from government institutions and banks, and so a new type of livestock trade emerged. Some arrangements for borrowing money took the shape of animal exchange. Animals were usually sold by those wealthy traders for double their relative price in the market, based on a fixed date of payment. People who bought at such prices and under such circumstances usually sold the animals at current prices to raise cash. This was being replaced by ambitious private money lenders, who were mainly traders and wealthy farmers.

Marketing decisions were related to a household's demand for cash and to environmental factors such as the high cost of production, disease, unstable government policy, household characteristics, and labour capacity. Producers sold for several reasons, the most important of which were to generate cash to buy feed or to reduce costs and eliminate the risk of disease. Access to credit could have provided families with a further source of income especially during critical cash flow periods. With cash available, they could increasingly rationalize their livestock operations and market the animals more effectively.

Differences in marketing strategies existed between small- and large-scale producers.

The latter were able to receive higher prices for their animals and could benefit from sale on credit. Furthermore, large-scale producers were able to supply good quality lambs when they let them suckle for a longer period, and most of the time they did not wean them at all. The data suggested that the credit arrangements between producers and traders took place because they fulfilled a genuine need for traders, often faced with low cash availability. For this reason, large-scale farmers benefited from such sales when they received higher prices for their animals. Larger-scale farmers could generate a net cash revenue surplus, and so faced few cash-flow problems. Greater wealth provided opportunities to buy livestock inputs including veterinary drugs and water.

In contrast, levels of annual cash income of small flock-owners were very low and often insufficient to cover their main cash expenditure requirements. In that situation, those households were incapable of introducing any production technologies. The size of a flock in a household was governed by the mobility of the labour force. Small-scale producers saw mobility as profitable management, but they were prevented from being mobile for reasons of lack of cash, transport and labour, having only a small flock, and needing to spend more on their children's education.

8.3. Recommendations

The study findings are of importance to economic and social development planning concerning Bedu welfare. There needs to be field-oriented rather than office-oriented contact between farmers, traders and policy makers if the situation in both production and marketing in the Badia is to improve. Before government designs and implements extensive changes in the livestock sub-sector, policy makers should be aware of Bedu behaviour, which can appear irrational to an outsider, but is based on solid economic beliefs. Any policy should be stable and should give farmers more time and space to plan in response to government policy changes. Changes in consumer demand and government policy require this type of production to be competitive to cope with new international trade developments. This gives us insight into the importance of addressing farmers' needs if the marketing system is to be appropriate for them. It is suggested therefore that improvement in the infrastructure of outlet markets would encourage farmers to market more animals through these channels.

There is a lack of comprehensive knowledge about the Bedu lifestyle and their strategies and behaviour toward livestock production. Bedu livestock production not only still contributes significantly to the gross national product of the country, but is also a major sector on which the welfare of many people, in terms of food, cash and labour, depend. Livestock development programmes should give full consideration to the integration of household activities and the role which livestock play in the survival strategy of households. Due to the widespread distribution of small livestock holders, a greater number of farmers and more low-income households would benefit if the importance of livestock development policy was stressed to small livestock holders. Given the basic need to address food and income security among the Badia inhabitants, emphasis on small livestock producers seems appropriate, since livestock continue to play an important role in the household.

Availability of financial resources, access to market information, intensity of use of veterinary clinics and drugs and introduction of regulations to keep rangelands free from plastic waste, are considered to be the most important factors to assist the development level of Bedu farm households. These factors are closely related, so that their interplay determines the direction in which pastoral households will develop. Development of livestock production obliges planners to approach this issue in a more widely-integrated manner, since livestock are found in nomadic and semi-nomadic hands.

A programme which educates livestock suppliers about daily and weekly prices in outlet markets would enable them to plan their marketing more efficiently. The media should play a leading role in updating farmers and traders about changing conditions in the markets; or perhaps mobile phones could be used for the purpose. Reporting on arable agriculture through the media is already practised in Jordan and this could easily be extended to cover general trends in livestock marketing.

Acknowledgements

I would like to acknowledge the BRDC and its staff for facilitating research and data, expertise, and practical help; and Dr. Oughton, Dr. Atkins, and Dr. Cox for offering guidance. Financial support from the HCST, BRDC (BRDP), is gratefully acknowledged.

References

Abbott JC (1987). Agricultural marketing enterprises for the developing world. Cambridge: Cambridge University Press.

Abbott JC (1993). Agricultural and food marketing in developing countries. Oxford: C. A. B. International.

Abdullahi A (1990). Pastoral production systems in Africa: a study of nomadic household economy and livestock marketing in central Somalia. Germany: The Institute of Agricultural Economics and Social Science in the Tropics and Sub-tropics, University of Hohenheim.

Abu Jaber KS (1987). The Badia of Jordan: the process of change. Amman: Jordan University.

Abu Jaber KS and Gharaibeh FA (1981). Bedouin settlement organisational, legal and administrative structure in Jordan; the future of pastoral peoples. Proceedings of Nairobi Conference, Kenya, IDRC-175e pp. 294-301.

AFCR (2005). Arab Forum Competitiveness Report. Accessed at www.forumblog.org/blog, January 17th 2007 <http://www.forumblog.org/blog, January 17th 2007>.

Alcantara C (1992). Introduction: markets in principle and practice. In: Alcantara C (ed), Real markets: social and political; issues of food policy reform. London: Frank Cass.

Bekure S and Tilahun N (1983). Livestock marketing studies. Proceedings of the IDRC/ILCA workshop held at ILCA, Addis Ababa, Ethiopia 21-24 March. Pastoral systems research in Sub-Saharan Africa: International Livestock Centre for Africa, Addis Ababa pp. 327-350.

Blench R (1995). The Hashemite Kingdom of Jordan national programme for range rehabilitation and development baseline survey of socio-economic and animal production data. Volume 1.1, Main Report, Version 1.0, Rome: IFAD.

BRDP (1994). Jordan Badia Research and Development Programme, HCST, Amman.

Briggs J (1985). An exploratory study of farmers' choice of crops in central Sudan. Trans. Inst. Br. Geogr 10:170-180.

Campbell DA, Jones K and Roe A (1995a). On-going monitoring of production, health and economic of selected flocks in Programme area, BRDP, Jordan.

Campbell DA and Roe A (1995b). Results of a preliminary survey of livestock owners in the Badia research and development programme area, BRDP, Jordan.

Chatty D (1986). From camel to truck: the Bedouin in the modern world. New York: Vantage Press.

CIA (2006). CIA World Factbook. Accessed at www.cia.gov/cia/publications/factbook/goes/jo <http://www.cia.gov/cia/publications/factbook/goes/jo>.

Cossins N (1983). Production strategies and pastoral man. Proceedings of the IDRC/ILCA workshop held at ILCA, Addis Ababa, Ethiopia 21-24 March. Pastoral systems

research in Sub-Saharan Africa: International Livestock Centre for Africa, Addis Ababa pp. 213-219.
Department of Statistics (2007). Agriculture Surveys. Accessed at www.dos.gov.jo/agr <http://www.dos.gov.jo/agr>, January 17th 2007.
Directorate of Statistics (1991). Results of the livestock census Oct. 24, 1991. Directorate of Statistics, Amman.
Drucker RF (1958). Marketing and economic development. Journal of Marketing 22 (3):252-259.
EIU (1993). Country profile: Jordan. London: The Economist Intelligence Unit.
Ekins P (1986). The living economy: a new economics in the making. London: Routledge & Kegan Paul.
Ekins P and Max-Neef M (1992). Real-life economics, understanding wealth creation. London: Routledge.
Ensminger J (1992). Making a market, the institutional transformation. New York: Cambridge University Press.
Fialor SC (1994). The maize marketing in Ghana: an alternative approach to the analysis of market performance. Ph.D Thesis, Newcastle Upon Tyne.
Galaty GJ (1981). Opening addresses. In: Galaty J, Aronson D and Salzman P (eds), The future of pastoral people: International Development Research Centre, Conference on the Future of Pastoral, Ottawa pp. 27-30.
Galaty J and Bonte P (1991). Herders, warriors, and traders pastoralism in Africa. Boulder, New York: Westview Press.
Ignatius, D, (2005). Washington Post, Accessed at www.relocalize.net/node, January 17th 2007 <http://www.relocalize.net/node, January 17th 2007>.
IMF (2007). Eliminating feed subsidies. <http://www.imf.org/external/np/loi/2002/jor>.
Jabbur S and Conrad L (1995). The Bedouins and the desert: aspects of nomadic life in the Arab East. Albany: State University of New York Press.
Kay S (1978). The Bedouin. New York: Crane Russak.
Kerven C (1992). Customary commerce: a historical reassessment of pastoral livestock marketing in Africa. London: Overseas Development Institute.
Kinsey J (1982). The role of marketing in economic development. European Journal of Marketing 16 (6):64-77.
Kohls R and Uhl J (1985). Marketing of agriculture products. London: Collier Macmillan.
Leybourne M (1993). Links between the steppe and cultivated area through migration: The socio-economic organisation of production of the semi-nomadic agro-pastoral society of the Syrian steppe. PhD Dissertation, France.
Livingstone L (1984). Pastoralism: an overview of practice, process and policy. Rome: FAO.
Marx E (1984). The changing Bedouin. London: Transaction Inc.
Masri A (1986). A strategy for increasing sheep production in the Hashemite kingdom of Jordan. Technical Paper 5.6, microfiche No. 5677. Rome: FAO.
Mitchell M (1985). Agriculture & policy: methodology for the analysis of development country agricultural sectors. London: Ithaca Press.
MoA (1990). Report on livestock production. Ministry of Agriculture, Amman.
MoA (1991). Report on livestock production. Ministry of Agriculture, Amman.
MoA (1992). Report on livestock production. Ministry of Agriculture, Amman.
MoA (1993). Report on livestock production. Ministry of Agriculture, Amman.
MoA (1994). Report on livestock production. Ministry of Agriculture, Amman.
MoA (1995). Livestock farmers' files. Ministry of Agriculture, Mafraq Branch, Jordan.
MoA (2006). The national strategy for agricultural development. Ministry of Agriculture, Amman, Jordan. www.moa.gov.jo <http://www.moa.gov.jo>
MoA (2007). The national strategy for agricultural development 2006. Ministry of Agri-

culture, Jordan.
MoP (1989). Report on the Badia of Jordan. Ministry of Planning, Amman.
MoP (1993). Economic and social development plan 1993-1997. Ministry of Planning, Amman.
Nesheiwat K (1991). Socio-economic aspects of the traditional Hema system of arid land management in Jordan. Rome: FAO.
Oakeley R (1996a). The nature and significance of livestock trading in the north east Badia of Jordan. Amman: BRDP.
Oakeley R (1996b). The use of supplementary feed animal in the Jordan Badia; a study of livestock producer perception and understanding sheep and goat feed alternatives, and their relative values. Amman: BRDP.
Oakeley R (1997). Production system for small ruminants in the north Badia. Amman: BRDP.
Oughton E (2008). Survival and change in the northern Badia of Jordan: the role of women in the Bedouin household. [in publication].
Randhawa N (1990). Agricultural policies and their implementation and administration in Jordan. Rome: FAO.
Roe A (1996). Input into the Badia livestock production system. Paper presented at the Workshop on Population and Livestock Related Studies in the Eastern Badia of Jordan (September 21-22, 1996), (BRDP), Amman.
Roe A (1997). Pastoralists of the lower Dana reserve area: attitude to and prospects for integration into the reserve management strategy. Dana Nature Reserve Project Socio-economic Study, Amman.
Sachs W (1992). The economist's prejudice. In: Ekins P and Max-Neef M (eds), Real-life economics, understanding wealth creation. London: Routledge, pp. 5-10.
Salzman C and Galaty J (1990). Nomads in a changing world: Issues and problems. In: Salzman W and Galaty J (eds), Nomads in a changing world. Naples: Istituto Universitario, Orientale pp. 3-48.
Schultz JL (1974). Primitive and peasant economies. In: Biggs, H and Tinnermeier, R (eds), Small agricultural development problems. Colorado: Fort Collins pp. 61-68.
Shaffer JD (1980). Food System organisation and performance: toward a conceptual framework. American Journal of Agricultural Economics 62:310-318.
USDHHS (2006). Order of the centres for disease control and prevention. USA: Department of Health and Human Services.
Wikipedia Encyclopedia (2006). Accessed at www.en.wikipedia.org/wiki/Economy-of-Jordan <http://www.en.wikipedia.org/wiki/Economy-of-Jordan>, October 2006.

CHAPTER 16

Livestock subsidy policy change in Jordan and implications for Bedouin households in the north-east Badia region

Victoria Schoen
2 Waddenhall Cottages, Petham, Canterbury, Kent, CT4 5PX. E-mail: victoria.schoen@talk21.com

This paper is based on fieldwork undertaken in 1994/5, before the removal of the livestock feed subsidy, and in 1998, about two years afterwards. Initial interviews were conducted in the households of 109 Bedouin sheep-owning families, and many of the same families were interviewed again in 1998. The interviews centred around milk production, consumption, processing and sales. The families were divided into three groups or clusters based on variables that described the degree of family dependence on livestock and milk production. In 1998 it was seen that the cluster of subsistence producers were least affected by subsidy removal while the cluster that had been most dependent on their livestock for income had been most severely affected by the change; there was less money for family expenditure and significant numbers of sheep had been sold. But the degree of overall change had also been limited by a somewhat desperate hope that circumstances would improve. However, since 1998 livestock numbers have been reduced, in response to drought as well as continuing high feed prices, but returns from the sale of milk have remained much the same as before.

Keywords: Milk, milk products, subsidy

1. Views from the literature on Bedouin household dependence on livestock

Livestock have traditionally played a central role in the household economy of pastoralist peoples. This is well documented in the literature and is borne out by fieldwork completed for this research in the desert region of Jordan in the 1990s. However, for the Bedouin in the study region, from 1981 livestock production and hence Bedouin lifestyles became increasingly tied to a subsidy on feed first introduced by the Jordanian government because of severe drought. The subsidy was eventually withdrawn over the period 1993-1996 as part of a package of Structural Adjustment measures.

With milk and dairy products playing an important role in the economy of the Jordanian Bedouin, such a change in agricultural policy, which potentially raises the cost of milk production and the opportunity cost of home consumption, would be expected to have a detrimental effect on household diets and livelihoods. The current research investigates the effects of the policy change on the milk consumption and sales habits of the Jordanian Bedouin.

The important roles which livestock and livestock products play in the lives of pastoralists has been well documented in the literature (see, for example: Jabbur 1995, Livingstone 1977, Sandford 1983). This importance is especially relevant to the Bedouin context. Writing about the Bedouin in the study area, Al-Oun (1997) states that, 'Some families…have livestock raising as their main employment and livelihood. Several have never learned any other occupation or trade, and caring for livestock best suits their way of life' (p 45). Also with reference to the Jordanian Bedouin, Abu Jaber et al (1987) explain that, 'In the domestic sphere, herding alone or in combination with cultivation constitutes the backbone of the [Bedouin] economy' (p 42). The only regular source of income to the population comes from remittances from family labour working outside the area, for example in the army or civil service. Camels and sheep also have a cultural value in Bedouin society, being regarded as a form of wealth, nobility and prestige.

The important role played by milk in the Bedouin diet is recognized in the literature. Kay (1978) states that, 'Milk and milk products from their flocks form the mainstay of the Bedouin's diet' and 'Many Bedouin meals consist solely of a bowl of milk or yoghurt' (p 18). The position of milk in the Bedouin diet relative to that of meat is highlighted by Abu Jaber et al (1987) who find that, 'In a society where camels, sheep and goats are the mainstay of the economy, meat is not the staple dish and in fact seems to be rarely eaten, being served for guests and for special occasions'. As a consequence, 'The Bedouin subsists on cereals and milk products' (p 68). For this reason, milk and dairy products rather than meat form the main commodity focus of this study.

2. The role of milk and dairy products in households in the study region

Fieldwork in the study area to demonstrate the role of livestock in the livelihoods of the population under study comprised a survey of 109 Bedouin households carried out over the summer of 1994 and the spring of 1995. Of these 109 households, 85 % were found to own some livestock. The majority owned both sheep and goats with flock sizes averaging 216 animals and herd sizes averaging 29. Livestock numbers were found to be limited mainly by availability of cash to buy feed but also cash for livestock purchase and household labour availability.

The milking season in the region occurs from March through to August and September and, because of the very high local temperatures and lack of refrigeration, any milk not consumed or sold soon after morning milking is converted to dairy products. In some households, surplus milk is sold. A variety of dairy products is made in the home ranging from butter through to soft cheese and of the 93 livestock owning households in the sample the five main dairy products were usually made by over 70 % of households. With such a high proportion of households making their own dairy products, it was predicted that these items would play a major role in diets.

To investigate this issue further, respondents were asked about the daily frequency of milk and dairy product consumption compared with the consumption of other food items, and about the percentage of their diet they believed to come from milk and dairy products at certain times of the year. Responses did suggest that dairy products are the most frequently consumed dietary items. In addition, it was found that during the milking season, 84 % of the sample takes more than 41 % of their diet from milk and dairy products and 36 % of the sample take 81-100 % of their diet from this source. Even during the winter season, milk and dairy products remain significant in the diet.

In many households, sales of milk and dairy products contribute towards household income. Sales of these goods involve different marketing channels and different intermediaries and it was unusual to find a household selling both simultaneously. In the sample, 51 of the 93 households with livestock in 1994/5 were selling or intending to sell milk to a cheese maker with quantities sold varying from three to more than 200 litres/d. For dairy products, only 23 of the 93 households had made any sales in the study period.

In many households sales of milk and dairy products contribute a significant proportion of total household income. Actual values for income were difficult to obtain. However, households were usually able to assign approximate percentages to the amount of income coming from various sources and this was sufficient information to highlight the importance of milk and dairy product sales. Fifty-nine households (63 %) in the sample received income from the latter source in the 1994/5 year with the percentage of total household income coming from this ranging 5-65 % (mean 24 %).

The timing of milk and dairy product sales is also important in maintaining a regular cash flow over the seasons. Research by Al-Oun (1997) shows that returns from milk sales early in the year provide an income to the household before the bulk of their livestock sales occur (May to November). Livestock sales are at their lowest during the winter period (November to March) thus providing an incentive for households to sell milk early in the year to provide an income at this time also.

As milk and dairy products play an important role in the economy of the Jordanian Bedouin, a change in agricultural policy which potentially makes milk production more expensive and raises the opportunity cost of home consumption would be expected to have a detrimental effect on household diets and livelihoods. The current research investigates the effects of the policy change on the milk consumption and sales habits of the Jordanian Bedouin.

3. Effects of the livestock feed subsidy policy in Jordan and reasons for its withdrawal

The Jordanian government introduced a subsidy on livestock feed in 1981 because drought in that year led to reduced availability of natural pasture for grazing. However, the subsidy continued into other years, irrespective of rainfall conditions, and became difficult to remove for political reasons. For sheep and goats, the subsidy was paid on a per head-of-livestock basis on barley and wheat bran. The quantity of subsidized feed available each year per head of livestock, and the rate of subsidy per tonne of feed, was dependent on various factors such as rainfall and fodder conditions, the budget made available by the Ministry of Finance for this purpose, and the size of the national herd (World Bank 1993a).

The subsidy allowed livestock producers to increase the size of their flocks so that, during the years of the subsidy, livestock numbers, mainly sheep and goats, increased dramatically. Estimates of exact numbers are difficult because producers would tend to inflate these when any livestock census was carried out. However, World Bank (1993b) estimates suggest that numbers increased 1.5-2.5 million animals between the years 1983 and 1991 alone.

The reduction in feed costs and increase in livestock numbers led to increased wealth amongst livestock producers and this allowed them to acquire vehicles and to transport feed and water to their animals rather than having to take the animals to the resources, as had occurred previously. Producers no longer had continually to move on in search of pasture so the combined effects of an increased flock size being concentrated on fixed areas of land for greater periods of time led to overgrazing of the rangeland and damage to, and inhibition of, new growth. In addition to this, as a result of subsidies being given to producers of cereals, the area under cultivation expanded, reducing the area of natural rangeland and further concentrating the effects of increased animal numbers (World Bank 1993a). With an increasing number of livestock utilising a declining area of productive rangeland, the percentage of total feed requirements of the livestock population met by natural grazing fell dramatically. By 1993, 50 % of feed requirements were being imported (World Bank 1993b).

The feed subsidy also caused distortions in feed markets; when barley and wheat bran were available at subsidized prices they were the cheapest purchased feeds available (IFAD 1993). Because their prices were held artificially low, there was no incentive for producers to use other feed products. Producers looked for the cheapest feed sources available and ignored the need to balance diets and vary them according to the life cycle stage of the animal. This led to declines in productivity (Oakeley 1996).

Because of these distortions, as well as the increasing cost of the policy, withdrawal of the feed subsidy was included in the 1994 Agricultural Sector Adjustment Programme. The subsidy on barley was removed slowly with the sale price to livestock producers gradually doubling JD60-120 over the period 1993 to 1997.

4. Subsidy dependence in the study region and predicted effects of its withdrawal on livestock production systems

The dependence of producers on the livestock feed subsidy is illustrated by the research of Oakeley (Oakeley 1996). A survey in the study area early in 1996, prior to the complete removal of the feed subsidy, found that all 75 livestock-owning respondents claimed to feed subsidized barley to their livestock and the majority of these also fed subsidized wheat bran. Alfalfa and maize were the next most popular dietary items but only 29 % and 16 % of respondents, respectively, purchased these.

This survey also investigated the likely reaction of respondents to a removal of the feed subsidy. 40 % claimed that they would sell all their animals. Although this was most likely an over-reaction, the drastic forecast response does indicate the degree to which producers had come to rely on subsidies for feeding their livestock. It also indicates the type of response, reducing stock numbers, which any cost increases may force them to adopt.

Works by the (World Bank 1993a) and Oakeley (1997) gave extremely pessimistic predictions about the likely effects of the subsidy removal on Bedouin households[1]. Using data presented by Campbell (1996) and Roe (1996), Oakeley (1997) calculated that the feed subsidy removal would result in a fall in gross margin per sheep JD13.50-1.62: 'Unfortunately the proposed changes will not facilitate the life of the rural producer', '...the removal will, in the short term, create hardships for especially the smaller full time private farmers' (World Bank 1993a, pps 4-6), and 'Less efficient producers will inevitably go out of business...' (Oakeley 1997, p31). Dietz et al (1993) concluded that some small and medium flock owners '...might not be able to increase productivity and might have to give up in the medium term' (p 54).

5. Methodology for assessing household response to the subsidy removal

Results of the survey carried out in the summer of 1994 and spring of 1995 were used to determine indicators for assessing household response to the policy change. The research study predicted that the effects of the subsidy withdrawal on households would vary according to their dependence on the livestock sector for diets and income in the pre-removal period. Hence the sample of 93 households with livestock was divided into groups or 'clusters' according to their dependency on livestock at this time[2]. It was expected that groups of households that displayed similar characteristics with respect to the production, consumption and sales of milk and dairy products would respond in a similar way to the removal of the livestock feed subsidy. In a return visit to the region in 1998, following the subsidy removal, it was possible to revisit a sample of households (58 in total) from each of the clusters to update the response indicators identified in 1994/5.

Variables on which to cluster the data were selected according to their suitability for describing the degree of the household's dependence on the livestock and livestock product sector. These are shown on the left side of Table 1. Households from the 1994/5 survey were clustered using the SPSS Professional Statistics programme and results produced by Ward's method using the squared Euclidean distance[3] were chosen at the three cluster solution. Clusters 1, 2 and 3 contained 39, 41 and 13 households respectively, of which 27 (69 %), 24 (58 %) and 7 (54 %), respectively, were revisited in 1998).

Table 1 provides the mean values of each variable for each cluster. These data describe three different household types: subsistence producers producing milk and dairy products mainly for their own consumption; milk sellers, being large sellers of milk; and milk and dairy product sellers who sell large quantities of milk as well as *jameed* (dried yoghurt) and *semneh* (ghee).

Table 1. Mean values for each of the variables, by cluster (1994/5 sample of 93 households)

Variable	Cluster 1 (n = 39)	Cluster 2 (n = 41)	Cluster 3 (n = 13)
Total number of sheep	16.21	284.02	148.15
Total number of goats	13.59	18.22	49.77
Area of barley planted (dunums)	34.15	64.37	153.31
Household size	8.28	10.76	12.31
% diet from milk and dairy products in the milking season	60.77	59.76	76.15
% diet from milk and dairy products in the summer / autumn	33.59	30.49	46.92
% diet from milk and dairy products in the winter	28.46	25.61	34.62
Quantity of milk sold per day (units of 3 kg / 3 litres[a])	0.44	25.08	22.64
Quantity of jameed sold in 1994 (kg)	2.82	5.46	108.69
Quantity of semneh sold in 1994 (17 kg tins)	0.31	0.61	32.00
% household income from livestock production	26.41	62.51	62.31
% household income from milk and dairy product sales	0.64	27.07	16.92

[a] Households and cheesemakers refer to sales of milk in units of 3 kg. 3 kg of milk is approximately 3 litres

5.1. Cluster 1, subsistence producers

Households in Cluster 1 may be termed subsistence producers as they own relatively few livestock (mean of approximately 30) and have a very small marketed surplus of milk and dairy products. Income from livestock production forms about 26 % of total (gross) income with the majority of income coming from other sources.

However, milk and dairy products do form an important part of the diet, particularly in the milking season. Household size is not large, reducing labour availability for managing livestock but also limiting requirements for milk and dairy products for home consumption. Some barley is grown at home to feed the small numbers of livestock.

5.2. Cluster 2, milk sellers

Cluster 2 contains households with very large numbers of stock. The mean size of a sheep flock is 284 animals, with a small goat herd (18 animals) probably to supply the household with milk over a longer milking season. Income from the large flock contributes 62 % of the household total and, when combined with income from milk and dairy product sales, these households obtain approximately 90 % of their (gross) income from the sector. They are thus heavily dependent on livestock production and milk sales.

These households are large milk sellers with an average of 75 litres of milk sold per day, yet there is still sufficient milk available for them to be highly dependent on milk

[1] However, two possible opposing environmental effects were thought possible. On the one hand, the World Bank (1993a) predicted that higher feed prices would eventually lead to a reduction in livestock numbers in the medium term so reducing grazing pressure, perhaps leading to a level lower than that prevailing before the subsidy removal. On the other hand, removal of the livestock feed subsidy meant that livestock producers would need to look for alternative, cheaper feed sources. The World Bank (1993a and b) was concerned that this might encourage the further planting of cereals for feed purposes; clearing of land for cultivation purposes increases the risk of soil erosion and degradation.

[2] See Sharma 1996 and Hair et al 1995 for further detail on the technique of cluster analysis.

[3] See Sharma 1996 and Hair et al 1995 for descriptions of these measurements

and dairy products for their diets, particularly in the milking season. With such large numbers of livestock and the resultant large milk supply, it becomes difficult to make dairy products for sale so quantities of *jameed* and *semneh* sold are minimal. Household size is approximately 11 persons, thus providing labour for flock management but simultaneously boosting home requirements for milk. Some barley is planted, but the ratio of livestock to area of barley planted is far lower than for households in Cluster 1.

5.3. Cluster 3, milk and dairy product sellers

Households in Cluster 3 are similar to households in Cluster 2. Livestock numbers are large, though not as large as for households in Cluster 2, and they also take a significant proportion of their (gross) income (79 %) from the livestock sector. However, as they take 21 % of their income from other sources and produce more of their own feed requirements, they are not likely to be as severely affected by the removal of the subsidy as households in Cluster 2.

The main difference between households in Clusters 2 and 3 is that the latter are large sellers of dairy products. With a slightly larger household size in Cluster 3 (12.31) compared with Cluster 2 (10.76), there is perhaps more labour available to convert milk to products. Despite the large quantities of dairy products sold, there are still large daily sales of fresh milk, with a mean of 68 litres. These households are also very dependent on milk and dairy products for their diets, particularly in the milking season.

The large sales quantities of dairy products and the high dependence of this Cluster on milk and dairy products for diets is explained by the fact that the majority of these households were interviewed whilst living in a *bayt ash-sha'r* (Bedouin tent) in more remote desert areas. Often there was no outlet for fresh milk in these areas and the distance from any town limited the opportunity to buy alternative foodstuffs.

Return visits were made to households in each of these three clusters in 1998 to investigate how they had been affected by, and how they had responded to, the subsidy removal. It was anticipated that households in Cluster 1 would be least affected, being the least dependent on livestock for diets and employment prior to the policy change while households in Cluster 2 with the largest stock numbers and a high dependence on livestock for diets and income would experience the greatest impact.

6. Household response to the policy of subsidy withdrawal

The subsidy on barley and wheat bran was finally removed by 1997 (Oakeley 1997). A change in policy allowed private traders to become active in the livestock feeds market and feed prices from these traders to producers were lower than those offered by government supply centres. However, the prices from private traders were still significantly above pre-1993 levels. The price of barley in 1998 varied JD95-100 and wheat bran JD75-90 per tonne (household interviews, 1998).

In addition to higher feed prices, in 1998 livestock producers were also being severely affected by lower prices for livestock sales. It was frequently suggested that the higher cost of keeping livestock had encouraged widespread sales of both adult and young animals and households generally believed that prices were still falling in 1998. As a result, they were having to sell increasing numbers of animals in order to obtain sufficient income to pay for feed for those remaining, encouraging producers often to claim that, 'the sheep are eating the sheep'. Data collected from livestock markets in 1998 suggested prices to be 40-50 % lower than prior to the subsidy withdrawal.

As predicted, households in the three clusters responded to these changes in different ways. Cluster 1, the subsistence producers, owned livestock mainly to supply the household with its milk and dairy product requirements. Prior to the subsidy removal, a very small percentage of gross income came from this source, and often feed costs had been met from other income sources. When the subsidy on feed was removed, households did not adjust by selling their stock as they still needed to consume milk and dairy products. Even if it was now less expensive to buy milk and/or dairy products than to keep livestock to produce these at home, households often preferred their own production and were prepared to increase the amount spent on feed from other salaries rather than switch to being net purchasers of food products. There was no significant change in the number of milk and/or dairy product sellers in this group, and no significant change in the percentage of gross income drawn from the sector, which remained at around 26 %.

Cluster 2 households, the milk sellers, were far more responsive to the policy change. They were originally the most dependent on the livestock sector for income so it was likely that they would be the households most affected as they did not have other sources of income which could be used to absorb the increase in costs, at least in the short term. Certainly, increased spending on livestock feed forced them to reduce expenditure on other items, usually food for the household, and despite this, they often needed to reduce their livestock numbers in order to feed those which remained. This is the only cluster where livestock numbers in fact showed a significant decline, of 18 %.

The decline in stock numbers led to reduced milk sales while consumption of milk and dairy products remained steady. There is a tendency for home consumption to be the first priority when milk output declines which suggests a trend towards subsistence production. Only three households in the cluster completely lost their ability to sell milk and/or products and were unable to replace this with income from other sources, but other households did reduce the amount that they sold.

Only three households in Cluster 2 were no longer dependent on livestock production for at least a part of their income following the subsidy removal. This suggests that the remaining households were worse off as they had frequently seen a decline in their livestock numbers and in milk sales, with very little diversification in income sources. Even when the flock size falls, the household head often still considers that he is unavailable for alternative work as, to a certain extent, 100 sheep require as much male labour presence as do 50. Hence, there was little shift of labour out of the livestock production sector and those households remaining were made worse off by higher feed prices and lower livestock prices.

Theoretically, a smaller flock may be managed better than a larger one leading to higher productivity. However, in this case, with management systems changing very little and wealth still measured in numbers of stock, households were not happy about reducing flock sizes and usually had not done this through choice (see Case Study 1 presented in Box 1).

Cluster 3 households, the milk and dairy product sellers, on average increased their livestock numbers (although this increase is not significant when tested statistically at the 5 % level of confidence). The sample is really too small to draw general conclusions. This is partly caused by the nature of households in this group who were often first interviewed in a *bayt ash-sha'r* in the desert region and hence were impossible to trace in 1998. One household in Cluster 3 that did not follow the trend to increase livestock numbers is described in Box 2.

Sources of income for households in Cluster 3 were slightly more diversified than those of households in Cluster 2 prior to the subsidy removal and this may have given them the reserves they needed to increase livestock numbers while prices were low, or indeed to provide an income to the household to delay sales of stock until prices improved.

Box 1: Case Study 1, Qasim village – Cluster 2 household

This household owned 50 sheep and eight goats in the interview in 1995 and in 1998, sheep numbers had fallen to 25 while goat numbers had increased to 12. The reduction in sheep numbers was more serious than the year-to-year figures suggest: the head had sold 120 sheep in 1997 as a direct result of the feed price increases. The remaining animals provided the sole income for a family of seven people. The head had tried to reduce the livestock dependency on purchased feed by extending the time spent out on pasture, and he had reduced the amount of feed available to each head of livestock as there was not enough money in the household to buy sufficient. Whereas the household used to be able to save money from sales of lambs to provide an income throughout the year, it was now unable to save anything and was having to sell breeding stock in order to meet household expenses.

As a result of the diminished feed ration, milk yields had fallen and there was no longer enough to sell milk to a cheese maker. Hence this was no longer a source of income for the family. The proportion of the family's diet coming from milk and dairy products increased as they had stopped selling milk and because they had very little cash to buy other goods from the shops. The proportionate increase in significance of milk and dairy products in the diet did not necessarily mean that total consumption had increased, just that these products were now more important to them relative to purchased items.

The current household situation was untenable in the long term: the family was not getting enough food and there was no money in the house to buy more. The head thought that if feed and livestock prices did not improve very soon, he would sell the rest of his animals and go to find work in Mafraq.

Box 2: Case Study 2, ar-Rifa'iyyat village – Cluster 3 household

In this village, one household visited had responded to the increase in the feed prices by selling stock, and intended to sell the remaining animals at the end of the 1998 milking season. In the last interview in 1994, the head claimed to own 100 sheep. This figure stood at 50 in the July 1998 interview and the intention was to sell everything by the end of August 1998. The owner had calculated that it cost JD60 to keep one sheep for a year, including all feed and veterinary costs. The lamb price was only JD40-50 so losses were being incurred on all animals.

Any money coming into the household from the sale of sheep and lambs was being used to pay for feed so the family did not have enough money to buy the food that they would choose or new clothing. Before the subsidy removal, money available to the household for such purposes amounted to approximately JD100 per month, whereas with higher feed prices, monthly disposable income had fallen to about JD60. The family also reported that they were unable to save anything out of their income.

In the 1994 interview, the household had sold both jameed and semneh but because of the more urgent need for cash income in 1998, they were selling all surplus milk fresh to the cheese maker as this brought a more rapid income. Despite this the head observed that a greater return could be made on milk when sold as dairy products.

Having sold all the livestock, the head planned to operate solely as a livestock trader next year. Milk would be bought from neighbours for the production of dairy products.

Sales of sheep milk showed no significant change over the period for Cluster 3 households but there was a decline in sales of *jameed* and *semneh*. Quantities of these products traded and the number of households involved are small, so again results are inconclusive but they may point to a general trend to sell milk fresh for more immediate returns rather than to produce products which might have taken longer to sell and which offered very little price advantage.

7. Discussion of results

Of the three Cluster types, households in Cluster 2 are most obviously identifiable as the worst affected by the subsidy removal, having seen a decline in their livestock numbers and milk sales, which had often not been overcome by diversification into other sectors. The fate of households in Cluster 3 is really indeterminate from the 1998 data but was perhaps dependent on the direction of livestock prices in 1999. Those increasing numbers through further investment were hoping for better prices for lambs when they sold from their increased breeding base, while those waiting to sell older stock to reduce their flock size were also hoping for better prices at the time of sale.

For the sample as a whole, livestock numbers per household fell by only 6 % (145-135) following the subsidy removal and this followed through to minor changes in consumption of milk and dairy products, and an insignificant change in the quantity of sheep milk offered for sale. Given the predictions regarding the fate of livestock producers following the policy change, it is perhaps surprising that livestock numbers on average did not fall further. However, the return household visits in 1998 occurred only one to two years after the subsidy had finally been removed and there was certainly still a feeling of 'wait and see' amongst many of the households interviewed. It takes time for households to adjust and often they will try to hold out as long as possible before making any adjustments. Many spoke of their hopes that feed prices would fall further and livestock prices would recover; it is likely that some believed the situation would eventually settle down and they would be able to continue as before. The main problem was to find some way of surviving the short term.

The majority of households in Cluster 3 had more livestock in 1998 than in 1994/5 and sometimes the only reason for this was that they were waiting for prices to improve. If stock numbers of Cluster 3 type households were large and these did sell at a later date, then the eventual aggregate reduction in numbers may have been far greater than 6 %, depending on whether the stock were bought for meat or by other households for breeding. Cluster 2 households may have eventually been forced to sell more of their livestock.

Up until the time of the return visits in 1998, households had been attempting to absorb the increase in feed costs and had managed to forestall large reductions in flock sizes. Some had been feeding the animals less, some had taken credit from feed traders, many had reduced expenditure on other goods, and some had made slight downward adjustments to their flock size. In difficult times, the household can subsist off the products of the sheep. A small marketed surplus of milk provides the cash to buy feed for the livestock.

The situation worsens when the market surplus of milk disappears and there is no longer a regular supply of income from this source. It is then necessary to sell young stock to obtain the cash to feed the remaining livestock, and when sales of these do not supply enough income to meet cash needs (especially when livestock prices are depressed), it becomes necessary to start selling the breeding stock. However, this may be a slow process as the day-to-day cash needs of Bedouin households are not high. By reducing their expenditure on food and clothing they have managed to divert any cash to pay for livestock feed to maintain their flocks. But it is unlikely that this situation is tenable in the long term as it would become increasingly difficult for the household to meet emergency cash demands, such as to pay for medical expenses or vehicle repairs.

8. Conclusion

Whatever adjustments occurred in the 1994/5-1998 period, it is likely that households would have needed to improve the efficiency of their livestock production in order to succeed in the longer term. It may be that some Cluster 2 and 3 households eventually became Cluster 1 households, maintaining livestock solely for subsistence purposes.

Eventually Cluster 1 households may have sold all their livestock and become dependent on markets and gifts from other, more successful, livestock producers for their milk and dairy product requirements.

Those households least able to adapt to the policy change would eventually have been forced out of the livestock sector and would have needed to find alternative employment elsewhere. According to some authors (for example: Helland 1980, Jahnke 1982, Toulmin 1983) the removal of some households and livestock from the land makes it easier for those remaining in livestock production to feed their animals from the natural pasture. In this case, the movement of some Bedouin families from the livestock sector would not be detrimental, if other opportunities existed.

To 1998, there was no large scale movement of households out of the Badia region. Development efforts need to be directed towards providing the opportunities which will prevent this from occurring in the future by looking for ways to help those households no longer involved in livestock production. However, the Badia is a remote region and the population has a history of livestock herding. Opportunities for diversification, especially for the older generation, are limited. Unless future development in the Badia region creates employment opportunities for the younger generation, it may have to be accepted that, in the longer term, younger household members will be forced to migrate in search of employment while only the most dedicated and most effective livestock producing households or household members remain in production.

While change may have occurred as households have adapted, in 1998 the short term situation remained extremely difficult for some households and there may have been no other option but to re-introduce some kind of targeted and time bound financial support to assist them while input and product markets settled.

9. Developments since 1998

Since 1998, numbers of livestock in the Badia region have fallen as a result of a lack of rain and hence natural grazing. Despite increases in livestock prices, it has become uneconomical for some producers to retain their animals. The price of barley has changed little, at JD90 per tonne from government sources and JD110 per tonne from private traders.

Returns from milk sales have remained much the same over the period despite a new cheese factory being developed in one village in the region, Tall ar-Rimah. This factory operates for only three months in the year so has had little effect on employment in the region or household income sources. Many households are still dependent on a salary from the army, a government department or work in the private sector.

Acknowledgements

The assistance of Dr. Awadh Al-Laimoon in gathering the information for this section is gratefully acknowledged. Dr. Al-Laimoon also provided most of the interpretation in the fieldwork undertaken for this study.

References

Abu Jaber KS, Gharaibeh FA and Hill A (1987). The Badia of Jordan, the process of change. Jordan: University of Jordan.

Al-Oun SS (1997). Livestock marketing in the Jordanian Badia. PhD, University of Durham.

Campbell DA (1996). Review of animal production in the Badia. Paper presented at the Workshop on Population and Livestock Related Studies in the Eastern Badia of Jordan (September 21-22 1996), JBRDP, Amman.

Dietz M, Hannover W and Lindauer G (1993). Assessment of the potential impact of agricultural adjustment measures, as proposed under ASAL (Draft), on farmers' incomes in Jordan.

Helland J (1980). Some issues in the study of pastoralists and the development of pastoralism. In: Helland J (ed). Five essays on the study of pastoralists and the development of pastoralism, African Savannah studies, Occasional Paper No. 20. Bergen: Sosialantropologisk Institut Universitetet I Bergen.

IFAD (1993). The Hashemite Kingdom of Jordan: national programme for range rehabilitation and development formulation report volume 1. Main report, International Fund for Agricultural Development.

Jabbur JS (1995). The Bedouins and the desert: aspects of the nomadic life in the Arab east. Translated from the Arabic by Conrad L I, New York State University of New York Press.

Jahnke HE (1982). Livestock production systems and livestock development in tropical Africa. Kieler Wissenschaftsverlag Vauk.

Kay S (1978). The Bedouin. New York: Crane, Russak and Company Inc.

Livingstone I (1977). Economic irrationality among pastoral peoples: myth or reality? Development and Change 8:209-230.

Oakeley R (1996). The use of supplementary animal feed in the Jordan Badia: a study of livestock producer perceptions and understanding of sheep and goat feed alternatives, and their relative values. Jordan Badia Research and Development Programme.

Oakeley R (1997). Economic and policy trends affecting the production and marketing systems for small ruminants in the north east Badia. Jordan Badia Research and Development Programme.

Roe A (1996). Inputs into the Badia livestock production system. Workshop on Population and Livestock Related Studies in the Eastern Badia of Jordan (September 21-22 1996). Amman: JBRDP.

Sandford S (1983). Management of pastoral development in the Third World. Overseas Development Institute, London: John Wiley and Sons.

Toulmin C (1983). Economic behaviour among livestock-keeping peoples: a review of the literature on the economics of pastoral production in the semi-arid zones of Africa. Development Studies Occasional Paper No. 25: School of Development Studies, University of East Anglia.

World Bank (1993a). The Hashemite Kingdom of Jordan environment ASAL and ASTAP recommendations. Working Paper no. 4 World Bank.

World Bank (1993b). The Hashemite Kingdom of Jordan rangelands and livestock: ASAL and ASTAP recommendations. Working Paper no. 3 World Bank.

CHAPTER 17

Socio-cultural, physical, and architectural attributes of bayt ash-sha'r *in transition in Jordan's Badia*

Majd Al-Homoud
German Jordanian University. E-mail: majd.alhomoud@gju.edu.jo

Salem Al-Oun
Business Administration Department, Faculty of Economics and Administrative Sciences, Al Al-Bayt University

Amal Abed
Department of Architecture, Jordan University of Science and Technology (JUST), Jordan

The authors interviewed 317 people from 34 villages in the north-east Badia and observed and photographed the entire evolving range of Bedu 'homes' from the traditional *bayt ash-sha'r* (hair tent) through transitional village homes to large modern houses, or villas. In spite of using entirely different materials for the construction of these various homes the spatial separation, shape, internal divisions and functions of the original hair tent were strongly reflected in the more recent transitional forms, made of mud and basalt rocks, and even in the more sophisticated villas, though the growing wealth of some families coupled with their increasing exposure to external influences (as well as the intrinsic qualities of the new materials) have gradually changed external shape and internal divisions of space and furnishings in the newest houses. However, some traditional public and cultural functions of the original *bayt ash-sha'r* seem still to require its presence, a dilemma often resolved by erecting the *bayt ash-sha'r* beside the new house.

The structure and usage of the different forms of the *bayt ash-sha'r* are described in meticulous detail, making a permanent record which will also help preserve its ancient Arabic nomenclature for a time when, perhaps, hair tents are no longer used.

Keywords: *Bayt ash-sha'r*, house, home, tent

1. Introduction

[1] لبيت تخفق الأرياح فيه   أحب اليّ من قصر منيف

A tent with resulting cool breeze /Delights me more than palace high[2]

Home is a social, cultural, and physical package that reflects users' changes in their way of living by directly transforming external and internal layout to fit their new needs. It is the closest representation of the relationship between the built environment and cultural values. Change in the way of life can be directly manifested in the built environment through form and layout, and the built environment is an image that reflects cultural values and way of life as well.

This study investigates Bedu (here defined as mobile, semi-settled and settled pas-

---

[1] Source: ميسون بنت بحدل

[2] Source: Nicholson, R. A., Literary History of the Arabs. (1997).

Figure 1. Settlements in the study area (Source: Al-Oun 1997)

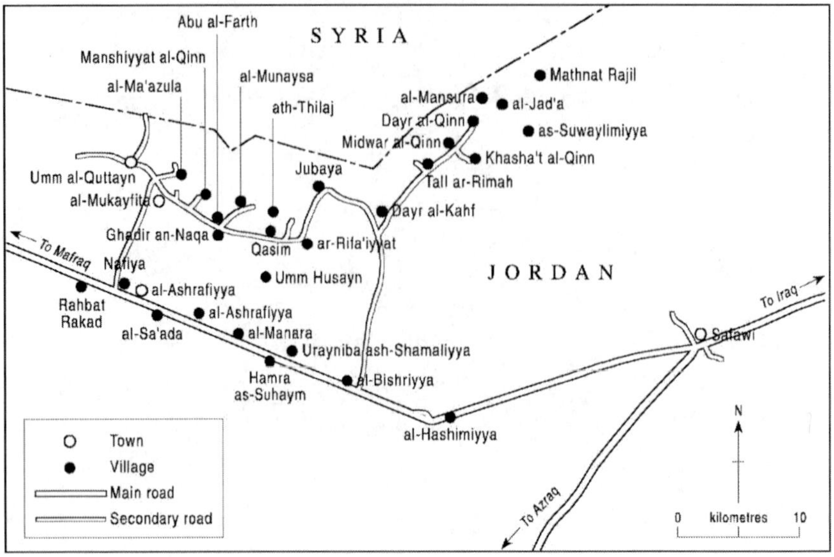

toralists) home layout prototypes as a result of the influence of lifestyle, with a view to understanding the development of the traditional home layout in the north-east Badia of Jordan. The traditional Bedu home is a *bayt ash-sha'r* (house of hair) where *bayt* means house or home, and *ash-sha'r* is related to the hair material the home is made of (Weir 1990). A *bayt ash-sha'r* is a long, horizontal black tent made of goat and camel haircloth woven by women. It is a symbol of Bedu identity and is part of their tradition and culture (Jabbor 1988, Weir 1990). A *bayt ash-sha'r* is very well adapted to desert life; it is an effective solution to shelter, privacy, environment, and construction (Al-Ozaizat 1980, Weir 1990).

The research methods used in the present study include quantitative (field survey) and qualitative aspects (field interviews and field visits). In the field survey, a structured questionnaire was administered by the interviewer in November 2000 to 317 respondents. The sample came from the 34 villages of the Jordan Badia Research and Development Programme (BRDP[3]) area. A proportional systematic random sampling technique was used to select the subjects. In the field interviews, a semi-structured questionnaire was administered by the interviewer to elicit data from 16 key informants from the north-east Badia. Analysis of variance and chi-square tests were used to measure the difference and association of the hypothesized relationship at the different variable levels.

It is hoped that the present investigation will contribute to a better understanding of the housing market in Jordan through promoting housing research for special groups of specific socio-cultural values and traditions in general, and for Jordanian Bedu societies specifically.

1.1. Research settings

The research setting included the 34 villages of the original BRDP area, located in the north-east Badia of Jordan (Figure 1). These villages are located east of Mafraq, the main city in the north-east Badia. The density of homes ranged from four to 395 per village

[3] The Badia Research and Development Programme became the Badia Research and Development Centre in May 2005, with a remit throughout the drylands of Jordan.

based on statistics from the 1993 demographic and socio-economic survey of BRDP. The survey revealed three representative types of home: the traditional *bayt ash-sha'r*, a tent of any size usually made of goat hair; a transitional home, made of mud, rough stone, concrete, or concrete blocks; and a modern home (villa), made of concrete blocks, concrete, or white stone and regular basalt stone.

Jordan is divided into four main regions distinguished by topography and climate. They are the highlands, the Jordan Valley, Aqaba Gulf, and the arid and semi arid lands of the Badia (Al-Oun 1997, Kana`n et al 1992). The Badia covers 72 600 km^2, which is 81.3 % of the total area of Jordan and is administratively subdivided into three sectors: the northern, central and southern Badia (Figure 2) (Al-Oun 1997). The study area is part of the north Badia, located in Mafraq governorate.

1.2. Cultural heritage of Bedu in Jordan

The Badia is the home of the Bedu. The number of Bedu living in the Badia has sharply declined since the introduction of the new Ottoman land laws in the mid-eighteenth century, which abolished the communal ownership of land that was a basic component of their nomadic lifestyle. The decline continued under twentieth century central governments that encouraged the Bedu to settle down in order to control them better (Al-

Figure 2. Jordan showing the different Badia sectors (Source: Al-Oun 1997)

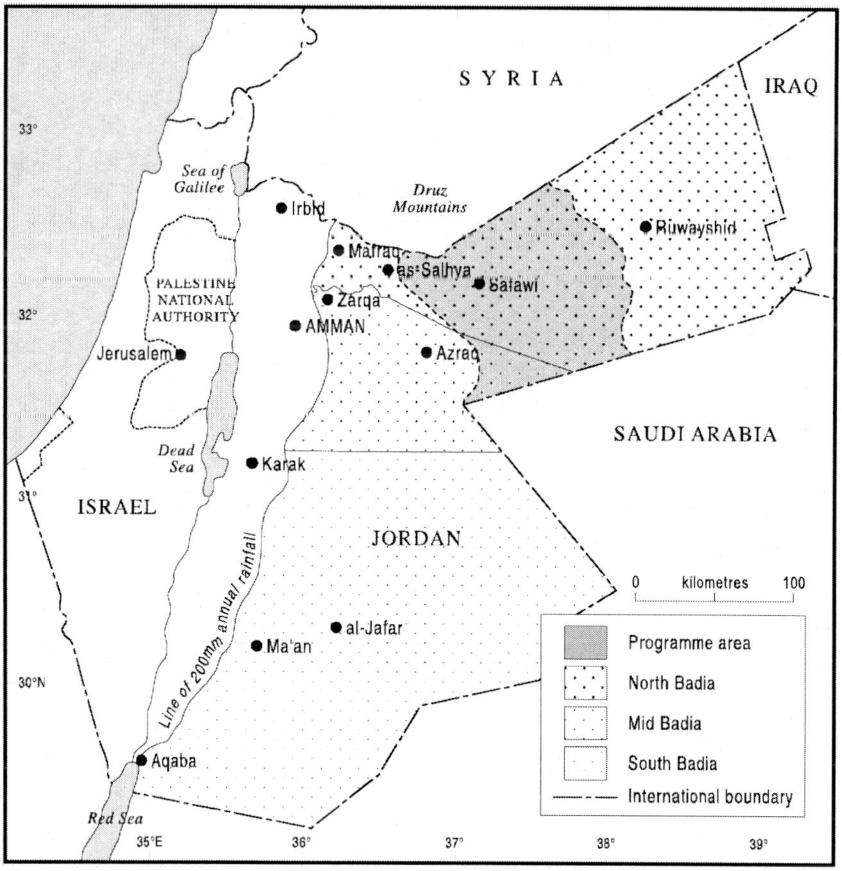

Oun 1997, Ismael 1983). Despite the difficulties the Bedu face, they are proud of their way of life and consider themselves the noblest class in the Arab society. Their pride stems from a paradoxical admiration of noble Bedu virtues as a model of pure Arabic-Islamic culture (Al-Oun 1997, Jabbor 1988).

Bedu society is based on tribal organization that is characterized by collectivism and fierce loyalty to family, clan and tribe (Al-Oun 1997, Al-Ozaizat 1980). Their structure is based on *'asheirah*, which is subdivided into clan (*hamoulah*), which is in turn divided into main families (*fakhdh*) (Al-Oun 1997). The Jordanian *'asheirah* of the north-east Badia include *Bani Khaled, As-Sirhan, Al-'Isa, Ahl Al-Jabal*, and *As-Sardiyyah* (Al-Mousa 1996, Al-Oun 1997). Bedu society is patriarchal; all members of a tribe claim that they have descended from a common male ancestor. The tribal chief is the *sheikh* who descends from a noble family and is characterized by generosity and bravery. When he dies, his role does not always pass automatically to his eldest son; it is rather open for any suitable member of the family who can gain the approval of the elders. The *sheikh*, as a leader of the tribe, has a considerable power limited by customs, precedence, and the advice of the council of elders called the *majlis*. He rules by virtue of his personality and the respect it engenders, by negotiation, consensus, and arbitration (*shoura*), rather than by dictate (Jabbor 1988). Bedu society regulates most life affairs by the customary law of ancestors, which are viewed as a strict code of rules. Customs inherited from ancestors stress the values of tribal obedience and honour, and defend it against all other groups through cunning and revenge (Weir 1990).

Bedu life depends on four basic elements (Jabbor 1988): the Badia as it is related to Bedu environment; the camel that is used as a means of transportation; the Bedu individual who is the son of the Badia where he grew up and lived all his life; and the *bayt ash-sha'r* where the Bedu live. The importance of these elements is linked to nomadic pastoral life, which characterizes Bedu life in Jordan, although Bedu have been shifting steadily to semi-nomadic and fully settled pastoral individuals (Abu-Jaber et al 1987b). Shifting to a settled life does not mean that the Bedu altered the basic patterns of their life; rather, they still have a hybrid life combining traditional and modern values. The traditional Bedu economy is still largely dependent on livestock and on its milk products, meat, and wool (Al-Mousa 1996, Al-Oun 1997), though other sources of income have a growing importance even in the most remote areas.

Bedu lifestyle has shifted to a settled one for the following reasons: factors that affect Bedu lifestyle development, including governmental strategies and policies for settlement; economic development; improved transportation and communication media; intermingling with urban societies; less dependence on sheep and other livestock and land cultivation; education and job opportunities; and settlement in fixed homes while keeping *bayt ash-sha'r* in some cases (Al-Fawal 1983, Brandenburg 1998, Ismael 1983). These factors reduced the average number of nomadic Bedu as a proportion of the total Arab population from 10 % in 1960 to 1 % in the late 20th century (Abu-Jaber et al 1987b).

2. Socio-cultural attributes of *bayt ash-sha'r*

The built environment is either a container of material objects or an empty frame to be filled as a livable space or a medium for communication that reflects an individual's activity and embodies specific community meanings (Al-Homoud 1995, Al-Homoud et al 1997). The built environment reflects cultural and socio-cultural components, and is a document that describes our past, an image that reflects our present, and a sign that calls us into the future (Richardson 1989). This suggests that there is a broad and complex relationship between the built environment and the socio-cultural values, where the best manifestation of the built environment is architecture, represented in its simplest form by the home (Al-Homoud 1995, Al-Homoud et al 1997).

Figure 3. Women's space occupancy in the *bayt ash-sha'r* (Source: Al-Homoud and Shauqour[4], 2007)

The dimensions of the socio-cultural world affect lifestyle, which is defined as the way of life (Rapoport 1969, Tognoli 1987). However, each cultural group is distinguished from others by lifestyle components. For example, Bedu lifestyle is distinguished clearly by tribal and clan loyalty, traditional customs, traditional collectivist society, and identity. Home identity, whether collectivistic or individualistic, is communicated in different ways: spatially and socio-spatially (Duncan 1985). Bedu were transformed from a nomadic pastoral society to a settled life (Abed 2001, Al-Mousa 1996). Lifestyle components in the present study include occupation, education, family origin, income, social status, gender, life cycle, ethnicity, role, hobby, aesthetic taste, cultural customs, identity, consumption behaviour, and family type and size (Erwin 1994, Rapoport 1991, Strauss 1976, Weisner et al 1981).

Bedu lifestyles are geared towards using the internal spaces traditionally, and this is mainly affected by traditional male and female household occupations. Householding is considered the traditional female occupation in Bedu culture. A housewife usually spends most of her time in the home to perform the daily work such as baking, cooking, spinning, and weaving (Figure 3) Because of the diversity of a housewife's work, it becomes difficult to assign one space for each activity; therefore, overlapping of such activities in one space becomes imperative. This overlapping appears clearly in the functional space of *bayt ash-sha'r* (*haderah*) where the housewife cooks, stores food, and washes clothes. Moreover, mobility dictates the overlapped use of one space for more than one activity (non-specialized space) (Abed 2001, Al-Homoud et al 2000b).

Traditionally, livestock ownership is the most common household occupation, involving travelling and searching for grazing land and water resources for animals. As a result, it requires homes that are mobile, light, and easy and quick in construction so as to suit their mobility, such as *bayt ash-sha'r*, as suggested by Weir (1990). Occupation usually has a direct effect on income; the traditional occupation, livestock ownership, offers a seasonal income, but since it depends on climatic circumstances, traditional Bedu income fluctuates from year to year depending on the rainfall. The *bayt ash-sha'r* suits fluctuations in income, because it is made out of hair taken from the livestock, which is woven and sewn by family women, as explained by Jabbor (1988) and Weir (1990)[5]. Income fluctuation does not allow Bedu to save money to escape droughts; therefore, they are not able to allocate

[4] All photos with this similar source were originally taken in 2002 by Dr. Majd Al-Homoud and were reproduced in 2007 by both Dr. Al-Homoud and architect Muhammad Shaqour, a graduate from the Institute of Islamic Arts and Architecture at Al Al-Bayt University, Jordan.

[5] See p 290 for an explanation of the meaning of all the words related to tent construction.

enough money for home structure since the first expenditure priorities are food, drink, hospitality, and clothing. Occupation and income affect social role and status; rich Bedu have large *bayt ash-sha'r* used to host guests as an emphasis of their traditional hospitality. This readily accommodates expansion or enlargement in home size according to guest numbers (Kellett 1999) (Abed 2001, Al-Homoud et al 2000b).

Traditionally, Bedu prefer polygamy, thereby requiring additional areas, because each wife needs a special space. Such characteristics are found in the *bayt ash-sha'r*, as its size can easily be enlarged through adding another number of *shigag* to form a new *raba'h*. This readily accommodates the Bedu family,which is big because a large number of sons and daughters are considered a source of pride and are needed as a work force. Bedu have a strong tradition of women's virtue and honour. Thus *bayt ash-sha'r* spaces are divided in terms of use into two parts reflecting gender segregation: one for men, and the other for women and the rest of the family. The men's quarter tends to be larger than that of the women as an expression of the honour of receiving guests, while the women's quarter is used to house the rest of the daily living activities of the family in addition to receiving female guests. Couples, daughters, and sons of less than ten years sleep in the *maharam* and adult males sleep in the *madheif*. Specifically, men's space is used for men's activities that include receiving guests, sitting, eating, and sleeping (Abed 2001, Al-Homoud et al 2000b).

Privacy is mainly regulated through the Bedu cultural values, beliefs, and ideas. Bedu women, for example, can participate in the family discussion, listen to male opinions, and volunteer or sometimes are asked for their opinions; if the guest is a stranger, women will participate from behind the scene, otherwise they can join the males of the family in the discussion. Women, within the Bedu code of honour, are allowed to receive male guests in the *madheif*, if the male householder is not available, especially if the guests are travelling for long distances. Bedu use cultural beliefs for privacy regulation; for example distances that separate *bayt ash-sha'r* from all sides and, particularly, from the front side, are a physical expression of the need for privacy. Another form of such beliefs is the location of the entrance, which is situated at the corner of the *bayt ash-sha'r*, as suggested by Al-Mousa (1996). In addition, privacy is achieved through hierarchical zones in a *bayt ash-sha'r*, through *haram bayt ash-sha'r* (to be explained later) (Abed 2001, Al-Homoud et al 2000b). Bedu use *haram bayt ash-sha'r* for privacy, territory, for children playing, and for animals. Horses are placed at the front, donkeys are tied at the back, and goats are delimited within a space, called *seyrah*, that is situated at a distance from the *bayt ash-sha'r* (Al-Homoud et al 2000b, Al-Ozaizat 1980).

The *bayt ash-sha'r* was considered a basic necessity for Bedu, and there are different reasons to explain this necessity, such as hospitality, psychological comfort, openness, mobility, durability, occasions, flexibility, lightness, simplicity and beauty, authenticity, continuity, social communication, heritage, identity, assurance, external work, and climatic suitability. The *madheif* is always opened as an expression and a symbol of Bedu hospitality; closing it brings shame on the householder (Abed 2001, Al-Homoud et al 2000b). Further, as a communal act, Bedu prepare *nezalah*, a dinner meal to which new neighbours are invited for more acquaintance (Abed 2001, Al-Homoud et al 2000b).

3. Physical and architectural attributes of the bayt ash-sha'r

A *bayt ash-sha'r* has different sizes that include *kharboosh, gatbah, shawa, mthoulath, mrouba', mkhoumas, msoudas, msouba'*, and *mthouman*, based on the number of *wassit* (to be explained later). It is made out of different materials that include goat and camel hair, cotton, and *kheish* (Figure 4). A *bayt ash-sha'r* is the basic and original Bedu home because it suits their traditional life that is based on constant moving. It is also considered an effective shelter that provides privacy, and it is handy for the construction and deconstruction that is required by continuous travelling. A *bayt ash-sha'r* has two major internal

Figure 4. Traditional *bayt ash-sha'r* (Source: Al-Homoud and Shauqour, 2007)

Typical external *bayt ash-sha'r*	Typical internal *bayt ash-sha'r*

space uses: *madheif* and *maharam*, furnished with simple furniture such as carpets and pillows. Its external spaces are related to the territory of the *bayt ash-sha'r* (*haram bayt ash-sha'r*) where the Bedu keep their livestock (Abed 2001, Al-Mousa 1996).

The *bayt ash-sha'r* has been rarely used since the 1940s as a result of Bedu interaction with urban centres in Jordan. In the present study, only 5 % of Bedu live in *bayt ash-sha'r*. The noticeably limited percentage of *bayt ash-sha'r* is as a result of Bedu settlement that has reduced their dependence on *bayt ash-sha'r* functionally, though it has stayed significant socially and symbolically. About one-sixth of the *bayt ash-sha'r* in the sample has the size *gatbah* or *mthoulath*, and two-thirds of the sampled *bayt ash-sha'r* are made out of goat and camel hair. The limited availability of *bayt ash-sha'r* among the subjects in recent years is related to their limited use for *ta'zeib* and travelling with the livestock (Al-Homoud et al 2000b).

It is useful to present as comprehensive an image as possible about the traditional Bedu home layout, *bayt ash-sha'r*, by describing its types, structure, spaces, environmental consideration, development, and furniture (Al-Ozaizat 1980, Weir 1990).

3.1. Types

The *bayt ash-sha'r* type is an indication of its size, which can be described in terms of its height, width, and length. Usually these dimensions have a positive relation to each other: when width and length increase, the height will directly increase. The height consists of four *shgag*, the width ranges 6-12 *shigag*, the exact number of which correlates with the financial and social status of the householder, and the length which is measured with *wassit* or *raba`h* number. Accordingly, the *bayt ash-sha'r* of the clan sheikh is the largest one and is usually of the *msoudas* or *msouba'* type (Al-Homoud et al 2000b). Generally, *bayt ash-sha'r* types are categorized by different clans into two classifications according to their length.

Figure 5. Mkhoumas *bayt ash-sha'r* type (Source: Al-Homoud and Shauqour, 2007)

Figure 6a-b. Elements of a *bayt ash-sha'r* (Source: Al-Homoud and Shauqour, 2007)

a. *Wasit* in *bayt ash-sha'r* | b. *shigag* and *tareigah* of the top of *bayt ash-sha'r*

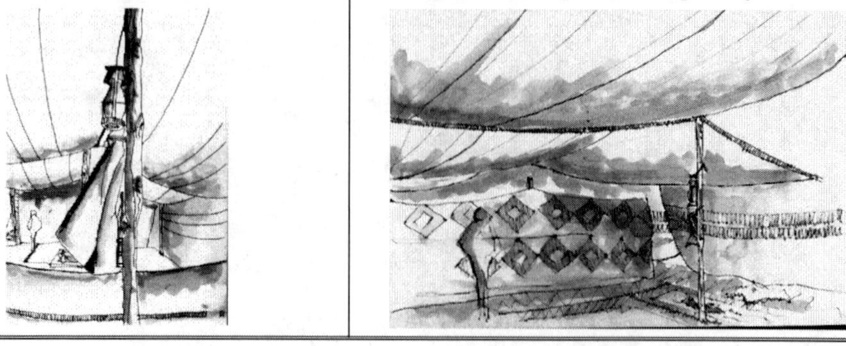

The first classification divides the *bayt ash-sha'r* into nine types according to *wassit* and *raba`h* number. These types are: the *kharboosh* which is built without a middle post (*wassit*) and has one space (*raba`h*); the *gatbah*, built with one *wassit* and two *raba`h*; the *shawa*, built with two *wassit* and three *raba`h*; the *mthoulath*, built with three *wassit* and four *raba`h*; the *mrouba'*, built with four *wassit* and five *raba`h*; the *mkhoumas*, built with five *wassit* and six *raba`h* (Figure 5); the *msoudas*, built with six *wassit* and seven *raba`h*; the *msouba'*, built with seven *wassit* and eight *raba`h*; and the *mthouman*, built with eight *wassit* and nine *raba`h* (Al-Homoud et al 2000b, Al-Mousa 1996, Al-Ozaizat 1980).

The second classification divides *bayt ash-sha'r* into two types according to a specific *wassit* number that divides a *bayt ash-sha'r*: the *mgouran* consists of two *wassits* or less and includes *shawa, gatbah,* and *kharboosh* types; the *mdoubal* has three *wassits* or more and includes *mthoulath, mrouba', mkhoumas, msoudas, msouba'* and *mthouman* types (Al-Homoud et al 2000b, Al-Mousa 1996, Al-Ozaizat 1980).

3.2. Structure

In general a *bayt ash-sha'r* is a long, horizontal tent, made of *sha'r* (hair), and having a simple structure that enables the Bedu to fold it, pack it up and move it easily (Al-Rawi 1949, Jabbor 1988). It consists mainly of four components: columns, sides (walls), top (roof), and *tanab*. The roof is defined as a rectangular cloth sewn and supported by the centre (*wassit*) (Figure 6a) and wooden edge posts anchored by guy ropes (*tanab*) and cleats (*wedd*) (Figure 7a-b). The framework is covered with *shigag* (longitudinal strips), 60-80 cm in width (Figure 6b), which are sewn together to run the length of the *bayt ash-sha'r*. Most roofs are made of 6-8 *shigag*, divided equally between both sides of the ridge. So, the *bayt ash-sha'r* width averages 3.6-6.4 m (Al-Homoud et al 2000b, Al-Ozaizat 1980).

The walls are pinned to the edge of the roof and are categorized into four types (Al-Homoud et al 2000b):

- The *rwaq* is a back wall that consists of two main decorative strips made of goat and camel hair connected with each other by longitudinal strips called *dafa'iyah* made of goat and camel hair too (Figure 8a). The *sfalah* is the lower *shigag* of the back wall that connects to the ground. Made of goat and camel hair, it is usually tied to the ground with a *sekkah*;
- The *hijal*, also called star, is the front wall also made of goat and camel hair or *kheish*;
- The *ruffah* is a side wall that consists of vertical *shigag* and is also made of camel and goat hair;

Figure 7. Structural system of the *bayt ash-sha'r* (Source: Al-Homoud, 2006)

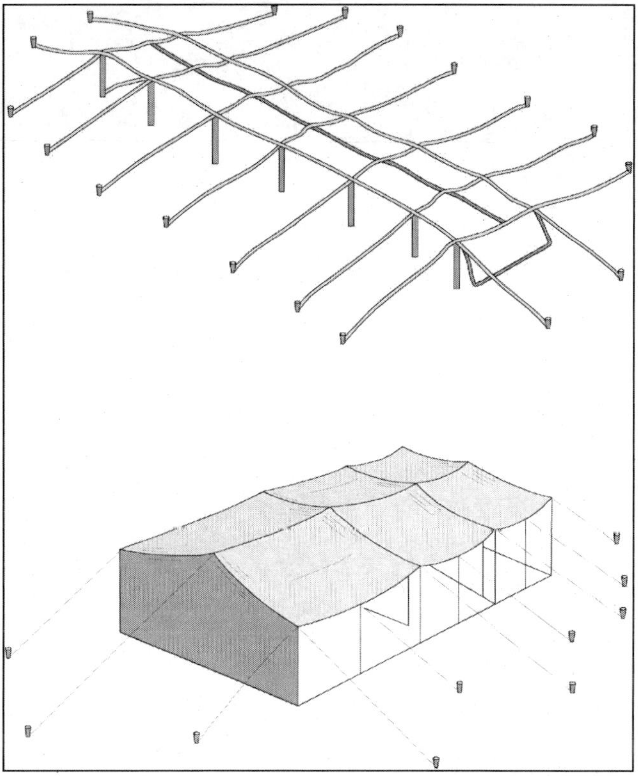

- The *sahah* is an interior wall used to define *bayt ash-sha'r* space divisions and is made from a woven decorated and colourful curtain made of goat and camel hair or cloth (Al-Ozaizat 1980, Weir 1990) (Figure 10). The top consists of several long pieces made out of goat and camel hair and called *shigag*, connected to each other with perpendicular long strips called *tareigah*. The *tanab* is the guy rope of the *bayt ash-sha'r*. It is anchored to the ground by cleats (*wedd*) (Figure 8b). Traditionally the *mayjama* hammer is used to drive the *wedd* into the ground.

Figure 8a-b. *Bayt ash-sha'r* constructional components (Source: Al-Homoud and Shauqour, 2007)

a. The Longitudinal Strip, *dafa'iyah*

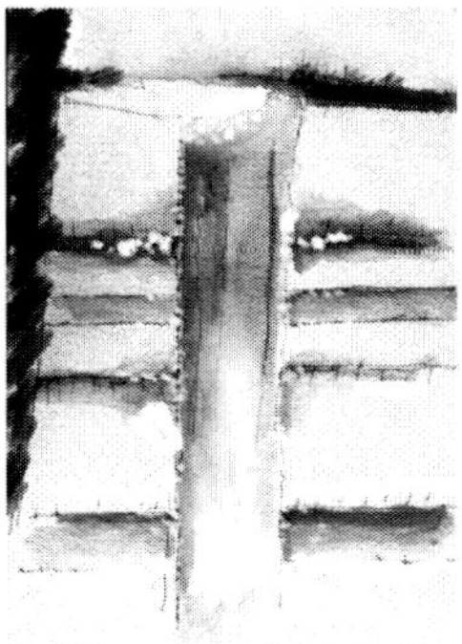

b. *wedd* in the ground

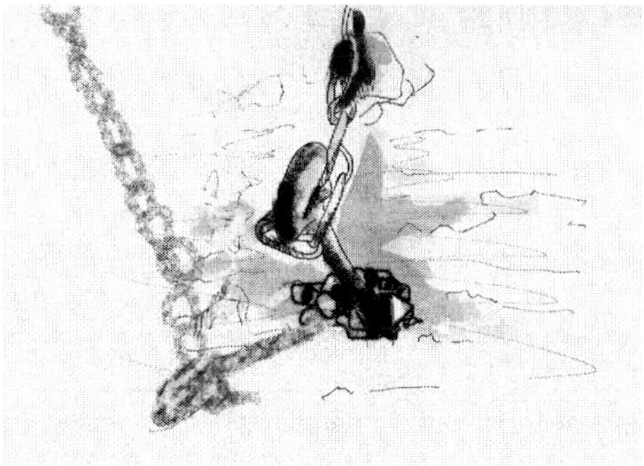

Figure 9. *Bayt ash-sha'r* structural components: top view (Source: Al-Homoud, 2006)

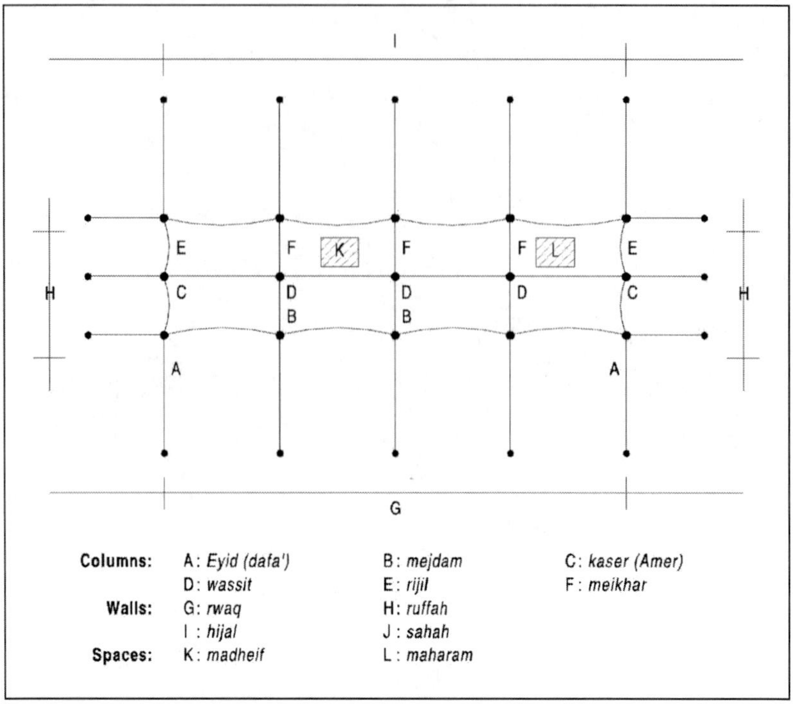

Figure 10. View through the *madheif* towards the *maharam* showing the *sahah* (Source: Al-Homoud and Shauqour, 2007)

The posts (*al-'amad*), held upright by guy ropes, differ in height; the central one (*wassit*) is 2.2 m while the others are 1.5 m. The distance between the posts is about 4 m. The *al-'amad* are classified into six types: the *wassit* is situated in the middle of the *bayt ash-sha'r*; the *kaser* or *'amer* is in the middle of the *bayt ash-sha'r* from the side elevation; the *mejdam* is the middle column from the front elevation of the *bayt ash-sha'r*; the *meikhar* is situated in the middle of the back elevation; the *eyid* or *dafa'* is the first post situated at the front corner of the *bayt ash-sha'r*; the *rejil* is the last post from the side elevation that is situated at the back corner of the *bayt ash-sha'r* (Al-Homoud et al 2000b, Al-Ozaizat 1980, Weir 1990) (Figure 9).

Figure 11a-c. *Bayt ash-sha'r* joints and details (Source: Al-Homoud and Shauqour, 2007)

a. *Shadhfah* and *mikrabah*

b. *Ameirah* and *khorb*

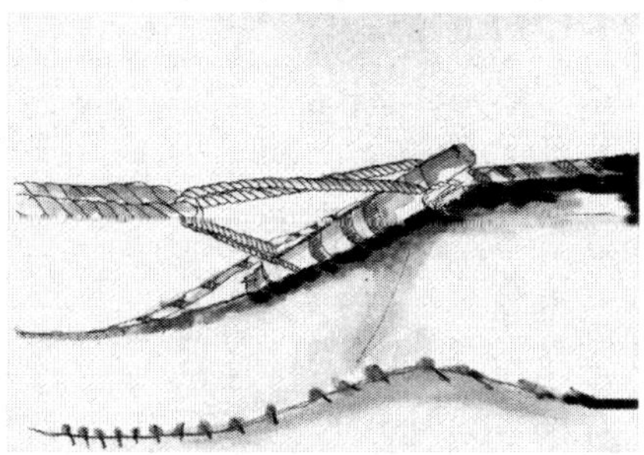

c. Top of *wassit* showing *lugmah* (*gatab*)

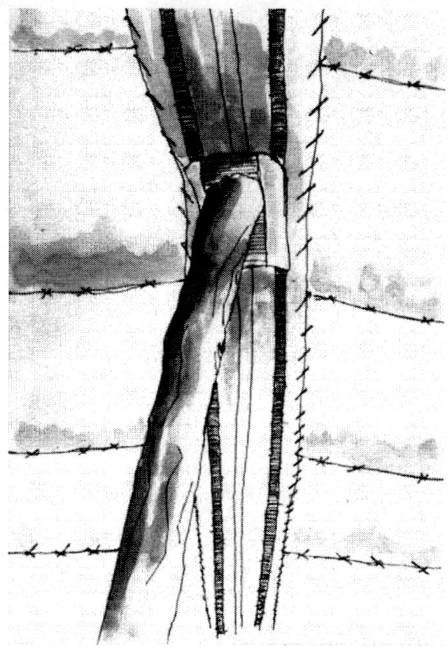

The main structural details of the *bayt ash-sha'r* include connections to the topside. The walls are connected to the roof through *makhalah* and *khillah*. A *makhalah* is a cloth made of goat and camel hair and connects the wall with the roof through metal pieces (*khlal*) that are tied together with a string (*jareer*). The roof is connected to the *bayt ash-sha'r* rope through a *shadhfah* (*mikrubuh*), a wooden piece that connects the roof to the ground through a tied v-shaped rope called *'ameirah/khorb* that connects the *tanab* to the *shadhfah*. When called *khorb*, it is usually covered with leather (Figure 11a-b). The *mikrabah* is connected with a rope called the *tanab* and ends in a metal cleat (*wedd*) that anchors it in the ground. Posts are connected to the roof through the *lugmah* and *gatab*. The *lugmah* is a small square piece made of goat and camel hair fixed at the top of the *kaser* columns. The *gatab* is a square piece put at the top of the *wassit* (Figure 11c). In some cases a wooden piece called *wawiyeh* is placed above the *wassit* to protect its roof fibre from wearing away. The interior walls are connected to the external walls by a rope (*mirsah*). The interior divider (*sahah*) that divides the interior spaces into two main spaces, is connected to the external back and front sides by a rope, also called *mirsah* (Figure 10) (Al-Homoud et al 2000b).

3.3. Spaces

The *bayt ash-sha'r*, like any home, consists of internal and external spaces. As indicated before, the assigned posts divide the internal area into two different quarters (*raba`h*) with a one third/two thirds proportion. The first quarter, called *madheif* or *maga'd*, is the men's space situated on the southern part of the *bayt ash-sha'r* (Figure 12a). It is centred around the *nigrah*, a fire and coffee place which symbolizes the hospitality and generosity of the Bedu (Figure 12b) (Al-Homoud et al 2000b, Al-Ozaizat 1980). The second quarter

Figure 12a-b. *Bayt ash-sha'r* special components (Source: Al-Homoud and Shauqour, 2007)

a: Men's space (*madheif*) | b. *Nigrah*

Figure 13a-b. Internal private domains of *bayt ash-sha'r* (Source: Al-Homoud and Shauqour, 2007)

a. View towards *maharam* | b. *haderah*

Figure 14. External layout of *bayt ash-sha'r* (Source: Al-Homoud and Shauqour, 2007)

is *maharam*, a women's domain (Figure 13a). It is divided into two components to satisfy privacy and the functionality of the home. The privacy component provides sleeping spaces and is divided into several specialized parts for males, females, couples, and additional members such as a married son. The functional component is the last space in *maharam*, and it provides a kitchen space called *haderah* used to cook food and store food and water; food is stacked at the back in sacks and containers (Figure 13b) (Al-Homoud et al 2000b, Al-Ozaizat 1980, Al-Rawi 1949, Weir 1990).

The external space that surrounds the *bayt ash-sha'r* is called *haram bayt ash-sha'r*. It is delimited from the front by the *mhathal* or *mhaja'* (goat space), 30-50 m away from *tanab* of the *bayt ash-sha'r*. Additionally, the *haram* is measured by the projections of the *bayt ash-sha'r's tanab*; when a *bayt ash-sha'r* is larger the distances will increase (Al-Homoud et al 2000b, Al-Mousa 1996). In general, these boundaries are defined by the Bedu cosmos, as there are no physical fences surrounding a *bayt ash-sha'r* (Figure 14) (Al-Homoud et al 2000b, Al-Ozaizat 1980).

3.4. Environmental considerations

The Badia has a harsh climate that helps define the ways in which the Bedu specify the features of the *bayt ash-sha'r*. These features include materials, moveable pieces, orientation, and site topography as follows (Al-Homoud et al 2000b):

- Materials: the *bayt ash-sha'r* in winter is pitched from *shigag* (made out of goat and camel hair), which provide warmth in the cold desert nights and are rainproof, and heating inside is provided by the *nigrah*. However, in summer the *bayt ash-sha'r* is made out of cotton, that provides coolness.
- Moveable pieces: these include lifting-up *al-riwaq* and raising the posts (*mijdam*) of the front elevation (east) in order to ventilate the *bayt ash-sha'r* in summer, as mentioned by Weir (1990) (Figure 15). In winter, the *hijal* (east front) is pulled down and posts, *mijdam*, are pulled back 2 m in order to increase the roof slope towards the ground, thereby helping the drainage of the rain from the top. In addition, a small ditch (*wany*) is dug around the *bayt ash-sha'r* to keep away rain water that falls from the top.
- Orientation: a *bayt ash-sha'r* is usually built with the front facing east, as also suggested by Jabbor (1988). Accordingly, Bedu open the *hijal* in order to receive winter sunshine and to avoid the prevailing west, winter winds.

Figure 15. Longitudinal section: lifting *al-riwaq* to allow air to flow inside (Source: Al-Homoud, 2006)

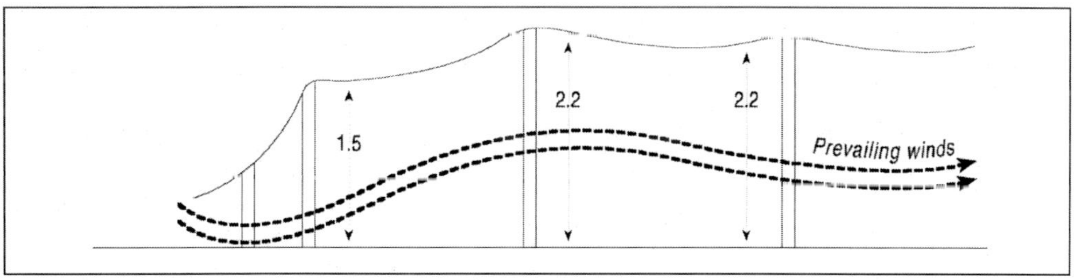

Figure 16. Topography surrounding *bayt ash-sha'r* (Source: Al-Homoud and Shauqour, 2007)

- Site topography: Bedu usually prefer to occupy flat land in order to facilitate their surveillance of the surrounding spaces (Figure 15). In addition, they look for low lands that provide water, range and grass for their livestock. However, in winter Bedu leave the floodable areas, which are dangerous for them (Al-Homoud et al 2000b).

3.5. Development

The development of *bayt ash-sha'r* design can be discussed in terms of transition from a mobile to a fixed type. There are four elements that are developed: materials, size, structure, and finishes (Al-Homoud et al 2000b):

- Material: *bayt ash-sha'r* in the past used to be woven by family members, but nowadays the parts are bought from factories and women sew them together. In addition, some people gave up camel and goat-hair *bayt ash-sha'r* to replace them with *kheish*, carpet, and synthetic materials that are more affordable and lighter. Tensile structural fittings are made out of metal (Figure 17a-b).
- Size: the size of the *bayt ash-sha'r* has increased, especially the width, to 130 % of the architypal dimension. Pickup trucks, compared with camels in the past, are used nowadays to transport *bayt ash-sha'r*, making it possible to lift bigger ones.
- Structure: lengths of metal pipes permanently fixed in the ground have replaced the wooden posts. This change has limited the concept of mobility of the *bayt ash-*

Figure 17a-b. *Bayt ash-sha'r* structural details development: *khorb* and *wedd* made out of steel (Source: Al-Homoud and Shauqour, 2007)

Figure. 18a-b. *Bayt ash-sha'r* tensile structural development (Source: Al-Homoud and Shauqour, 2007)

a. Metal pipes used instead of the wooden columns (*wassit*) in fixed *bayt ash-sha'*
(Original Source: Abed (2001)

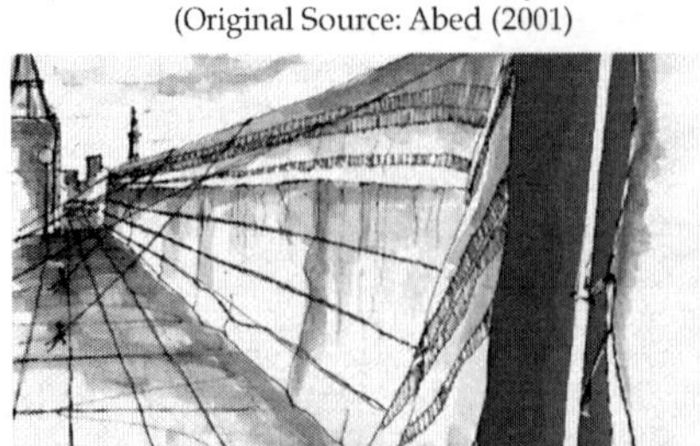

b. Tiled floors inside fixed *bayt ash-sha'r*
(Original Source: Al-Oun (2001)

Figure 19a-b. *Bayt ash-sha'r* finishes development (Source: Al-Homoud and Shauqour, 2007)

a. Electricity inside fixed *bayt ash-sha'r* | b. Fireplace inside fixed *bayt ash-sha'r*

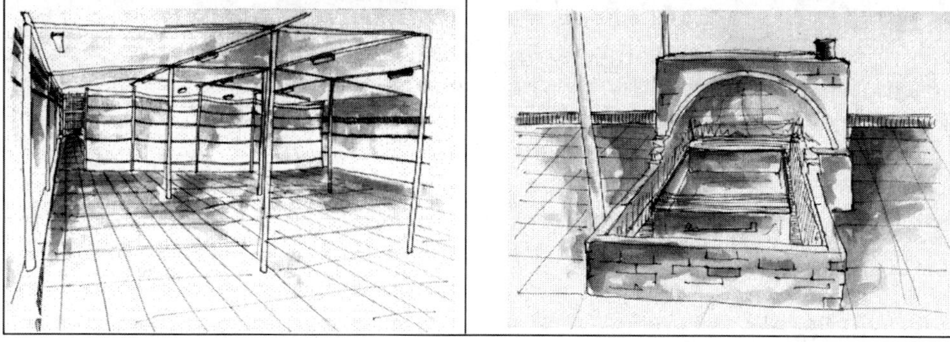

sha'r and lending and borrowing it. These changes have mainly affected *bayt ash-sha'r* shape by converting its side angles to vertical angles and by freeing it from *tanab* and, therefore, from the need for a large lot area (Figure 18a).

- Finishes: Bedu now use concrete, tiles, carpet, and *mokeit* to cover the ground inside fixed *bayt ash-sha'r* types (Figure 19a-b). In addition, they use electricity (Figure 19a) and install fireplaces instead of the *nigrah* (Figure 19b).

4. Transformational changes in Bedu homes

Thus the Bedu home has been transformed from a mobile structure to a fixed structure with new materials, new forms, larger spaces and areas, and diversity of outdoor facilities. This transformation is expressed in the attitude towards the use of internal home layout from traditional to specialized. New attitudes towards the use of internal spaces can be gauged from the findings of the authors' interviews: around one-tenth of the *bayt ash-sha'r* receive guests from both genders who are not members of the *'asheirah,* in one room, but mostly, because of traditional gender segregation, they tend to offer a separate space for each gender. Almost the entire sample indicated that family members eat in one space. About two-thirds of the respondents indicated that couples and sons and daughters who are more than ten years old, sleep in separate rooms; in general, the aim of this separation is to provide privacy for family members. In addition, almost all the respondents use the kitchen for cooking, one-third use the storage for keeping food, and one-fifth have a washing room. About one-third of the sample offer a playing space for their children inside the home, and the rest let their children play in the outdoor spaces (Al-Homoud et al 2000b).

Although new life circumstances have led to the design of the outwardly different transitional house, the Bedu have kept the *bayt ash-sha'r*'s traditional spatial layout, constructing it in the vicinity of their modern home if land and money are available. The internal layout is maintained traditional, so the transitional homes lack dining, office, and child play rooms that are typically available in the universal modern home. Bedu felt the need to live in modern homes that provide comfort, cleanliness, and environmental control. However, they tend to be tied to the traditional layout. Although they are able to receive their guests in a structured house, they prefer to receive them in *bayt ash-sha'r,* which can easily be made much bigger to accommodate larger numbers of guests. This can be explained by their belief that the *bayt ash-sha'r* is a symbol of hospitality and generosity. So, social and ceremonial activities, which are done in the guest-receiving spaces, are carried out in *bayt ash-sha'r,* while private activities, which are represented by sleeping space, women receiving, cooking, storing, and washing, are done in structured rooms. This finding is supported by Al-Ozaizat (1980) who stated that the *bayt ash-sha'r* is mainly used for receiving guests, because it is always open.

The Bedu try to maintain a *bayt ash-sha'r* next to their modern home (Figure 20) because of its cultural value and meaning, their strong emotional attachment to it, and its significance and necessity for clan assemblies for problem solving. The *bayt ash-sha'r* has remained essential for Bedu cultural integrity in the *madheif*, which limits *bayt ash-sha'r* use to special occasions and receiving guests and assures the symbolism of hospitality and other traditional functions. Generally, it can be noticed that the uses of the *bayt ash-sha'r* have changed from functional to symbolic ones (Al-Homoud et al 2000b).

Thus, today, given in more detail below, the *bayt ash-sha'r* is used as an expression of cultural processes and adaptation, in the following ways (Abed 2001, Al-Homoud et al 2000b):

- Being a symbol of hospitality and being able to accommodate a large number of people, it is used for special occasions. As a result, it is built only as a *madheif* to receive guests.
- It is used as a place for judgeship and to solve clan problems, besides being a symbol of dignity, authenticity, and assurance. Thus there are several legal norms connected with *bayt ash-sha'r* in many judgeship cases; in making peace, a *bayt ash-sha'r* is built in the middle of the violator's *bayt ash-sha'r* territory; anger cases are solved inside a *bayt ash-sha'r*; lying cases (*bish'ah*) are conducted in a *bayt ash-sha'r*; honour cases are solved by putting a white cloth on the front elevation of *bayt ash-sha'r*; moving (*jalwah*) cases would make the family move away for some time and it takes place in a *bayt ash-sha'r*; people seeking help (*dakheil*) enter the *bayt ash-sha'r* and the owner must build a special one for them.
- A *bayt ash-sha'r* is used for the family gathering because of its openness and symbolic heritage.
- It is used for moving away with the livestock (*ta'zeib*) especially in the spring season.

4.1. Modern lifestyle as a socio-cultural attribute and the modern Bedu home

In the present study, lifestyle has changed as a result of urbanization, which has affected Bedu living circumstances. Bedu responded to urbanization with transformation in their lifestyle patterns. Factors such as schools, provision of infrastructure services like roads,

Figure 20. A *bayt ash-sha'r* adjacent to a modern home (Source: Al-Homoud and Shauqour, 2007)

electricity, water, telephone, and land ownership encouraged settlement. Bedu basic features of settlement are emphasized through ownership of land and construction of fixed homes (Al-Mousa 1996). However, their inherited social values and manners and codes of honour were not lost (Jabbor 1988).

The qualitative and quantitative analyses of the present study show that lifestyle change may be perceived through attributes such as occupation of male household, education, income, family structure (lifecycle, size, and type), expenditure priorities, attachment to the local community, strong loyalty to *'asheirah* origin, expenditure priority of hospitality and food, and collectivistic identity (Abed 2001, Al-Homoud et al 2000b).

Education levels are sharply rising from illiteracy to higher education. The present study indicates that education has encouraged Bedu to settle and to live in fixed homes, while householders who have an official occupation and a university degree are more exposed to urban values. Education affects Bedu lifestyle by making them settle in villages so as to obtain education for their children. Bedu encourage their children to go to school because a higher level of education is closely linked to one's social position among clan members as indicated by Brandenburg (1998). Higher education encourages the Bedu to have new home layouts that exemplify new ways of living. Furthermore, education makes individuals open-minded to adapt to new themes, ideas, and beliefs while choosing to live a traditional lifestyle. Education has made Bedu interact with urban settings and encouraged them to borrow some city features like concrete blocks and white stone building materials. These features are increased when the family is large and when most of its members are adults who bring income. Nowadays, although they cost more money, Bedu prefer to use new and modern things, a necessity for the younger generation. Bedu women nowadays are educated and work outside the house, but they are still protected by codes of honour and restricted with chastity, a finding which supported Abu-Jabr et al (1987a). Work and education make women more able to take care of their children, to participate in the household, and to provide for some family needs (Abed 2001, Al-Homoud et al 2000b).

A university degree makes the Bedu more qualified to have a permanent job, which in turn encourages settlement because it offers financial stability that enables the householder to increase his expenditure on food and drink to cover new types of expenses that may include home structure. Occupations have changed from livestock management to employment in governmental and private sectors (Al-Mousa 1996). This does not mean that all Bedu have left their traditional occupations or livestock pastures (Abed 2001, Al-Homoud et al 2000b).

To adapt to new life circumstances, settled Bedu tried to change and develop their jobs from livestock ownership to farming, and other occupations such as labouring, trading, and driving. Changes in employment required permanent home structures, in addition to the traditional one, built out of cheap material such as mud or rough basalt stone. Despite their cheapness, mud and stone represented beauty for Bedu because they are part of their environment. Moreover, mud and rough basalt stone do not require a skilled labour force; family members themselves can build the home. It is noticed that permanent houses are normally limited to one floor and a small number of rooms, depending on the financial capacity of the householder. However, the availability of a *bayt ash-sha'r* is a necessity for Bedu who look after livestock as additional work to their permanent jobs. Therefore, the new homes, consisting of a number of rooms, are built adjacent to the *bayt ash-sha'r* with size depending on the number of rooms and floors within the specific lot area. The transitional layout emphasizes that Bedu do not consider home construction as an expenditure priority, except for having the extra rooms whose number depends on the financial circumstances of the householder. In more recent times, the market has provided several job opportunities for Bedu who have university degrees, in addition to job opportunities in the Jordanian army. These occupations offer Bedu financial stability that enables them to order their expenditure priorities to include new expenses, and encourages them to live in structured modern homes. Income evolved from being seasonal (informal) to formal with a regular monthly salary. Income affects home size; when the family has a high income, a more specialized home structure is required. Income spending includes new services like education and health care, which require more expenditure, and impact on home location and mobility, which supports the findings of Abed (2001), Al-Homoud et al (2000a), Al-Mousa (1996) and Brandenburg (1998).

Additionally, family structure, such as family lifecycle, family type, and family size, affect the home layout variable. When the family has children, it usually settles in a village that offers education, and therefore, lives in a structured house. However, families are changing from being extended and large, with an average size of 8.7 individuals for each household, to small nuclear ones as indicated by Al-Fawal (1983) and Al-Hunayti (1997). Usually, the large family size is associated with financial strength, which enables the householder to develop his home by increasing the number of spaces. This supports Al-Rawy's (1949) finding that good economic status is connected with large home size and more interior spaces. When the family size increases, Bedu enlarge the house size in response to their immediate family expansion, in the absence of general planning. Nevertheless, the *bayt ash-sha'r* remains the ideal solution for poor families that tend to be large and include more than one wife within the household.

It can be noticed that the large size of the Bedu family is influenced by values that link family size to wealth and high status. Accordingly, Bedu have tried to build new houses in a new style to emphasize their status. In addition, aesthetic taste affects modern internal spaces; when the householder is attached to modernity he tends to use the internal spaces in a specialized way. Furthermore, the number of wives affects space requirements, if not separate detached structures, and this requires an increase in home spaces and lot area. Large family units with adults, need more privacy, which is usually achieved through providing more specialized spaces for each activity and for each of the members, who may be separated from each other by hierarchical zones. Meanwhile, when most family members are children, small spaces can satisfy their needs; for example, the family may use one bedroom for all the children as indicated by Weisner and Weible (1981).

Bedu are a collectivistic society, therefore *'asheirah* members have a strong relationship that insures homogeneity through home layout. As a result, the new lifestyle may direct Bedu to live in new house prototypes, whilst keeping the traditional internal living spaces of the *bayt ash-sha'r*, or keeping a *bayt ash-sha'r* in addition to, and next to, the newly constructed house whether they are modern or transitional. Family size and wealth strengthen the householder's *'asheirah* relationships, leading him to receive large numbers of guests; thus, more spaces are needed to secure family privacy. When the householder has a stable occupation, it secures income stability that helps him to participate in problem solving, leads to more involvement in the local community and strengthens *'asheirah* relations as a response to his high social role. The social role of the householder within the *'asheirah* and participation in occasions that strengthen kinship ties lead to an increase in the number of guests; thus, it becomes a necessity to have separate specialized spaces, a finding supported by Akbar (1982), Kurdi (1982), and Rapoport (1969). This is seen in all types of Bedu homes: in *bayt ash-sha'r*, the *madheif* is usually separated from the family space by *sahah*; in the transitional home, the *bayt ash-sha'r* is mainly designed for receiving guests, and other home facilities are placed in the fixed structure; in modern homes, a separate structure for receiving guests, called *madhafah*, is located at a distance from the family home, and is placed in such a way as to have a direct access from the main entrance.

At another level, territoriality is linked to privacy. It is noticed that territoriality has a significant relationship with lifestyle. Social role positively affects territoriality, where social role is an indication of a householder's prestige, power, and wealth represented through home ownership within the *'asheirah*. When the householder has a high social role, as mentioned before, larger home size is required and physical elements separate home domain from the surroundings through barriers such as fixed fences and through vast space provision. The traditional hierarchical zones of privacy are translated into costly physical properties featuring wide spaces, surrounded by a fixed fence (or wall) and a gate. Furthermore, different homes are built within a certain lot area and are surrounded by one fence. This phenomenon is explained by the fact that such householders have strong kinship ties, being either brothers or cousins, and this cancels the need for individual or physical territoriality. This gives them a strong sense of collective honour and loyalty to defend their spaces against others.

4.2. Modern Bedu house physical attributes

In the present study about two-thirds of the subjects live in houses built of concrete blocks, and almost all of the subjects live in houses with concrete roofs. Almost all the sampled houses consist of one floor, for multi-level homes are not a traditional concept in Bedu home layout. One-third of the sampled houses consist of four rooms that are mostly ordered in a square shape. The number of rooms reflects the number of spaces in a *bayt ash-sha'r*: *madheif* and *maharam*, the latter being divided at least into two parts (sleeping space and kitchen), making *bayt ash-sha'r* consist of at least three spaces. However, the square home shape is a transitional shape in Bedu culture that has emerged out of the need for more space. Half of the respondents' houses are built within a lot of 1-5 *dunum*[6], which is mostly left open as a result of the traditional Bedu practice of wide-open space; some Bedu nowadays try to make use of such space by cultivating it with fruitful trees, including olives (Al-Homoud et al 2000b).

Bedu are adapting to changes and it is clear that their lifestyles are affected too; as a result, their home layouts seem to be in transition. This transition appears clearly in the use of fixed houses whose characteristics can be summarized in terms of:

- The use of new materials such as reinforced concrete, concrete blocks, and white stone;
- The increase in the number of rooms and floors;
- The appearance of new outdoor facilities, such as parking lots, wells, and a garden;
- The appearance of new shapes for plan layout;
- The increase of internal home spaces and specialization in space use.

These fixed houses can be categorized to two types of layouts: transitional and modern (Al-Homoud et al 2000b):

The transitional house: The transitional home consists of three different types or evolutionary phases. First, are 2-5 rooms that are open to a platform; rooms are made out of mud, rough stone, concrete, or concrete blocks. If there are two or three rooms, they are arranged longitudinally to form a rectangle, and if there are more, they are arranged to form a square consisting of two rows of rooms. Second, the *bayt ash-sha'r* are of any size and material. Third, are separate small structures made out of concrete blocks used as toilets. In this prototype, the internal space is a mix of specialized and non-specialized spaces in terms of use, where the rooms are used for private activities, and the *bayt ash-sha'r* is used as a *madheif* for hosting activities (Figure 21a-b).

---

[6] One dunum equals 0.1 hectare.

Figure 21a-b. Bedu transitional home layout (Source: Al-Homoud and Shauqour, 2007)

a. Typical external transitional layout | b. Typical internal transitional layout

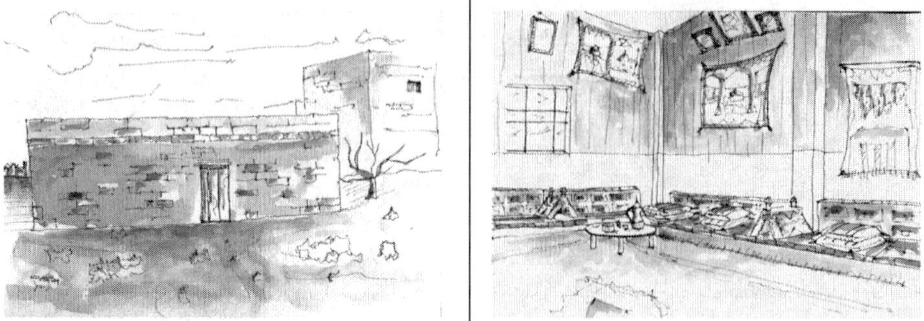

The mud or basalt stone home first appeared in the 1940s. It was built from mud bricks or basalt stone, and its roof was made of reed covered with a mixture of mud and chaff. Its internal spaces were limited to two or three rooms that were arranged longitudinally to form a rectangular shape. It usually consisted of one floor like the *bayt ash-sha'r* layout. Consequently, this category lacked specialization in terms of use, as the same space was used for living, sleeping and eating. There were three different patterns for this home prototype:

- A building that consisted of one room used to receive guests, sleep in, and prepare food, in addition to an outside toilet which consisted of an open space surrounded by a wall
- A building that consisted of two rooms open to a platform: one room was used to receive guests and for men's different activities, and the other was for the rest of the family's activities. The platform was usually used for family seating. This type also had an outside toilet
- A building that consisted of three rooms open to a platform: one room was used for receiving guests and the other for family uses. Here also there was an outside toilet. The number of rooms later increased to become four or five, forming buildings of two parallel longitudinal rows of rooms with a final square shape, which supports previous findings (Abed 2001, Al-Mousa 1996, Merrill 1981).

Figure 22a-b. Modern house (villa) layout (Source: Al-Homoud, 2001)

a. Villa | b. Villa internal layout

The modern home (villa): The modern home has more than five rooms and one, two or three floors that are articulated in plan shape. A home of this type is made out of reinforced concrete, concrete blocks, regular basalt stone, or white stone (Figure 22a-b). The internal space is usually specialized in terms of use. Sometimes, a modern home may include, adjacent to it, a bayt ash-sha'r that is used as a madheif for symbolic reasons. This type appeared in the 1970s, when the mud or rough stone home types began to disappear. The modern home is referred to as a villa, identified as a large and single family home that has specialized use of spaces for different activities, eg separate spaces for sleeping, living, and receiving guests, a kitchen that contains a storage area, and an indoor bathroom (Abed 2001, Al-Mousa 1996, Merrill 1981).

4.3. Characteristics of transitional and modern Bedu homes

Generally speaking, Bedu maintain the basic design principles of *bayt ash-sha'r* in their modern and transitional homes internally and externally. Bedu fixed homes can be described internally in terms of ceremonial and community spaces, private spaces, and functional spaces, and externally in terms of plan layout, home size, home material, outdoor facilities, home direction, and topography (Al-Homoud et al 2000b).

4.3.1. Principles of internal home design

Internal layout describes the specialization and overlapping of uses of space and the specific users. The internal home layout includes ceremonial and community space, private spaces, and functional spaces (Al-Homoud et al 2000b):

Ceremonial and community space: Such space in Bedu homes is represented by the guestroom. In Bedu homes there are two separate guestrooms (*madhafah*), one for men and another for women, and the men's *madhafah* is mostly larger than that for women. Further, the guestroom has a rectangular shape and is usually furnished with colourful traditional Arabic furniture, which consists of: *wisayed*, pillows used by the guest to lean on when they bend their body; alternatively they may use *shdad*, a seating put on a camel's back; and *sfouh*, the carpet that the guests sit on (Figure 23a). Sometimes guestrooms have a one-metre wide and one-metre high platform, called *tawaty*, that is furnished in the Arabic style. Guestrooms are rarely furnished in modern style with davenports and chairs. Sometimes, the guestroom is furnished in both traditional and modern styles. In some cases, there are two guestrooms, each furnished in one style. Generally, Bedu, whether living in transitional or

Figure 23a-b. Ceremonial and community spaces in the Bedu home

a. *Madhafah* layout
(Source: Al-Homoud and Shauqour (2007)

b. Sketch for modern *madheif* layout
(Source: Khalfawi (2001)

modern homes, are attached to the traditional Arabic furnishing style (Figure 23b). The guest-room is usually large in size as a symbol of Bedu hospitality. Additionally, some homes have a veranda or terrace space at the entry level where family members may gather.

Private spaces: Bedrooms for different family members are not available in modern Bedu homes; however, one bedroom for each gender is usually found if children are more than ten years old. Sometimes male children sleep in the *madhafah* or in the living room if they do not have their own bedroom. In transitional homes, private spaces are mainly found in the built room and not in *bayt ash-sha'r*.

Functional spaces: The functional spaces inside modern homes include kitchen, storage, and washing room. In some villages, especially in Umm al-Quttayn, there are some homes with two kitchens; one is inside and the other is outside. Storage space usually overlaps with the kitchen. If it is a specialized space, it will be an old and separate outside structure mostly made out of mud or basalt stone. Sometimes, in transitional home layouts, a *bayt ash-sha'r* made out of *kheish* (*gatbah* or *kharboosh*) is used as firewood storage or as a kitchen. A room for doing the washing is rarely available in the Bedu home; laundry is done in the available space beneath the staircase or in the bathroom.

4.3.2. Principles of external home design

External layout may be described for transitional and modern homes in terms of plan layout, size, material, outdoor facilities, orientation, and topography (Al-Homoud et al 2000b):

Plan layout: The most common plan layout is the square shape with some articulation, where the internal rooms are ordered in two rows of rooms. Sometimes, two or three different structures are scattered on the site plan. For example, a two-room linear structure is added to another three-room linear structure within the same lot area. These particular scattered structures are common in transitional home layout.

Home size: A Bedu home seems to be mainly of one floor. Sometimes it extends to two floors, noticeably those homes located on the Jordan-Iraq highway, where the ground floor is made up of commercial stores and the first floor is used as the residence. In modern homes, the number of rooms can exceed five. However, the average number of rooms in transitional homes is two to four on one floor.

House materials: Concrete block and reinforced concrete seem to be the common construction materials in most villages for both modern and transitional homes, although old homes are built of mud or rough basalt stones (Figure 24a-b). Still, in Rahbat Rakad village, a number of modern homes are built out of white limestone. Houses are surrounded with low rough walls that define the boundary of the lot and are made out of basalt stones.

Figure 24a-b. Mud, basalt, and concrete homes

Mud home (Source: Al-Homoud and Shauqour, 2007)	Basalt and concrete home (Source: Al-Homoud, 2001)

Figure 25. External vicinity of neighbouring Bedu homes: new towns, traditional layout (Source: Khalfawi[7], 2001)

Outdoor facilities: The outdoor facilities of a modern Bedu home include: open space that forms 70 % of the whole lot area; a separate large room used as *madhafah*; a *bayt ash-sha'r* that is pitched in spring and summer seasons as well as for special occasions; a garden planted with orchards or olive trees; a well used for collecting rainwater for daily uses and agriculture; a fence consisting of stone or concrete walls; parking in the empty space around home. Generally, the availability of parking and garden are an indication of a large lot area for modern home prototypes.

The outdoor facilities of a transitional Bedu home include: *seyrah* used to keep livestock; scattered structures that might include a small home for newly married sons, as well as rooms for services such as kitchen and toilet; rough basalt stone fixed fence, trees (semi-fixed fence), or an open space (non-fixed fence); and a *bayt kheish* used for baking or storing goods.

Home orientation: Modern and transitional Bedu homes face east or south for climatic reasons as *bayt ash-sha'r* used to. This orientation is dominant regardless of street access.

Topography: Bedu prefer to live on flat rather than hilly or mountainous land, and favour wide-open space and views. Generally, modern homes in the villages of the research area are distanced from each other, as the case used to be in the *bayt ash-sha'r* layout (Figure 25).

4.4. Summary and conclusions

As indicated earlier, Bedu responded to urbanization and their lifestyle changed from a traditional towards a modern one, which encouraged Bedu to live in permanent structured homes with specialized internal layouts, providing more physical and spatial comfort compared with *bayt ash-sha'r*. The relationship between lifestyle and home layout can be summarized and explained as follows.

The Bedu's first phase of settlement encouraged them to seek settled jobs, such as farming, driving, and labouring, besides their traditional occupation of livestock management. These occupations increased their income and helped them to adapt to new life demands, such as the construction of fixed homes made out of available material on site, such as mud and rough stone. These homes were limited to a small number of rooms arranged in a linear or square shape that, unconsciously, resembled the *bayt ash-sha'r*. When a Bedu family increased in size, more space was required, and the *bayt ash-sha'r* remained as an important entity to satisfy the family's traditional needs. The fixed rooms were mainly used for the family's private activities, and the *bayt ash-sha'r* was used as

[7] Khalafawi is an architect who graduated from the Department of Architecture at Jordan University of Science and Technology in 2002.

madheif or for other living activities, or for both. It can be concluded that those fixed rooms adjacent to the *bayt ash-sha'r* accommodated traditional and specialized internal layouts that met all Bedu basic needs of the transitional living style.

Recently, Bedu have had more access to higher education, which has enabled them to compete for secure jobs in the governmental and private sectors. Accordingly, their income has become more stable, and home construction has become an item on their expenditure list. Consequently, Bedu give more importance to a home image commensurate with their social and financial status. This has led to building new forms of multi-floor houses with a larger number of spaces, in addition to accommodating specialized inside facilities such as kitchen, washing room, storage room, and toilet. Outside, new facilities have become available at home, such as parking spaces, yards, and children's playing areas. It has become necessary to assign or specialize certain spaces to provide the family with more comfort and privacy.

Although new home prototypes have met some of the Bedu needs, the *bayt ash-sha'r* remained part of their home layout though it has varied in size. In traditional home prototypes, the *bayt ash-sha'r* is used functionally, while in modern home prototypes it is used symbolically and for ceremonial events. In transitional home prototypes, *bayt ash-sha'r* use depends on the available number of fixed rooms and family size. The *bayt ash-sha'r* remains constant in Bedu minds, and this appears clearly in the following aspects:

- Home direction: Modern homes generally face east, although this direction is sometimes opposite to street entry, which is an unconscious replica of *bayt ash-sha'r*'s orientation
- Land topography: Bedu inhabit flat lands similar to those of *bayt ash-sha'r* and avoid hilly or mountainous regions
- Home fencing: Bedu avoid constructing fences because they look for openness and favour observation from a distance, exactly as in the case of the *bayt ash-sha'r*. Usually the extended family encloses its lot within a fence, and no internal fences are erected within the extended family's boundaries or around individual units
- Large lot area: Bedu prefer to inhabit large open spaces so as to use them for family extension in the future, or for agriculture
- Home internal space components: Bedu internal home components consist of two large rooms for receiving guests, one for men and the other for women, in addition to family living and sleeping spaces
- Overlapping use of spaces: this is clear through the use of the women's guestroom as a living and sleeping room, the use of the kitchen as a storage area, the use of the bath as a washing room, and the use of the outdoor space for children playing, parking, and as a yard. These similarities clearly illustrate the Bedu mentality which enforces attachment to the *bayt ash-sha'r*, which is considered the most important component of their tradition.

Generally, Bedu society is connected to their traditional non-material cultural values (beliefs, norms, attitudes, feelings, goals, and identity) as indicated by Richardson (1989). This is emphasized by Rapoport (1969) who suggests that the home is a result of choices that are generated from possibilities offered by cultural values. In traditional societies, the non-material cultural values provide choices, while home material cultural value is low in its criticality. Therefore, physical changes are slow because if they do not fit within the non-material cultural values, they will be rejected as beliefs and norms are the only constants that provide or allow choices for change. This accounts for the slow rhythm of change the Bedu home has faced and the fact that Bedu now live in modern homes with a traditional *bayt ash-sha'r* atmosphere, where this phenomenon reflects Bedu attachment to *bayt ash-sha'r* because they insist on developing their homes within their traditional cultural values.

Acknowledgments

Our thanks go to BRDC for their interest in this research and for their partial financial support. Great thanks are also forwarded to the Deanship of Research at JUST for their partial financial support. Great thanks also are extended to BRDC employees at the Safawi field centre for their assistance – Ameenah Ash-Shurafat, Muhsen Ash-Shurafat, Odeh Al-Meshan, Mowafaq As-Sirhan, and Mohammed Al-Sumairan – as well as to all employees in the Safawi field centre who facilitated hosting us during the fieldwork.

References

Abed A (2001). Change in traditional Bedu home layout as a function of lifestyle: from bayt ash-sha'r to villa. M.Sc. Thesis, Jordan University of Science and Technology.

Abu-Jaber K, Gharaibeh FA and A H (1987a). The Badia of Jordan: The process of change. Amman, Jordan: University of Jordan.

Abu-Jaber K, Gharaibeh FA and A H (1987b). The Badia of Jordan: The process of change. Publication of the University of Jordan, Amman, Jordan.

Akbar JA (1982). Court yard houses: a case study from Riyadh, Saudi Arabia. In: Akbar, JA (ed), The Arab City: Its Character and Islamic Cultural Heritage. Arlington, Virginia: The Arab Urban Development Institute.

Al-Fawal M (1983). Social structure in Bedu societies. Cairo: Dar Al-Fekr Al-Arabi.

Al-Homoud M (1995). External space enclosure affords neighborhood-level social interactions. Ph.D. Dissertation, Texas A&M University, USA.

Al-Homoud M and Al-Oun SS (2000a). Interviews conducted by the researchers. Al-Homoud, Al-Oun, and Abed.

Al-Homoud M, Al-Oun SS and Abed A (2000b). Field interviews undertaken by the authors.

Al-Homoud M and Tassinary L (1997). External space enclosure affords neighborhood-level social interactions. Proceedings of The Third International Convention on Urban Planning, Housing and Design. ICUPHD '97. Cities for the 21st Century: Centre for Advanced Studies in Architecture, National University of Singapore, Singapore pp. 1-9.

Al-Hunayti D (1997). Social and economical directions of development in the Jordan Badia. MA Thesis, University of Mosul, Mosul, Iraq.

Al-Mousa H (1996). Desert habitat in the north Badia: a study in the urban structure. MA Thesis, Jordan University, Amman, Jordan.

Al Oun SS (1997). Livestock marketing in the Jordanian Badia. Ph.D. dissertation, University of Durham, UK.

Al-Ozaizat Y (1980). Arab and their culture. Amman, Jordan: Matba't Alquwat Al-Musallaha.

Al-Rawi A (1949). Al-Badeyah. 2nd edition, Baghdad, Iraq: Matba't Al-A'ni.

Brandenburg C (1998). Development anthropology: a primary survey. In: Dutton R, Clarke J and Battikhi A (eds), Arid land resources and their management: Jordan's desert margin. London: Kegan Paul International pp. 247-265.

Duncan J (1985). The house as symbol of social structure: notes on the language of objects among collective groups. In: Altman I and Werner C (eds), Home environments, New York, U.S.A: Prenum Press pp. 133-151.

Erwin R (1994). Lifestyle. Wilson Quarterly 18.

Ismael F (1983). Change and development in desert societies. 2nd edition, Alexandria, Egypt: Dar Al-Ma'refah Al-Jame'yah.

Jabbor J (1988). Bedu and Badia: images from Bedu life of Badeyat Al-Sham. Amman, Jordan: Dar Al-Elm Lelmalayeen.

Kana`n R and Estatiyyah M (1992). The guidance of climatic design. Ministry of Energy, Amman, Jordan.

Kellett P (1999). Cultural values and housing behavior in spontaneous settlements. Journal of Architecture and Planning Research 16:205-224.
Kurdi TM (1982). Influence of Arabian tradition on the old city of Jeddah: house form and culture. The Arab city: its character and Islamic cultural heritage. Arlington, Virginia: The Arab Urban Development Institute.
Merrill H (1981). Between desert and town: a case study of encapsulation and sedentarization among Jordanian Bedouin. Ph.D. Dissertation, University of Pennsylvania, USA.
Rapoport A (1969). House form and culture. N.J.: Prentice-Hall, Englewood Cliffs.
Rapoport A (1991). Housing: symbol, structure, site. In: Taylor, L (ed), Housing and Culture, New York: Rizzoli pp. 14-15.
Richardson M (1989). The Artifacts as abbreviated act: a social interpretation of material culture. In: Hodder, I (ed), The meaning of things: material culture and symbolic expression. London: Unwin Hyman.
Strauss A (1976). Life styles and urban space. In: Proshnasky H, Ittelson W and Rivlin L (eds), Environmental psychology, people and their physical settings. New York: Holt, Rinehart and Winston pp. 528-537.
Tognoli J (1987). Residential environments. In: Altman, I and Stokols, D (eds), Handbook of environmental psychology, 1, New York: Wiley Sons pp. 655-690.
Weir S (1990). The Bedouin. London: British Museum Publications.
Weisner T and Weibel J (1981). Home environments and family lifestyles in California. Environment and Behavior 13:417-460.

Glossary of Arabic terms

English pronunciation	Word in Arabic	Meaning in English
ameirah/khorb	عميرة/ خرب	Tied rope of 'v' shape that connects the *tanab* with the *shadhfah*. When called *khorb*, it is usually covered with leather.
'asheirah	عشيرة	'Bedu tribe', it refers to a group of people living in one wide region and having common socio-cultural value and rule systems that distinguish their character.
Badia	البادية	The arid and semi-arid land that is inhabited by the Bedu; its annual rainfall does not exceed 200 mm.
Bedu (Badu)	بدو	Traditionally, nomadic pastoralist tribes who are today a mix of mobile, semi-settled, and settled.
bayt ash-sha'r	بيت الشعر	'Bedu tent', refers to the traditional mobile Bedu home which described as a long, horizontal black tent made of camel and goat hair/cloth and pitched with the support of a line of central poles in the middle, while the front, back, and sides are supported by lower poles.
bish'ah	بشعة	The burning of the tongue as a punishment in Bedu judgeship for cases of lying.
dafa'iyah	دفاعيه	Longitudinal strips that connect two *shigag* in *rwaq* and made of goat and camel hair.
dakheil	دخيل	Person who enters a *bayt ash-sha'r* and asks for help and protection.

English pronunciation	Word in Arabic	Meaning in English
eyid/dafa'	ايد	Front column of the side elevation that forms the front corners of a *bayt ash-sha'r.*
fakhdh	فخد	'Family name', refers to a group of families that have the same ancestry, which does not exceed the fifth grandfather, and live in neighbouring regions.
gatab	قطب	Square wood piece that connects the *wassit* with the roof and protects the roof fiber from wearing out.
gatbah	قطبة	A *bayt ash-sha'r* that is pitched with one *wassit*, and consists of two *raba'h.*
haderah	حادره	End space in the *maharam* space of a *bayt ash-sha'r* that is used as a kitchen.
hamoulah	حمولة	'Bedu clan', refers to a group of *fakhdh* that have the same ancestry (refer to one grandfather) and they common rules and values.
haram bayt ash-sha'r	حرم بيت الشعر	Area that surrounds *bayt ash-sha'r* and that usually defines the territorial zones of influence of the family, marked by *tanab* as the first zone of influence and by the outdoor facilities as a second zone of influence.
hijal	الحجال	Front wall of *bayt ash-sha'r* that is made of goat and camel hair, and consists of several pieces called *shigag* that are patterned horizontally and made out of alternative white and black strips of colour.
jalwah	جلوة	Family moving outside the *'asheirah* for some time as a punishment by Bedu judgeship.
jareer	جرير	Thin rope used to connect *khlal* of the back wall together to keep track of these little iron pieces.
kharboosh	خربوش	The smallest *bayt ash-sha'r* in size that is pitched without a mediating *wassit*, is made out of *kheish*, and consists of one *raba'h.*
kaser/'amer	كاسر/عامر و جمعها كواسر	Middle column of the side elevations of *bayt ash-sha'r.*
kheish	خيش	Cheap material of *qinnab* that is used to make *bayt ash-sha'r.*
khillah	خلة	Metal pieces that are used to connect the *bayt ash-sha'r* roof with the back side, and are all connected with a rope that is called *jareer.*
lugmah	لقمه	Square cloth piece made out of goat and camel hair, and located between the *qatab* and the roof to decrease bearing load effect by the pole.
madhafah	مضافة	Guestroom in new Bedu home prototypes.
madheif/Maga'd	مضيف/ مقعد	Men's main receiving space in *bayt ash-sha'r* and main guest receiving space; usually open 24 hours as a symbol of hospitality.
mhathal/mhaja/ seyrah	مهظل/ مهجع/ صيره	Goat space that delimits *haram bayt ash-sha'r* as a territorial zone of influence.
majlis	مجلس	Council of clan elders who advise *Sheikh* and help him to take the right decision.

English pronunciation	Word in Arabic	Meaning in English
makhalah	مخله	Upper longitudinal strip made out of goat and camel hair that connects the top with the side.
mayjama	ميجامة	Hammer used to drive the *wedd* into the ground.
mdoubal	مدوبل	A *bayt ash-sha'r* that is pitched with three *wassit* or more; it includes *mthoulath, mrouba', mkhoumas, msoudas, msouba'*, and *mthouman*.
meikhar	ميخر	Middle columns located in the back elevation of *bayt ash-sha'r*.
mejdam	مجدم	Middle columns located in the front elevation of *bayt ash-sha'r*.
mgouran	مقورن	A *bayt ash-sha'r* that is pitched with two *wassit* or less, it includes *Sahwa, Gitbah*, and *kharboosh*.
maharam	محرم	Women's space in *bayt ash-sha'r*. It is divided into several parts to satisfy family privacy and functional facilities. It is usually divided into rooms separated by *sahah* depending on the family size.
mirsah	مرسه	Rope that connects the internal divider (*sahah*) with the external rope (*tanab*).
mkhoumas	مخومس	A *bayt ash-sha'r* that is pitched with five *wassits*, and consists of six *raba'hs*.
mokeit	موكيت	New material used to build *bayt ash-sha'r*.
mrouba'	مرويع	A *bayt ash-sha'r* that is pitched with four *wassits*, and consists of five *raba'hs*.
msouba'	مسوبع	*bayt ash-sha'r* that is pitched with seven *wassits*, and consists of eight *raba'hs*.
msoudas	مسودس	A *bayt ash-sha'r* that is pitched with six *wassits*, and consists of seven *raba'hs*.
mthoulath	مثولث	A *bayt ash-sha'r* that is pitched with three *wassits*, and consists of four *raba'hs*.
mthouman	مثومن	A *bayt ash-sha'r* that is pitched with eight *wassits*, and consists of nine *raba'hs*.
nezalah	نزلة	Dinner meal to which are invited new neighbours for acquaintance.
nigrah	نقرة	The middle ground enclosure hole used in the *madheif* for fire and coffee making; it symbolizes the hospitality and generosity of Bedu.
raba'h	ربعة	Internal space in *bayt ash-sha'r* located between two *wassit*s and making a room. It is seen by some Bedu as an indication of *madheif*.
rejil	رجل	End column of the side elevation of *bayt ash-sha'r*, forming the back.
rwaq	رواق	Back or front wall of *bayt ash-sha'r* made of goat and camel hair and consisting of several pieces called *shigag* connected with dafa'yeh; the front faces east and the back west. The front is called *wajh al-bayt* and the back *khalf al-bayt*.
ruffah	رفه	Side wall of *bayt ash-sha'r* made of goat and camel hair, and consisting of several pieces called *shigag*

English pronunciation	Word in Arabic	Meaning in English
		that are patterned vertically and made out of various colours.
shadhfah/mikrabah/ 'Asfoorah	شظفه/ مكرابه/ عصفورة	Wooden piece that connects the roof with the main structured ropes (*tanab*).
sahah	ساحه	Interior divider of the two major interior spaces, *madheif* and *maharam*; it is made of sheep wool patterned and coloured in design.
sekkah	سكة	Ropes to tie the lower part of the *rwaq* (front or back), the *sfalah*, to the ground, especially in bad weather conditions so as to avoid wind.
sfalah	سفلة	The lower *shigag* of the back wall that connects to the ground. It is made of goat and camel hair. It is usually tied to the ground with *sekkah*.
sfouh	سفوح و مفردها سفح	The carpet used in the *madheif* as the main element to sit the guests and usually made of either camel hair or sheep wool. If used in men's space, it is usually made of two colours only.
shawa	صهوة	A *bayt ash-sha'r* that is pitched with two *wassits*, and consists of three *raba'h*
shdad	شداد	The seating put on the back of the camel and usually used as a holder for guests in place of a *wasayed*.
sheikh	شيخ	The head of a tribe; the highest social position among the Bedu.
shigag	شقاق	Longitudinal pieces connected with each other to make the *bayt ash-sha'r* roof and walls; made of goat and camel hair.
tanab	طنب و جمعها طناب	Guy rope that is used as a structured tensile element for pitching *buyt ush-shu'*; usually made of camel hair, it is named after each pole it is connected to.
tareigah	طريقه وجمعها طرايق	Long strips that connect between *shigag* in the roof of *bayt ash-sha'r*; they are made of goat and camel hair, and connected to *shigag* from the outside the *bayt ash-sha'r*.
tawaty	طواطي	A platform, one metre high and wide, furnished in Arabic style; it is usually found in the guestroom (*madheif*).
ta'zeib	تعزيب	Goat feeding.
wany	وني	Channel that surrounds a *bayt ash-sha'r* to collect rain water draining off the roof and to prevent it entering the *bayt ash-sha'r*.
wasayed	وسايد و مفردها وسادة	Pillow for the guest to lean against in a relaxing pose. It is in Arabic or Bedu style. They are usually made of hair and piled two or three high depending on the guest's comfort and request. It is usually colourful, and used in women's space.
wassit	واسط	Column that is located in the middle of a *bayt ash-sha'r* (central pole), and usually made of expensive wood (*zan*).

English pronunciation	Word in Arabic	Meaning in English
wawiyeh	واوية	Wooden sockets used above the *wassit* to protect its roof fibre from wearing out.
wedd	ود أو وتد	Metal or wooden peg that is driven into the ground and attached to the *tanab*.

CHAPTER 18

Bedouin livelihoods, the role of women in the Jordanian Badia

Rula Dajani
Directorate General for Humanitarian Aid – ECHO, Regional support office – P.O.Box 142584, Amman, European Commission Delegation in 11184, Jordan. E-mail: salmadj@hotmail.com

Research was undertaken to investigate gender divisions, roles and decision-making processes in Bedouin society in three villages in part of the north-east Badia of Jordan, helping to fill a gap in research knowledge in the area. Bedouin women were found to contribute significantly to the household economy. Their roles can be divided into unpaid domestic and livestock-related work, and paid work which, typically, is agricultural, seasonal and poorly paid. Unpaid work or work that does not directly yield an income for the family (including work with sheep; men sell the sheep and thus are seen as responsible for their income) is not regarded as productive work. A 'power game' showed that the women are aware of power differentials, in favour of men, but also showed that power relationships are changing and that women foresee or would like more change to occur though not to the extent of reversing tradition. Younger and unmarried women are more restricted in their movements outside the family home; they are more bound by tradition and the local interpretation of religious mores. The removal of the livestock subsidy in 1996 followed by three years of drought undermined the economic viability of sheep keeping and encouraged some women to speak out (and be listened to) about not keeping sheep if they merely added to women's work while losing money. With income from sheep becoming unreliable and with more educational opportunities being taken up by women, more women are securing paid, even salaried, work, essential to maintain household security, and in consequence are seen as making productive contributions to the household, which promotes an increase in their influence in family discussions about the household economy.

Keywords: Gender roles, power game, status, decision-making

1. Introduction

The scarcity of information on the status of women and the domestic economy within the Middle East and particularly rural and Bedouin women is justification for this study. Understanding the key roles of women within their communities and the decision making patterns at the household level is among the first steps towards understanding the economic and survival patterns within specific communities such as the Bedouin society. In order to realize the different forces that affect a country's economy and structure, it is essential to start from the bottom up. Individuals and assets at the household level and related mobilization patterns of available resources are key indicators of communal activities that make up collective industries.

The livestock industry has always been the main income-generating activity for Bedouin communities in the Middle East. Whether nomadic or urbanized, Bedouin groups have relied mostly on livestock for household consumption and trading. Although much credit has been generally attributed to men, the bread-winners in such communities, women were the back-stage operators of the whole show. Their daily routine activities have revealed their contributions in different ways including caring for the animals; producing related products for household consumption and sale in the markets. One of the reasons for having a second and third wife in a Bedouin family was merely related to enlarging the base of the livestock work force. A Bedouin mother trains her little children including daughters on herding and looking after the family livestock from a very early age. During my research a Bedouin woman once told me that a sheep can be more expensive than a child. It requires feeding three times a day while a child can eat less if the family is going through financial difficulty and sometimes the family would cut-off their food intake to feed their animals. While normally women are the ones who prepare family meals and care for the livestock, a study of them makes one appreciate their role in managing the survival of both the household members and their livestock.

1.1 Aim of the research

The research aimed at investigating:

- Gender division of roles within the Bedouin community with respect to livestock activities;
- Productive and reproductive roles of Bedouin women;
- Decision making processes within Bedouin households;
- Status of Bedouin women with respect to the social and economic factors.

The findings of this research were used to draft recommendations for future studies and potential interventions in the area where the research was conducted.

1.2. Methodology and identification

As research using qualitative methods is based on discovery not proof (Denscombe 1998, Maykut et al 1994) it was the methodology chosen for this small-scale social research. The analysis of social and economic factors impacting people's livelihoods, behaviours and attitudes can never be classified in a quantitative manner. It will always be subject to interpretation and to the level of deep knowledge of the community researched. Therefore, rich and detailed description achieves a higher level of understanding of any community which the researcher does not belong to compared with statistical information. Although reliable figures do provide a great deal of confirmation for social analysis, they were not suitable for this study where only three villages were chosen for research and thus could only have given an indicative model of the whole Bedouin community.

The qualitative research methods used in this study included:

1.2.1. Focused group discussion

Discussions were carried out with household members and groups of women throughout the research. General discussions were conducted at the beginning of the fieldwork to develop a clear understanding of the Bedouin community, which included women, men and children. More focused discussions were organized with women's groups as the research proceeded. Some of the specific topics discussed included productive and reproductive chores, division of roles, decision making structures, mobility patterns, etc.

1.2.2. Individual interviews

Individual interviews provided reliable information as they normally produced specific relevant information from selected individuals who were considered 'to know' (Denscombe 1998). Still, it required the researcher's investment in time and social-bridging with the targeted individuals since it was realized that generally people would give out information only to those they trusted. Interviews were organized with women and men after courtesy visits had already been carried out and more interviews were conducted to cross-check some findings towards the end of the fieldwork.

1.2.3. Participant observation

The use of this tool was found to be essential for all the qualitative studies as it opened the door for more questions and investigation. In this case it helped the researcher to develop a critical eye for the significant details of even the personal life of the person being observed. Observation needs practice and sensitivity to the finest details and non-communicated words. My earlier involvement with different communities during early years of employment with CARE international had allowed me to develop my observational senses and use them where relevant. Although this method is highly affected by personal perception it helps to confirm responses obtained from direct or indirect enquiry. Participant observation was applied throughout my research.

1.3. Villages and household selection

The northern Badia, located in Mafraq governorate, is the area served by the Jordan Badia Research and Development Programme (BRDP), which became a national Centre (BRDC) in May 2005, and therefore is the area where this research was undertaken. Out of 34 villages covered by BRDP, three villages were chosen for the research including Salhia Village which was not covered by BRDP but was included within the scope of a project funded by the UK's Department for Overseas Development (DFID). The selection of villages was based on the following criteria:

- Selected villages must not have less than 300 inhabitants;
- Selected villages should vary in location relative to the main Mafraq Safawi Road;
- Selected villages must have different tribes for purposes of comparison;
- Selected villages must not share common features; variation to be ensured.

The three selected villages which complied with the above criteria were Dayr al-Kahf (13 k off the main road), Abu al-Farth (3 k of the main road) and Salhia Village (located on the main road).

In such small-scale qualitative research, it is not possible to adhere to the principles of probability and randomness in selecting people and events.

According to Denscombe (1998), in qualitative research it is difficult statistically to represent people, texts and events in a 'normal instance'. Special instances, ones that are extreme, unusual, best or worst might be included. Since such studies are based on discovery not proof, they could result in a wide spectrum of descriptions rather than narrowly focused information.

The most appropriate household selection technique was the 'snowballing sampling' which provided 'handpicked' people and information to feed into this study. This strategy allowed me to select people and events relevant for investigation and led to unfolding situations. Other methods including random, systematic, stratified and quota sampling were discarded.

The main criteria observed during the selection of the 30 households of this study were:

- At least half of the selected households included different female generations;
- Selected households had diversified sources of income among the total number of households including livestock production, employment, income generating activities and seasonal work.
- Selected households must differ in their familial organizational structures. Nuclear, extended and female headed households were included.

Eventually, the final number of visited households was 12 in Dayr al-Kahf, five in Abu al-Farth and 13 in Salhia, a total of 29 households. Irregular visits were paid to random households. Those households were not counted in the end as the information gathered from visits paid to them was repetitive in most cases.

1.4. Limitation of field work

I was considered as an outsider and insider to the community throughout my research; an insider for being a Jordanian female conducting research in a community within a Jordanian society; an outsider for not being a Bedouin female but rather regarded as somehow a liberal city female commuting back and forth to Amman, the capital city.

Having to commute daily using BRDP transportation, I was always accompanied by a male driver. This by itself imposed a question regarding the appropriateness of such an act according to Bedouin standards. But at the same time it facilitated my interaction with Bedouin men during discussions in which the Bedouin driver already known to the community was always invited to participate.

Related to logistic matters, although using the BRDP transportation facilitated my travel within the Badia, it did not allow room for spontaneous visits. A strict schedule had to be followed and presented to the BRDP logistics department in advance for organizational purposes.

Although I had always provided clear explanations about the purpose of my study, some families could not accept the mere fact that I simply wanted to understand the livelihood patterns of Bedouin families and women activities. Some families could not accept my explanations as valid reasons.

The usage of a recording machine had to be discarded following the sense of unease when used during focused discussions and individual interviews. People became always much more conservative when the device was in use, therefore I opted to document all discussions on paper in addition to memory registration of events and non-verbal behaviour.

1.5. Conclusion

Jordan's strategic location in the heart of the Middle East has played a major role in shaping its economic, political and social structure. Since its creation, Jordan has been a buffer zone due to the nature of its political and economic involvement in regional and international affairs.

Inevitably, the people of Jordan, including the small minority of the 4 % Bedouin population, were hit by the changing economic and political situation in Jordan during the late eighties and early nineties, linked to the political and economic implications of the Gulf War and changing relations with Gulf countries, the Structural Adjustment Program (SAP) and the devaluation of the Jordanian currency. More importantly, the Bedouin's livelihoods which had relied on the livestock industry, and proved to be profitable until the mid nineties, could no longer be dependent on this type of sole activity after the removal of animal feed subsides and the successive years of drought during the late nineties. Those factors have imposed changes on Bedouin economic and social lifestyles and livelihoods.

For these reasons, important efforts of BRDP/BRDC since its establishment in 1992 have been directed towards researching the Bedouin community and understanding the real situation in the Badia in order to help the Bedouin to adapt their livelihoods through the evolving difficult circumstances. BRDP/C has been regularly in contact with the local people in the area, researching different political, economic and social aspects of their lives, through technological transformation and by seeking to harness the natural resources available within the area.

BRDP/C early realized the importance of the single unit of the community, the household and their inter-relations, in understanding the mechanisms of the economic and social behaviour of the Bedouin. Therefore, more emphasis was put on studying the social factors of the Bedouin's livelihoods, which are inevitably incorporated into all other aspects of their lives. The realization of the importance of the human factor in any development effort made it crucial to decode individual and collective practices, gender and power relations, tribal considerations and other relevant aspects of Bedouin lifestyle and social organization. The micro level analysis of the household leads to a better understanding of the complex macro issues that formulate general trends concerning specific aspects of a community's life.

2. Women's productive roles within Bedouin households

The fact that attention has only recently been given to women and their productive contributions in sustaining their families' survival, has made the data available on women's status scarce at the time of my research. There is little information on Arab women in general and Bedouin women in particular. Maghadam (1993) confirms the scarcity of data regarding women's productive activities and their contribution to national development. Moreover, Maghadam pointed out that women's performance in the informal sector and at the household level is suffering real challenges to become recognized as being a considerable contribution to the overall economy for a given society. Therefore, one could expect to witness the diminished roles women have been credited with in Bedouin society in which generally men receive most of the recognition due to their outdoor interactions and visible contribution to the household budget.

Still, the findings of my research contradict this perception. Bedouin women in Jordan contribute substantially to the household economy and, similarly, to the livestock industry. Their assumed roles contribute greatly towards generating income and perhaps should be considered as the main input in this industry. If it were not for their daily tending and caring for the livestock, there would be little chance for the family to generate income from selling sheep, producing dairy products for household consumption and manufacturing items for sale in the market. If girls do not sometimes leave school to substitute the role of their mothers in travelling around with the herds or caring for them at their homes, then there is little chance to earn income from livestock activities. Norms and traditions expect boys to complete their education and enrol in the army to generate another source of income for the family. Bedouin society relies heavily on women's domestic activities for household survival but unconsciously denies this fact.

This is attributed partly to the general conception of women as domestic labour in the Middle East. If, recently, urban females have managed to demonstrate their role in the formal market, Bedouin women have not yet been released from the household boundaries and its related chores. These chores are inherited from generation to generation. Mothers pass on their duties and skills to their daughters at an early age. It is common to find a Bedouin girl at the age of 13 who has dropped out from school and is being trained to become the carer for the family and its belongings including the livestock. Yet, social traditions and norms confine most of the female activities to indoor chores with the exception of livestock herding which could be done while the family is travelling from one area to another.

The female activities can generally be classified into two categories in the northern Badia: unpaid domestic and agriculture-related activities, and the informal seasonal work and poorly paid work. In order to understand the pattern of women's contribution in the Bedouin household, more light has to be shed on these two main categories.

2.1. Domestic and agriculture-related activities

Unpaid activities include all regular domestic chores ranging from cooking, cleaning, washing clothes, serving visitors and caring for little children. Caring for animals falls within this category including all related activities, except selling and buying sheep, for all female members of the family. Market-related activities are a strictly male responsibility which proves gender segregation of chores by location where the activity is being conducted. The female livestock-related activities are generally considered domestic and are thus placed within the range of unproductive roles for women in the Badia. When most of the women interviewed were asked about how they perceive the nature of these activities, few were able to identify those as productive, although all women agreed that they were crucial to the survival for their families. All men similarly thought that these activities were part of the female domestic responsibility and could not recognize them as productive. The reason for this, as one man explained, is that these activities do not generate income; the trading of the livestock is the only way that money is brought back to the family and therefore was the only productive activity, in his opinion.

2.2. Seasonal and part-time work

The picking of summer vegetables and olives in winter is a new role in the Badia. In recent years, due to the deterioration of the livestock industry and the need to identify other sources of income, women have started taking up part-time and seasonal work on farms. Tomato farms have absorbed a large amount of Bedouin female labour as well. Men find less demand for their labour on farms with the exception of box fetching. Some Bedouin men consider such jobs down-grading and farm owners prefer to hire foreign cheaper labour to do the job. The demand for older Bedouin women is high due to their higher commitment and assertiveness, according to farm owners. Still some younger or newly married females seek seasonal work when their economic situation is desperate so they face less opposition from the family to take up such jobs.

The fact that this type of activity requires travelling and dealing with strangers normally makes it difficult for young and unmarried women to embark on such work. Older women have more freedom to take up these activities leaving indoor and domestic ac-

Figure 1. Typical mobility patterns of elderly and married women

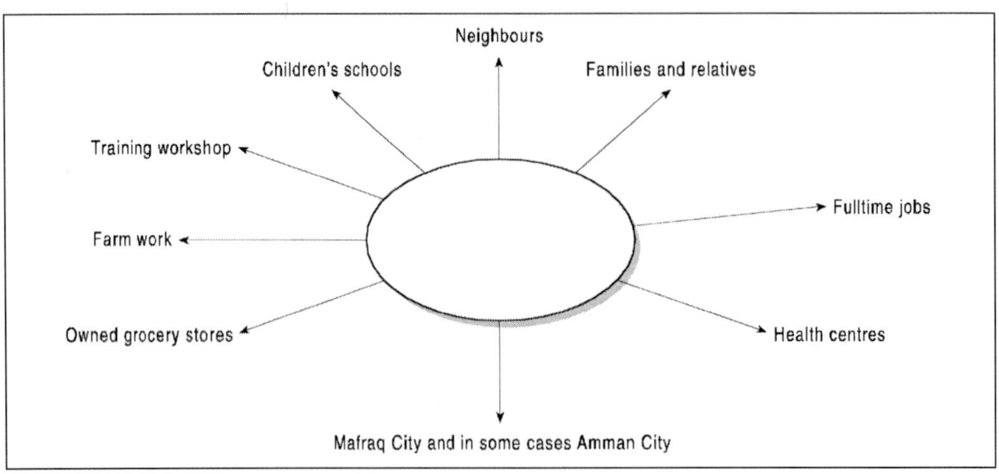

Figure 2. Typical mobility patterns of young and newly married women

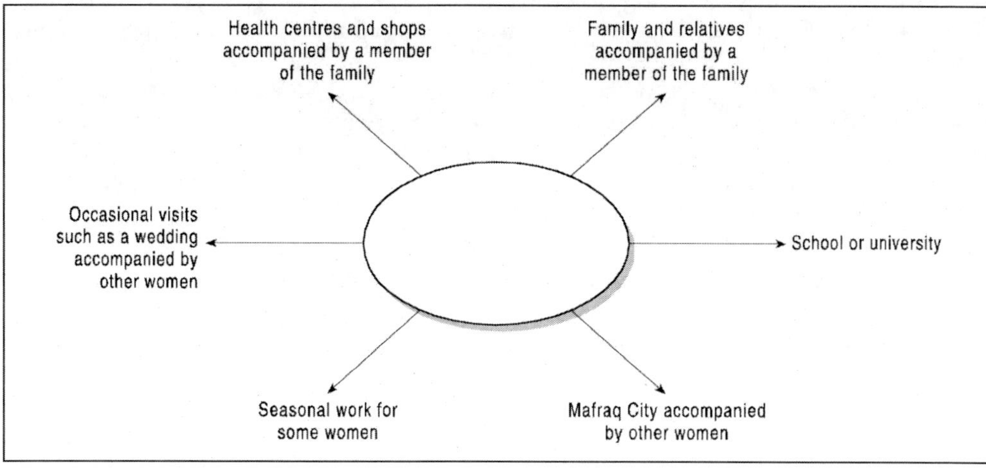

tivities to be managed by their daughters and daughters-in-law. Although such farm activities are physically very tiring, as women work for an average of 9–10 hours, sometimes in very hot and very cold weather, they are forced to accept the low wage offered which does not exceed JD50-60 per month. I came across two cases where young daughters were hired as labourers to cover their personal expenses which the household budget could not otherwise accommodate. In Salhia village, more women were involved in seasonal activities due to the location of the village on the main road, facilitating their travel. In the other two villages researched, only women in desperate financial situations sought this type of work during winter and summer.

The tendency towards the Bedouin community's acceptance of older women taking up seasonal work is related to the flexibility of their mobility patterns. Interviewing 40 women from different age groups, it was clear that younger women and unmarried females are more restricted by traditional and social norms in their outdoor activities.

Two patterns demonstrate the accessibility options to women in the Badia (Figures 1 and 2). Together they confirm the stricter rules that apply to the access of younger/unmarried women to people outside the household, which normally requires the presence of other members of the family.

2.3. Conclusion

There is a lack of data regarding women's active participation in different fields of economic activity, not to mention the scarcity of data on women's unpaid family labour, according to Mujahid (1985), Dixon-Mueller (1985), Maghadam (1993), United Nations (1995) and Haider (1995). The social values and family structures, in addition to educational and marital status, all contribute towards women's economic inactivity in the Jordanian labour market. The sectors of public services and education are the most common fields for women's productive participation in Jordan (CARDNE 1998, Flynn et al 1999). The agricultural sector also accounts for a large percentage of women's productive participation in the informal market which is still uncounted in most cases (Shukri 1996). Women's unpaid farm and domestic work occupy most of the time of rural and Bedouin women, leaving little space for them to participate in other officially recognized jobs. It is important here not to forget the fact that this informal participation in the market needs unskilled labour, which is the case of this stratum of the female population in Jordan. Because women in these areas lack qualifications and skills to compete in the formal market, they see big opportunities to secure their income from working as cheap labour. But since their work brings little financial reward to the entire family, their productive participation continues to be undervalued by the family. Yet, it is clear that without Bedouin women's various contributions towards their household survival, Bedouin families would suffer far more from the negative impact of the deterioration of the livestock market. Bedouin women's domestic and agriculture-related tasks, seasonal work and self-employment alternatives, in the last decade, have proved that they can pull Bedouin households through the difficult economic conditions that have been affecting the Bedouin population in Jordan.

3. Women's decision-making role within Bedouin households

It is not possible to illustrate the productive role of the Bedouin women without relating the issue to their decision-making capacity within the household and the community at large. The more recognition Bedouin women receive for their role in maintaining their household survival, the more power they exert over decisions made within their families. Their negotiation capacity grows stronger with the level of contribution they are able to sustain to improve the economic condition of their families. The challenge is, though, that in many cases their contribution could be seen as routine domestic chores and thus unproductive. But income-earning capacity is not the only criterion upon which Bedouin women's decision-making is judged. Their status within the household, their age, their belongings and their natal tribal history all determine the level of power they achieve. A wealthy Bedouin woman at an advanced age with powerful tribal background is expected to enjoy a more powerful status within her husband's family compared to a newly married female who comes from a poorer and less powerful family.

Social mores continue to limit the freedom of even well-placed women. According to Afsher (1993): 'Even the most powerful women are thus constrained by the ideological frameworks which delineate the patriarchal boundaries of their lives. Within these constraints women have differing options and embark on different strategies but those who step out and attempt to break the bonds are likely to spark off problems and be signalled out as scapegoats should troubles arise.'

Sen (1990) and Kabeer (1991) introduce the household model as a functioning bargaining unit in contrast to Becker's (1965) new household economic model presented as a total unit of inputs and outputs, subject to market prices. Becker's model has been challenged (Kabeer 1994, Senauer 1990) given that the household comprises more than one member with different preferences, resulting in constant conflict in positions regarding various matters. Therefore Sen's consideration of the household as a bargaining unit, in which family members are involved in decision-making, serves my research analysis better. According to Afshar (1998) power is a source of oppression in its abuse and a source of emancipation in its use. This is particularly true in the Bedouin context. Male-dominant power over household members including women, exemplifies the usage of power as a tool of oppression, subordination and negative control. On the other hand, women's ability to exercise their power over household matters and in their personal lives, can promote higher levels of self-reliance and self-confidence that could lead to self-sufficiency in some cases and, accordingly, would be regarded as emancipatory power.

It is essential here to highlight the role of religion and traditions in shaping power relations within Bedouin families since much of the social practices are related to human interpretation of Islamic teaching rather than the law itself. Generally, traditions play an important role in most societies affecting the practices and behaviour of individuals, especially in patriarchal societies, such as the Bedouin society (Esposito 1982). The patriarchal nature of Islam reinforces the pervasive belief among Bedouin that women are subject to men's control due to the basic teachings of Islam which place women under guardianship of their fathers, brothers or husbands who are responsible for maintaining women's well-being in such matters as food and shelter (Callaway et al 1994). The patriarchal society of the Bedouin in general tends to observe traditions with the same level of caution with which it observes religion.

While the Quran has clearly stated that men and women are equal: - 'O mankind! Lo! We have created you from a male and a female, and have made you nations and tribes that ye may know one another. Lo! The noblest of you, in the sight of Allah, is the best in conduct. Lo! Allah is knower, aware.' (49:26:13) Quran, Surat al-Hujurat (al-Hujurat Chapter) - some believe that Islam has placed men on the top of the tree but that this is due to the extra amount of responsibility they have to shoulder, rather than their superiority to women. As the Quran also says: - 'Men are protectors and maintainers of women, because Allah has made one of them to excel the other, and because they spend (to support them) from their means.' (4:5:34) The Quran, Surat an-Nissa (Women's Chapter).

The fact that some Muslim societies are less exposed to Islamic literature and scholarship, but instead rely on their personal interpretation of the words of the Quran, leaves the door open for misunderstanding and inaccurate adoption of Islamic teaching in their daily lives.

The issue of women in Islam is highly controversial and it is not within the scope of this work to explore the status of women in Muslim societies, but it has been my intension to stress the fact that traditions and religion may overlap and result in confusing positions between what is from God and what may be thought to be. This can impose false limitations and restrictions on vulnerable groups within the communities, Bedouin women being among those, within the wider Jordanian community. But Jordan, like many other Middle Eastern countries, has been through a progressive period of change during the last 30 years. Modernization and change continue to occur throughout the region, affecting some groups more than the others. Naqshabandi (1995), stated that although the Jordanian constitutional law and the personal status law have been amended several times and granted women increasing advantages compared to the past, in matters such as employment, marriage, inheritance and others, still the law has many pitfalls that hinder women's full participation and integration in public and political life. For Bedouin women in Jordan, it is even more challenging to gain recognition for their important role and allow them genuine participation in decision-making even at the household level.

3.1. Power game

To analyze women's perception of the division of power within the household; I practised a 'power game' on groups of women during the field work in each of the three villages. Following the division of women into groups, they were asked to have a close look on eight cards and formulate a story out of them depending on how the participation interpreted the development of power relations between men and women in the Badia.

Included were five cards with different arrangements of male and female characters. One card showed a female tied with strings that were controlled by a male. The second showed the opposite, a female controlling a male who is tied to her hands with strings. Two other cards showed a female and male person standing next to each other, with one having the male considerably larger than the female and vice-versa. The fifth card showed a male and female person standing on the same level and displayed as the same size. The sixth card included a male on his own and participants were asked to draw a female figure in relation to the male according to their own wishes. The seventh card presented a case opposite to that of the sixth card. The remaining card was left blank so participants could add whatever female-male power relation they desired in order to complete their story. The arrangement of cards within each group then depended on how women perceived power relations between men and women, not how they ought to be. Although some groups were not sure whether the cards were intended to show how power relations were divided at the present time or could possibly be in the future, they tried their best to come up with a logical version of the power story.

I obtained seven results from three discussion groups when group meetings were held in each of the three villages. After each group had presented their version of 'power story', a discussion of the reasons behind each group's arrangements of the cards followed. In addition, we discussed their perception of each card, their preferences and choices regarding the blank card and to what extent these cards reflected the current situation regarding male-female relationships in the Badia. The following presents some of the key discussion points:

- All groups without exception placed the card which illustrates the male being in control of the female as their first card describing the beginning of the 'power story' in Bedouin society. They said that this case was more representative of male-female relationships in the past, when men were completely in charge of the material and human resources of the households they belonged to, including female members of the household.
- Five groups placed the card which showed the male larger than the female as their second choice, while the other two groups chose to draw a female character released from strings, but still with a male character displayed larger than the female, as their second placement. They argued that the relationship between men and women did not develop very fast but, however, when women were somehow freed from being tied to the strings then they acquired a better status since they managed to stand at the same level with men.
- Six groups placed the card that showed both male and female characters at the same level as the third card representing the 'power story'. Only one group disagreed with this placement and thought that it was better to draw both female and male cards that showed them mutually tied to strings by each other. They then placed the card that showed them on the same level, freed from strings. This group argued that the relationship did not immediately develop from having the female smaller than the male, which they regarded as similar to being tied to strings, to standing in equality on the same ground with men. This group explained that there must be a stage in between when both males and females were equally controlled by each other, then a following stage when both were set free from each other's control and became totally equal.
- All groups thought that the fourth card must be the one which shows the female larger in size than the male, which indicates that she has acquired a more powerful status than the male. Most groups said that they cannot say that the female can be more important than the male in any stage of the 'power story', but they had to place the cards in this sequence to develop the story. They argued that in reality this story is hardly true but the most logical way to see it is when they become equal. Then, perhaps, women's status may become more recognized and their contribution towards the welfare of the household more evident, but the role of the women can never be more important than that of the men.
- The fifth card that completed the 'power story' was the one that showed the male tied to strings under the female's control. Almost all groups said that this is an unrealistic situation. One group said it might be applicable to a few cases when the male is desperately helpless, in situations of major physical disability or economic unemployment. One participant in a Salhia group thought that these two cases are not the only ones when the female is in control of the male. She said that although

her husband is not suffering from any physical disability and he is economically sustaining his family, still her husband's relationship with his second wife can be expressed by this card. She thought that her husband's second wife was totally in control of her husband, his earnings and the running of the daily routines of their lives. She added that mostly her husband's second wife's opinions affected her (the first wife's) household routines as well. But in general almost all participants agreed that this case was rare in the Badia. One said that if this situation was true, it would present a deviation in Bedouin society from both religion and tradition.

- Five groups could not suggest another female-male relationship regarding the two cards showing one of the opposite sexes on their own. Two groups suggested leaving the cards as they were because there are cases when the husband is dead and therefore the female can be in charge of household survival. The same applies to the male-only card. They added that this is not a common case because most men remarry after they divorce their wives, but it is still possible.
- The blank cards were left blank in four cases because the groups did not see any possible male-female relationship to add. Three groups added the card that showed, in two cases, one male with two females and, in another case, three females tied to him. They decided that these cards might go in the middle of the 'power story' or at the end when the husband decides to marry more than one wife and would be in control of all his wives.

The 'power game' confirmed that the traditional perception of men controlling women is common in the Badia. Still, there were some arguments that women nowadays can gain more control over their lives and assume an active role in the decision-making within their households.

3.2. Women, power and decision-making in the Badia (case study)

For the reason that few female researchers of Bedouin women's lives have managed to conduct fieldwork in the past two decades, little material is available to provide us with a reliable background to the nature of this topic. As most of the research work has been undertaken by male researchers, who have minimal access to women's private domain, the picture that has been given is likely to be misleading and guided by men's opinions.

Dawn Chatty is one of the few female researchers who have been able to unveil the private sector of the world of the Bedouin women. Her own views regarding women's power do not conform to what has been suggested by previous or current male researchers. After 20 years of conducting fieldwork, she argued in her book 'Mobile Pastoralists' (1996) that Harasiis Bedouin women in Oman have played a complementary role to that of men rather than a dependant and subordinate one. As Chatty expressed it: 'In the context, the private domestic or informal universe of women was regarded as generally subordinate to the public, extraordinary or formal universe of men. Power was generally seen to be attached to the public and formal sphere of life, and women were subsequently viewed as powerless, subjugated... I conclude that Harasiis gender relations are not based on a concept of inferiority and superiority but are founded on a complementarity of duties, roles and needs... women do not take a formal role in the gatherings of men. They do nonetheless participate informally from the background sidelines and their opinions can inform the discussion.' (p 143, 145, 147).

Lancaster's work with Rawala Bedouin (1981) clarified the fact that Rawala Bedouin women, as much as other Muslim women, are kept within the private domain of Bedouin life, but he is not of the belief that women are the property of men. On the contrary, women in Bedouin settings form the survival element for Bedouin families. Women can easily assume men's roles and undertake their duties whenever needed, as well as performing their own household survival role which is essential to all. On the other hand, men cannot assume women's roles in the way that women can easily assume both roles.

What Lancaster (1981) tried to point out was that because of the importance of the women's role in the Rawala Bedouin community, they were treasured and brought into the public sphere only in emergencies which required their involvement. As long as stability prevailed, women undertook their natural roles as mothers and family managers but when instability occurred, women's natural role continued, or might be delegated to other women, and new roles might be assigned to them (or taken by them) to face the upcoming situation.

Women in Bedouin areas enjoy a powerful status within the household even with the existence of men. According to the study undertaken by Oughton and Adas (1999), concerning how Bedouin households in northern Jordan make a living, women were found to hold a relatively strong position regarding household decision-making matters. Their economic contribution and ownership of assets in the form of livestock, land and jewellery enhance their power even further. During the course of this study, one case illustrated how a Bedouin wife did not allow her husband to buy livestock by threatening to leave the house if he did. This wife thought that owning livestock at this particular period would generate debts and cause a worsening of their financial status. I encountered a similar case during my work in the Badia.

3.3. Cases of decision-making regarding livestock ownership (Karima's case)

Karima is young women married to Ahmad, a school principal, and has four young children. They live in Abu al-Farth village. While her husband has a full-time job as a school principle in their village, he trades a few livestock in his free time after work. Karima has always discouraged her husband from keeping livestock since the profit they make out of selling is very little and in some cases they have spent more money than they gained from selling them, leaving them with a loss. After several unsuccessful attempts from Karima's side to convince her husband to abandon this trade, she refused to look after the few head of sheep her husband bought because she thought that they were not helping in securing their survival, but rather consuming more money in unprofitable activity.

Karima said: 'I have four children and these few sheep are part of the family. My husband makes sure that he secures their food before ours. When I calculated their viability, I found out that it would be cheaper to sell them and get our dairy products from the grocery shop or buy fresh milk from neighbours and process it myself. When I suggested this to my husband, he did not like the idea and said that he has nothing much to do in his spare time and these sheep keep him busy. In fact, I was doing all the work and he only took them for a walk whenever he felt like it. I decided, after several attempts to put him off this trading activity, not to look after them because I was already overwhelmed with domestic and child caring duties. When my husband tried for a few days to do the task I was assuming caring for livestock, he discovered that it needs hard work, therefore he gave up and sold the last 100 sheep we had and never bought any after that' (Karima, third interview on Nov 8, 1999 in Abu al-Farth village).

In this particular case the wife used her power to push for what she thought was a sensible thing to do. She wanted to make it easier for herself, by reducing the amount of work she had to undertake, and for the benefit of her household. Although her husband was against it, he found himself in a weak position and incapable of handling the situation on his own after she had stopped looking after the animals. In this case, the husband took the practical action of selling the sheep and not buying any more, a decision that was strongly influenced and based on the wife's refusal to perform any duty related to livestock. While this cannot be considered as a common case, since married women normally follow their husband's will, in this particular situation Karima had the power to influence her husband's decisions since their family was not dependent on livestock as a source of income to the household but rather it was more her husband's preference. Therefore, Karima's opposition to this activity did not negatively affect her husband's earning but instead, from her point of view, saved them the risk of losing more money and saved her a considerable amount of livestock maintenance duties.

4. Main conclusions and recommendations

There is no doubt that issues of Muslim women in general and Jordanian women in particular are receiving more recognition and being given more attention by researchers, scholars, state officials and development practitioners.

Jordanian women's political economic and social status is formed from three sources of influence: Islam, the law of the state and traditional culture.

The overall aim of this study was to investigate Bedouin livelihoods, and women's roles in securing their household survival and in their participation in the decision-making process at the household level. Based on this aim and their related study objectives, stated earlier, I present below the major conclusions of this study. But my conclusions should just be understood in the context of the social and economic changes that have occurred in Jordan and, more specifically, within Bedouin society over the last two decades. Social and economic changes provide a background to understanding changes in women's economic power and their social status in the Badia of Jordan.

4.1. The changing livestock industry and Bedouin women's economic contribution towards household survival

Bedouin attitudes towards livestock were negatively affected by the deterioration of the livestock industry, which was mainly caused by the removal of the animal feed subsidies, the low levels of rainfall in the Jordanian Badia during the late nineties, problems of market access to Saudi Arabia and sheep imports from other parts of the world including Eastern Europe. A large proportion of livestock owners were adversely affected (in general, owners of large and some medium-sized herds managed to sustain their livestock numbers and, in some cases, generate profit, while owners of small herds were more negatively affected). The rise in animal feed prices coincided with the degradation of the grazing areas due to the lack of sufficient rainfall during the three-year period 1997/8-2000/1. Therefore, livestock owners were forced to substitute natural grazing with purchased fodder, whose prices rose drastically in the second half of the nineties. Since many livestock owners could not maintain their livelihood and afford their livestock expenses at the same time, some were forced to sell their herds to reduce herd numbers, and to avoid negative consequences from trying to maintain their herd size. As a result, Bedouin people were forced to explore other sources of income since the livestock industry did not maintain its credibility as a dependable source of cash. Livestock keeping became a subsistence-oriented measure mainly for providing household consumption needs. Traditional practices which involve livestock were being altered due to the deterioration in livestock rearing conditions. The findings of my research show that Bedouin norms of slaughtering seven sheep in gratitude to a dead member of the family had been affected by these economic hardships. Fewer sheep were being slaughtered, perhaps only one or two depending on the financial status of the family. In addition, practices regarding sheep slaughtering during Ramadhan and the Eids were being modified and the number of sheep slaughtered reduced, and perhaps disregarded in some cases, to save the household excess expenditure.

Bedouin women showed less interest in contributing efforts towards raising livestock unless it proved still to be profitable for their households. A large number of women interviewed expressed their desire to keep a few animals in order to secure household consumption of dairy products. Most of the women interviewed kept a small number of animals, which they would retain despite the economic hardships. Some women had shifted their interest from raising sheep to poultry for household consumption and, in some cases, as an income generating activity. The new generation of young women had already developed a negative attitude towards livestock activities and preferred in most cases to perform domestic duties instead of the livestock related chores.

With regard to sustaining household survival, Bedouin women's contributions have become more substantial and measurable in contemporary Bedouin household. Bedouin women have developed strong attitudes towards seeking work opportunities in the informal market, such as seasonal farm work and part-time employment. An increasing number of Bedouin women work in full or part-time jobs, seasonal paid work and self-employment activities.

Older women were found to be more eligible to seek this kind of work since they had more flexible and less restrictive mobility patterns than younger women in general within the Bedouin society.

4.2. Allocation of Bedouin women's productive roles

In the past, young women were associated with outdoor activities related to livestock, such as herding and grazing, but recently they have become more associated with indoor domestic activities since the livestock industry is not as strong as it used to be. Regarding older Bedouin women, they are forced these days to seek employment and income generating activities, if they need to secure income, outside the boundaries of their households rather than being confined to their households as the case used to be in the past. Although older Bedouin women were never too involved in domestic activities in the past, still they were more attached to their household territories (tents). Today it is more a matter of adopting outdoor activities as a necessity to secure their household's survival, leaving younger women to manage domestic work.

Linked to the previous point, the mobility pattern of Bedouin women has changed accordingly. To achieve the freedom that some Bedouin women have to work outside their households necessitated a less restrictive mobility pattern to emerge, particularly regarding older women's groups. Younger Bedouin women are more associated with domestic activities and have more restrictive mobility patterns that limit their movements to specific places and impose certain limitations on making long distance visits while being accompanied by a member of their households.

4.3. Power relations

Although Bedouin women make major contributions to securing their household needs and managing their livelihoods, the control of all the resources involved in the processes remains within men's hands. Therefore, the division of power between men and women in the Badia, which is related to the control one has over resources, still weighs more to the advantage of men.

Bedouin women's perceptions of the division of power between men and women are mostly traditional, seeing male-dominance as a naturally given phenomenon supported by Islam and Bedouin traditions. The findings of the 'power game' conducted with Bedouin women showed that most women believe in male superiority regarding the share of power and control over the household resources. Nevertheless, Bedouin women identified some cases in which women's power was manifested more clearly such as in cases of female-headed households, men's physical disabilities or absence, which require major contributions from the female side and therefore entitle women to have more power, varying in each particular case. Although the picture of power in the Badia is still male-oriented, Bedouin women are more entitled to participate in decision-making and share more power with men at the household level. The more productive and financially rewarding contributions women make, the more significant their inputs in making decisions concerning their household survival become. Although Bedouin women's contemporary decreased involvement in livestock-related activities has caused a devaluation of the importance of their productive roles, their other productive contributions to household survival in the form of full-time, self and seasonal employment have added another dimension that maintains recognition of their productive roles and is strongly linked to their share of power within their households. Age plays an important role in the sharing of power among women from different age groups within Bedouin households. From my fieldwork, it became clear that as a woman grows older, her influence over other members of the household, whether men or women, increases. Older women are normally considered wiser and more experienced than younger women and thus more eligible to impose their opinion and therefore influence the decisions taken by the male head of the household. Women continue to be under the supervision of their fathers, elder brothers and mothers as long as they are unmarried and living with their natal families.

4.4. Female-headed households

The emergence of female-headed households is relatively new in Bedouin society. The significance of this phenomenon lies in the fact that Bedouin women whose husbands die or divorce them are now entitled to remain single. In the past, widowed or divorced women accepted to go back to live with their parents-in-law or natal families in order to be under the guardianship of the male head of those households. Due to the economic and social changes in contemporary Bedouin households, it is no longer unusual for a woman to run the household on her own, depending on her ability to do so. The same applies for divorced women who go back to live with their natal families and practically, in most cases, become the managers of the households.

4.5. Changing social attitudes towards Bedouin women's education and marriage

Bedouin societies are still strongly affected by religion and traditions, as they have always been. Hence, the economic changes in the Bedouin structure play a major role in altering people's social attitudes towards education and marriage. The Bedouin have developed a positive attitude towards educating their children, whether males or females, in response to their settlement in villages, the deterioration of the livestock industry and the increased possibility to secure household income through formal employment. The Bedouin regard education as a means to qualify their children to have better employment opportunities in the future. They regard this as being particularly important because the livestock industry has become less reliable as a means of securing Bedouin livelihood in recent years. I found out that a large number of the young women who do not attend school or university but have the opportunity and permission to pursue their education, are the ones who decided to quit because of some lack of ambition. These cases were quite different from others in which families imposed their decisions on their daughters to stay at home for various reasons such as long distance, domestic responsibilities and conservative attitude. The cases of young women's commitment to livestock herding leading to difficulties of attending school are no longer evident.

Regarding marriage, a negative correlation between livestock ownership and marriage has developed in recent years. Many Bedouin families now tend to avoid marrying their daughters to livestock owners since it involves more work and a less secure income unless the potential husband comes from a relatively wealthy family. Bedouin families prefer to marry their daughters to government and army employees since they earn monthly salaries, which entitle them to have pensions when they retire and allow them to afford marriage expenses by taking loans which can be repaid over a long period of time. On the other hand, educated women who have secure jobs certainly attract more marriage proposals. Men perceive women's direct financial contribution into the household budget as purely productive, while their productive contributions towards their household's maintenance in forms of domestic and child-rearing activities have always been underestimated.

5. Recommendations for further research

The overall picture of Bedouin women's status within their households has changed over the previous two decades due to changes in the social and economic structure of Bedouin society. Yet, a comprehensive assessment of their roles within their households and the larger community needs to be further researched and analyzed in order to identify the focal points of their roles and contributions towards their families' survival and livelihoods. Although my research has been able to reveal specific areas regarding Bedouin women's productive contributions, their family livelihoods and their share of decision-making power at the household level, there are yet more interesting areas to be investigated which would add more depth and enable a better understanding of their status within their households and the larger society. For example, specialized research on female-headed households and the Bedouin women's roles in maintaining their households in Bedouin society will provide more evidence of Bedouin women's evolving roles in domestic management. A thorough analysis of young Bedouin women's roles in sustaining their household's survival could be another area of investigation. A continuous assessment of the impact of Bedouin women's changing roles and contributions on their lives, is recommended. Further research should identify whether Bedouin women's changing status works as an advantageous factor or extra burden.

References

Afsher H (1993). Women in the Middle East: perceptions, realities and struggles for liberation. Basingstoke: Macmillan Press.

Afsher H (1998). Women and empowerment: illustrations from the Third World. Basingstoke: Macmillan Press.

Callaway B and Creevey L (1994). The heritage of Islam: women, religion, and politics in West Africa. London: Boulder Lynne Rienner Publisher.

CARDNE (1998). Final report: The impact of structural adjustment programs on the rural poor in Egypt, Jordan and Yemen. Regional Center in Agrarian Reform and Rural Development for the Near East, Amman, Jordan.

Chatty D (1996). Mobile pastoralists: development, planning and social change in Oman. New York: Colombia University Press.

Denscombe M (1998). The good research guide: for small-scale social research projects. London: Open University Press.

Dixon-Mueller R and Anker R (1985). Women's work in third world agriculture: concepts and indicators. Geneva: International Labor Office.

Esposito J (1982). Women in Muslim family law. New York: Syracuse University Press.

Flynn D and Oldham L (1999). Women's economic activities in Jordan. WID tech report, International Centre for Research on Women.
Haider R (1995). Gender and development. Cairo: American University in Cairo Press.
Kabeer N (1991). Gender, production and well-being: rethinking the household economy: discussion paper. Brighton: Institute of Development Studies.
Kabeer N (1994). Reversed realities: gender hierarchies in development thought. London, New York: Verso.
Lancaster W (1981). The Rawala Bedouin today. Cambridge: Cambridge University Press.
Maghadam V (1993). Modernizing women: gender and social change in the Middle East. Cairo: American University in Cairo Press.
Maykut P and Morehouse R (1994). Beginning qualitative research: a philosophical and practical guide. London: Falmer Press Teachers' Library series.
Mujahid G (1985). Female Labor Force Participation in Jordan. In: Abu Nasr J and Khoury N (eds), Women, employment and development in the Arab World: Mouton Publisher.
Naqshabandi B (1995). Women's changing political participation in Jordan. PhD thesis, University of Durham.
Oughton E and Adas W (1999). Report on the Baseline Study of Households. Jordan Unpublished Report, Amman: BRDP.
Sen AK (1990). Gender and cooperative conflicts. In: Tinker I (ed), Persistent inequalities: women and world development. Oxford: Oxford University Press.
Senauer B (1990). The impact of the value of women's time on food and nutrition. In: Tinker I (ed), Persistent inequalities pp. 150-161.
Shukri S (1996). Arab women: unequal partners in development. Avebury: Aldershort.
UN (1995). Women: looking beyond 2000. New York: United Nations Publications.

CHAPTER 19

Common infectious diseases and risk factors in the north-eastern Jordan Badia

Laila F. Nimri
Department of Medical Laboratory Sciences, Jordan University of Science and Technology, Irbid, Jordan.
E-mail: nimri@just.edu.jo

The objectives of this study were to determine the common infectious diseases prevalent in the area by identifying the predominant causative pathogens, their antibiotic susceptibilities, clinical outcomes, and the risk factors. The study was conducted over 2-year period in the north-eastern Jordan Badia, a rural area inhabited mostly by recently settled Bedouin and small-scale farmers of low socioeconomic status. Four types of clinical specimens were collected between September 1999 and September 2001 from outpatients that were physically examined by a physician at four major health centres in the Jordan Badia. The patients were enrolled in the study based on the symptoms and the clinical diagnosis.

Diarrhoeal diseases caused by a variety of enteric pathogens, brucellosis, community-acquired bacteraemia and urinary tract infections are prevalent in that region and were the focus of this study. The reporting of newly emerging parasites eg *Cryptosporidium parvum* and *Cyclospora cayetanensis* in this rural area increases awareness about these parasites. The recovered pathogenic bacterial species showed a wide range of resistance to commonly used antimicrobial agents, which was most probably due to the empiric treatment used in invasive and recurrent infections and reflects an uncontrolled use of antibiotics. Risk factors that correlated significantly with contracting certain diseases included socioeconomic status, level of education, drinking unpasteurized milk, drinking unchlorinated well or tank water, contact with animals, and a low level of personal hygiene. To our knowledge, this is the first clinical study on brucellosis, the aetiologies of diarrhoea, bacteraemia and urinary tract infections in this rural area of Jordan. The findings of this study do emphasize the need for further longitudinal epidemiological studies in which the issue of indigenous infections and modes of transmission should be further investigated.

Keywords: Infectious diseases, brucellosis, diarrhoea, bacteraemia, urinary tract infections

1. Methods

1.1. The study area and patients

The study was conducted over a 2-year period in the north-eastern Jordan Badia, a rural area inhabited mostly by recently settled Bedouin and small-scale farmers of low socioeconomic status. Four types of clinical specimens were collected between September 1999 and September 2001 from outpatients that were physically examined by a physician at four major health centres in the Jordan Badia. The patients were enrolled in the study based on the symptoms and the clinical diagnosis. The total inhabitants that receive services from the four centres exceed 12 000. The centres are: Sa'idiyya, which receives patients from 12 villages, Umm al-Quttayn, patients from 12 villages, Dayr al-Kahf, patients from eight villages, and Sabha, patients from eight villages.

All patients consented to the collection of the specimens and the study was approved by the University ethical review committee for human investigations.

All relevant information for each patient was recorded by a nurse from the centre using a structured questionnaire and from the medical records. The demographic and clinical data collected for each patient included sex, age, immune status, acquisition of a bacteraemia, underlying disease, a history of hospitalization within the previous month, previous use of antimicrobial therapy, response after the introduction of presumptively appropriate antimicrobials and the final outcome of the disease.

1.2. Processing the specimens

Each specimen was processed based on the type of the specimen (Nimri 2003a, Nimri 2003b, Nimri 2007, Nimri et al 2004a, Nimri et al 2004b, Nimri et al 2004c).

Blood specimens were immediately inoculated under strict aseptic conditions in broth media or diphasic blood culture bottles. Plates were incubated at 37°C and were checked daily for growth. Gram staining of subcultures on enriched and selective media for a wide variety of pathogens was undertaken for suspected colonies.

Organisms from the positive cultures yielding pathogens were identified by microbiological and biochemical methods, including the API 20E system.

Susceptibility tests: Susceptibility testing of the isolates to at least eight out of 15 antibiotics was performed on Mueller–Hinton agar by a standard disc diffusion assay, following National Committee for Clinical Laboratory Standards (Wayne 1999) guidelines. Antibiotics tested were ampicillin, carbenicillin, ceftriaxone, clindamycin, chloramphenicol, doxicillin, gentamicin, kanamycin, lincomycin, methicillin, penicillin, tetracycline, erythromycin and vancomycin for the Gram-positive isolates and trimethoprim-sulfamethoxazole. 'Non-susceptible' refers to resistant and intermediately resistant. Multidrug resistance refers to resistance of the isolates to two or more antibiotics.

2. Diseases

2.1. Community-acquired bacteraemia

It has been recognized that the presence of living microorganisms in the blood of a patient carries with it considerable morbidity and mortality (Reimer et al 1997). Despite all the advances in medical practices in recent years, the bacteraemia continues to be a serious problem that needs immediate attention and treatment. In addition, it is still one of the main causes of mortality despite the existence of numerous antimicrobial agents and an increase in means of support.

Patients with a community-acquired bacteraemia have diseases with a differing spectrum of pathogens and prognosis. A variety of factors such as the type of microorganism, age, the underlying disease and where the bacteraemia was acquired, can change the prognosis of the infection (Cisterna et al 2001). Blood was withdrawn from 215 patients presenting with fever and a presumed diagnosis of a bacteraemia. Some patients had other symptoms depending on the underlying disease, eg cough, chest pain and joint pain; rash was observed in 12 cases.

The primary focus of infection was defined as a culture-positive site and/or a clinically evident site of infection concomitant with the bacteraemia. The results are given in Table 1.

Positive blood cultures were recorded for 126 out of the 215 cases (58.6 %), 4 % of these cultures yielded double isolates. No pathogens were recovered from 41.4 % of the cultures. The mean age of the patients was 44 years (range, 10 months to 87 years), children less than 14 years old accounted for (34.9 %), while 38 % were older than 50 years. The incidence in males and females was 44.4 % and 55.6 %, respectively.

Table 1. Bacteria recovered from blood cultures of 126/215 bacteraemia patients and rates of resistance to different antibiotics tested

Organism	No. of cases (%)	Antibiotic resistance (%)*
Staphylococcus aureus	40 (31.7)	25 (62.5)
Brucella melitensis	20 (15.9)	5 (25.0)
Streptococcus pneumoniae	14 (11.1)	8 (57.1)
Streptococcus spp	13 (10.3)	6 (46.2)
Escherichia coli	10 (7.1)	8 (80.0)
Coagulase-negative *staphylococci*	4 (3.2)	1 (25.0)
Klebsiella spp	4 (3.2)	3 (75.0)
Yersinia pseudotuberculosis	4 (3.2)	1 (25.0)
Salmonella enteritidis	3 (2.4)	3 (100)
Haemophilus influenzae	3 (2.4)	1 (33.3)
Enterococcus spp	3 (2.4)	1 (33.3)
Enterobacter cloacae	2 (1.6)	1 (50.0)
Neisseria meningitidis	2 (1.4)	1 (50.0)
Unusual bacterial spp	4 (3.2)	1 (25.0)

* Resistance to one or more antibiotics as mentioned in results.

Staphylococcus aureus (31.7 %) was the most common bacterial agent isolated, followed by *Brucella melitensis* (15.9 %), *Streptococcus pneumoniae* (11.1 %) and other pathogens such as *Staphylococcus epidermidis* and non-typhi *Salmonella (S. enteritidis)* which were associated with soft-tissue infections and abscesses.

The increasing frequency of *Staphylococcus aureus* bacteraemia, coupled with increasing rates of antibiotic resistance, have renewed interest in this serious, common infection (Finkelstein et al 1984, Petti et al 2002). The isolates were assessed as clinically relevant based on physical examination, symptoms and finding of the origin of the bacteraemia in these patients.

The recovered pathogens showed a wide range of resistance to commonly used antimicrobial agents, and multidrug resistance was documented in 44.4 % of the isolates. This wide range of resistance of the isolates is most probably due to the empiric treatment used in invasive and recurrent infections and reflects an uncontrolled use of antibiotics.

Results indicate that therapeutic options for a community-acquired bacteraemia have not yet become seriously limited by the prevalence of acquired antibiotic resistance. Although almost any localized infection can disseminate into the bloodstream (Raimundo et al 2002, Weinstein et al 1997), the most probable common origins of the bacteraemia based on physical examination, signs and symptoms supported by positive cultures in most of the cases were urinary (15.9 %), respiratory (14.3 %), no source of the bacteraemia identified (a primary bacteraemia) (13.5 %), gastrointestinal (12.7 %), soft-tissue infection, eg wounds and superficial skin infections (7.9 %), or meningitis (1.6 %). The most probable primary sites of infection were superficial skin wounds or skin abscesses. Characteristics permitting recognition of patients with such strains would aid infection control efforts and choice of empiric therapy pending culture and susceptibility results (Rezende et al 2002).

Risk factors for infection could not be identified in 34.1 % of the patients. Studies have shown that despite the best efforts of clinicians, the source of bacteraemia or fungaemia cannot be determined in one-quarter to one-third of the patients (Anthony et al 2000). No mortality was reported in association with the bacteraemia in these patients.

In conclusion, this study reports on the relatively different predominant aetiologic agents of the bacteraemia in a rural area, which are most probably due to the lifestyle and health practices in that area. It demonstrates the importance of characterizing patients with such strains to permit recognition of groups at increased risk for bacteraemia, and characterizing the antibiotic susceptibility distribution of isolates to develop appropriate policies for the treatment and prevention of these diseases in the community. Furthermore, results confirm that appropriate empirical antimicrobial treatment increases the survival of the bacteraemia patients since no mortality was reported in these cases.

2.2. Brucellosis

Brucellosis affects human populations in many developing countries including the Middle East, and Latin America where it is still endemic. According to the World Health Organization (WHO 1997), half a million new human cases each year are reported worldwide. These numbers greatly underestimate the true incidence of the human disease (Young 1995) because the clinical picture of human brucellosis is extremely variable (Zerva et al 2001) and because it is under-declared to local authorities and misdiagnosed by clinicians. The diagnosis of human brucellosis based on the clinical picture alone is not definitive since the symptoms mimic other diseases and are fairly non-specific. Either the isolation of the organism in culture or detection by molecular methods is necessary to confirm the clinical diagnosis.

Brucellosis has been prevalent in Jordan for years, where 7 842 cases of human brucellosis were registered at the Ministry of Health during the last 10-year period. The increase in the incidence of human cases was linked to the lambing season in Jordan (Abu Shaqra 2000, Dajani et al 1989); these reports are in agreement with the findings of the current study. Bedouin are at higher risks of contracting brucellosis, due to the lifestyle, environmental and social conditions of this population (Levy et al 1998).

This study was initiated by the recent increase in the number of human cases diagnosed in the northern Jordan Badia in order to help assess the status of the disease in that area. For this purpose, a total of 215 peripheral blood specimens were collected from 50 healthy control subjects and 165 patients. These patients had compatible signs and symptoms that were clinically diagnosed to indicate brucellosis and were seropositive. The specimens were cultured in broth media, and were tested by PCR assay using genus-specific primers from the conserved region of the 16S rRNA sequence, which showed high specificity for the *Brucella* spp.

Table 2. Epidemiological, clinical, and serological results of 165 cases of human brucellosis

Characteristics:	Values
No. of patients studied	165
Male: Female	1:1.85
Mean age in years (range)	45.5 (5-86)
Clinical and laboratory diagnosis:	
Mean duration of the chronic cases in years (range)	8.5 (1-16)
No. of patients with fever	165/165 (100 %)
No. of patients with muscle pain and anorexia	140/165 (84.8 %)
No. of relapsed patients	58/165 (35.2 %)
No. of positives in Rose Bengal agglutination test (≥1/160)	140/165 (84.8 %)
No. of patients with positive blood cultures	20/165 (12 %)
No. of patients with positive PCR	120/165 (72.7 %)
No. of relapsed patients with positive PCR	48/58 (82.8 %)

Table 3. Comparison of the PCR results with blood culture and serology for the 165 cases

Result of tests	Sero + Cult +	Sero + Cult -	Sero - Cult +	Sero - Cult -	Total
PCR +	20	100	0	0	120
PCR -	0	20	0	25	45
Total	20	120	0	25	165

Serology (sero): Rose Bengal agglutination test (titer, 1/160); Cult: blood culture

The mean age of the patients was 46 years (range, 6 to 86 years). The information recorded in the questionnaire indicated that 40 % of the 165 patients had contact with animals (eg sheep and goats) either raised nearby where they live or at work (eg farmers). Most probably, 45 % acquired their infections by consuming contaminated unpasteurized milk or homemade dairy products (eg soft cheese), 19.4 % acquired their infection from contact with animals, 14.5 % acquired their infection possibly from one or other of these two sources.

The other 35 (21.2 %) patients acquired their infection from unknown sources, since they either drink pasteurized milk, do not drink milk, or drink powder milk, or had homemade cheese.

The mean duration of symptoms before diagnosis of the brucellosis in 140 (84.8 %) of the cases was 35 days (range, 10–60 days). The duration for the other 25 (15.2 %) was more than 60 days. Thirty-five percent of the previously treated patients had symptoms suggestive of a relapse within a year after concluding treatment. These patients had symptoms consistent with brucellosis, with an increase in antibody titers and no other infections (Table 2). They complied with the treatment according to the information provided in the questionnaire.

Positive results of the Rose Bengal agglutination test with titers ranging between 1/1 60 and 1/1280 units were recorded in 140 (84.9 %) of the patients. The 25 clinical cases that were seronegative were excluded from the group of certainly infected patients.

Brucella was isolated in blood cultures from 20 of the 165 (12 %) specimens who did not receive antibiotic treatment at the time the specimen was collected. Isolates were identified by standard methods as B. melitensis. These specimens were positive by both PCR and serology. The results are shown in Table 3.

Diagnosis of *Brucella* was established by PCR in 72.7 % of 165 cases and none of the control groups was positive. All of the PCR positive cases were seropositive. Forty-eight of 58 (82.8 %) of the relapse cases two months after completing the treatment with an increase in the previous serological titers were positive by PCR. The PCR test proved to be useful in the early detection of relapses as was reported by another study (Morata et al 1999). The diagnosis of these relapses is difficult by conventional methods.

Relapses after concluding a correct treatment are known to occur in brucellosis (Ariza et al 1995, Young 1995). This problem is related to the ability of *Brucella* spp to evade some of the basic mechanisms of the host's immune system (Morata et al 1999). The relapse rate of 32.5 % of the patients after a combined treatment could be explained in part by lack of compliance in some patients to the full course of treatment.

This is common in such rural communities where Bedouin show up for treatment several weeks after having symptoms and discontinue treatment once the fever subsides. It might also be explained by the fact that *Brucella* organisms localize in the reticuloendothelial system after surviving the intracellular control mechanisms by phagocytes and polymorphonuclear leukocytes (Solera et al 1997, Young 1995). A combination regimen of two or three drugs is more effective than a single-drug regimen. The relapse rate may be as high as 50 % in cases of single-drug regimens (Solera et al 1997).

Longer courses of therapy are required to cure relapses and focal forms of the disease such as meningitis and osteomyelitis.

The positive PCR results obtained in the 82.8 % of the relapse cases in this study is especially important indicating that the assay could be a useful tool to confirm a relapse in cases of a treated brucellosis (Nimri 2003a). However, due to the extremely high sensitivity of the technique, the ability to amplify the DNA of non-viable, non-culturable *Brucella* DNA from dead or phagocytized cells, especially the remains of DNA present in the circulating mononuclear cells in certain patients who have concluded successful treatment (Morata et al 1999), should be considered when interpreting the results.

The criteria to indicate a cure of brucellosis is not definite, since negative blood cultures do not exclude the presence of the disease. Another study reported recovering the bacteria in 53.4-95 % of the blood cultures of patients but successful isolation of the organism decreased over time (Yagupsky 1999). Other specimens from bone marrow and liver tissue or lymph nodes may improve recovery rates (Young 1995), but these specimens are invasive and their use not always feasible. Antibodies may remain elevated for a long time after conclusion of the treatment (Ariza et al 1995), which was noticed in some of the chronic cases in this study that were negative by PCR.

The results of this study are in agreement with several previous studies (Isabel et al 1997, Morata et al 1999, Morata et al 2001, Romero et al 1995) that reported on the usefulness of PCR and recommended its use for the diagnosis of initial episodes, focal complications, post treatment follow-up and relapses.

The peripheral-blood-based PCR assay described is highly sensitive and specific, and it avoids the risks to laboratory personnel associated with handling the micro-organism.

2.3. Diarrhoeal diseases

Diarrhoeal disease is a frequent illness in developing countries, and contributes to the death of 4.6–6 million children annually in Asia, Africa and America (Torres et al 2001). Morbidity is also important in poor countries. Despite the improvement in health services in the Jordan Badia, diarrhoeal diseases still represent an important cause of infectious morbidity in children as well as in adults. Diarrhoea is the third most common diagnosis in outpatients seen at the health centres in the area, exceeded only by respiratory and urinary tract infections. Causes of diarrhoea in areas of endemicity involve a wide variety of bacteria, viruses and protozoa. Poor quality food hygiene, water and sanitation are common in communities with high levels of diarrhoeal disease. Underlying conditions, such as malnutrition, which increase the risk of contracting diarrhoea, are also common in this area. These factors combine to facilitate the spread of enteropathogens, and epidemics are common in such settings (Gasco´n et al 2000). The present study investigated the occurrence of enteric pathogens in patients with persistent or acute diarrhoea during a period of two years in order to elucidate the aetiologies of diarrhoeal diseases and risk factors in the Jordan Badia population.

Apart from well-described enteropathogens, such as *Shigella* and *Salmonella* spp, particular attention was paid to identifying other bacteria and intestinal parasites not previously detected, in order to characterize local strains of these pathogens and assess their prevalence and pathogenicity in this population.

Stool specimens were collected from 180 patients who presented with acute or persistent diarrhoea and other symptoms, and from 100 non diarrhoeal controls. All samples were examined for parasites and bacterial pathogens by culture and PCR. Bacterial isolates were tested for their susceptibility to common antimicrobial agents. Pathogens and potential enteropathogens were identified from 77.8 % of the patients, with more than one pathogen being recovered from 37.2 % patients. Potentially pathogenic parasites were observed in 50 % of patients; those that were associated significantly with diarrhoea were *Giardia lamblia, Blastocystis hominis, Cryptosporidium* spp, *Entamoeba histolytica* and *Cyclospora cayetanensis* (Nimri 2003b).

Parasites such as *Ascaris* and *Hymenolepis nana* might be responsible for some of the symptoms, but were not considered to be the causative agent of diarrhoea. Uncommon parasites reported for the first time were *Cyclospora cayetanensis, Isospora belli* and *Dientamoeba fragilis* (Table 4). The presence of non-pathogenic parasites (eg *Entamoeba coli, Endolimax nana*) in the non-diarrhoeal control subjects indicates a common source of infection in this population.

There was a relatively low yield of enteropathogenic bacteria from the patients with diarrhoea, although *E. coli* O157:H7, *E. histolytica* and *S. dysenteriae* were associated with severe symptoms and visibly bloody specimens. Pathogenic bacteria were isolated from 40 % of patients of which 62.5 % were resistant to at least one antibiotic, and 30.6 % of these were multiresistant. The most common enteropathogenic bacteria found were *Shigella* spp, *Campylobacter jejuni*, diarrhoeagenic *Escherichia coli* strains, *Yersinia enterocolitica*, and *Salmonella* spp.

This study reports the isolation of several uncommon potential bacterial pathogens for the first time in Jordan. They included *Citrobacter koseri, Chryseobacterium meningosepticum, Eikenella corrodens, Enterobacte cloacae, Hafnia alveli, Morganella morganii*, and *Serratia* spp. These organisms were isolated either individually or in association with other microorganisms from patients, but not from the control subjects. They were considered to constitute a possible aetiology of diarrhoea when they were the predominant organism recovered.

The aetiological agents could not be determined for 40 (22.2 %) of the 180 patients investigated. The causative agents might be rotavirus, or non-infectious causes, such as chemical exposure, oral antibiotics, stress, food allergies or dietary indiscretion. The problem of pinpointing an associated organism is difficult, even in outbreaks of gastrointestinal illness (Payment et al 2002).

Persistent diarrhoea was reported in 21 % positive cases, they had diarrhoea over 14 days or longer and it was associated with infections by *E. histolytica*, Enteroaggregate *E. coli* strains and polymicrobial infections.

Thirty-five percent of the positive cases had more than one and up to three associated pathogens. These cases involved a member of the Enterobacteriaceae plus another pathogen (20 cases), rotavirus plus another pathogen (10 cases) such as *Giardia, B. hominis, H. nana* or *EAST; E. histolytica* plus *Cryptosporidium*; LT ETEC plus *Giardia*. These

Table 4. Pathogenic and potential pathogenic microorganisms found in 180 patients with gastroenteritis

Organism	No. of cases (%)
Blastocystis hominis	54 (38.6)
Giardia lamblia	36 (25.7)
Entamoeba histolytica	18 (12.6)
Cryptosporidium spp	15 (10.7)
Cyclospora cayetanensis	13 (9.3)
Diarrhoeagenic *E. coli*	20 (14.3)
Shigella spp	15 (10.7)
Campylobacter jejuni	10 (7.1)
Yersinnia enterocolitica	8 (5.7)
Salmonella spp	7 (5.0)
Uncommon bacterial spp	12 (8.6)
More than one pathogen	67 (47.9)

infections were strongly associated with prolonged disease and more severe clinical presentation, eg diffuse abdominal pain, dysentery or persistent diarrhoea, anorexia that eventually caused a reduction in nutrient intakes (odds ratio = 5.9; 95 % confidence interval = 1.3–37.3), and with a reduced functional ability at presentation (P = 0.02).

Polymicrobial infections are also common in such communities and individuals harbouring such infections may suffer exacerbated morbidity, making children even more vulnerable. Thus, these infections pose a serious threat to the health and development of children in low-income countries. Use of an antibiotic is indicated only for more severe cases or in the presence of fever, dysentery or severe dehydration (Duchini et al 1999).

There is increasing evidence that polymicrobial infections in which microorganisms present specific pathologies may act in a synergistic or inhibitory fashion, impacting on either tissue, host cell destruction or the maintenance of health (Bowler et al 2001). Among these, bacteria–bacteria, virus–virus, parasite–parasite or virus–bacteria interactions have been well documented.

Risk factors that correlated significantly with contracting diarrhoea were socioeconomic status, level of education, use of unchlorinated well or tank water, and a low level of personal hygiene.

Malnourished children (primarily undernourished), which accounted for 25 % of the cases, presented more risk of contracting persistent diarrhoea caused by multiple pathogens.

Co-infections were observed in eight (8 %) of the control subjects involving mainly the above mentioned parasites and bacteria or parasites and rotavirus, but none of these was associated with diarrhoea or any other symptoms.

From the results obtained, it is clear that many enteropathogenic bacteria and parasites are endemic in the Badia region of Jordan. These pathogens are all water- or food-borne, and a multivariate analysis of the demographic data revealed that factors such as the low socioeconomic status, low level of education, use of unchlorinated well or tank water, and a low level of personal hygiene correlated significantly with diarrhoea, and might have led to the transmission of these pathogens (Nimri et al 2004a). These factors reflect the living conditions, lifestyle and environmental conditions of the population. The burden of gastrointestinal illness is not limited to those living in poverty. Even in developing countries, the level of gastrointestinal disease associated with the faecal–oral route of transmission could be decreased significantly by implementing relatively simple strategies, such as better wastewater treatment and hygiene education.

The highest infection rate (42 %) was recorded in children less than five years old, which is not surprising since the rate of gastrointestinal illness is always highest in children between one and four years. Basic health strategies could alleviate a great deal of unnecessary suffering and loss of productivity, reduce the number of lives lost to these diseases, and result in significant savings in health care costs (Payment et al 2002).

The high prevalence of gastrointestinal disease is a strong indication that gastrointestinal pathogens circulate through this population with ease and that the environment, the food, and water are vehicles of transmission for these microorganisms. The polymicrobial infections registered in many children indicate a common source: poorly chlorinated water or tank water among them; risk factors that were significantly associated with the infections.

2.4. Urinary tract infections

Upper and lower urinary tract infections (UTIs) are among the most common bacterial infections and frequently recurring problems encountered by clinicians in community practice (Abdul-Ghani et al 2002, Millette-Petit 1988).

The approach to these infections remains a difficult and sometimes controversial issue. One area in particular is accurate diagnosis. The objectives of this study were to determine the incidence of community-acquired UTI, and the clinical characteristics, risk factors, causative organisms, and the antimicrobial susceptibilities of the uropathogens to commonly used antibiotics.

Urine cultures and urine analysis were performed on 250 urine specimens collected from patients having symptoms consistent with UTIs. Symptoms reported by patients included dysuria (burning pain on passing urine), urgency. Some patients had fever and low back pain. Quantitative bacteriological cultures were performed according to standard procedures. Identification of isolates was based on gram reaction, morphology and biochemical reactions. Antibiotic susceptibility of the isolates was assayed with the disk diffusion method.

Of the 250 urine specimens cultured, 168 (67.24 %) were positive for counts of >105 organisms per ml. All infections were due to a single species. Mean age of the patients was 43 (rang, 2 - 84 years). *Escherichia coli* was the predominant uropathogen (46.4 %) isolated in acute uncomplicated infections. It was followed by *Candida* spp (14.9 %) *Klebsiella, Proteus* spp and *Pseudomonas* spp. The only gram-positive species isolated were *Staphylococcus saprophyticus* (4.8 %) and *Enterococcus faecalis* (2.4 %). *Staphylococcus saprophyticus* accounted for (7.8 %) of the female cases and was associated with uncomplicated infections. Results are presented in Table 5.

The lowest incidence (10 %) of UTIs was seen among the 13-20-year-old age group, and 12 years (20 %). Frequent infections of the lower urinary tract accounted for 40.5 % of the cases. The most resistant pathogens were found to be E. coli and Proteus species. Resistance rates to commonly used antibiotics were consistent with those from major recent studies reported in the literature.

The aetiological role of *Candida* species in the pathogenesis of UTIs can be hypothesized if the CFU number of yeasts is higher than 10^4/ml and bacteria are present in low numbers or are absent. The patients who had candiduria in the current study were children with urogenital abnormalities or elderly patients with diabetes. The major predisposing factor associated with candiduria in these patients (64 %) was previous antibiotic therapy.

Men and women with frequent UTIs were 30 %, and 70 % respectively. This is not surprising since in community-acquired UTIs, women are significantly more likely to experience these infections than men (Foxman 2002, Stamm 2001). The significant lower infection rate observed in males might be partially attributed to circumcision. This is a

Table 5. Pathogens that were isolated from 168 patients having symptoms of UTIs

Organism isolated	Number (%)
Escherichia coli	78 (46.4)
Candida spp	25 (14.9)
Klebsiella spp	15 (8.9)
Proteus spp	12 (7.0)
Pseudomonas spp	10 (6.0)
Staphylococcus saprophyticus	8 (4.8)
Enterococcus faecalis	4 (2.4)
Enterobacter spp	4 (2.4)
Serratia marcescens	3 (1.8).
Non-typhoidal *Salmonella*	3 (1.8)
Stenotrophomonas maltophilia	2 (1.2)
Citrobacter spp	2 (1.2)
Plesiomonas shigelloides	2 (1.2)

cultural and religious practice in all male infants in Jordan. The question of circumcision is an area of long-term interest in the study of UTIs. Data published suggest that uncircumcised males have a higher incidence of urinary tract infection, however, this continues to be a point of controversy (Friedman 1998). Significant risk factors for developing a community-acquired UTI as determined by multivariate analyses were frequent antibiotic exposure, female gender, age and pregnancy.

In conclusion, the incidence of urinary tract infections is high and influenced by the age and sex of the patient and previous use of antibiotics. Overall, susceptibility testing demonstrated decreased usefulness of common antibiotics and demonstrated a need for a re-evaluating of common antibiotics used in the therapy for UTIs (Nimri 2007).

3. Discussion

The prevalence of infectious diseases in the Badia region is due to several factors that might affect the epidemiology of these diseases. These factors include the socioeconomic status, where some are still living in tents or very small crowded houses, and the lack of health education, which influences personal hygiene, and certain practices such as walking barefooted and drinking unpasteurized milk (Nimri et al 2004a).

Brucellosis is common in such rural communities due to practices such as drinking unpasturized milk. The patients show up for treatment several weeks after having symptoms and discontinue treatment once the fever subsides. This leads to chronic infections and increases the morbidity in those patients. The health centres in the area offer free treatment for the patients who visit the centres but cannot follow up on these patients once they stop coming to the centres.

Diarrhoeal illnesses still remain an important cause of infectious morbidity in children and are exceeded only by respiratory tract infections (Torres et al 2001). Underlying conditions, such as malnutrition etc, that modify the risk of contracting diarrhoea are common in this area. The results indicate the prevalence of polymicrobial infections among children in rural areas and emphasize the need to develop exploratory approaches to examine polymicrobial interactions and think beyond the one organism-one-disease concept.

The knowledge of the etiological agents of diarrhoea in children and the associated risk factors should stimulate future research involving inter-microbial interactions. This knowledge will also assist in defining the environmental/physiological conditions that contribute to the microbial synergy or inhibition within the same host and lead to an increase of susceptibility. Such studies will provide useful information to investigators in the development of more effective diagnosis, prevention and treatment strategies.

In Jordan, as part of a general improvement in the quality of life, local programmes of diarrhoeal disease control were instituted by the Ministry of Health, including promotion of breast-feeding, oral re-hydration therapy and specific health education. However, the infection rates are expected to decrease in the study area as a result of a general improvement in the quality of life. In addition, local programmes for the control of the diarrhoeal disease were instituted; including promotion of breast-feeding, oral re-hydration therapy and specific health education. The high prevalence of gastrointestinal disease is a strong indication that gastrointestinal pathogens circulate through this population with ease and that the environment, the food and water are vehicles of transmission for these micro-organisms.

Factors that might affect the epidemiology of diseases in that area include the socioeconomic status, where some are still living in tents or very small crowded houses, and the lack of health education, which influence personal hygiene, and certain practices, such as walking barefooted and drinking unpasteurized milk.

Results indicate that therapeutic options for community-acquired bacteraemia have not yet become seriously limited by the prevalence of acquired antibiotic resistance resulting from the uncontrolled use of antibiotics.

The study reports on the relatively different predominant aetiological agents of bacteraemia, which are most probably due to the lifestyle and health practices in that area. It demonstrates the importance of characterizing patients with such strains to permit recognition of groups at increased risk for bacteraemia, and characterizing the antibiotic susceptibility distribution of isolates in order to develop appropriate policies for the treatment and prevention of these diseases in the community. Furthermore, results confirm that appropriate empirical antimicrobial treatment increases the survival of the bacteraemic patients since no mortality was reported in these cases.

References

Abdul-Ghani MA, Glicberg F and Sahagian H (2002). High incidence of resistant pathogens in community-acquired bacteriuria from patients in the Jerusalem area with lower urinary tract infections. Harefuah 141:1032-1035, 1090-1091.

Abu Shaqra QM (2000). Epidemiological aspects of brucellosis in Jordan. Eur J Epidemiol 16:581-584.

Anthony RM, Brown TJ and French GL (2000). Rapid diagnosis of bacteremia by universal amplification of 23S ribosomal DNA followed by hybridization to an oligonucleotide array. J Clin Microbiol 38:781-788.

Ariza J, Corrediora J, Pallare's R, Fernandez-Viladrich P, Rufi G, Pujol M and Gudiol F (1995). Characteristics of and risk factors for a relapse of brucellosis in humans. Clin Infect Dis 20:1241-1249.

Bowler PG, Duerden BJ and Armstrong DG (2001). Wound microbiology and associated approaches to wound management. Clin. Microbiol. Rev 14:244-269.

Cisterna R, Cabezas V, Gomez E, Busto C, Atutxa I and Ezpeleta C (2001). Community-acquired bacteremia. Rev Esp Quimioter 14:369-382.

Dajani YF, Masoud AA and Barakat HF (1989). Epidemiology and diagnosis of human brucellosis in Jordan 1989. J Trop Med Hyg 92:209-214.

Duchini A and Rodgers VD (1999). Diarrhea in the international traveler. Curr Treat Opt Gastroenterol 2:251-257.

Finkelstein R, Sobel JD, Nagler A and Merzbach D (1984). Staphylococcus aureus bacteremia and endocarditis: comparison of nosocomial and community-acquired infections. J Med 15:193-211.

Foxman B (2002). Epidemiology of urinary tract infections: incidence, morbidity, and economic costs. Am J Med 113(1A):5S_13S.

Friedman AL (1998). Urinary tract infection. Curr Opin Pediatr 10:197-200.

Gasco´n J, Vargas M and Schellenberg D (2000). Diarrhoea in children under 5 years of age from Ifakara, Tanzania: a case-control study. J Clin Microbiol 38:4459-4462.

Isabel MA, Queipo-Ortuño MI, Morata P, Oco N, Manchado PP and Colmenero JD (1997). Rapid diagnosis of human brucellosis by peripheral-blood PCR assay. J Clin Microbiol 35:2927-2930.

Levy A, Fraser D, Vardi H and Dagan R (1998). Hospitalizations for infectious diseases in Jewish and Bedouin children in southern Israel. Eur J Epidemiol 14:179-186.

Millette-Petit JM (1988). Urinary tract infections in older adults. Nurse Pract 13(29):21-24.

Morata P, Queipo-Ortuño MI, Reguera JM, García-Ordoñez MA, Pichardo C and Colmenero JD, (1999). Post treatment follow-up of brucellosis by PCR assay J Clin Microbiol 37:4163-4166.

Morata P, Queipo-Ortuño MI, Reguera JM, Miralles F, Lopez-Gonzalez JJ and Colmenero JD (2001). Diagnostic yield of a PCR assay in focal complications of brucellosis. J Clin Microbiol 39:3743-3746.

Nimri LF (2003a). Cyclospora cayetanensis and other intestinal parasites associated with diarrhoea in a rural area of Jordan. Int Microbiol 6:131-135.

Nimri LF (2003b). Diagnosis of recent and relapsed cases of human brucellosis by PCR assay. BMC Infect Dis 3:5.

Nimri LF (2007). Candida spp. and other predominant uropathogens and their resistance to common antibiotics. In progress.

Nimri LF and Batchoun R (2004a). Community-acquired bacteraemia in a rural area: predominant bacterial species and antibiotic resistance. J Med Microbiol 53:1-5.

Nimri LF, Elnasser Z and Batchoun R (2004b). Polymicrobial infections in children with diarrhoea in a rural area of Jordan. FEMS Immunol Med Microbiol 42:255-259.

Nimri LF and Meqdam M (2004c). Enteropathogens associated with cases of gastroenteritis in a rural population in Jordan. Clin Microbiol Infect 10:634-639.

Payment P and Riley MS (2002). Resolving the global burden of gastrointestinal illness: a call for action. American Academy of Microbiology, Washington, DC.

Petti CA and Fowler VG Jr (2002). Staphylococcus aureus bacteremia and endocarditis. Infect Dis Clin North Am 16:413-435.

Raimundo O, Heussler H, Bruhn JB, Suntrarachun S, Kelly N, Deighton MA and Garland SM (2002). Molecular epidemiology of coagulase-negative staphylococcal bacteremia in a newborn intensive care unit. J Hosp Infect 51:33-42.

Reimer LG, Wilson ML and Weinstein MP (1997). Update on detection of bacteremia and fungemia. Clin Microbiol Rev 10:444-465.

Rezende NA, Blumberg HM, Metzger BS, Larsen NM, Ray SM and McGowan JE Jr (2002). Risk factors for methicillin-resistance among patients with Staphylococcus aureus bacteremia at the time of hospital admission. Am J Med Sci 323:117-123.

Romero C, Gamazo C, Pardo M and Lo'pez-gon I (1995). Specific detection of Brucella DNA by PCR. J Clin Microbiol 33:615-617.

Solera J, Martinez-Alfaro E and Espinosa A (1997). Recognition and optimum treatment of brucellosis Drugs 53:245-256.

Stamm WE (2001). An epidemic of urinary tract infections? N Engl J Med 345:1055-1057.

Torres ME, Pi'rez MC and Schelotto F (2001). Etiology of children's diarrhoea in Montevideo, Uruguay: associated pathogen and unusual isolates. J Clin Microbiol 39:2134-2139.

Wayne P (1999). Performance Standards for Antimicrobial Disk and Dilution Susceptibility Tests for Bacteria Isolated from Animals. Approved standard M31-A, National Committee for Clinical Laboratory Standards, NCCLS.

Weinstein MP, Towns ML, Quartey SM, Mirrett S, Reimer LG, Parmigiani G and Reller LB (1997). The clinical significance of positive blood cultures in the 1990s: a prospective comprehensive evaluation of the microbiology, epidemiology, and outcome of bacteremia and fungemia in adults. Clin Infect Dis 24:584-602.

WHO (1997). World Health Organization Fact sheet N173. World Health Organization, Geneva, Switzerland.

Yagupsky P (1999). Detection of Brucellae in blood cultures. J Clin Microbiol 37:3437-3442.

Young EH (1995). An overview of human brucellosis. Clin Infect Dis 21:283-290.

Zerva L, Bourantas K, Mitka S, Kansouzidou A and Legakis NJ (2001). Serum is the preferred clinical specimen for diagnosis of human brucellosis by PCR. J Clin Microbiol 39:1661-1663.

CHAPTER 20

Delivering primary healthcare in the north-east Jordan Badia

Neil Spicer
Health Policy Unit, London School of Hygiene and Tropical Medicine Keppel Street London WC1E 7HT.
E-mail: neil.spicer@lshtm.ac.uk

This chapter is based on a sociological study that involved interviewing semi-nomadic and sedentarized Bedu communities of the north-east Jordan Badia. The chapter explores the use of healthcare services by these groups focusing on the following key issues: the most commonly used Arab medicines are reviewed and contemporary patterns of the use of healthcare services are described; changes in communities' health and illness beliefs are explored, with a focus on parents' accounts of their children's health, including the ways economic activities are linked to perceptions of health and illness and the extent to which families perceive they can exert control over their children's health; key geographical, organizational, economic and social factors mediating access to healthcare services experienced by settled and semi-nomadic communities are identified; and the particular problems of healthcare accessibility among households practising pastoral migration are discussed.

Keywords: Healthcare, pastoral migration, sedentarization, Arab medicines, health beliefs, accessibility

1. Introduction

Delivering accessible, appropriate healthcare services to rural groups in the context of geographically and socially inequitable development programmes in low income countries continues to confound progress in achieving the World Health Organization's aim of achieving 'Health for All in the Twenty First Century'. Until recently Jordan's economic and social development centred almost entirely on urban areas, whilst government health and other services were relatively undeveloped in rural parts of the country. In the 1980s Jordan's development policy placed considerably greater emphasis on rural social development programmes including the expansion of healthcare services, schools, surfaced roads, piped water and electricity to domestic residences in most rural areas. Indeed, the Jordanian Ministry of Health asserted that the national health policy is based on 'the principle that all citizens have the right to health services' and that 'health services are available, accessible and acceptable to all communities, and ... seeks to ensure the equitable distribution of these services'(DSME 1992). Honey and Kharmeh (1989) noted that the majority of rural settlements in Jordan populated by one thousand or more residents, and many with far fewer, were provided with government clinics by the late 1980s. Within the north-east Jordan Badia the first rural clinic was established in 1981; the majority were subsequently introduced in the early 1990s (Spicer 1999). By the late 1990s the region was provided with a range of comprehensive health centres in some of the larger towns, primary clinics in many villages and a number of health outposts in smaller villages. Government vaccination programmes for DPT (diphtheria, pertussis and tetanus), polio, hepatitis B, tetanus (for mothers), BCG (tuberculosis) and measles are administered through government clinics and schools. Private and charitable healthcare is concentrated in the governorate administrative centre, Mafraq City, where there is a large government hospital together with a number of pharmacists and Arab medicine apothecaries.

Nomadic livestock pastoralism, a method by which limited resources in arid regions can be exploited, has historically dominated the economy of the north-east Jordan Badia. The region has undergone significant socioeconomic changes. Estimates suggest that 90 % of the population of the region was fully settled in permanent houses by the late 1990s, mostly within around thirty villages concentrated near the Syrian border (Findlay et al 1998). In 2003 the total population of the region was around 21 000, based on projections (Maani et al 1998). Whilst livestock production continues to be an important part of the region's economy, only a small minority of families practise fully nomadic pastoralism. Instead, most families engaged in pastoralism adopt 'semi-nomadic' patterns of production involving members of extended families migrating for part of the year whilst other family members tend to settle permanently and are employed in the non-agricultural sector. Indeed, the study suggests that 40 % of semi-nomadic families contained members employed by government agencies or military forces. In contrast 78 % of fully settled families contained members with these occupations (Spicer 2005).

The transitional economy of the study area enables semi-nomadic and settled households to be compared. This chapter explores a number of issues relating to patterns of use of healthcare services among semi-nomadic and settled households in the region:

- The most commonly used Arab medicines are reviewed and contemporary patterns of the use of healthcare services are described;
- Changes in communities' health and illness beliefs are explored, with a focus on parents' accounts of their children's health. Two key changes are discussed: the ways economic activities are linked to perceptions of health and illness; and the extent to which families perceive they can exert control over their children's health;
- Key geographical, organizational, economic and social factors mediating access to healthcare services experienced by settled and semi-nomadic communities are identified;
- The particular problems of healthcare accessibility among households practising pastoral migration are discussed.

This chapter is based on a study that involved interviewing semi-nomadic and sedentarized[1] Bedu communities living within the north-east Jordan Badia. Results of this study have previously been reported in (Spicer 1998, 1999, 2000, 2005). A multiple method approach was adopted: this comprised structured interviews conducted with 175 rural families (125 sedentarized families randomly selected from eight villages and 50 semi-nomadic families) and in-depth qualitative interviews with six families, each over a number of sessions (three sedentarized families and three semi-nomadic families). The interviews focused on patterns of use of healthcare facilities and Arabic medicines, experiences of problems accessing services, and changing beliefs about health and illness. A further eight in-depth qualitative interviews were carried out with clinicians working in government clinics in the region. These interviews complemented the family interviews in exploring issues of accessibility and patterns of healthcare utilization. A limitation of the study is that male heads of households were interviewed. Mothers are principally responsible for childcare in the north-east Badia, although fathers and other family members participate in decision-making about health and sometimes accompany children to use health services. Hence, whilst interviews with adult males provide a useful set of perspectives on health-related decision-making, it is acknowledged that these perspectives are likely to differ from those of women. It is also not possible to explore gender differences in experiences of accessing health services.

---

[1] The term sedentarization refers to the process whereby nomadic groups settle permanently in locations such as villages and towns.

It is recognized in the literature that a number of geographical, economic, social and cultural factors mediate patterns of healthcare utilization in non-industrialized countries. Geographical factors include travel times, and the costs and availability of transport to health services. The costs of using health services and buying medicines may also influence patterns of use (Phillips 1990). Improving healthcare accessibility to nomadic groups in low income countries can be particularly problematic due to their mobility, coupled with limited rural health services in more geographically marginalized, sparsely inhabited areas (Hampshire 2002, Spicer 1999, Swift et al 1990). Indeed, health services in many low income countries tend to be neither accessible nor appropriate in addressing the health problems of the majority of populations, and deficits in rural health service provision can be particularly acute. Social and cultural factors are also important influences on healthcare use, including how service users relate to providers, as well as the knowledge communities have about health and healthcare. In many societies the unequal distribution of power and resources within households exaggerates difficulties women may have in accessing healthcare (Hampshire 2002, McCray 2004, Ojanuga et al 1992, Puentas-Markides 1992). Moreover, it is increasingly recognized that mothers' levels of education, duration of marriage and maternal age may influence their use of children's health services (Abbas et al 1986, Akhtar et al 1994, Hobcraft 1993).

Understandings and explanations of health, illness and disease among non-medically trained people are often referred to as 'lay' health and illness beliefs. Such beliefs influence the ways communities use biomedical services and 'traditional medicines', as well as reflecting wider sociocultural contexts (Atkinson et al 1995, Nettleton 2006, Puentas-Markides 1992). Studies conducted in high-income countries, for example, reveal variations between different socioeconomic groups' accounts of health and illness. Poorer groups' accounts may closely reflect material constraints and economic and social pressures. Healthcare may only be sought by poorer people if individuals' experience of illness disrupts their usual economic and social obligations; this may result in relatively low levels of use of healthcare services. Health may also be equated by poorer people simply with an absence of disease, and with diseases being perceived as caused by factors outside the community's control. In contrast, wealthier groups may conceive health more broadly in terms of an individual's feeling of well being, and that health can be maintained through their own actions (for example: Blaxter 1990, for example: Calnan 1987, Curtis et al 1996, Donovan 1986, Nettleton 2006).

2. Arab medicines and the adoption of healthcare services in the north-east Jordan Badia

There is insufficient space in this chapter to describe in detail the range of Arab medicines employed by Bedu communities in the study area historically and currently. Instead, some of the more important Arab medicines are briefly reviewed. Current patterns of healthcare utilization are then summarized, and the implications of Bedu cultural values for these changes discussed.

Herbal medicines, used to treat an array of symptoms including colds, coughs, intestinal pains, menstrual pains, diarrhoea, fatigue, insomnia, swollen glands, headaches and for cleaning cuts, were the most commonly used Arab medicines in the past. Accidents and injuries, particularly broken bones, cuts and bites, were common hazards; treatments tended to be practical and effective, including bone setting that was a particularly successful treatment amongst the Bedu (Kay 1978), and using salt, oil or herbs to clean wounds. Various techniques employed in the event of snake, spider or scorpion bites show that the Bedu had some awareness of the concept of the circulatory system and blood poisoning. Reading and quoting from the Quran was, and continues to be, a part of everyday life. It was also used in the event of illness; an individual who was literate would be asked to read the Quran in an attempt to cure some illnesses. Another common treatment involved branding the skin with a hot iron, although this tended only to be used for particularly serious illnesses that could not be dealt with in other ways, for example, unrecognized or serious pains, and also for gangrene, or if other treatments had failed. Reputable Arab practitioners were commonly sought in the past to administer these and other treatments including herbalist doctors, wizards, witches and shamans. Shamans would often be approached to treat illnesses believed to be linked to spiritual etiologies such as mental/neurological illnesses, and indeed tuberculosis and serious internal pain; Sajdi (2005), provides a detailed account of shamanism in southern Jordan. Interviewees' accounts suggested that the use of these and other Arab medicines varied considerably amongst users and often depended on the knowledges and experiences of a family and other members of a tribal network. Often there was no definitive set of conditions that corresponded to a particular treatment, and sometimes a number of treatments would be used concurrently or consecutively for a particular illness episode.

Biomedical services in the form of basic government healthcare clinics were introduced in the north-east Jordan Badia in the 1980s and 1990s and are now widely accepted by the communities of the region for the majority of minor and serious adults' and children's illnesses. Preventive medicine including children's vaccinations and ante- and post-natal maternal and child health services are also widely accepted and used. Overall, 98 % of settled families participating in the study indicated that they had used some form of modern medicine, and the vast majority (96 %), use government services if a child has a minor illness. A very low proportion of the respondents, 2 % (only three families from a total of 126 settled families), reported that they would do nothing in the event of minor illness, or that they would simply use Arab treatments administered by family members rather than using formal healthcare, unless the illness became more serious. The majority of respondents (72 %) indicated that they would tend to travel directly to an urban hospital if a serious illness was suspected. Comparatively few settled families (7 %) stated that they would first use private health services for serious illness, and even fewer would use them for minor episodes. Levels of children's vaccinations are also high; the vast majority, 98 % of sedentarized and 92 % of semi-nomadic families, indicated that their children had received a number of vaccinations. The majority of rural families (81 %) also stated that they continued to give herbal medicines to their children, albeit on a supplementary basis to formal healthcare, but would rarely if ever use other Arab medicines. As with settled families, semi-nomadic households widely use government health clinics, although they face a number of key accessibility issues that are discussed below. Arab medicines continue to be used by this group, although very few participants in the study indicated that Arab medicines play a significant role in curing children's illness, and the numbers stating that they would not do anything in the event of a minor illness were negligible (only two out of fifty mobile families).

Bedu cultural values were described by interviewees as precipitating the rapid adoption of biomedical services among rural communities living in the region. Two Bedu cultural values appear to be particularly significant: *adaptability* and *reputation*. Adaptability continues to be as important in the north-east Jordan Badia and in other parts of the Middle East as it was in the past; the Bedu have pragmatically adapted their production methods to changing social, economic and political circumstances that are seen as advantageous to their livelihoods including adopting new technologies and vehicles, and diversifying into other agricultural and non-agricultural occupations (Chatty 1986, Dutton et al 1998, Lancaster et al 1990). This, interviewees explained, is reflected in the willingness of communities of the north-east Badia rapidly to adopt government health clinics. The Bedu cultural value of reputation mediated patterns of use of Arab medicines in the past. The reputation of a particular Arab medicine or practitioner based on experience of their efficacy was the primary rationale for their use. Hence, the use of Arab practitioners and medicines was not based on unconditional faith in them. Indeed, Arab practitioners and medicines would only remain reputable if they were effective in treating particular illnesses. Closely associated with the wide-scale adoption of government services is the perception that biomedicine is now considered more efficacious than Arab medicines in treating most illnesses; reflecting positive experiences of their efficacy, biomedical health services have gained positive reputations. The few Arab medicines that continue to be popular, in particular herbal medicines that are often used before seeking formal healthcare, or concurrently with formal healthcare, are used on the basis that they are considered effective in treating a number of specific illnesses (Spicer 2000, 2005). This contradicts prevailing assumptions among healthcare planners that rural communities in non-industrialized countries are resistant to adopting and utilizing biomedical services whilst continuing to rely on traditional medicines (Phillips 1990, Zaidi 1994).

3. Changing health beliefs in the north-east Jordan Badia

Historically, beliefs about health and illness were closely linked to systems of nomadic pastoralist production among Bedu communities of the north-east Jordan Badia; there were parallels with poorer socioeconomic groups in high-income countries described in a number of earlier studies (Blaxter 1990, Calnan 1987, Donovan 1986). Interviewees' accounts of health and illness emphasized two key changes; firstly, that changing patterns of economic activity have had an important influence on how health and illness are perceived, and secondly, there have been considerable changes in the extent to which the Bedu are able to exert a degree of control over health and illness. This chapter will now explore these two issues.

Pastoral production was highly time-consuming and physically demanding; pressures were greatest during key activities such as lambing, milking and shearing. Carrying out these activities was critical in terms of maintaining a family's livelihood. This had a major influence on pastoralists' definitions of health and illness in the past; families rarely reflected on issues of health, and would seek treatment only if an illness was serious and disrupted key economic activities. Until recently healthcare services were concentrated in urban areas, exacerbating the problems of seeking healthcare and the potential disruption this would cause. The association between health and illness was therefore dichotomous; health, or being healthy, was defined as an absence of disease. Provided individuals could undertake their work adequately, they were considered healthy. Individuals who were not able to fulfil their obligations were considered ill. Historically, among pastoralist communities of the north-east Badia, children aged five years and above, and indeed those over two years of age, sometimes participated in family economic activities. Hence the health of children tended to be viewed in a similar way to that of adults; illness was acknowledged if a child was not able to fulfil its economic role. A strong sense of family and tribal solidarity reinforced this perception of health and illness. Family groups were reliant on the contribution of all members, and closely cooperated with their wider tribal groups. Collective needs tended to be emphasized over individuals' health and wellbeing as this was the basis of a group's survival (Spicer 2000, 2005).

The transition from nomadism to semi-nomadic/sedentarized economies was associated by interviewees participating in the study with considerably greater emphasis on individual children's health and wellbeing. Importantly, government healthcare services are widely used for both serious and minor children's illness in contrast with the past when children's illnesses were only acknowledged and acted upon if they were not able to work. A total of 98 % of sedentarized and 94 % of semi-nomadic families indicated that they would seek government healthcare services for minor children's illnesses. In part, these patterns can be explained by increasing accessibility of healthcare services discussed below. The wide-scale reduction in children's participation in family pastoral production associated with the adoption of non-agricultural employment and children's enrolment in primary and secondary schools are also key factors influencing such changes. Among families participating in the survey 96 % of sedentarized families and 64 % of semi-nomadic families stated that all of their children of school age were enrolled in schools. Many families explained that this was an important motivation for sedentarizing; the purpose of education was to enable their children to enter into non-agricultural employment, and hence to strengthen the families' economic security in the longer term. As a result, their families tend to acknowledge the importance of maintaining their children's health (Spicer 2000, 2005).

Interviewees' also described most children's and adults' illnesses as being attributed to factors beyond their control reflecting the hardships of living in a harsh climate with few natural resources. Hence, it was perceived that they could do little to prevent illnesses, and although a range of Arab medicines were available, their efficacy in treating illnesses tended to be limited. Historically nomadic Bedu lived in tents in which it was difficult to maintain hygienic conditions. Poisonous snakes, scorpions and spiders were common, together with hazards such as altercations with rival tribes. Water tended to be in short supply for most of the year, and water quality poor. Droughts limited animal grazing resources and resulted in variable food security. In the past Bedu communities also had limited understandings of the body, health and disease. Few specific diseases were recognized, and there tended to be limited understandings of the causes of most diseases, although specific symptoms were sometimes understood to be caused by environmental factors such as extreme heat and cold as well as poor hygiene and contagion. Chest-related conditions were sometimes seen as being caused by a person becoming cold, and intestinal pains were sometimes connected to a person eating unclean or decaying food. Contaminated water was rarely identified as a cause of illness. Measles and smallpox tended to be recognized as highly contagious, although they were frequently conflated. The two diseases were treated with particular circumspection and fear since there were no known treatments and the death of adults and children tended to be inevitable; suspicion that an individual was infected generally resulted in them being isolated or abandoned by a family or wider group. Mental and neurological illnesses were not clearly understood although they were sometimes connected to spiritual causes. For example, djin (spirits) were often perceived to cause such illnesses, although it was recognized that little could be done to prevent them (Spicer 2000, 2005), and sufferers were frequently stigmatized.

Interviewees participating in the study widely described parents as having considerably more control over their health and that of their children today than they did historically. Importantly, preventative healthcare has become widespread in its use since its introduction in the 1990s, particularly children's vaccinations and ante- and post-natal maternal and child health services. Indeed, as suggested above, the vast majority of sedentarized and semi-nomadic families stated that their children had received a number of vaccinations. All interviewees who expressed a view emphasized that vaccinations are a very important part of maintaining their children's health. Very few interviewees indicated that they felt nothing could be done to prevent children's illnesses (4 % of sedentarized and 6 % of semi-nomadic interviewees), and a minority indicated that they did not know of any means of preventing children's illnesses (10 % of sedentarized and 44 % semi-nomadic families). The majority of interviewees (86 % of sedentarized and 50 % of semi-nomadic interviewees) described a number of practices in addition to vaccinations that they frequently employed in order to maintain the health of their children, as well as prevent illness. Common health-related practices include maintaining hygienic conditions in domestic settings, avoiding exposing children to extremes of temperature, using clean drinking and washing water, ensuring food was clean, providing nutritious diets, boiling milk and carefully monitoring children's health. Interviewees felt that government social marketing campaigns through the mass media, especially the radio, have been important sources of health information, as had health information displays within healthcare clinics and seminars run by doctors in village settings. Hence they were better informed about preventative practices in domestic settings and the range of facilities present at government healthcare services. Changing parental roles, particularly among settled families not involved in agriculture, were also connected by interviewees to improved control over children's health. Mothers are the primary child-carers in Bedu society currently, as they were historically, although other female relatives, and indeed fathers and other male relatives, were described as contributing to health-related decision-making, as well as sometimes accompanying children to use health services. Most mothers from semi-nomadic families continue to be, as they were in the past, engaged in pastoralism-related activities, hence they had limited time for childcare, monitoring their children's health or dealing with illnesses. Conversely, sedentarized mothers' roles primarily relate to childcare and domestic activities and hence they tend to have more time to monitor their children's wellbeing and employ health and illness practices expediently if the need arises (Spicer 2000, 2005).

4. Factors mediating the accessibility of rural health services in the north-east Jordan Badia

The sedentarization of formerly nomadic families, together with the extended provision of health clinics in rural Jordan, means that the majority of rural families of the north-east Badia have relatively few geographical barriers to utilizing basic healthcare (Spicer 1998, 2000). Nevertheless there are some local variations in geographical accessibility among settled communities, who also experience a number of other barriers to effectively accessing healthcare services including factors relating to the organization of healthcare delivery, as well as economic and social factors. The chapter will now review these key factors of accessibility.

It is estimated that by 1993, around 78 % of the families from the study area[2] were settled in villages with health clinics, and another 12 % lived in villages without clinics, albeit located within 10 km of formal clinical services. All sedentarized interviewees living in villages provided with clinics indicated that distance and travel time did not represent barriers to them using government clinics. However, most families living in villages without a clinic, or living beyond a reasonable walking distance of a clinic, experience geographical accessibility problems; a key problem is that a relatively small proportion of settled families own private vehicles. Moreover, women are generally not permitted to drive vehicles by their families, even though mothers are most often responsible for accompanying their children to utilize medical services. Thus, those people living in villages without health clinics often find it necessary to rely on the public transport system, which, at the time of the study, consisted of infrequent and unpredictable bus services between the smaller villages and the larger settlements. Community support continues to be important in the north-east Badia, and is an important enabling factor facilitating healthcare utilization, especially amongst settled groups. Those who do not have available private transport frequently request that their neighbours take them to health clinics when necessary.

'Coupling constraints', such as facility opening times/days relative to times when

[2] Estimate based on population figures generated as part of the BRDP baseline survey (1993).

potential users are able to visit, as well as waiting times and queues for consultations, are recognized as key influences on accessibility to health services, and therefore have implications for healthcare utilization. Such time-related 'organizational' factors may create inconvenience and economic cost to users, and render the delivery of services ineffective in terms of the abilities of target populations to use them within their time-budgets (Carlstein et al 1978, Phillips 1990). The larger rural clinics in the north-east Badia are permanently staffed with qualified doctors. Smaller clinics are not permanently staffed; a nurse is available most days of the week, and a doctor is in situ one or two days per week. As a result, 64 % of interviewees stated that the main reason for not using them was that they were open part time. The limitations of the facilities provided at the clinics, and the quality of services compared to other health centres, were also widely seen as disincentives to using them. The study also suggests the opening hours of the smaller health clinics vary considerably, and hence many families are not aware of facility opening times. This adds an additional dimension to the problems of effectively utilizing basic health services; for more serious illnesses, these clinics were frequently bypassed (Spicer 2000).

Whilst at the time of the study the Jordanian government health system provided subsidized medical care for all users, in line with broader structural adjustment programmes since the late 1980s, the state has attempted to recover part of the cost of delivering healthcare by charging outpatients prescription fees for medicines, and inpatients using urban hospitals a fee for medicines and care on a pro rata daily basis. However, a user fee concession system operates in Jordan. A number of types of 'card' are issued, linked to military and civilian government employment, which reduce the cost of treatment/medication, or even exempt payment for some groups. This system is widely perceived by many interviewees participating in the study as highly unfair since the availability of a health concession card, and therefore the cost of medical care, is not commensurate with a family's income. In particular, few families whose main income was derived from agriculture were eligible for a concession card. The study suggests that user fees have the potential significantly to reduce utilization rates in the region; the majority (63 %) of families without health concession cards stated that the cost of treatments had a negative impact on their willingness to approach healthcare clinics in the event of illness.

The study suggests also that the concession card system is inconsistently applied, both in terms of who is eligible and who actually receives a card in practice. The system of qualification for concession cards is erratic, often depending on personal contacts service-users have with bureaucrats administering the system. A number of respondents stated that they expected to be entitled to a concession card, but the complex and confusing bureaucratic system had stopped or delayed them receiving one. Many families, for example, complained that they thought that they should qualify for a 'poverty' card intended for low-income families, but despite having applied, had not received one. These findings accord with DeJong's (1995) study, which suggested that loopholes in the Jordanian system are often exploited; some families obtain concession cards whilst not being eligible, or they retain cards even after the death of an eligible individual. In addition to inconsistencies obtaining concession cards, it was widely reported by interviewees that there is considerable irregularity in the levying of user charges; whether or not doctors charge patients for medications is highly discretionary in practice, regardless of card possession. A number of respondent families felt that fees for medicines are often levied on an arbitrary basis, which, coupled with the high turnover of staff in rural clinics, makes the costs of health service utilization highly unpredictable. Health staff participating in the study acknowledged that this problem exists. One doctor, for example, suggested that many of his colleagues administered free medication, particularly orally administered medicines for simple infections and colds, if put under pressure to do so by service users. This, he argued, was wasteful since it was not always necessary to dispense such medications (Spicer 2000).

Studies have suggested that health professionals' attitudes and behaviour may affect the utilization of health services among different groups, including their attitudes to patients from other ethnic or linguistic groups. Social practices may also be an important influence on healthcare utilization including gender relations within societies and attitudes to reproductive issues: in some contexts these may make it difficult for women to use health services (Hampshire 2002, McCray 2004, Ojanuga et al 1992, Puentas-Markides 1992). The study suggests that whilst there were some tensions between communities and staff working in rural health clinics, the majority of families felt that this would not directly affect their willingness to use health services. Very few rural families indicated that they did not trust specific medical staff, treatments and medicines, or had a sufficiently poor relationship with staff such that it would negatively affect their willingness to use a service. However, whilst ethnicity, tribal affiliation and language did not appear to be problematic issues, socioeconomic divisions appear to impact negatively on perceived quality of service. For example, many families practising agricultural occupations emphasized that the clinical staff, who originate from urban/educated backgrounds, often act derisively to them, and discriminate against them by providing poor quality care. Some families stressed that rural doctors are usually under-qualified, or recent graduates posted to a rural clinic to gain experience. They are therefore seen as inexperienced and hence less able to provide good quality healthcare than their more experienced colleagues who were unwilling to work in rural areas. Moreover, problems of staff retention mean that high staff turnover in many of the basic health clinics of the region made it difficult for service users to develop positive, trusting relationships with service providers.

DeJong's (1995) study in Jordan suggested that *al-wasta* (family or tribal 'connections') may be used to avoid or reduce payments for government medicine if a family has connections with employees of the health service. A number of respondents to this study indicated that *al-wasta* may be used to reduce waiting times for heath services or might result in them receiving better quality care or the prescription of better quality medications. Conversely those families without social connections complained that they might have to wait longer for a consultation, or that the doctor would be less conscientious. Most respondents were opposed to *al-wasta* as a principle, but if these opportunities could be used to their advantage, they acknowledged that they would exploit them (Spicer 2000).

5. Pastoral migration and accessibility to rural health services in the north-east Jordan Badia

Contemporary patterns of semi-nomadic migration involving seasonal residence in villages and vehicle ownership have improved levels of access to health services for part of the year for pastoralist families. During the periods they were located in or around settled parts of the north-east Badia region pastoralist families tended to experience similar problems accessing health services to settled families discussed above. However, 62 % of interviewees from this group indicated distance/travel time presented them with difficulties in using clinics during seasons that they migrated to remote parts of the region. The key issues of accessibility that result from pastoral migration are now explored.

Migration within the region is based on maximizing the commercial profitability of sheep production; natural grazing for animals is sought as much as possible, although artificial supplementary animal feeds are also used. Until 1996 the Jordanian government supported the livestock sector by subsidizing the purchase of artificial animal feeds and vehicles, as well as constructing deep mechanical wells available to pastoralists. Most pastoralists rapidly adopted such means of improving productivity. Many also diversified into livestock marketing and related activities as a means of improving their incomes (Spicer 1998, 1999).

Pastoral migration can be highly dynamic geographically. A pastoralist family may change its location frequently, often travelling considerable distances to locate natural grazing or to collect water (Spicer 1999). Return visits to towns and villages are also common in order to visit family, market animals and collect materials. Access to health services therefore depends on economic activities and may vary considerably throughout the year, as well between years depending on variations of rainfall. Geographical proximity of health services is an important factor influencing patterns of utilization among pastoralist families; many families indicated that when they were located in remote parts of the region travel to use health services was extremely difficult or impossible. Moreover, whilst there are relatively high levels of ownership of vehicles among pastoralist families, especially trucks that are used for transporting animals and water, the majority interviewed suggested that their vehicle is used predominantly for economic activities; it is rarely available for personal activities such as transportation to clinics.

Among families currently practising semi-nomadic pastoralism there continue to be parallels with the ways health and illness were perceived historically. Pastoralist production determines the amount of time a family has available for travelling to and using health services. Especially at critical times in the production cycle such as lambing, shearing and milking, the health of family members would not be considered a high priority. At these times they would often delay or not approach health services for more minor illnesses in order to avoid disrupting pastoralist activities. Many families suggested that they would tend only to use health services if there were other reasons for travelling to settled parts of the region such as to collect animal feed or other supplies. Furthermore, they would tend to travel greater distances to larger clinics where they could guarantee staff would be present and would have sufficient equipment to attend to their needs in order to avoid the potential coupling constraints of smaller clinics. Children under six years of age are particularly vulnerable since they frequently travel with their parents until they reach school age. Many are not vaccinated against many common diseases including measles and polio; 32 % of families who migrated seasonally indicated that they had children who had not been vaccinated for some diseases or had not completed the full course of a vaccination. Conversely, no settled families indicated that they had children who were not vaccinated for some diseases or had not completed the full course of a vaccination. The survey suggests that potential disruption to economic activities appears to be a more significant accessibility problem than other costs of using healthcare, including user fees.

6. Conclusions

The study identifies a range of factors that have promoted the rapid and wide-scale adoption of healthcare services including increased accessibility to services linked with sedentarization. Bedu cultural values, namely adaptability and reputation, also appear to have precipitated the adoption of healthcare services.

Socioeconomic changes have had a profound influence on beliefs about health and illness among Bedu communities of the north-east Badia. Historically, systems of nomadic pastoralist production meant that individuals, including a child, might only be considered ill if they were unable to perform their usual economic roles within a family. Hence, illnesses were rarely acknowledged. Furthermore, harsh living conditions and limited understandings of health and illness meant that the communities of the north-east Badia believed they could do very little to prevent illness. There have been important changes in these respects. Children are now widely enrolled in schools, and hence tend not to be involved in family economic activities. There is considerably greater emphasis on individual children's health and wellbeing. Family members, especially mothers, carefully monitor children's health and act promptly if illnesses are suspected, even relatively minor illnesses. Families also recognize that they are able to exercise greater control over their health and that of their children in a number of ways including children's vaccinations and employing an array of activities within the household to maintain children's health, such as maintaining hygienic conditions and nutrition. Participants in the study also acknowledged that government social marketing programmes provided them with important health information that improved their health-related decision-making. Among sedentarized families who no longer practice livestock production, mothers have more dedicated time for childcare, and if necessary employ health and illness practices expediently.

Nevertheless, the study also shows that there continue to be a number of geographical, organizational, economic and social factors mediating effective accessibility to rural healthcare services in the region. Low levels of private vehicle ownership, coupled with limited public transport services and the fact that mothers, who are most often responsible for accompanying their children to utilize medical services, tend not to be permitted to drive, present considerable local variations in accessibility of government healthcare services among settled families. 'Coupling constraints', such as part-time and erratic opening hours, also constitute significant accessibility problems to basic healthcare among both settled and semi-nomadic groups. Economic factors undermining healthcare accessibility include what is seen by many families as an unfair user-fee concession system; the costs of using health services are not commensurate with a household's ability to pay. Those families whose main income was derived from agriculture, including pastoral production, tend not to be eligible for reduced user fees. Disruption to pastoralist production is also seen as an important problem influencing healthcare seeking, especially during critical times in the annual production cycle when family members are too busy to travel to distant health facilities. Geographical and economic factors of healthcare accessibility are therefore interlinked. There are also a number of social factors influencing healthcare utilization. Many pastoralist families believe they may be discriminated against since they are from poor, rural backgrounds, resulting in reduced quality of care. Social connections (*al-wasta*), are seen as advantageous in terms of improving some groups' access to healthcare; many families who lack social connections believe they experience increased waiting times and reduced quality of care.

The study therefore demonstrates the importance of healthcare planners developing detailed understandings of what are complex and interlinked factors of accessibility and the motivations behind households' willingness and ability to use health services if they are effectively to deliver appropriate healthcare interventions in transitional economies.

References

Abbas A and Walker G (1986). Determinants of the utilization of maternal and child health services in Jordan. International Journal of Epidemiology 15(3):404-407.

Akhtar R and Izhar N (1994). Spatial inequalities and historical evolution in health provision: Indian and Zambian examples. In: Phillips D and Verhasselt V (eds), Health and development. London: Routledge.

Atkinson S and Farias M (1995). Perceptions of risk during pregnancy amongst urban women in northeast Brazil. Social Science and Medicine 41(11):1577-1586.

Blaxter M (1990). Health and lifestyles. London: Routledge.

Calnan M (1987). Health and illness: the lay perspective. London: Tavistock.

Carlstein T, Parkes D and Thrift N (1978). (eds), Human activity and time geography. New York: Halsted Press.

Chatty D (1986). From camel to truck. New York: Vantage Press.

Curtis S and Taket A (1996). Health and societies: changing perspectives. London: Arnold.
DeJong J (1995). Households, health and crisis: coping with economic upheaval in Jordan 1988. Journal of International Development 7(3):443-465.
Donovan J (1986). We don't buy illness, it just comes: health, illness and health care in the lives of people in London. Aldershot: Gower.
DSME (1992). Jordan population and family health survey, Jordan Department of Statistics and Ministry of Health.
Dutton R, Clark J and Battikhi A (1998). (eds), Arid land resources and their management: Jordan's desert margin. London: Kegan Paul International.
Findlay A and Maani M (1998). Population fertility trends. In: Dutton R, Clark J and Battikhi A (eds), Arid land resources and their management: Jordan's desert margin. London: Kegan Paul International.
Hampshire K (2002). Networks of nomads: negotiating access to health resources among pastoralist women in Chad. Social Science and Medicine 54:1025-1037.
Hobcraft J (1993). Women's education, child welfare and child survival: a review of the evidence. Health Transition Review 3(2).
Honey R and Kharmeh S (1989). Rural policy in Jordan's 1986-1990 development plan. Journal of Rural Studies 5(1):75-85.
Kay S (1978). The Bedouin. New York: Crane, Russak and Company Inc.
Lancaster W and Lancaster F (1990). Desert devices: the pastoral system of the Rwala Bedu. In: J Galaty and Johnson D (eds), The world of pastoralism: herding systems in comparative perspective. London: The Guilford Press.
Maani M, Hunaiti H and Findlay A (1998). Demographic change and population projections 1976-2013. In: Dutton R, Clark J and Battikhi A (eds). Arid land resources and their management: Jordan's desert margin. London: Kegan Paul International.
McCray T (2004). An issue of culture: the effects of daily activities on prenatal care utilization patterns in rural South Africa. Social Science and Medicine 59:1843-1855.
Nettleton S (2006). The sociology of health and illness (2nd edition). Cambridge: Polity Press.
Ojanuga DN and Gilbert C (1992). Women's access to health care in developing countries. Social Science and Medicine 35 (4):613-617.
Phillips D (1990). Health and health care in the Third World. Harlow, Essex: Longman Development Studies.
Puentas-Markides C (1992). Women and access to health care. Social Science and Medicine 35(4):613-617.
Sajdi R. Desert land <http://www.acacialand.com/>.
Spicer N (1998). Sedentarisation and accessibility to health services. In: Dutton R, Clark J and Battikhi A (eds). Arid land resources and their management: Jordan's desert margin. London: Kegan Paul International.
Spicer N (1999). Pastoral mobility, sedentarisation and accessibility of health services in the northeast Badia of Jordan. Applied Geography 19(4):299-312.
Spicer N (2000). Health care in sedentarising communities: a case study in the Jordan Badia. University of Glasgow.
Spicer N (2005). Sedentarization and children's health: changing discourses in the northeast Badia of Jordan. Social Science and Medicine 61(10):2165-2176.
Swift J, Toulmin C and Chatting S (1990). Providing services for nomadic people. UNICEF Staff Working Papers No. 8: UNICEF, New York.
Zaidi S (1994). Planning the health sector: for whom, by whom? Social Science and Medicine 39(9):1385-1394.

CHAPTER 21

Developing a landfill suitability map in the northern Badia of Jordan using GIS

Odeh M. Al-Meshan
Badia Research and Development Centre, P.O. Box 902, Jubeiha 11941, Jordan.
E-mail: al_meshan@yahoo.com

Ian Foster
Coventry University, UK & Rhodes University, South Africa

This paper is an example of the logical development of a decision-support tool, based on GIS technology combined with the careful collection of field date from surveys and other sources integrated with international regulations and best practice, as an aid to improving the management of solid waste in Jordan. Because such a high proportion of Jordan is desert, with a very low population density, there has been a tendency in the past to assume that waste can be disposed of with ease and safety almost anywhere. In consequence, today there are examples of many types of problems stemming from lack of planning care. As this paper makes clear, there are only relatively small parts of the Badia suited to solid waste disposal. The approach used in this paper can be used to develop other planning tools so long as the appropriate data sources are available.

Keywords: GIS, decision-support, mapping

1. Introduction

GIS has been widely recognized as a powerful tool for managing spatial data (Burrough 1996). For example, it has been used for environmental modelling (Burrough 1996, Engelen et al 1999), groundwater vulnerability applications (Al-Adamat 2002, Padgett 1994), wellhead protection and groundwater modelling (Adams et al 1992, Hammen et al 1993, Vieux et al 1998), water resource management (Baban 1997), locating wind farms (Baban et al 2001), urban planning (Yegorov et al 2000), and catchment area management (Yusof 2000).

GIS applications have been highlighted in several studies during the last two decades as a tool to assist in locating new landfill sites (Atkinson et al 1995, Baban et al 1998, Basagaoglu et al 1997, Bonham-Carter 1994, Davies et al 1991, Dikshit et al 2001, Hussey et al 1996, Kao et al 1996, Lin et al 1999, Lindquist 1991, Richason et al 1988, Siddiqui et al 1996, Tchobanoglous et al 1993). GIS is used in this study as a tool to aid effective planning by informing the decision-making process to analyze and screen out specific criteria for preliminary assessment (Berry 1987, Burrough 1996).

Multi-criteria evaluation (MCE) techniques and the application of GIS for spatial decision making in relation to landfill sites have been demonstrated by several authors (Bonham-Carter 1994, Kao et al 1996, Malczewski 1999, Siddiqui et al 1996). These authors have demonstrated the practical use of GIS and multi-criteria for supporting the landfill selection process by manipulating spatial data from various sources that ultimately provide support for a wide range of planning processes and decision-making applications.

The landfill suitability map produced in this study was developed in two phases.

Table 1. A set of landfill siting criteria used for developing the landfill suitability map in this study

Criterion type	Constraint factors	Buffer distances and specifications Buffer 1[a] No landfill sites should be located:	
(1) Exclusive	Built-up areas	< 3 km and more than 10 km	
	Farms	< 300 m	
	Wells	< 500 m	
	Ponds	< 400 m	
	Wadis	< 200 m	
	Floodplains	Not within 100-year flood frequency event on flood plains	
	Faults	< 90 m	
	Roads	< 500 m and > 2 km	
	Slope	> 20 %	
	Political borders	< 1 km	
(2) Restrictive	Constraint factors	Buffer 2[b] Acceptable site locations:	Buffer 3[c] Preferred site locations
	Built-up area	3 km to 5 km	5 km to 10 km
	Farms	300 m to 400 m	> 400 m
	Wells	–	> 500 m
	Ponds	400 m to 500 m	> 500 m
	Wadis	200 m to 275 m	> 275 m
	Flood plains	<100 m	> 100 m
	Faults	90 m to 200m	> 200 m
	Roads	1 km to 2 km	500 m to 1 km
	Slope	20 % to 5 %	< 5 %
	Political borders	1 km to 3 km	> 3 km
	Oil pipeline	–	> 125 m
	Site capacity	Sites of less than 20 ha should be excluded	
	Qualitative criteria		
	Local community	Selected sites must be accepted by local community	
	Wind direction	Sites with upwind dominant direction must be avoided	
	Visibility	Sites must not be visible to the public	

[a]: buffer weight is 0; [b]: buffer weight is 5; [c]: buffer weight is 10.

First, GIS tools were used to create buffer zones corresponding to specific criteria. Within the GIS, calculation of distances and screening was used to eliminate areas that do not meet specific exclusionary criteria. In this phase 95.4 % of the study area was eliminated. The second phase used Weighted Linear Combinations (WLC) and then arithmetic operations to combine and overlay the input maps together to produce a single output map. Finally, 36 potential sites (2.95 %) of the study area were found suitable for landfill sites. Based on feedback from the local community within the study area, 11 had potential for future development. These sites were evaluated environmentally and economically and future geotechnical investigations were recommended prior to final development in order to avoid any potential pollution to groundwater.

This paper presents a strategy for developing a suitability map to identify potential landfill sites in the northern Badia of Jordan. The strategy is based on integration of the MCE derived from a survey using GIS techniques.

2. The study area – an overview

The study area is located in the eastern part of the northern Badia of Jordan and falls within the Mafraq district. It covers an area of about 1 500 km^2, north-east of Amman near the Syrian Border. The study area is located in a transition climatic zone between the wetter environment of the Jordan highlands in the west and the dry desert interior in the east. Figure 1 shows the distribution of built-up areas including 34 towns, villages and small settlements in addition to the main roads and also the old and current dumpsites within the study area.

The population of the study area was 15 655 people in 1993 (Dutton 1998). In 2005, the population was expected to have risen to 23 113, based on a 3.3 % annual growth rate reported by Maani et al (1998) for the study area, based on a field survey that was undertaken in 1993. The population lives in 34 towns, villages and small settlements.

The study area is fully served by roads, telecommunications and a water distribution network. Employment is mainly in the agricultural sector, using both rain-fed and groundwater irrigation for livestock rearing and small areas of cultivation. In general, the area is expanding in residential and agricultural activities.

Local relief in the area is quite modest; the altitude ranging 700-1 200 m asl. Basalt strata underlie the area. Throughout the region are wadis and mudflats as well as some small basalt hills that are exposed irregularly. In general the study area is flat to gently sloping (Ibrahim 1997, Sunna 2002).

Figure 1. The location of the study area and the distribution of the built-up areas (Source: modified from BRDC, 1994)

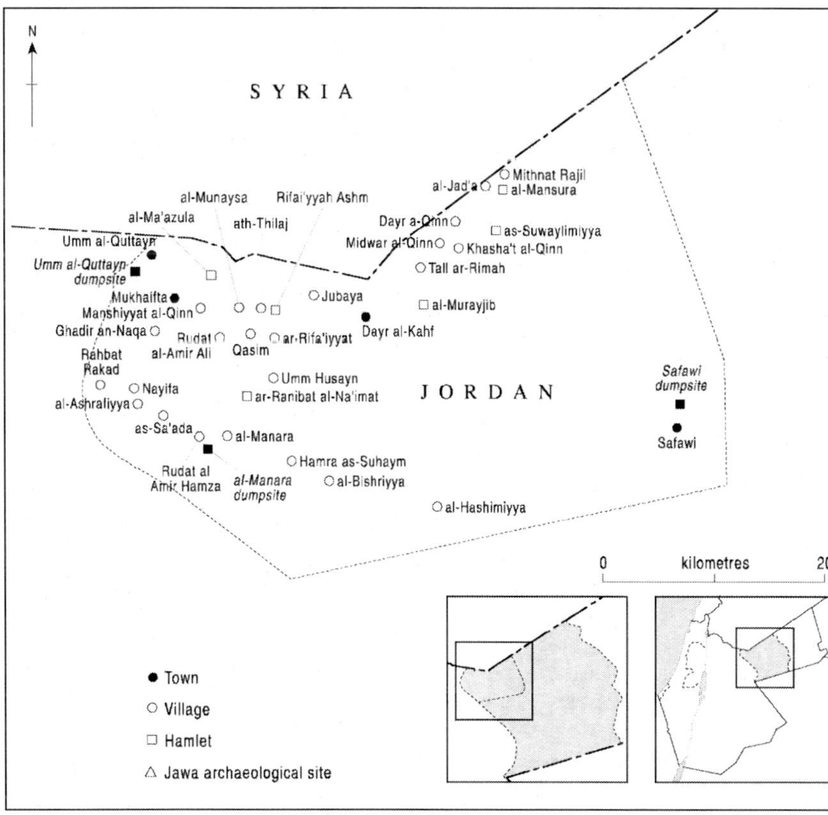

The annual rainfall in the study area is erratic both spatially and temporally, with a range of 50-200 mm with rare snowfall. The mean monthly temperature range is about 25-40 °C in the summer and 20 °C to about –5 °C in the winter (Allison et al 1998). There are two dominant wind directions in the study area; north-west in the summer season and south-west during the winter and spring seasons. These two dominant winds are reflected in the local unsettled weather coming from the Mediterranean (Kirk 1998). Potential evaporation in the study area is relatively high at 1 500 to 2 000 mm yr^{-1}.

The general geological history of the study area has been described by many researchers (Al-Homoud et al 1998, Allison et al 1998, Batayneh et al 2001, Bender 1974, Ibrahim 1997) and has been dominated by a series of tectonic events that caused six volcanic eruptive episodes that occurred during the Tertiary and Quaternary periods.

The unpredictable winter rain can lead to ponding in the *qa'a* (mudflats or playas) and rapid runoff in the wadis. The summer is dry.

Three major aquifers were identified in the study area (Dottridge 1998). The first one (upper aquifer) occurs both in alluvial and basaltic aquifers in unconfined conditions. This aquifer is one of the most productive aquifers in Jordan. Some 24 x 10^6 m^3 is extracted annually, partly used locally for agriculture and partly pumped to Greater Amman (MWI 2002). Groundwater levels are 300-500 m below ground level (Dottridge 1998). The second and third aquifers in the study area are confined and deep (more than 500 m) and the possibility of being polluted is very weak. The upper aquifer is the most important one to be considered in relation to landfill locations.

3. Existing thematic maps to the study area

A total of eleven data sets representing geographical attributes were made available at BRDC together with other sources as thematic layers to be used for GIS analysis in this study. These data sets were collected and their sources and characteristics are presented in Table 2. Thematic maps related to mineral deposits, future landuse planning, land ownership, land value and land cost were not available in a form suitable for GIS analysis.

4. Steps to develop the landfill suitability map

Analysis was carried out using Arc View GIS. Arithmetic spatial operations were carried out using GIS based on exclusionary criteria first and then applying further restrictive criteria (Table 1) to produce the landfill suitability map based on the following steps:

- Proximity analysis;
- Editing tables and maps;
- Performing map overlays;
- Surface analysis and modelling; and
- Raster vector conversion (vectorization)

(DeMers 2005, Hanna et al 1998, Hutchinson 2004, Malczewski 1999). Figure 2 shows the methodology for developing the landfill suitability model.

5. Proximity analysis and map overlays

Using the BUFFER command in Arc View, proximity analysis was carried out in this study to create buffer areas surrounding existing geographical attributes (constraint factors) in the study area corresponding to the quantitative criteria that are given in Table 1. In general, three kinds of surrounding buffers were used. These are: point buffers (wells and ponds), line buffers (roads, oil pipelines, wadis, political borders and faults)

Figure 2. Model flowchart showing the techniques used to produce the suitability map

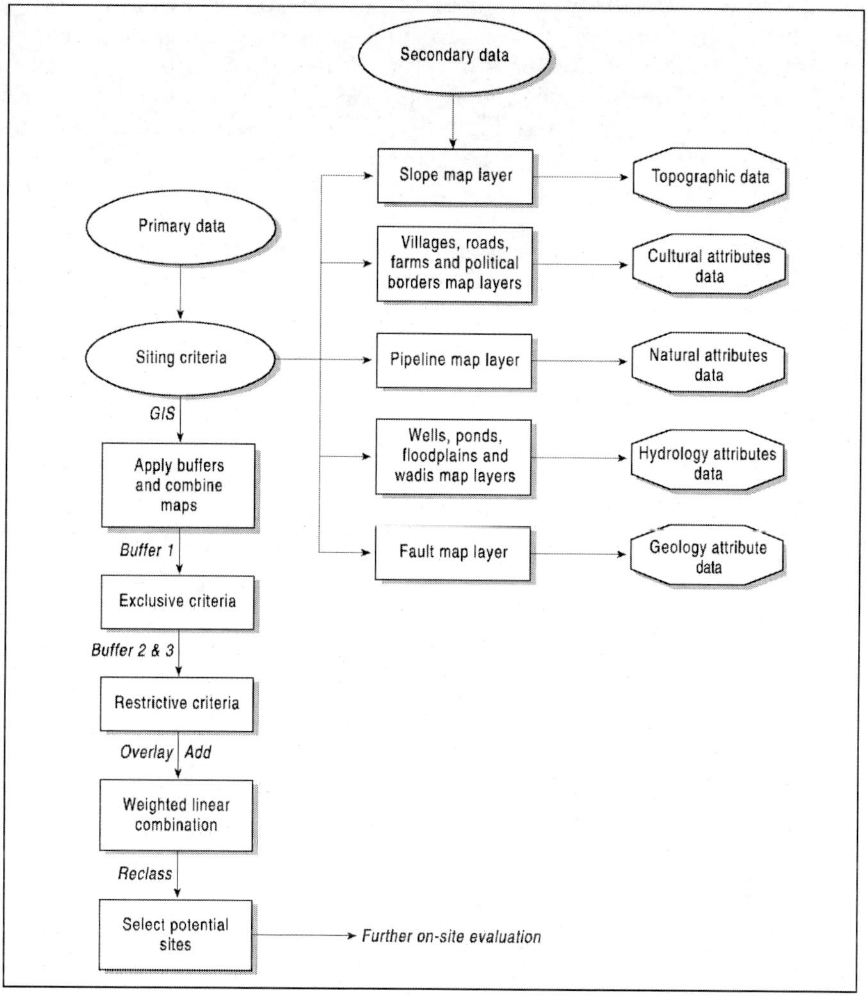

Table 2. The secondary data obtained from BRDC database for the study area and their characteristics and sources

Data sets	Date	Format	Scale	Data shape	Data source
Land use	2002		1:25 000	Polygon	Al-Adamat
Wadis (valleys)	1995		1:50 000	Lines	
Ponds	1995		1:50 000	Points	
Roads	1995	All	1:50 000	Lines	All
Oil pipelines	1995	Digital	1:50 000	Lines	BRDC
Floodplains	1995		1:50 000	Polygon	
Political borders	1995		1:50 000	Line	
Faults	2001		1:250 000	Line	
25 m contour lines	1995		1:50 000	Line	
Wells	2002	Sheet	Sheet	Point	Al-Adamat
Soil	1995	Sheet	1: 250 000	Polygon	MoA

and polygon buffers (villages, farms and flood plains). These create new polygons at a specific distance away from these attributes. Some geographical attributes, such as 100-year flood plain, wells and oil pipelines, were buffered into two ranges, either acceptable or not acceptable. Wells were buffered into two distances because the analysis of the questionnaires showed no difference between the second and the third buffers (ie 500 m). On the other hand, flood plains and oil and gas pipelines were also buffered into two ranges because the difference between the second and the third buffer was small (25 m) which means that, for the available GIS scale, there will be no clear area between these buffers.

Geo-processing tools in Arc View were also utilized in this study to produce these polygons and the final reclassified maps. Five command tools are available in Arc View GIS 8.3. These are: CLIP, DISSOLVE, MERGE, UNION and DELETE. Spatial analyst functions were also available in Arc Map, where all layers were converted to a raster environment, using the CONVERT command in order to run calculations. After that, data for all polygons were processed using the ZONAL STATISTICS command.

Analysis was carried out in two phases. The first phase applied the exclusionary criteria. The second phase differentiated between preferable and acceptable sites. Maps were produced in phase one to show areas that were excluded from further consideration, and also areas that were retained for further analysis. Tables were produced to summarize the areas excluded and areas which were left.

In the first phase, areas where landfills cannot be located for a variety of reasons (see exclusionary criteria in Table 1) were identified. The database was edited (using EDITOR-GIS) so that all the area located within buffer one (Table 1) was assigned a 0 weight and then screened out from any further analysis by using the DELETE command in Arc View. For example, locations less than 500 m from wells, less than 3 km and more than 10 km from built-up areas, and less than 500 m and more than 2 km from roads were excluded from further analysis.

Proximity to roads was analyzed for economic reasons (transportation cost) while proximity to settlements was analyzed to maximize public health and community satisfaction. The area also had to be located away from wadis, wells, farms, political borders, oil pipelines, ponds and steep slopes.

The percentage slope was extracted from the topographic map based on a 25 m contour interval using a Triangular Irregular Network (TIN) for creating and classifying slope angle. Table 2 shows the areas that were excluded in this phase.

By applying the exclusive criteria (Table 1) on the constraint factors, 143 450 ha (95.6 %) of study area were screened out, so that only 6 546 ha (4.4 %) of the study area

Table 3. Areas excluded after applying exclusionary criteria in phase one

	Exclusionary criteria		
	Area excluded (ha)	Percentage	Notes
Built-up area buffer	70 957	47.2	Built-up areas included
Roads buffer	107 916	71.8	
Wells buffer	4 478	3	
Valley buffers	311 5	20.7	
Fault buffers	2 266	1.5	
Political border buffer	5 354	3.6	
Farm buffers	1 243	0.8	Farms included
Steep slope (> 20 %)	7 011	4.7	
Flood plains	375	0.25	No buffers
Pond buffers	98	0.06	

Table 4. Siting factors/constraints and associated data layers chosen for analysis

Index	Data layers (attributes)
Geology	Faults
Cultural	Built-up areas, agricultural areas, political borders and roads
Topography	Slope
Hydrology	Ponds, valleys, floodplains and wells
Natural resources	Oil pipelines

remained as potential areas for landfill sites. Table 2 shows the results of applying the exclusionary criteria including areas that were excluded and the area that remained.

In the second phase of the analysis, proximity analysis was carried out on areas that remained from phase one. Each of the eleven attributes was buffered into two buffer zones corresponding to the restrictive criteria in Table 1 (buffers 2 and 3). MCD analysis algorithms were integrated with Arc View to analyze and rank all remaining areas. This was done by assigning weights for each criterion.

Following the second phase of the analysis each buffer distance (criterion) was assigned a weight by using the EDITOR command in Arc View. The aim was to recreate a distribution of values with a hypothetical range from zero to ten, where 0 indicates unacceptable area (exclusive criterion), 5 is acceptable and 10 is preferable.

Then by using weighted linear combination (WLC) all constraints/factors were combined using map algebra (Chrisman 1997, DeMers 2005).

Map overlay was used to combine two or more layers to create a new layer using the UNION command in Arc View with successive layers where each polygon is superimposed on another (Bernhardsen 1992, Martin 1996). The final result was a single layer containing all areas that satisfy all suitability conditions in the study area.

6. Suitability map indices

Five landfill suitability indices were used to create the suitability map. The indices were: geology, topography, hydrology, natural resources and cultural resources (Table 4). Each of these component indices has one or more data layers. In fact, each layer represents constraints within the study area as denoted in Table 2.

The final suitability indices are made up of a combination of these multiple data layers based on the following equation:

[(Atkinson et al 1995, Oweis et al 1998)]:

$$LSI = \sum_{i=1}^{n} (Wi^{*}Si) \qquad (1)$$

Where: LSI: Landfill Suitability Index; W: data layer importance weight; S: criterion score; n: number of data layers.

The five landfill suitability indices and the impact of each index on the final selection are discussed below.

6.1. Geology suitability index

The geology index suitability map is based on the fault layer. The relative importance of faults is 9. Based on Table 1, a protective buffer of 90 m was set around the fault lines in the exclusionary phase of the analysis. In this step 2 266 ha (1.5 %) of the study area was excluded. The other two buffer zones (restrictive criteria) were drawn around the faults.

The second was from 90-200 m (2 832.8 ha: 1.9 % of the study area). The third was more than 200 m (145 240 ha: 96.6 %) in order to rank the suitability of land adjacent to faults from low (close to the faults) to high (far from the faults); the three buffer zones were assigned weights of 0, 5, and 10 respectively. These values were multiplied by the relative importance factor of 9 to create the final geology suitability map. Areas less than 90 m from the fault lines were excluded entirely from further analysis as potential hazards to groundwater pollution and also for construction considerations (Figure 3).

6.2. Topographic suitability index

The final topographic index suitability map is based on the slope map layer. The relative importance (weight) of slope from the survey is 6. Aspect was not considered in this study because it has no effect on the siting process. Slope was derived initially from the 25 m contour map (1:50 000 top. Point map), using 3 D ANALYSIT. This was achieved by creating a Triangular Irregular Network (TIN) in GIS. The TIN was then converted to a grid data set in a raster format for arithmetic operations.

The result showed that most of the study area has a slope of less than 20 % (95.3 % of the area). Sites having slope angles more than 20 % (7 011 ha: 4.7 %) were excluded in phase one (exclusionary criterion). Areas that have a slope range of 20-5 % (71 689 ha: 47.7 %) were considered acceptable and given a sore of 5, and finally areas having a slope less than 5 % (71 633 ha: 47.6 %) were considered preferable and assigned a score of 10. The final topo-

Figure 3. Geology index suitability map

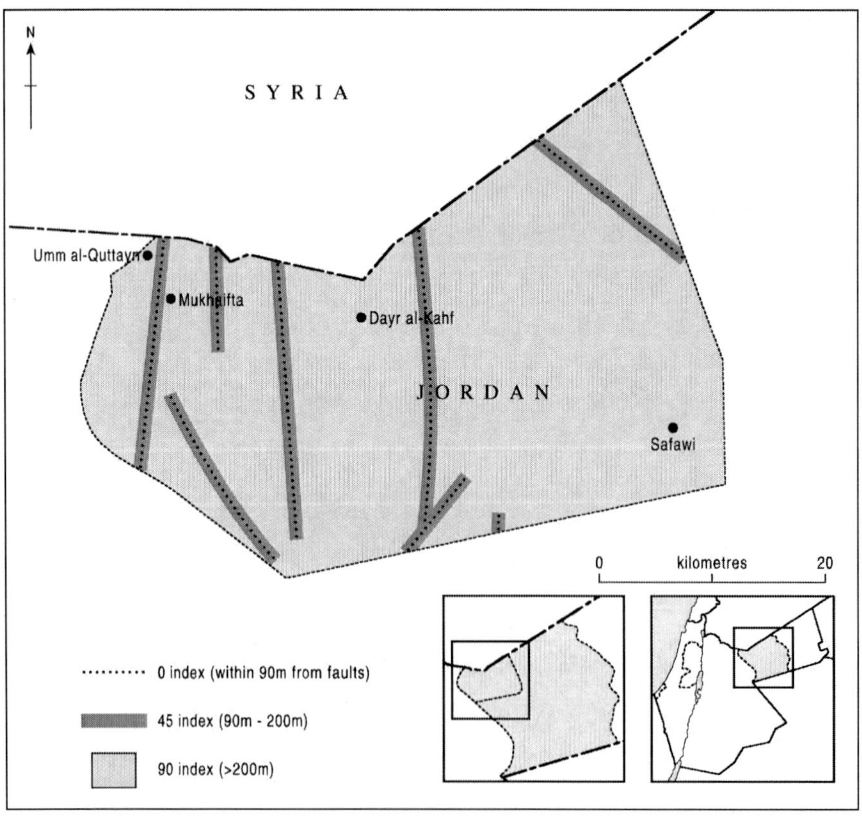

Figure 4. The Slope index suitability map

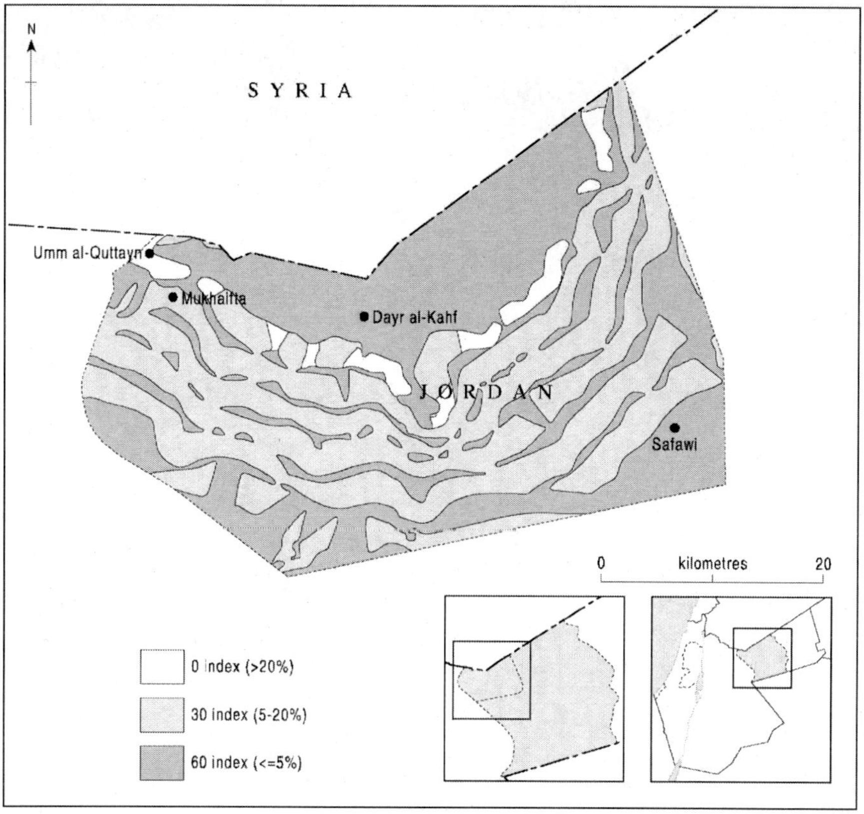

graphic index map was attained by multiplying the slope by the relative importance derived by the survey. The final output was then reclassified into three divisions (Figure 4).

6.3. Hydrology suitability index

The data sets representing water supply wells were buffered at 500 m and excluded. Areas with a 10 score were considered for further analysis. Most of the government wells are located within 200 m of the main roads. Since a 500 m distance has been specified from the main roads, the requirement for a safe distance from these wells is automatically satisfied.

One dam was found within the study area, located within the border of Dayr al-Kahf

Table 5. Percentage of areas resulted from applying exclusionary and restrictive criteria to the hydrology constraints

Constraints	Exclusionary criterion (% area excluded) (buffer 1)	Restrictive criteria (% area remained) buffer 2	buffer 3
Wells	3 % (wells included)	–	97 %
Valleys	20.5 %	7.7 %	71.8 %
Ponds	0.06 %	0.04	99.9 %
Flood plains	0.45 % (flood plains included)	–	99.55 %

Figure 5. Hydrology index suitability map

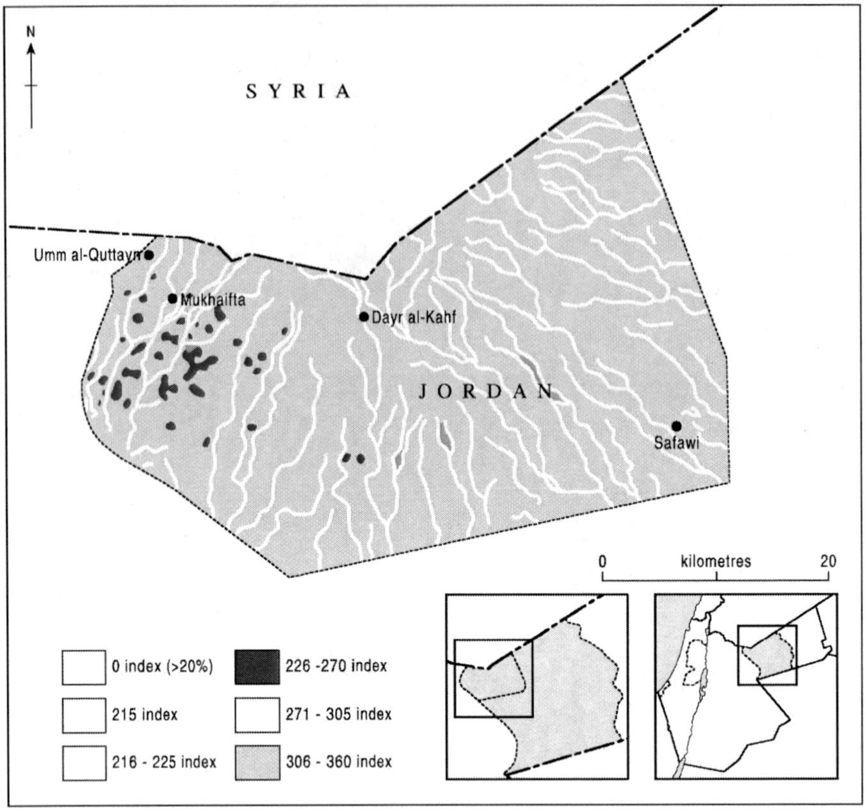

village. By applying the four exclusionary criteria in the first phase, an area of 36 100 ha (24 %) had been screened out (Table 5). The final hydrological suitability index map was a combination of these three map indices as shown in Figure 5.

6.4. Natural resources suitability index

The natural resources index is derived so that the site does not interfere with any natural resources that would be utilized by the local community. The natural resources analysis is based on the oil pipeline layer. Oil wells, conservation areas and gas wells were not considered in this study because none currently lies within the study area. Some mining activity locations (there are six volcanic tuff centres in the study area and some of them are currently exploited for agricultural and industrial purposes) were not available as thematic maps; however, these locations are automatically excluded because they form hills and have a steep slope (ie >20 percent). Only oil pipelines were available for analysis. Underground oil pipelines could be relocated if there are no available landfill sites, but the best option is to avoid siting on or near them. The pipeline was buffered with a 125 m wide area either side. By applying restrictive criteria to the oil pipeline (ie =/-125 m) a small area (ie 1 080 ha: 0.7 %) was excluded from further analysis (Figure 6).

The final hydrology suitability index is based on four data layers (Wadis, Ponds, Wells, and Flood plains; Table 5). Wadis were buffered at 0-200 m, 200-275 m and more than

Figure 6. Natural suitability index map

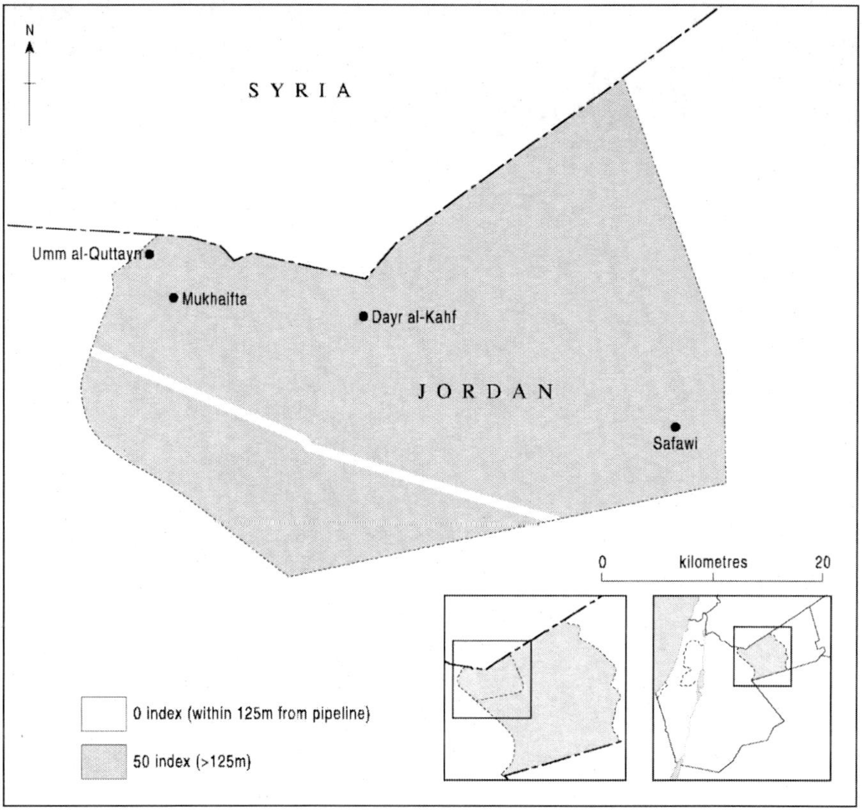

275 m respectively. The first buffer was allocated a zero score and excluded in the first phase. Ponds were buffered at 400 m, 400-500 m and more than 500 m respectively. The first buffer was allocated a zero score and excluded while the second and third buffers were allocated scores of 5 and 10 respectively and considered for further analysis.

6.5. Cultural resources suitability index

The suitability index representing cultural resources consists of four attributes (Table 4). The data set representing built-up areas was buffered to ranges: up to 3 km from built-up areas (excluded areas), 3-5 km and from 5-10 km. The weight assigned to the built–up areas from the survey is 9. Political borders were buffered to 1 km as an exclusion area, and 1-3 km and more than 3 km respectively. The weight assigned to the political border is 7. Agricultural areas (farms) were also buffered to 300 m (excluded areas), 300-400 m and more than 400 m respectively, where areas with a distance more than 400 m are considered preferable. The relative importance for agricultural areas is 7.

The final layers considered in the cultural suitability index dealt with the transportation (least cost) issue. Buffers were set around the different road types based on the ease of using them as access to the sites. The relative importance of the roads was 8. A buffer area was placed 500 m around the road routes as an excluded area for aesthetic considerations and also for any potential future expansion in the road construction. The areas beyond 500 m were buffered to 1 km, and then from 1-2 km. Areas more than 2 km

Figure 7. Cultural suitability indices map

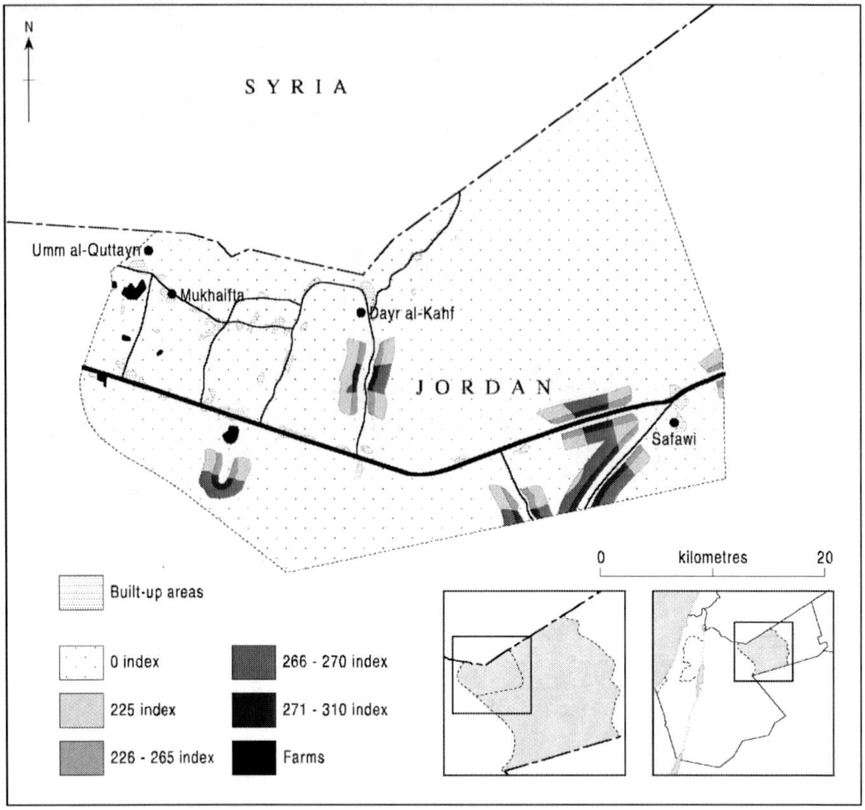

away from roads were excluded to allow reasonable access to the sites. The buffer weights were highest close to the road (500-1 000 m) and lowest far from roads (> 2 000 m) (Figure 7).

The final cultural suitability index was created from the mathematical combinations of the four layers to create a final composite map. By applying the cultural exclusionary criteria ca 141 000 ha (94 %) of the study area was excluded. Only ca 8 900 ha (6 %) remained for further analysis. Finally all areas with a score of zero were removed from the maps. The areas remaining were classified into two divisions; either less acceptable (score: 5) or preferable (score: 10). Roads accounted for the largest portion of the excluded areas followed by built-up areas then political borders when applying the exclusionary criteria as shown in Table 6 and Figure 7.

7. The final suitability map overlay and preliminary evaluation

The final landfill suitability index map was based on equation 1. Initially all map layers were converted to a raster model for calculation purposes and then combined. 35 classes were generated as a result of applying the three buffer areas around all the 11 constraint factors (based on Table 1). For example four classes were drawn around the settlements and villages; < 3 km, 3-5 km, 5-10 km and more than 10 km, using the Map algebra calculator (RASTER CALCULATOR). The individual areas and polygons were combined using the WLC, the weighting summation resulting from the overlay process

Figure 8. Potential landfill areas and their scores after applying the WLC

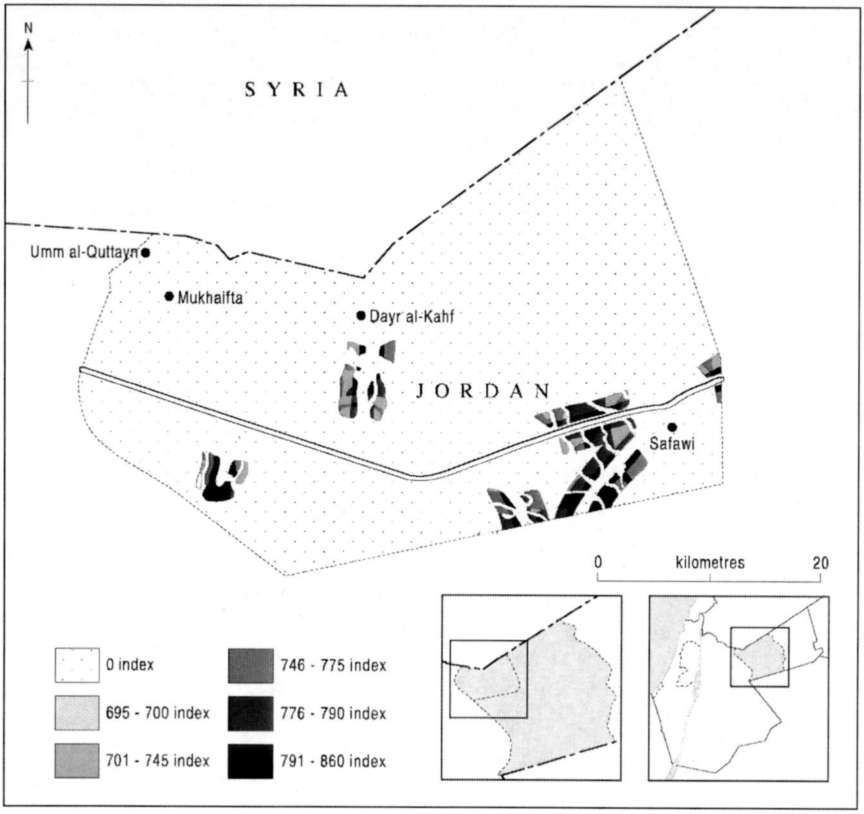

suitability map. This was achieved by using the ADD function where all map layers were calculated and summed. After more than seventy GIS-runs had been carried out on these data sets and criteria, 178 potential landfill sites were identified within the study area forming a suitability map including a varying range of suitability indices from 695 (least acceptable areas) to 860 representing the most acceptable or preferable areas (Figure 8).

After applying a site capacity criterion, the number of potential sites that had an area greater than 20 ha declined to 36 (4 437 ha: 2.95%), Figure 9. The suitability indices ranged from 755-860. The smallest site had an area of 29.9 ha while the largest site was 421.4 ha (Table 7).

Table 6. Percentage of areas resulting from applying exclusionary and restrictive criteria to cultural constraints

Constraints	Exclusionary criterion (% area excluded) (buffer 1)	Restrictive criteria (% area remained) buffer 2	buffer 3
Villages	(45 %) + (2.2 %: Villages polygons)	16.8 %	36 %
Farms	0.9 % including farms	0.2 %	98.9
Political borders	3.6 %	6.8 %	89.6
Roads	71.8 %	17.7 %	10.5 %

Table 7. The 36 acceptable and preferred sites remaining after the final analysis

		Restrictive criteria			
Site ID	Acceptable (ha)	Site ID	Acceptable (ha)	Site ID	Preferred (ha)
1	42.7	17	300.9	2	31.8
4	158.0	18	53.2	3	41.8
5	113.4	19	103.5	6	59.3
7	304.2	22	46.5	14	129.5
8	158.5	23	101.6	20	41.8
9	222.1	24	121.9	21	86.4
10	130.5	25	202.6	27	421.4
11	157.0	26	98.2	29	181.8
12	75.0	28	119.0	30	76.4
13	57.4	23	195.5	31	29.9
15	62.2	34	57.9	32	219.2
16	62.2	36	136.0	35	37.5
		Total	3 080.0	Total	1 357.0

8. Classification and re-colour of potential sites

Using the RECLASSIFY tool in GIS, these indices were reclassified again into two equal class ranges as a 'classes choropleth map'. The first class ranges from an index of 755-807.5 (acceptable areas). The second class ranges from 807.5-860 (preferable areas). The two classes were merged together using the MERGE tool in GIS and assigned different colours to represent different types. This was achieved by selecting the Unique Value from the Legend Type pull-down menu in GIS. The new numbers mean that the attributes are grouped into new ranges with a similar level of suitability (Figure 9). Berry (1987) claims that reclassification of the map into two quantitative values creates a new level of suitability map for general use. Adam and Gangopadhyay (1997) also suggest the use of this method of grouping when using map algebra to produce suitability overlay maps. They gave two reasons: first, to make it easier to understand the values generated within each index in order to produce the final suitability map when merging the maps, and to assign appropriate colours to the polygons; second, to ensure that each of the data layers was normalized and that those pixels with the lowest values have the lowest desirable ranges and those with the highest values have the highest desirable ranges. The final landfill suitability map in this study was assigned three different colours. Each colour represents a class and each class represents a criterion. Black was assigned for preferable areas and red for acceptable areas. Thus, the suitability indices map has been grouped into suitability ranks. Map cells with scores in the lower half of the score-range for all cells were assigned a second rank (755-807.5) while the rest were assigned a first rank (807.5-860).

Finally, the new map of landfill suitability indices shows three suitability classes. The first one (exclusionary criterion; class 1; score 0) which accounts for about 95.6 % of the study area map, the second class (restrictive criterion: score 5), (24 sites; 2.05 %; 3 060 ha) and the third class 12 potential sites (restrictive criterion: score 10), (12 sites; 0.9 %; 1 356 ha). White implies that an area must be avoided, red implies 'carefully go' (ie the area is conditionally acceptable) and black implies that the land is potentially suitable for disposing solid waste (Table 7 and Figure 10).

9. Summary and main conclusions

This study has illustrated the use of GIS as a decision support system for analyzing po-

Figure 9. Landfill index suitability map after excluding 20 ha

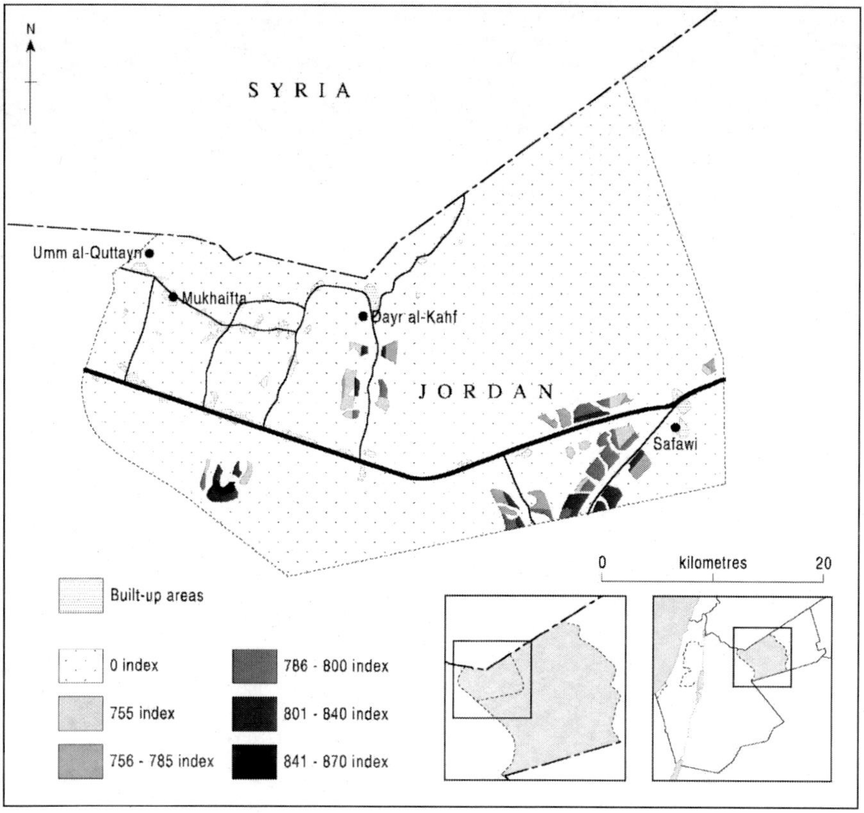

tential sites for landfills in the northern Badia of Jordan. A suitability map was developed through a progressive elimination process. In the first phase about 95.6 % of the study area was eliminated Built-up areas and roads had the greatest influence in the final selection. Hydrological features, in general, have a greater impact than the remaining natural features. The sites with potential cover about 4.4 % of the total study area.

In the second phase, the MCDA approach was used to produce a landfill suitability map using arithmetic operations. This was accomplished by developing criteria weights and then performing overlay procedures. The weighted linear combination (WLC) technique was used to combine eleven data layers together. It was then possible to identify potential solid waste landfill sites. A site capacity criterion was added to select only potential areas greater than 20 ha. This criterion reduced the number of potential areas to 36 sites (2.95 % of the total study area). Potential sites were divided into two classes: preferable and acceptable.

Overall, the strategy adopted in this chapter to develop the suitability map for potential landfill sites makes the process of decision-making easier and tries to be more objective. This strategy has created a model based on expert knowledge in Jordan regarding landfill siting criteria, and also enables decision makers, with limited experience, to make informed decisions about landfill site selection in other areas in the future.

It is recommended that in order to minimize time and effort and to make the process of selection as easy as possible, no more than two buffers are used (ie either exclusive or acceptable). Also, the constraint factors and the criteria that were developed in this study should be completely evaluated prior to any final landfill site selection.

Figure 10. The final outcome: the list and classification of the acceptable and preferred sites (details of the site capacities are given in Table 7).

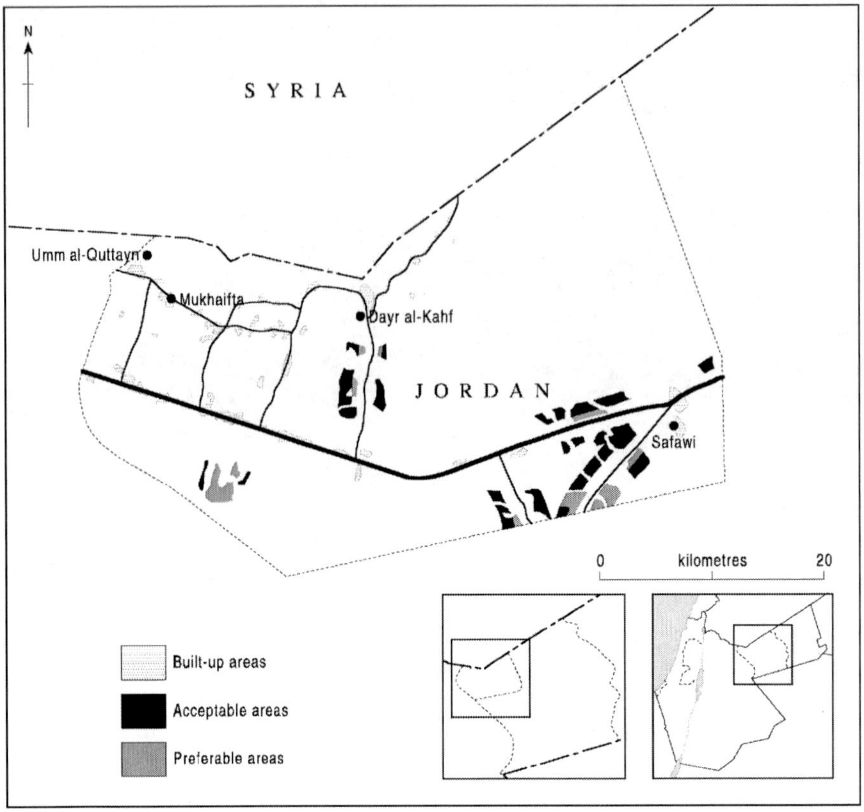

The approach used here should be viewed as a general automated model which could be applied in other regions with similar ecological characteristics. However, this will depend on the availability of both factors and data sets within the research area.

References

Adam NR and Gangopadhyay A (1997). Database issues In Geographic Information Systems (2nd edition). London: Kluwer Academic Publishers.

Adams B and Foster SS (1992). Land-surface zoning for groundwater protection. In the Annual symposium on groundwater pollution and aquifer protection in Europe: Paris (October). J IWEM pp. 313-319.

Al-Adamat RN (2002). The use of Geographical Information Systems (GIS) and remote sensing to investigate groundwater quality in the Azraq Basin, Jordan. Unpublished, PhD Thesis, Coventry University, UK.

Al-Homoud AS, Allison RJ, B S and White K (1998). Geology, geomorphology, hydrology, groundwater and physical resources of the desertified Badia environment in Jordan. GeoJournal 37(1):51-67.

Allison RJ, Higgitt DL, Kirk AJ, Warburton J, Al-Homoud A, Sunna BF and White K (1998). Geology, geomorphology, hydrology, groundwater and physical resources. In: Dutton R, Clarke J and Battikhi A (eds), Arid land resources and their manage-

ment, Jordan's desert margin. London: Kegan Paul International pp. 21-44.
Atkinson SF, Schoolmaster FA, Lyons DI and Coffey JM (1995). A Geographic Information Systems approach to sanitary landfill siting procedures: a case study, professional report. Journal of the Environmental Professional 17:20-26.
Baban MJ (1997). Potential applications of satellite remote sensing (SRS) and GIS in maximizing the use of water resources in the Middle East: examining Iraq as a case study. Remote Sensing and Geographic Information Systems for Design and operation of water resources systems. Proceedings of Rabat Symposium S3. IAHS publication no.242 pp.23-32.
Baban MJ and Flannagan J (1998). Developing and implementing GIS-assisted constraints criteria for planning landfill sites in the UK. Journal of Planning & Research 13 (2):139-151.
Baban MJ and Parry T (2001). Developing and applying a GIS-assisted approach to locating wind farms in the UK. Journal of Renewable Energy 24:59-71.
Basagaoglu H, E, C, MA, M and Usula N (1997). Selection of waste disposal sites using GIS. Journal of the American Water Resources Association 33 (2):455-464.
Batayneh A, Haddadin G, A, AZ and Abu eladas A (2001). Geophysical investigations for the location of a proposed dam in Al Bishriyya area, northeast Badia of Jordan. Journal of Environmental Geology 40 (7):132-144.
Bender F (1974). Geology of Jordan: contributions to the regional geology of the earth. Berlin: Gebruder Borntraeger.
Bernhardsen T (1992). Geographic Information Systems. Norway: ViakIT, Arendal.
Berry JK (1987). Fundamental operation in computer-assisted map analysis. Journal of Geographic Information Systems 1 (2):119-136.
Bonham-Carter GF (1994). Geographic Information Systems for geoscientists: modelling with GIS. Oxford: Pergamon.
Burrough PA (1996). Introduction to Geographic Information Systems. In: Burrough, PA (ed), Principles of Geographic Information Systems for land resources assessment. Oxford: Oxford University Press p. 194.
Chrisman N (1997). Exploring Geographic Information System. Canada: John Wiley & Sons, Inc.
Davies RE and JK, L (1991). Applying an expert system methodology for solid waste landfill site selection. Proceedings of annual conference in the urban and regional information system association 1, Ohio 40-53.
DeMers MN (2005). Fundamentals of Geographic Information Systems (3rd Edition). New Mexico: John Wiley & Sons.
Dikshit AK, Padmavathi T and Das RK (2001). Locating potential landfill sites using Geographic Information System. Journal of Environmental Systems 28(1):43-54.
Dottridge J (1998). Water resources quality, sustainability and development. In: Dutton R, Clarke J and Battikhi A (eds), Arid land resources and their management, Jordan's desert margin. London: Kegan Paul International pp. 67-80.
Dutton RW (1998). Population, environment and development. In: Dutton R, Clarke J and Battikhi A (eds), Arid land resources and their management, Jordan's desert margin. London: Kegan Paul International pp. 3-20.
Engelen G, Greetman S, P, S and Wessels C (1999). Dynamic GIS and strategic physical planning support: a practical application. In: Stillwell J, Geertman S and Openshaw S (eds), Geographic Information and Planning. Berlin, Heidelberg: Springer-Verlag.
Hammen JL and Gerla P (1993). Devising a wellhead protection strategy using Geographic Information Systems Technology: Proceeding of the symposium on Geographic Information Systems and water resources. AWRA Publications USA pp. 361-368.

Hanna K and Culpepper R (1998). GIS in site design. New York: Wiley.
Hussey V, Dodd VA and Dennison GJ (1996). Locating a landfill site for Dublin using Geographic Information Systems. Proceeding of the Institution of Civil Engineers-Municipal Engineer 115 (3):125-133.
Hutchinson S (2004). Inside ArcView GIS 8.3. Canada: Thomson Delmar Learning Inc.
Ibrahim K (1997). The geology of Al-Bishriyya (Al-Aritayn) Area, Map Sheet no.3354 II with special reference to geology at the economic zeolites deposits. Geology Directorate, Geological Mapping Division, Bulletin no.37, Amman, Jordan.
Kao JJ and Lin HY (1996). Multi factors spatial analysis for landfill siting. Journal of Environmental Engineering 122(10):902-907.
Kirk A (1998). A synthesis of climatic data with specific interest in the precipitation record. In: Dutton R, Clarke J and Battikhi A (eds), Arid land resources and their management, Jordan's desert margin. London: Kegan Paul International pp. 48-66.
Lin HY and Kao JJ (1999). Enhanced spatial model for landfill siting analysis. Journal of Environmental Engineering 125 (9):845-852.
Lindquist RC (1991). Illinois cleans up: using GIS for landfill siting. Geo Info Systems 1(2):30-35.
Maani M, Hunaiti H and Findlay AM (1998). Demographic change and population projection. In: Dutton R, Clarke J and Battikhi A (eds), Arid land resources and their management, Jordan's desert margin. London: Kegan Paul International pp. 215-246.
Malczewski J (1999). GIS and multicriteria decision analysis. New York: John Wiley & Sons Inc.
Martin D (1996). An assessment of surface and zone models of population. Journal of Geographical Information Systems 10(8):973-989.
MWI (2002). Jordan water strategy and policies. Report, Ministry of Water and Irrigation, Amman, Jordan.
Oweis IS and Khera RP (1998). Geotechnology of waste management (2nd edition). Boston: PWS Publishing Company.
Padgett DA (1994). Using "DRASTIC" to improve the integrity of geographic information system data used for solid waste management facility siting: a case study. The Environmental Professional 16:211-219.
Richason BF and Johnson J (1988). GIS application in the landfill siting process. Proceedings GIS/LIS'88, ACSM, ASPRS, AAG, URISA pp. 695-703.
Rimawi O and El-Naqa A (1995). Assessment of transmissivity from specific capacity data of the fractured basalt aquifers in the north-northeastern part of Jordan. Arab Gulf Journal of Scientific Research 13 (3):67-80.
Siddiqui MZ, JW, E and Vieux BE (1996). Landfill siting using Geographic Information System: a demonstration. Journal of Environmental Engineering 122 (6):515-523.
Sunna BF (2002). Integrated studies of the Azraq basin for optimum utilization of the natural resources, geology of Azraq basin: target areas. Vol. 1, part B, Unpublished Report. Higher Council for Science and Technology, Amman, Jordan.
Tchobanoglous G, Theisen H and Vigil SA (1993). Integrated solid waste management. New York: McGraw-Hill, Inc.
Vieux BE, Mubaraki MA and Brown DR (1998). Wellhead protection area delineation using a coupled GIS and groundwater model. J. of Environmental Management 54:205-214.
Yegorov YK and Zerkal OV (2000). Geological data management for urban planning using GIS. Proceeding of the 8th international congress for engineering geology and the environment. Vancouver, Canada IAEG Congress.
Yusof KW (2000). Developing a hydrological catchment and planning strategy for Malaysia using remote sensing and Geographic Information System. PhD Thesis, Coventry University, UK.

CHAPTER 22

A surface water management tool for arid lands using GIS – a Jordan Badia case study

Saad M. AlAyyash

Jordan Badia Research and Development Centre, P. O. Box 902, Jubeiha 11941, Jordan.
E-mail: s_ayyash@yahoo.com

Fidelia N. Nnadi

University of Central Florida, Orlando, Florida, USA

Rania Shatnawi

Applied Science University, Amman, Jordan

In Jordan as well as most of the developing countries located in the arid regions, water is a priority issue at national and individual levels. The limited rain that falls over wide areas of the arid lands of Jordan (the Badia) are of significant quantity but, due to lack of proper management, the water volume has not been estimated. Much water is lost due to the high evaporation rates. In this study, the Natural Resources Conservation Services (NRCS) Curve Number (CN), and unit hydrograph models were used to simulate the surface runoff over the study area using the ArcView® software. ArcView® micro language, called Avenue, was used to customize the Graphical User Interface (GUI) of the software. A management tool was developed that includes various steps of simulation used to estimate the runoff volume in the Badia. First the map data were accessed and displayed on the screen so the user can locate the catchment for further calculation. Other data files, which are the hydrological database that describe the hydrology of the project area, were also loaded.

Several programmes and management practices are in place to conserve water resources and make available new resources to meet the demand of various sectors of water users. With very high demand on the ground water resources, the limited volume of surface water becomes a promising option for new and innovative water supplies.

The Wadi Salma catchment (#11) was used to collect field rainfall-runoff data. Rainfall storms were recorded in this catchment December 2000-April 2001 as well as the runoff volumes. These data with the hydrological database were used in the model to calibrate the simulation parameters and calculate the runoff. Using initial abstraction (Ia) as 42 % of the potential soil storage, (S′) was found to be the best estimation for rainfall losses calculation. Using the Muskingum routing method, a travelling parameter (K), equal to the travelling time, and weighting parameter (x) of 0.18 were estimated for the study area. The simulation tool calculated the peak discharge and runoff volume with acceptable accuracy for the whole catchment; the difference in peak discharge estimation was within the range 4-6 % while the runoff volume was 7 % for the first storm and 10 % for the second storm.

The Wadi Salma catchment (95.63 km^2), located in the north-east Badia, was selected to establish rainfall-runoff records. Also field experiments were performed at various parts of the study area to define the hydrological response. Rainfall-runoff records showed high runoff coefficients of 13–20 % for Wadi Salma catchment. Rainfall simulation tests were used to estimate CNs, and high values (up to 94) were estimated for different land covers.

Keywords: surface water, arid lands, surface hydrology, wadi hydrology, GIS, ArcView, Jordan, Badia, Safawi

1. Introduction

Historically, arid lands have not invested in hydrological data acquisition such as flood volumes and rates, rainfall intensities and volumes, etc. Hydrological data collection such as rainfall intensities and flood volumes started in Jordan not long ago and it is concentrated in the humid parts of the country. In the last few years, the Badia started to receive more attention in documenting natural resources. This was as a result of the high population pressure in the limited urbanized areas of the country and the need to utilize the Badia for future development. Daily and annual rainfall recorders were installed on a very limited scale over the Badia. Flood volumes were not recorded in the Badia areas except in very few cases where limited time records were taken in small areas for specific projects (AlAyyash and Nnadi 2001)

In order to define the hydrology of an area, detailed and frequent records have to be established, and also field observations are required to define the hydrological parameters. For this purpose, a typical catchment in the study area was chosen for detailed study. In this catchment, a rainfall recorder and three flood recorders were installed. Infiltration tests in various parts of the catchments were also made in order to define the infiltration parameters. Because of the limitation of the rainfall-runoff data, rainfall simulation tests were applied to various land covers and an average CN for each land cover type was established.

2. Background

Jordan is a developing country situated off the south-eastern shores of the Mediterranean Sea between longitudes 35° and 39° east and latitudes 29° and 33° north (Figure 1). It is bounded in the north by Syria, in the east by Iraq and Saudi Arabia, in the south by Saudi Arabia and in the west by the River Jordan and the Dead Sea. Its area is about 90 000 km^2 and consists of various distinctive topographic units trending in a north-south direction. These units are the rift valley, the high lands and the arid lands-desert region.

The Badia, the desert region, which comprises more than 80 % of the country, is rich in natural resources such as minerals and range lands. These factors make it a potential area for future development. The Jordanian government is motivated to develop the Badia because the Jordan Valley and the high land areas have become over-populated.

The study area (Safawi area) is located in the north-eastern Badia (Figure 1), which receives more rain on average than the other parts of the Badia. The annual rainfall varies from over 150 mm in the north-west to less than 50 mm in the south-east with an average of 70-80 mm. The annual mean temperature is 15 °C. with a mean monthly range of 8-28 °C, January to July. Evaporation rates are in general very high, with an average daily potential evaporation range of 3-17 mm/day, January to July (Taqieddin et al 1995, AlAyyash and Nnadi 2001)

2.1. Geomorphology and Soil

The basaltic rocks cover 11 000 km^2 of the north-eastern Badia. Gentle slopes and relatively flat topography characterize the surface. Around Safawi area, the surface elevation is mostly 600-1 100 m asl; highest in the north-east (Figure 2). Geological variation is reflected in the geomorphology; faults have developed into wadis but the sub-surface rocks made the development of channel networks very limited except in some parts where the weathering rates are high. Large boulders are distributed in the north and east of the area where the younger basalt is found. On the other hand, smaller rocks are found in the western part of the Badia where older basalt and higher rainfall helps weathering.

Figure 1. Location map of Jordan and general topography

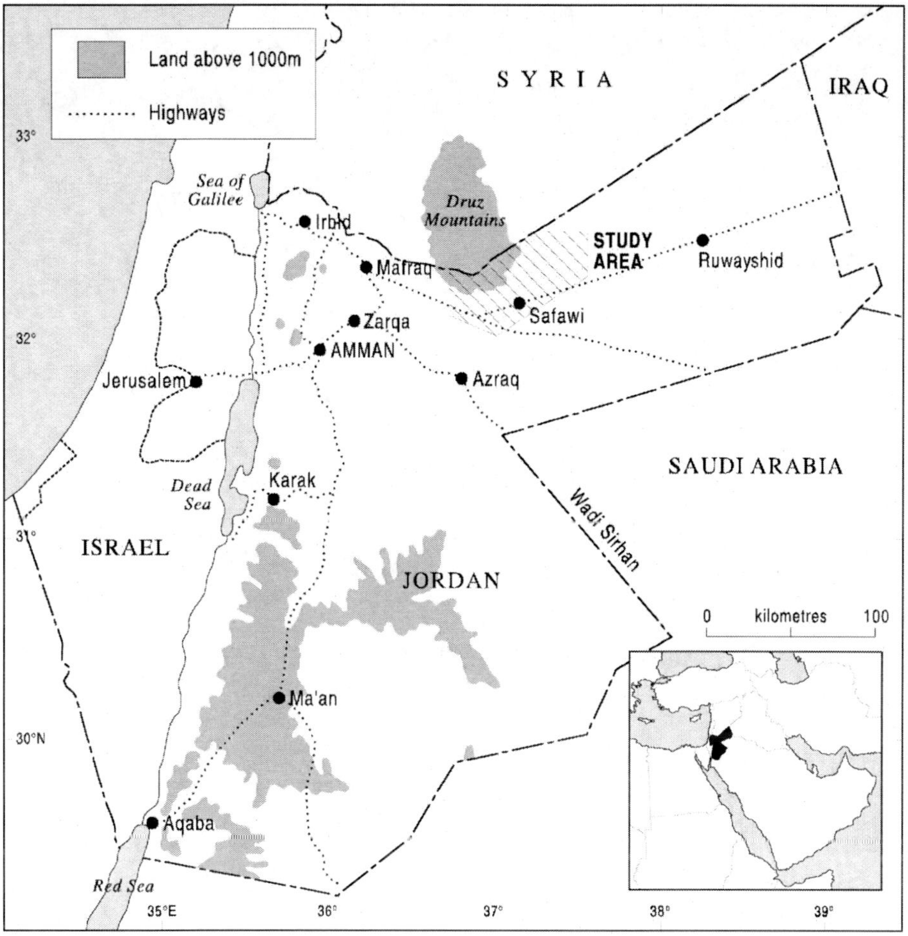

The poor channel network results in blockages across most of the water courses in the Badia. These blockages create natural pools in the depressions where most of the runoff ends and evaporates leaving salty fine soil. The depressions (*qa'a*) are widely distributed in the Badia and form a major part of the surface geomorphology. Another type of depression (*marab*) is formed in the course of major channels. The *marab* are formed from blockages that are not substantial enough to create pools but which impede runoff which has led to soil accumulation. Flood water that does not infiltrate into the soil finds its way out of the *marab*. This process results in renewing the *marab* soil; only the coarse soil particles have the chance to settle, resulting in good productive soil for rangeland vegetation and some seasonal crops such as barley and wheat.

The parent soil material and the climatic parameters control the soil types and properties. In the north-east Badia, the soil is derived from basaltic lava and its development is influenced by topography and weathering. Most of the soils are of depositional origin, which is dominant for arid lands. On steep slopes with high erosion, residual soils occur on the rocks. Soils include well-developed xerochrepts in the areas of older basalt flows and high weathering effect, and weakly developed xerothents on the more recent basalt flow (HCST 1992).

Figure 2. Geology and topography for the Safawi area

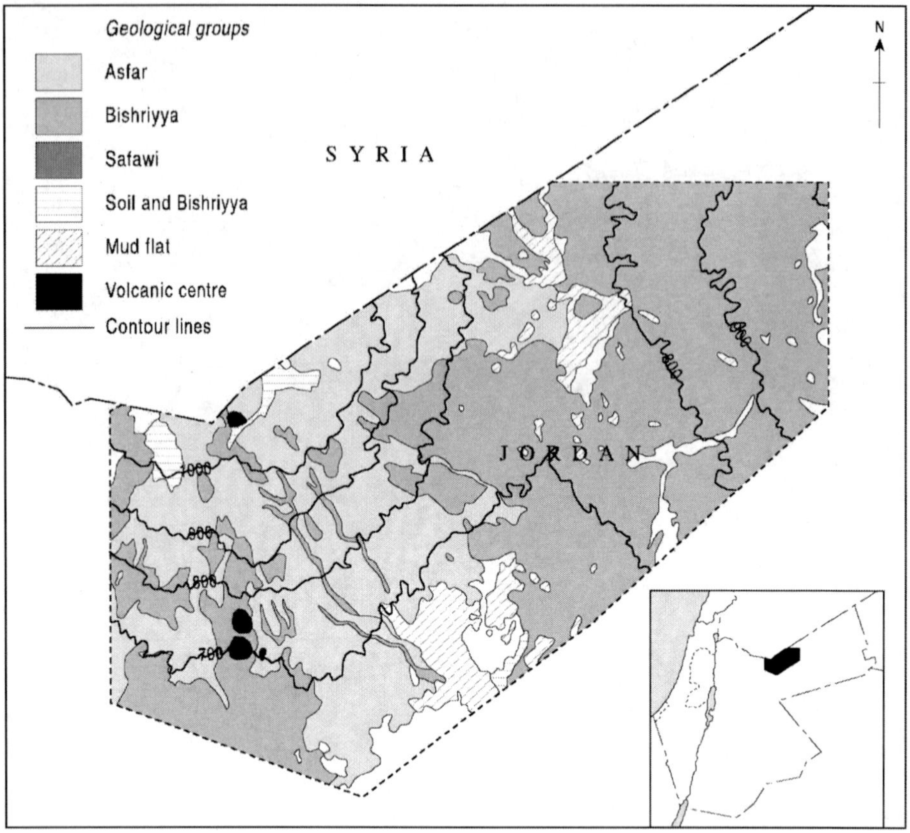

Figure 3. Daily rainfall data in Safawi weather station (F2)

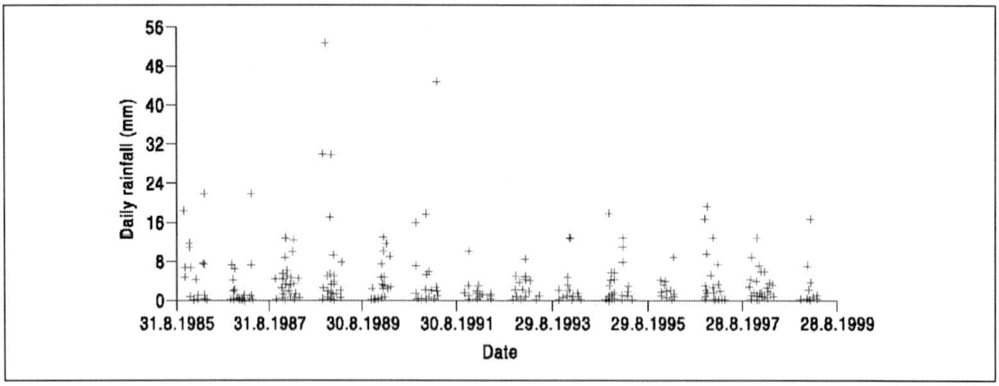

2.2. Rainfall and evaporation

Rainfall as thunderstorms, characterized by irregular intensity and duration, forms the greatest part of total precipitation in the Safawi area. The annual mean rainfall is 70-80 mm, and its amount decreases from 200 mm in the north-west to less than 50 mm in the south-east. Rainfall occurs mostly during winter months (October to April) with the highest intensities in the months December to February.

The daily rainfall data in Safawi Weather Station (F2), 1985-99, were plotted (Figure 3). The data show that more than 89 % of the storms yielded less than 8 mm and only about 4 % yielded more than 16 mm; these storms are the ones that could generate significant runoff. Figure 3 also shows that there are very few thunderstorms that generate more than 24 m of rainfall.

The evaporation rates are very high in the area, mainly during the dry months of June to September. The evaporation Class A Pan records for Safawi Weather Station (F2) show that the mean annual evaporation rate is 9.9 mm/d with a maximum rate of 17.7 mm/d in July. The total potential evaporation is around than 3 600 mm.

3. Hydrological simulation

The hydrological parameters which are required for rainfall-runoff simulation, can be defined in various ways depending on the simulation method. Understanding the approach used in calculating or estimating these parameters is necessary before using them in rainfall-runoff simulation. Parameters used in the simulation in this study include the CN and the time of concentration (Tc).

3.1. Curve number

In the NRCS-CN method, the infiltration losses and other losses such as depression storage are estimated from the watershed parameter, the CN. The CN is a function of the area cover, soil type, surface texture and depression areas. The relationship between the CN, potential storage (S′) and the Initial Abstraction (Ia) is given by the following equation:

$$S' = \frac{25400}{CN} - 254 \quad \text{[SI units]} \quad \text{Where } Ia = \alpha S' \tag{1}$$

The coefficient α defines the percentage of the potential storage (S′) that is contributing to the initial abstraction (Ia). The NRCS suggested a value of 0.2 for most land covers in the USA.

3.2. Time of concentration

The time of concentration is the time required for the excess rainfall from the most remote part of the watershed to reach the outlet of that watershed. The longest collection channel in the watershed is used in calculating the time of concentration. The NRCS (Wanielista et al 1997) used an equation that relates the time of concentration (T_c) and the lag time (T_l) to calculate the time of concentration as:

$$\hat{O}_c = 1.67\, x\, \hat{O}_l \tag{2}$$

And the lag time (T_l) is calculated as a function of the length (L), watershed slope (w_s) and potential storage (S′) as:

$$T_l = L^{0.8} \frac{(S' + 25.4)^{0.7}}{7070\, x\, w_s^{0.5}} \quad \text{Where: } T_l \text{ in hr, L in m, } S' \text{ in mm and } w_s \text{ in \%} \tag{3}$$

NRCS developed a dimensionless unit hydrograph from data all over the USA. By knowing the peak discharge (Q_p), the time to the peak (T_p), and with the dimensionless unit hydrograph, the unit hydrograph could be developed for specific durations. The time to peak (T_P), the peak discharge (Q_P), and the time of concentration (T_c) are calculated using empirical formulas.

NRCS used equations to calculate the time to peak (T_P) as functions of the time of concentration (T_c) and the duration of the unit hydrograph (D). The peak discharge (Q_P) is then calculated as a function of the watershed area (A) and the time to peak (Wanielista et al 1997).

$$T_P = \frac{D}{2} + 0.6 * T_c \tag{4}$$

$$Q_P = \frac{0.208 * A}{T_P} \tag{5}$$

Where: D and T_P are in hours, A in km^2 and Q_P in m^3/sec.

NRCS developed equations that relate the rainfall excess (R) to rainfall volume (P), the initial abstraction (Ia) and the saturation volume (S') as follow:

$$R = \frac{(P - Ia)^2}{P - Ia + S'} \qquad \text{If } P \geq Ia \tag{6}$$

$$R = 0 \qquad \text{If } P \geq Ia \tag{7}$$

When the storage volume of water in the soil approaches saturation value (S), and the rate of infiltration becomes constant, the rainfall excess rate then equals the precipitation rate.

If there is a pond within the sub-basin, the pond volume should be satisfied before any runoff occurs beyond the pond. A simple model is used to route the runoff hydrograph along the pond volume. This model uses two parameters: the pond volume (V_pond) and parameter (K) that represent the ratio of the area that feeds into the pond to the total sub-basin area. This factor (K) ranges from 0 (no pond) to 1 (the pond at the outlet of the sub-basin where all the sub-basin area feeds to the pond). Accumulated volume is calculated using this factor and the runoff hydrograph as follows:

$$V = \sum_{i=0}^{t} K\,Q_i \tag{8}$$

At the beginning of the runoff process, a K portion of the calculated runoff is routed into the pond until the pond volume is satisfied (V_pond = V). During this time, the sub-basin outflow equals $[(1-K)*Q_i]$. Once the pond volume is satisfied, the sub-basin outflow equals the calculated runoff (Q_i).

After the runoff hydrograph is generated for the sub-basin and routed through ponds, the resulting hydrograph is routed to the outlet of the next downstream sub-basin. The Muskingum method was used to route the hydrographs. The output hydrograph at any time ($O(t+\Delta t)$) is calculated as a function of inflow at that time ($I(t+\Delta t)$) and the previous time step for the inflow and outflow as:

$$O(t+\Delta t) = C_0\, I(t+\Delta t) + C_1\, I(t) + C_2\, O(t) \tag{9}$$

Where: $C_0 = \dfrac{-Kx + 0.5\Delta t}{K - Kx + 0.5\Delta t}$, $C_1 = \dfrac{-Kx + 0.5\Delta t}{K - Kx + 0.5\Delta t}$, $C_2 = \dfrac{-Kx + 0.5\Delta t}{K - Kx + 0.5\Delta t}$

The routing parameter, *K*, represents the travelling time along the routing channel, *x* is a parameter that defines the weighting factor of inflow and outflow when determining storage volume in the reach (it varies 0.0 - 0.5) and Δt is the time increment for the hydrograph. The parameter *K* can be estimated as equal to ß x T, where ß is a coefficient and T is the travelling time in hours and is calculated from Manning's equation as:

$$T = \frac{L}{V} = \frac{n}{3.6} \frac{L}{R^{2/3} S^{1/2}} \quad (10)$$

Where: V: Velocity (m/sec), n: Manning coefficient, R: Hydraulic radius (m), S: Main channel slope (m/m), L: Main channel length (km).

3.3 Rainfall simulation

A rainfall simulator was constructed and used to simulate rainfall events at various parts of the study area. A set of half-inch PVC pipes were designed in a parallel structure with a span of 15 cm. Holes of about 1 mm diameter were drilled on the PVC pipes at 10 cm intervals in order to simulate uniform intensity over the test sites. Figure 4 shows the plan of the rainfall simulator. The simulator applies water into an area of 1.5 m x 1.5 m from 2.0 m high. The hydrological responses for the experimental sites were documented and, based on the data, hydrological parameters were calculated such as the NRCS CN and runoff coefficient. Rainfall simulation tests were performed at various parts of the study area. As shown in Figure 2, the sites were chosen such that each of the geological soil groups was tested.

The rainfall simulator designed and constructed for this purpose was used to apply water on a 1.5 m by 1.5 m area and the intensity of the applied water was measured using the automatic rainfall recorder. During the testing process the different stages of soil response to the applied water were documented. The time when the water started to show above ground, or the ponding point, was recorded such that the amount of water applied to get to this point was obtained from the rainfall recorder. The time when

Figure 4. Rainfall simulator plan

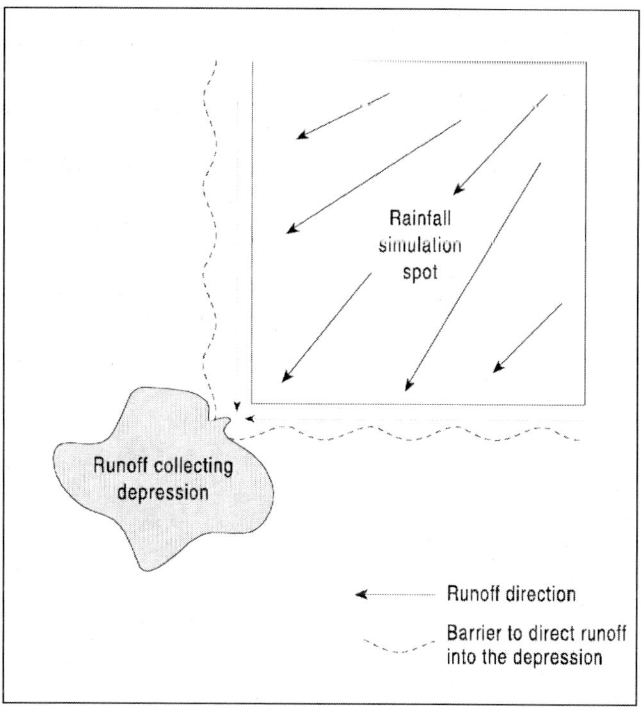

Table 1. Summarized rainfall simulation results and the calculated Curve Numbers

Site	Geological group	Initial abstraction, Ia (mm)	S (mm)	Derived CN	Average CN
1	Bishriyya	3.9	19.4	93	91
2	Bishriyya	6.5	32.5	88	
3	Bishriyya	5.1	25.6	91	
4	Bishriyya	3.9	19.4	93	
5	Asfar	2.9	14.4	95	93
6	Asfar	4.9	24.4	91	
7	Safawi	3.3	16.3	94	94

the water started to run over the ground, the runoff point, was also recorded and the water volume at that point was recalled from the rainfall recorder. Using these soil response points, the initial abstraction of the soil was estimated as the water applied to get to the runoff point. The CN was then calculated for that site using this value.

Tests at 4 simulation sites (sites 1, 2, 3, and 4) were performed in the Bishriyya geological group, which covers most of the study area, as shown in Figure 2. Sites 5 and 6 were in the Asfar group and site 7 in the Safawi group. The results of the rainfall simulation and the average CNs calculated are summarized in Table 1 for the test sites and the associated geological groups.

On sites 3, 6 and 7, the runoff from the test area was directed to depressions that were made at the corner of the test area. The depressions were lined with a plastic sheet to prevent infiltration. After each test, the water in the depression, which is the total runoff from the test area, was measured and the overall runoff coefficient was calculated as the ratio of the runoff depth to the water applied depth. The runoff coefficients for different geological surfaces in the study area are shown in Table 2.

4. Map data

Other types of data that were collected for this study include thematic maps for the study area. Some of them were in print format, such as the topographic maps of scale 1:50 000. Others were in digital form as ArcView shape files (*.shp) and ArcInfo format.

Topographic paper maps, scale 1:50 000, were used to identify the boundaries of the major catchments in the study area. Twelve major catchments with areas 28.70-209.09 km^2 were identified (Figure 5). Each one of these catchments was divided into smaller sub-basins (with areas less than 20 km^2) for modelling purposes.

From the topographic paper maps, the sub-basin boundaries, channel networks and depression data layers for each catchment were developed in ArcView file format to be used in the model. Figure 6 shows these data layers for the Wadi Salma Catchment (#11)

Table 2. Runoff coefficients for various geological groups

Site	Geologic group	Applied water depth (mm)	Runoff volume (cm^3)	Runoff depth* (mm)	Runoff Coefficient** (%)
3	Bishriyya	9.25	4000	1.8	19
6	Asfar	19.5	10500	4.7	24
7	Safawi	7.5	4750	2.1	28

* Runoff depth = runoff volume/area, (area = 1.5 m x 1.5 m);
** Runoff coefficient = runoff depth/applied water depth

Figure 5. Major catchments in the study area

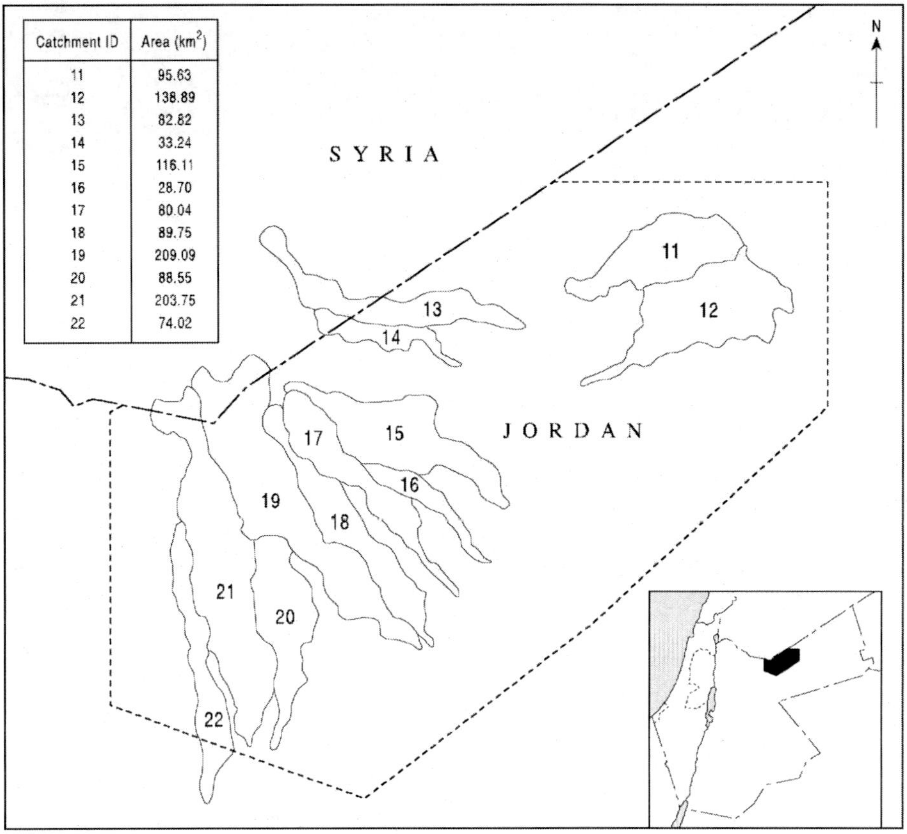

Catchment ID	Area (km^2)
11	95.63
12	138.89
13	82.82
14	33.24
15	116.11
16	28.70
17	80.04
18	89.75
19	209.09
20	88.55
21	203.75
22	74.02

The other data layers, already in digital format, included contour lines, soil units and the geological groups. They were clipped by the catchment boundaries using ArcView® GIS software. In this way the data layers for specific catchments were developed.

The next step was to develop a spatial and relational database for the study area such that the study area major catchments were defined and each catchment was divided into smaller sub-basins. The characteristics of the major catchments and the parameters of their sub-basins were stored in a database. The database has two types of tables, the general parameters of the catchments such as the areas, and the number of sub-basins. The physical and hydrological data associated with each sub-basin were also included in the database (Tables 3 and 4).

5. Management tool outline

The modelling tool includes various stages of simulation. The first step is starting the simulation with all catchments; the software then takes the user smoothly and logically from screen to screen on the computer seeking data entry or demonstrating results. The first screen shows the layout of the catchments in the study area and the main data table, which include the catchment identification numbers, their areas and the number of sub-basins in each catchment (from Figure 5). The second step involves the selection of catchment, upload of its data and display of its map. The sub-basin boundaries, channel

Table 3. Sub-basins' data table structure in the database

I. The sub-basin parameters:	Sub-basin identification number (ID). Sub-basin area (A). Curve number (CN).
II. The collecting channel (the longest streamline in the sub-basin) parameters:	Length (L_col). Slope (S_col).
III. The ponds parameters:	Pond Volume (V_pond). Pond factor (K_pond).
IV. The routing parameters:	Downstream sub-basin identification number (Down_id). Length of the travelling channel in the downstream sub-basin (L_trav). Slope of the travelling channel in the downstream sub-basin (S_trav).
V. Other parameters:	Manning coefficient (n).

Table 4. General table structure in the database

Catchment parameters	Catchment identification number (Catch_id) Catchment area (Area). Number of sub-basins on that catchment (Sub_Num).

network contour lines, and the depression layers data are displayed for the catchment once it is selected for analyses. In this stage of the modelling, three menus are added to the tool to allow the execution of the next steps of the hydrological calculation. These menus include: 'Select', which allows the user to select a new catchment for analysis; 'Modelling' that has items to control the various hydrological calculation steps such as rainfall data entry and project selection; and 'Results' which has items that control the way results are viewed and presented.

The third step defines the modelling control data, such as rainfall, and then calculates the runoff. The 'modelling' menu has five items: 'Enter Rainfall Data', which activates a save file dialog window to save the new table with '*rin*' extension, to differentiate it from other tables, and allows the user to enter the duration of the rainfall, the time, and the rainfall volume for each time increment up to the duration of the rainfall; 'Load Rainfall Data', which allows the user to choose the rainfall table and load it into the modelling system; 'New Project', which allows the user to define a new project and save it with '*out*' extension to differentiate it from other types of files such as the rainfall files; 'Load Project', which gives the user the option to retrieve saved projects for the selected catchment, which activates a dialog that allows the user to select the name of the project to be loaded, and add the runoff table of that project to the system; and then the 'Run Model' (disabled by default), which becomes active and thus, once the user defines a new project or loads an old project, the simulation is performed.

The last step is presenting the results. The 'Results' menu has items that control the way to view and present the model-generated data and the results of the simulation. The 'View Rainfall Data' activates a window with the rainfall data tabulated and also gives the user an option to change the data and update the rainfall data table; 'View Model Data' displays the project name, description and simulation parameters; and, finally, 'View Runoff Results' displays the runoff calculation results.

Figure 7. Runoff hydrograph for the rainfall storm of December 9th, 2000

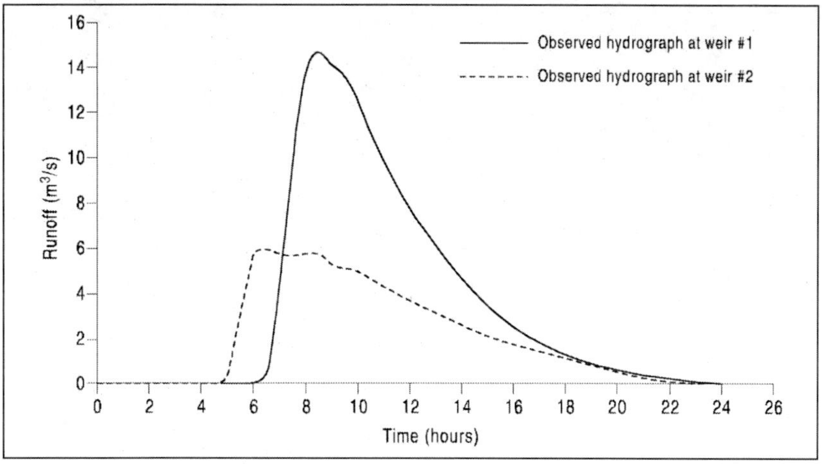

Figure 8. Runoff hydrograph for the rainfall storm of April 4th, 2001

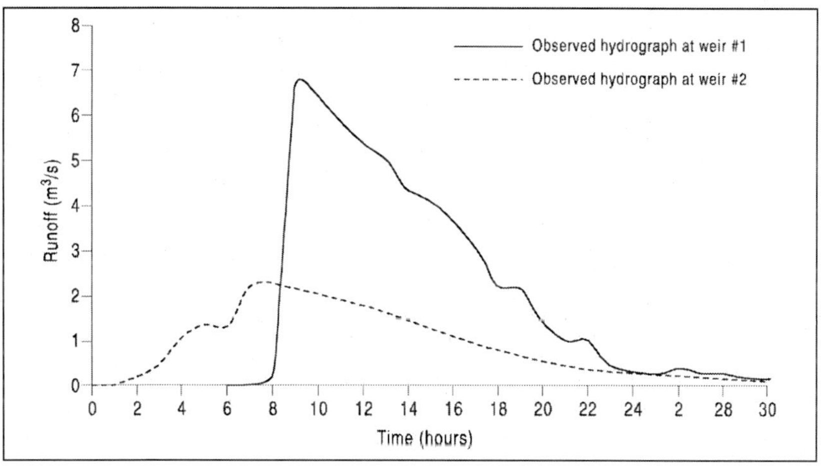

K and greater than *2*x*K*, five combinations of *K* and *x* parameters were used in Muskingum routing.

Figure 10 shows the simulated hydrographs for the storm of December 9th, using different combinations of the routing parameters. The absolute difference between the simulated hydrograph and the measured hydrograph for these combinations showed that using *K* parameter equal to the travel time (ß = 1.0) and *x* values as 0.18 gave the closest simulated hydrograph to the measured hydrograph.

The modelling parameters and the sub-basins data (Table 6) were used with the rainfall data to calculate the runoff hydrographs and volumes for the catchment sub-basins. The results of the simulated hydrographs at weirs #1 and #2 (Figure 6) were plotted with the measured hydrographs at the same locations as shown in Figures 11 and 12 for the storms of December 9th, 2000 and April 4th, 2001 respectively.

For the storm of December 9th, 2000 simulation (Figure 11) the results show that at weir #1, (the whole catchment), the simulated and measured peak runoff rate were 13.87

Figure 9. Estimating Ia and S′

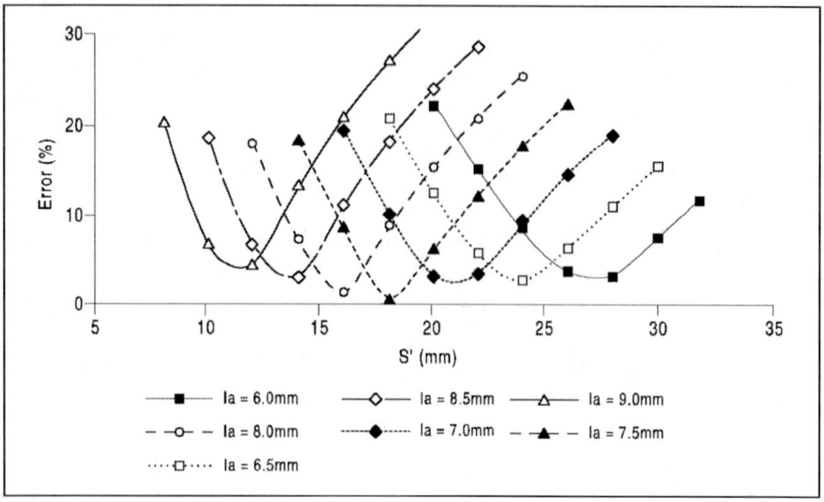

Figure 10. Simulated hydrographs for various routing parameters

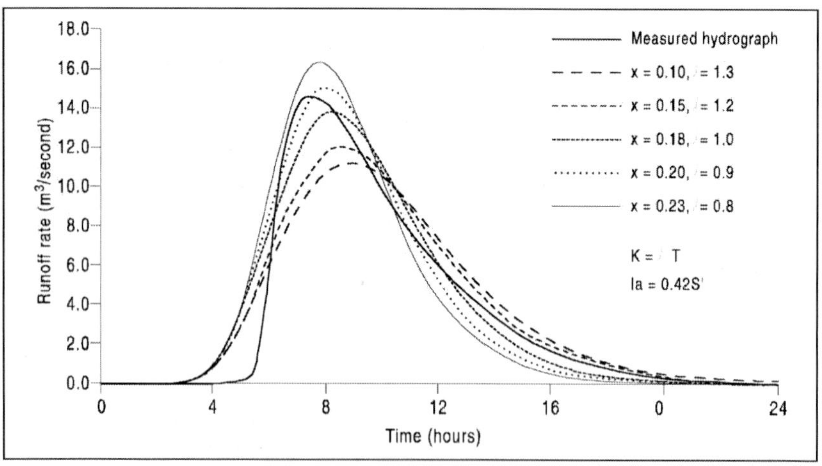

and 14.70 m^3/s respectively, while at weir #2 (outlet of sub-basin 1107), the simulated and measured peak runoff rate were 6.13 and 5.92 m^3/s respectively. On the other hand, the total simulated runoff volume (at weir #1) was 0.32 MCM which is very close to the measured runoff volume of 0.30 MCM, while for part of the catchment (at weir #2), the simulated and measured volumes were 0.10 and 0.18 MCM respectively.

For the storm of April 4th, 2001 simulation (Figure 12), the results show that at weir #1, (the whole catchment), the simulated and measured peak runoff rates were 7.07 and 6.82 m^3/s respectively, while at weir #2 (outlet of sub-basin 1107), the simulated and measured peak runoff rates were 2.81 and 2.27 m^3/s respectively. On the other hand, the total runoff volume (at weir #1) is simulated as 0.18 MCM and measured as 0.20 MCM while for part of the catchment (at weir #2), the simulated and measured volumes were 0.06 and 0.10 MCM respectively.

Table 6. Sub-basins data for catchment 11 (Wadi Salma). For key see Table 3

ID	A (km^2)	L_Col (km)	S_Col (m/m)	V_Pond (m^3)	K_Pond	Down_ID	L_Trav (km)	S_Trav (m/m)	n	CN
1101	19.26	7.62	0.0098	116000	0.8	1103	3.30	0.0152	0.05	93
1102	10.31	6.30	0.0127	0	0.0	1105	4.30	0.0119	0.05	93
1103	8.77	3.30	0.0152	0	0.0	1106	2.81	0.0053	0.05	93
1104	6.05	4.34	0.0150	32000	0.3	1107	3.12	0.0080	0.05	93
1105	13.78	5.04	0.0119	0	0.0	1108	3.20	0.0136	0.05	93
1106	4.48	2.81	0.0053	0	0.0	1107	3.12	0.0080	0.05	93
1107	4.69	3.12	0.0080	0	0.0	1109	4.00	0.0091	0.05	93
1108	7.75	5.15	0.0136	0	0.0	1109	4.00	0.0091	0.05	93
1109	13.26	7.13	0.0091	0	0.0	1110	4.04	0.0088	0.05	93
1110	8.25	5.68	0.0088	0	0.0	1110	0.00	0.0000	0.05	93

Figure 11. Measured vs simulated runoff of December 9th, 2000 storm

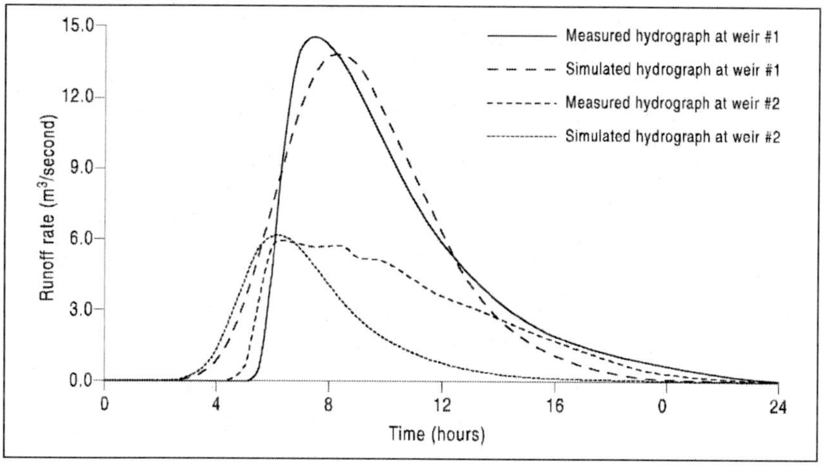

Figure 12. Measured vs simulated runoff of April 4th, 2001 storm

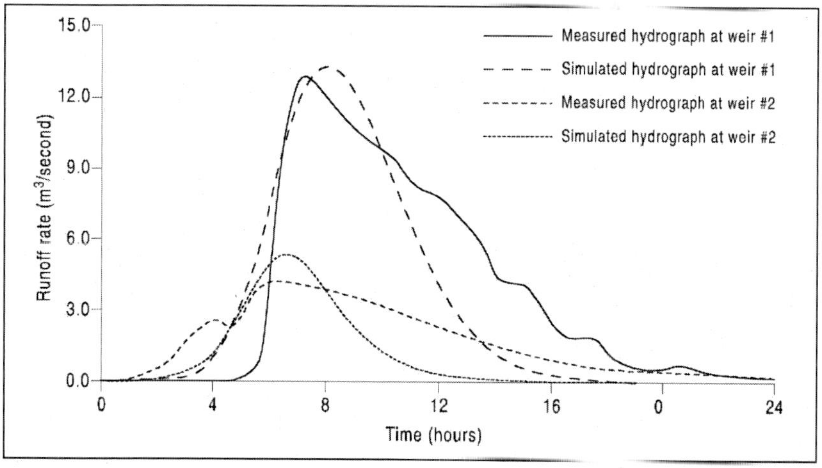

7. Discussion and conclusion

The management tool developed includes various steps of simulation used to estimate the runoff volume and peak discharge in the Badia.

The calculated values are with acceptable accuracy for the whole catchment (at weir 1); the difference in peak discharge estimation was 4-6 % and in runoff volume was 7 % for the first storm and 10 % for the second one. For parts of the catchment (weir 2), the peak runoff rates are estimated with an error margin of 4 % for the first storm and 24 % for the second storm, while the runoff volumes are estimated as 55-60 % of the measured volumes.

The results of the rainfall simulation showed varied hydrological responses for the different types of land covers in the study area. The runoff was generated faster in the Safawi Geological Group and relatively slower in the Bishriyya Geological Group. In general all the land covers were highly impervious which resulted in relatively high CNs. The calculated CNs were 91, 93, and 94 for the Bishriyya, Asfar and Safawi Geological Group surfaces respectively. The texture of the soil and the high rock and stone covers for the surfaces explain these high CNs. Considering the small depths of water applied, 7.5-19.5 mm, the results of rainfall simulation showed that a good portion of the applied water transformed into runoff; in some cases up to 28 % of the applied water ended up as direct runoff. These rainfall simulation results are close to these calculated from the observed rainfall and runoff records, where runoff coefficients were 13-20 %.

Given the geomorphology of the area, which is defined by its boulder cover and the distribution of mud flats, the estimated values of the runoff coefficients are relatively high. The first storm (with a higher rainfall) gave a higher runoff coefficient than the second storm. This comparison is expected since the infiltration rate drops very rapidly; once the initial abstraction is satisfied, most of the remaining rainfall is translated into runoff.

The overall assessment of the hydrological response of the study area indicates poorly drained land-cover and low infiltration rates; a relatively large proportion of the rainfall, after the initial abstraction, ends up as direct runoff.

Acknowledgements

This project was partially funded by University of Central Florida and BRDC. We thank the staff of BRDC field station in Safawi for their help in field data collection.

References

AlAyyash SM and Nnadi FN (2001). Surface water management in arid lands of Jordan with GIS application. Proceeding, Decision Support Systems For Water Resources Management Conference: AWRA, Snowbird, Utah pp. 419-423.

HCST (1992). Geomorphology and physical resources. Technical report, Higher Council for Science and Technology; Jordan Badia Research and Development Programme, Amman, Jordan.

Roberson JA, Cassidy JJ and Chaudhry MH (1998). Hydraulic engineering; 2nd edition. USA: John Wiley & Sons.

Taqieddin S, Al-Homoud A, Awad A and AlAyyash S (1995). Geological and hydrological investigation of a water collection system in arid Jordanian lands. Environmental Geology 26.

Wanielista M, Kirsten R and Eaglin R (1997). Hydrology, water quantity and quality control. 2nd edition, USA: John Wiley and Sons Inc.

CHAPTER 23

Water resources management for rain-fed agriculture in the north-east Badia, Jordan

R. S. Shatnawi
Department of Civil Engineering, Applied Science University, Amman-Jordan.
E-mail: rania_sha@asu.edu.jo, or rania.shatnawi@gmail.com

S. M. AlAyyash
Badia Research and Development Centre, Jordan

C. R. Cratchley
University of Wales, Bangor

J. A. Kay
c/o Dr Cratchley, University of Wales, Bangor

In the Badia there are a number of farms on *marab* which attempt to raise winter crops utilizing sporadic inundation by wadi flows. Outcomes vary greatly from year-to-year, as the winter rainfall varies greatly. Recent research has recognized important correlations between global climatic indices and the total amount and duration of winter rains in Jordan. In particular the effects of the El-Niño Southern Oscillation (ENSO) on Jordanian rainfall have been studied. The results of the study showed that there is a significant correlation between the global climatic indices (ENSO SST and SOI) and the Jordanian rainfall. The correlations have been used to develop a rainfall prediction model for Jordanian rainfall. The developed rainfall prediction model was linked with the traditional approaches that have been practised by the local people of the north-east Jordanian Badia. The link showed that the rainfall predictability in Jordan could be enhanced. The combined prediction model also offers the potential for forecasts to be made and confirmed at intervals during summer and the early months of autumn, generating forecasts with high reliability and thus contributing to water resources management in Jordan.

The research is an initial step towards providing the approach throughout broader areas of Jordan. This work in Jordan could be a precursor to programmes elsewhere in the Middle East that could utilize similar correlations and integrate them with locally used methods of forecasts to develop area-specific sets of forecasting criteria. Proactive management of rain-fed agriculture and natural pastures on the desert margin offers disproportionate benefits to the subsistence farmers currently struggling to survive in these inhospitable areas.

Keywords: Global climate indices, rainfall prediction model

1. Introduction

The Badia region comprises the eastern part of Jordan. It covers an area of 72 000 km^2, more than 80 % of the total area of Jordan. The Badia area is divided into three regions, north-eastern, central and south Badia. The word Badia refers to the place where the Bedouin live. The north-east Badia has an area of 25 600 km^2 that comprises 35.5 % of the country's total area.

The north-east Badia of Jordan is one of the most economically deprived areas in the country, where the rainfall is the most unpredictable element. Jordan experiences only two seasons, a rainy season and a dry season. The rainy season lasts from October through April, during which the country receives 85 % of its total annual rainfall. Average annual rainfall varies from over 300 mm per year in the north and north-west parts of the Badia to less than 50 mm per year in the south and south-east.

There are many depressions or pans in the Jordanian Badia, composed of fine-grained sediments, which are called *marab* or mudflats. *Marab* have formed as a result of regional and local faults and other geological processes that caused the formation of small depressions, where sediments have accumulated from standing water fed by ephemeral wadis discharging into mudflats (Ibrahim 1997). *Marab* are regularly flooded by water from the wadi flow; this water drains onwards into the wadi as the flood recedes. The excess water leaves the marab through a well-defined channel.

The *marab* are nearly flat with relatively deep soil and receive floods from the surrounding catchments. Flooding of the *marab* is short-lived, occurring because of short-term high water levels in the adjacent wadi. Flood water rarely accumulates on the soil surface because of the gentle slope of the *marab*, and rapid recession of the wadi hydrograph. This *marab* soil remains saturated for several days after each flood during the winter season. The estimated area of the *marab* in the Badia area is about 0.2 million ha, (Abu-Zanat 2002). In the last 20-30 years, the best areas of the *marab* have been cultivated with wheat and barley, relying on spate irrigation.

In the past the Bedouin used to be more nomadic than today, because they were travelling within and outside the country before the borders with the neighbouring countries were closed. Recently the lifestyle of the Bedouin has changed towards a settled style of life. This change in lifestyle has increased the demand on infrastructure, facilities and services in the Badia area. Also, the number of villages and settlements in the Badia area has increased which encourages the expansion of cultivated areas and coincides with a shift from nomadic pastoralism to farming-based systems, (Millington et al 1999). Only five percent of the Badia population is still nomadic. The rest is now settled and involved in agriculture, livestock production, the civil service, commerce, the armed forces, and mining and prospecting industries (Allison et al 1998).

The majority of the Bedouin farmers' income comes from the sale of their animals and animal products like meat and yoghurt. Their main concern is to have enough fodder for their livestock (sheep and goats), especially during the summer months. The cultivation of the *marab* is still primitive, as most of the farmers who cultivate their *marab* use traditional approaches, hoping to have sufficient winter flood events at intervals of the growing season to maintain soil moisture and thus to sustain the crop until maturity. The decisions they make are described later, and depend on both seed type and local rainfall. Achieving a better understanding of the duration and quantity of winter rains is a key to improving the agricultural economy of the Badia; this study has examined methods for better agricultural water management in the *marab* areas.

Table 1. Marab Hassan characteristics

Size (km^2)	3.5
Catchment area (km^2)	357
Mean annual rainfall (mm)	93.1
Median length of wet season (month)	4
Medium number of runoff events	6
Soil type	NAT (loamy texture)
Depth of soil profile (cm)	100-200
Flow pattern across the marab	Channel from the eastern wadi and sheet (overland) flow from the western wadi

Figure 1. Marab Hassan catchment area

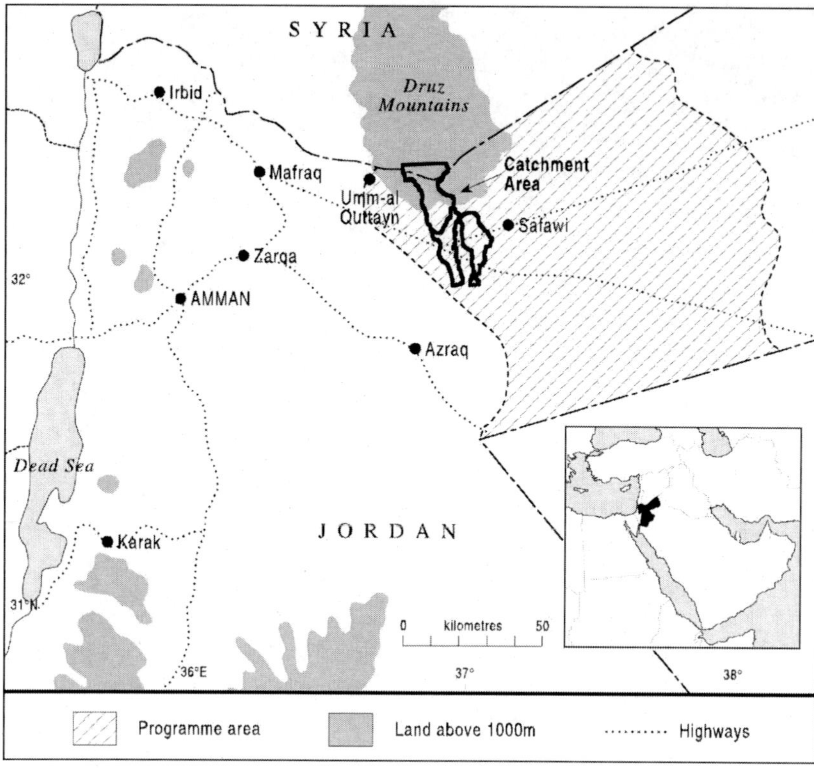

2. Study area

The study area is the Marab Hassan catchment with an area of 357 km² and is located in the north-western part of the north Badia as shown in Figure 1. The catchment consists of three wadis that flow from north to south until they reach a large mudflat, Marab Hassan. Marab Hassan has an area of 3.5 km² and has been cultivated since 1994. Average annual rainfall over the catchment area varies between 300 mm in the north-western part to 50 mm in the south-eastern part. The *marab* floods three to four times in a wet year.

Marab Hassan is located 35 km south-west of Safawi, and almost 10 km from Azraq. From aerial photographs of 1:25 000 scale, (Figure 2), 1:50 000 geological maps and ground truth inspections, it can be concluded that Marab Hassan has formed as a result of the existence of natural structural systems (faults) in the area. The *marab* takes almost a pear shape and has two inlets and two outlets, one major outlet and a minor one that only functions in cases of high floods. Table 1 summarizes the characteristics of Marab Hassan.

3. Indigenous approaches to *marab* cultivation in the Badia area

Traditional approaches to cultivation depend on surface water from the rainfall for irrigation ie they are rain-fed. Most of the rainfall in Jordan occurs between October and May, mainly concentrated between December and March. Usually farmers in the Badia area prepare their land for cultivation in late summer, and then they wait for the rainfall to come.

The farmers have two types of farming; wet and dry farming. In both kinds the *afeer* concept is used, which is the Arabic for seed broadcasting. Barley is the crop that is

Figure 2. Original aerial photograph of Marab Hassan (scale 1:25,000)

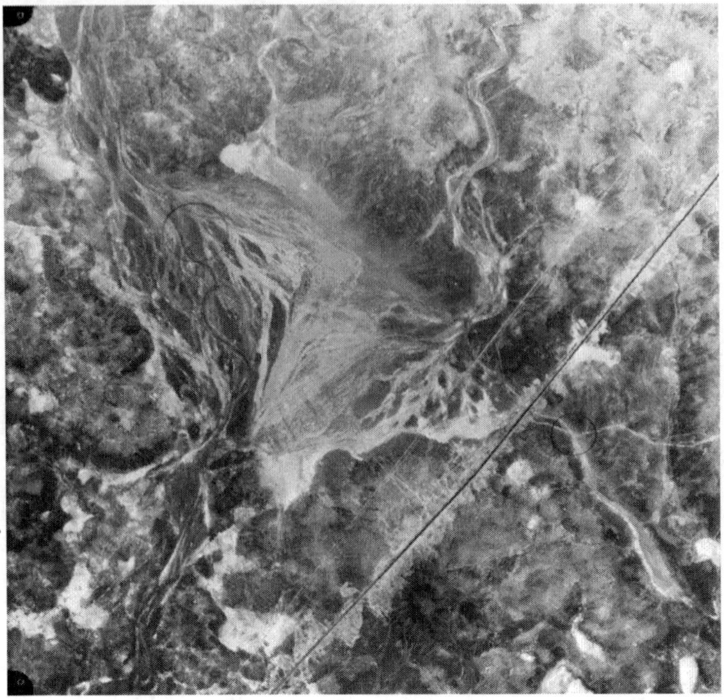

mostly cultivated, as it is known for its drought tolerance. If they have a local variety of seed they sow it before the *marab* receives any flood (ie dry farming). If the local variety of seeds is not available then the wet farming approach is used, ie they wait until the *marab* receives floodwater. Provided this happens before 20th November they sow the seed, otherwise they just leave the land bare without cultivation for that year.

To evaluate this approach the onset and end of the rainy season have been analysed. Three stations have been selected, two are outside the catchment and one is inside. The outside stations are Umm al-Quttayn and Dayr al-Kahf in the north-western part of the catchment. Al-Ariyyatayn is located inside the catchment area as shown in Figure 3.

3.1. Water availability and its importance to barley cultivation

Barley is the main crop that is cultivated on the *marab*. The optimum time to sow the barley seeds is between the last week of September and the middle of October. Rainfall has the highest contribution to grain yield during December and January (Oosterom et al 1993). Depending on variety and seasonal growing conditions, barley requires 90-120 days to reach maturity. Late seeding will increase the risk of lower yield. But, as mentioned before, what controls the seeding time is the availability of the 'local seeds' and/or the flooding of the *marab*. The most important factor that affects the barley is to have pre-sowing flood, or irrigation water in the case of irrigated cultivation, as soil needs to have adequate moisture content. Whatever amount of water the crop receives afterwards is not as critical as the pre-sowing amount. But, normally one or two floods later in the season around December and January are very helpful, especially for the grain yield. This requirement coincides with the 'traditional method' of cultivation that is *afeer*, when people wait to sow the seeds after the *marab* receives the floodwater, using non-local seeds.

Figure 3. Rain gauge distribution in Marab Hassan catchment area

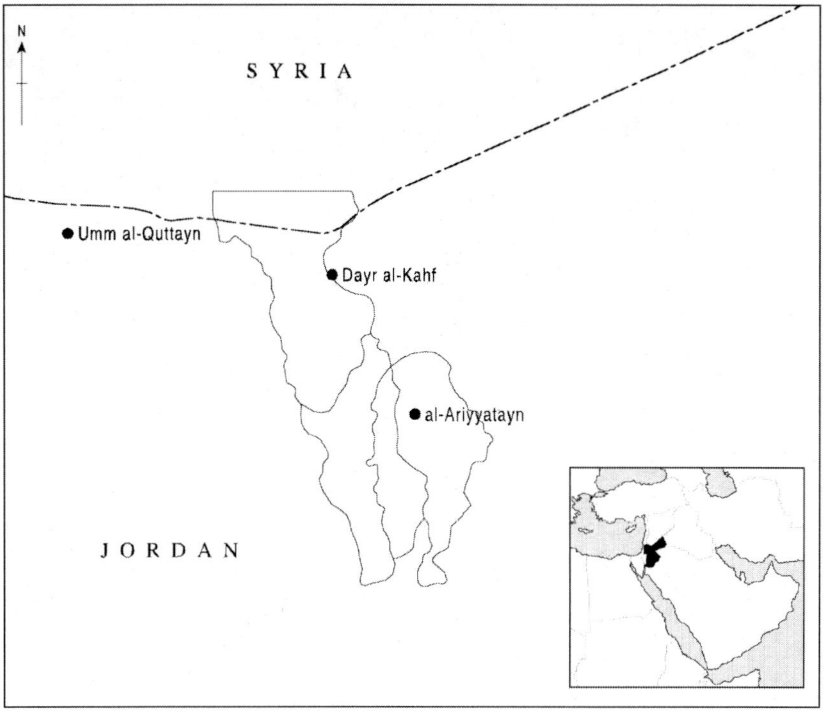

The present traditional farming practices carry a significant risk of the crop failing to reach maturity. However, it seems clear that there is the potential to improve the productivity of these areas, by improving water management through understanding the winter rainfall season better.

4. Predictive approach in rainfall forecasting

The economic and social impacts of rainfall variability are especially significant on the desert margin and this makes it important to understand the nature and causes of this variability. Increased world-wide awareness of rainfall irregularity and recent studies that explain this variability (in whole or in part) in terms of global climatic phenomena have raised the possibility of predicting year-to-year variations. A number of studies have confirmed that El-Niño Southern Oscillation (ENSO) has a significant impact on rainfall variability over different parts of the globe (Beltrando et al 1993, De Rojas 2000, Nezamosadat 1999, Nezamosadat et al 2000, Osman et al 2001, Shatnawi 2002).

4.1. Possible correlations

The Southern Oscillation (SO) phenomenon is defined as the fluctuation in atmospheric pressure fields within the Pacific region, and is associated with the El-Niño phenomenon, (Glantz 2001). Because the El-Niño and Southern Oscillation phenomena are closely interlinked, the names are sometimes combined and the term ENSO is used to address both events. According to this terminology El-Niño refers to the warm phase of the SO and La-Niña (which is the opposite of El-Niño) is the cold phase of the SO. The Southern Oscillation Index (SOI) is used to measure the strength of the El-Niño. SOI is the pressure difference between Tahiti in the east and Darwin in the west. High negative values of SOI represent an El-Niño or a warm event.

Recent research to forecast the duration and quantity of winter rains in the Badia has demonstrated correlations between winter rainfall and global climatic indices associated with El-Niño (Shatnawi 2002). The correlations are sufficiently reliable to allow long-term forecasts of winter rains to be made. Such long-term forecasting creates the potential for pro-active management of desert margin agriculture. Integration of the new methods with the traditional forecasting technique based on the timing and amount of early rains is advocated for adoption by the local community.

The methodology that has been used in this study considered testing two approaches in rainfall prediction in the Jordanian Badia, particularly in the Marab Hassan catchment area. The approaches are the predictive approach and the traditional approach (or the use of the ethnoclimatological element) that have been used in rainfall forecasting.

4.2. Rainfall-SOI correlation

The predictive approach depended on testing the correlation between the rainfall in the catchment area and a global climatic index, which is the El-Niño Southern Oscillation Index (SOI). To test for correlations, sets of monthly rainfall for three rain gauges as well as monthly global climatic indices were used. The gauges as shown in Figure 3 are Umm al-Quttayn, Dayr al-Kahf and al-Ariyyatayn. Umm al-Quttayn and Dayr al-Kahf are located outside the catchment area, while al-Ariyyatayn is inside the Marab Hassan catchment.

For each of the three stations the Average Rainfall Index (ARI) was derived using (1).

$$ARI = \frac{R - \bar{R}}{\sigma_R} \qquad (1)$$

Where: R is the seasonal rainfall of the station; $\bar{R}$ is the average seasonal rainfall for the same station; σ_R is the standard deviation of the seasonal rainfall values.

The hydrological year concept has been used in this study for both the rainfall and SOI. The seasonal (winter) rainfall was taken between September and May. The reason for considering the rainfall between September and May is that the rain-fed agriculture in the area uses water supply that is essentially a winter activity. This omits from consideration the very occasional and, usually, localised summer rains, which have minimal relevance to winter agriculture. SOI was averaged over three-month blocks for the previous hydrological year. The sequential seasons are the previous autumn (SON-), the previous winter (DJF-), the previous spring (MAM-) and the previous summer (JJA-), as shown in Figure 4. ARI was then linearly correlated with the sequences of SOI.

4.3. Results of correlation

Dayr al-Kahf did not show very significant correlation with the SOI, while Al-Ariyyatayn and Umm al-Quttayn did have significant correlation. Table 2 summarizes the correlation results.

Figure 4. Illustration of the previous and current hydrological year terminology

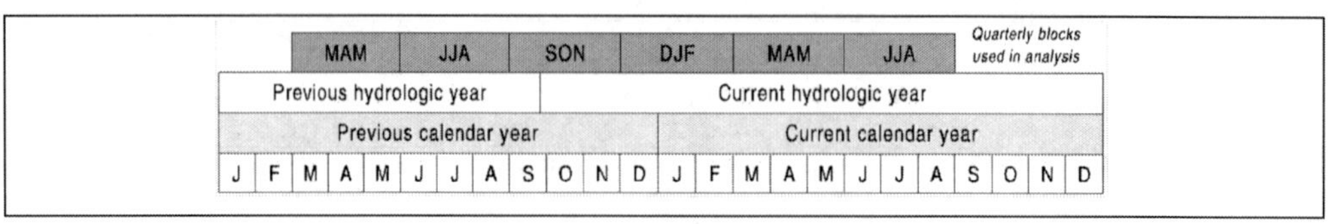

Table 2. Linear correlation results between seasonal ARI and seasonal SOI

SOI	ARI		
	Umm al-Quttayn	al-Ariyyatayn	Dayr al-Kahf
SON-	-0.11	-0.06	0.12
DJF-	0.03	-0.20	-0.03
MAM-	0.21	0.01	0.13
JJA-	0.27	0.29	-0.01

Figure 5. Scatter plot of the data showing the Umm al-Quttayn seasonal ARI and MAM- SOI, for the years 1953-1989. The coefficient of correlation is 0.21

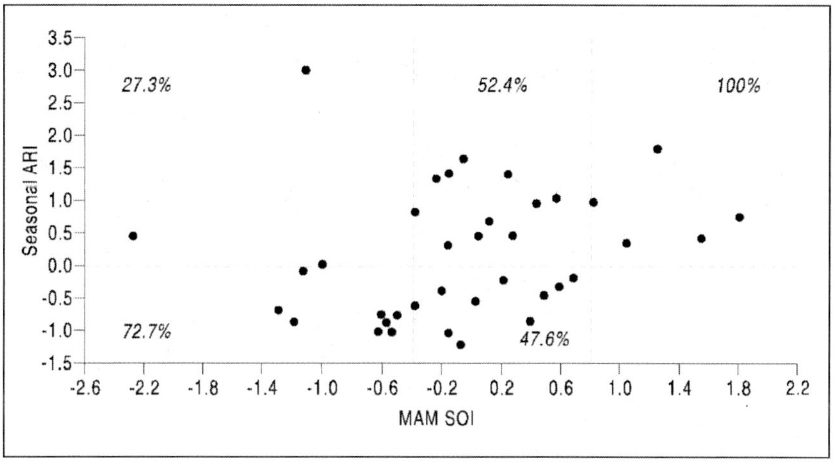

As shown in Table 2 ARI for Umm al-Quttayn is significantly correlated with MAM- and JJA- SOI. Because Umm al Quttayn has the most significant correlation it has been used in this study.

Scatter diagrams of SOI vs. ARI for Umm al-Quttayn were developed using MAM- and JJA- SOI as rainfall predictors. For Umm al-Quttayn, rainfall data are available between 1953 and 2000; the period 1953-1989 was used to build the correlation (rainfall prediction model) and the last 11 years (1990-2000) were used to verify it and check its efficiency.

The scatter diagram between seasonal ARI and MAM- SOI for Umm al-Quttayn is shown in Figure 5.

The following summarizes the procedure of developing the scatter diagram.

- The diagram is a result of plotting ARI vs. MAM- SOI, then dividing the plot into windows according to the selected ARI and MAM- SOI limits;
- The ARI limit that has been selected is 0. If ARI > 0 the year is considered a wet year, while if ARI < 0 the year will be considered dry;
- Three limits have been selected for MAM- SOI, which are representing cold phase (MAM- SOI > 0.8), warm phase (MAM- SOI > -0.4) and normal phase (-0.4 < MAM- SOI < 0.8);
- The diagram was used to develop the rainfall prediction model using MAM- SOI as a rainfall predictor;
- According to the diagram, the probability of having a wet winter in Jordan with a cold ENSO phase is 100 %;

- The probability of normal ENSO and a dry winter year is 47.6 %, while a likelihood of having normal ENSO phase and a wet winter in Jordan is 52.4 %. If the SOI is in its cold phase, the probability of a dry winter occurring is quite high (72.7 %).

Table 3. Summarizes the rainfall prediction model for Umm al-Quttayn using MAM-SOI as a rainfall predictor.

The developed model based on Figure 5 was tested and verified using rainfall and SOI records for the 11 years 1990-2000. It gave 82 % efficiency, as it predicted eight dry years and three wet ones, while in reality the number of dry years was six and the number of wet years was five, Table 4 shows the results of the model verification.

The maximum linear correlation coefficient (Cc = 0.27) for Umm al-Quttayn station is found between the seasonal rainfall and the previous summer JJA- SOI. This correlation coefficient (Cc) is statistically significant. The scatter diagram between the seasonal ARI and JJA- SOI has been developed as shown in Figure 6. This diagram is based on choosing one limit for ARI that is 0. A positive ARI value indicates a wet year, while a negative value represents a dry year. The JJA- SOI limit is –0.2; values > -0.2 indicate a cold phase and values < -0.2 point to a warm phase. The rainfall probability model is summarized in Table 5.

Table 5 is based on Figure 6, each showing that the probability of having a wet or a dry year depends on the value of JJA- SOI.

Table 3. Rainfall prediction model for Umm Al-Quttayn station using MAM- SOI

MAM- SOI category	Probability of rainfall year category	
	Dry	Wet
Cold	0.00	100
Normal	47.6	52.4
Warm	72.7	27.3

Table 4. Verification of the probabilistic model using MAM- SOI

Year	MAM- SOI	Predicted ARI	(Observed ARI) class	Evaluation
1990	1.41	Wet	(0.85) Wet	T
1991	0.10	Wet	(1.09) Wet	T
1992	-1.55	Dry	(0.49) Wet	F
1993	-1.62	Dry	(-0.24) Dry	T
1994	-1.34	Dry	(-0.02) Dry	T
1995	-1.75	Dry	(-0.46) Dry	T
1996	-0.80	Dry	(0.07) Wet	F
1997	0.63	Wet	(1.80) Wet	T
1998	-1.68	Dry	(-1.38) Dry	T
1999	-1.98	Dry	(-0.96) Dry	T
2000	-1.11	Dry	(-1.24) Dry	T

Table 5. Rainfall probabilistic model for Umm al-Quttayn station using JJA- SOI

MAM- SOI category	Probability of rainfall year category	
	Dry	Wet
Cold	38.0	62.0
Warm	62.5	37.5

Figure 6. Scatter plot of the data showing the Umm al-Quttayn seasonal ARI and JJA- SOI, for the years 1953-1989. The coefficient of correlation is 0.29

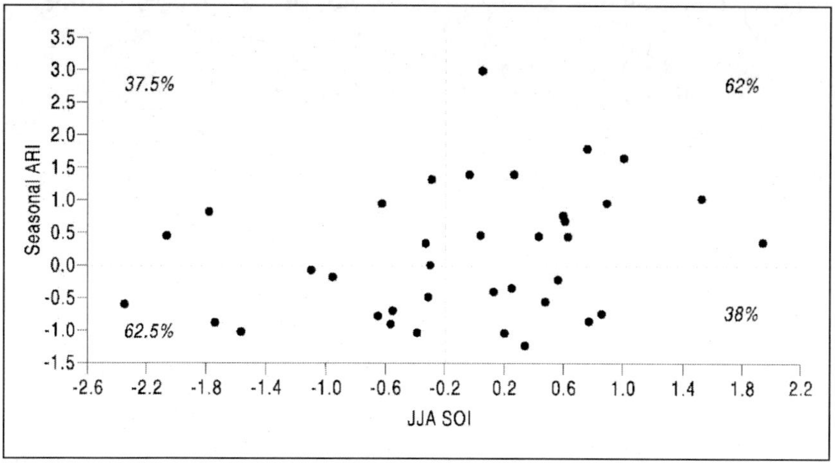

Table 6. Verification of the probabilistic model using JJA- SOI

Year	JJA- SOI	Predicted ARI	(Observed ARI) class	Evaluation
1990	0.31	Wet	(0.85) Wet	T
1991	-0.07	Wet	(1.09) Wet	T
1992	-0.64	Dry	(0.49) Wet	F
1993	-0.82	Dry	(-0.24) Dry	T
1994	-1.66	Dry	(-0.02) Dry	T
1995	-1.73	Dry	(-0.46) Dry	T
1996	0.01	Wet	(0.07) Wet	T
1997	0.88	Wet	(1.80) Wet	T
1998	-2.08	Dry	(-1.38) Dry	T
1999	1.12	Wet	(-0.96) Dry	F
2000	0.16	Wet	(-1.24) Dry	F

The model was developed using 1953 to 1989 data, and it was then verified using the more recent data from 1990 to 2000. The model anticipated six wet and five dry years, but the actual figures show that there were four wet and seven dry years, ie the model's efficiency is 73 %. The detailed analysis of the model's efficiency is shown in Table 6.

5. Traditional approaches of rainfall prediction

A survey was carried out using semi-structured interviews with local people in the Badia area about the rainy season as well as the traditional approaches in cultivation that have been used across the *marab* in the north-east Badia area. Interviews were conducted with those who have lived in the catchment area for a long time, typically more than 20 years.

The interviewees were asked about the number of flood events, frequency, when the events happened and the rainfall duration. The major findings from these interviews are the following:

- Early rainfall at the beginning of the season (October and November) is locally known as *hareef*, while rainfall at the end of the season (March, April and May) is known as *khareef*. *Hareef* and *khareef* both occur in a thunderstorm type of rainfall

that lasts for a short period of time, 10-45 minutes. This type is the one responsible for floods in the wadis in the catchment area as elsewhere in the Badia;
- Mid season rainfall is termed as *daim* and occurs December to February. *Daim* is particularly frontal type rain that affects the area for more than 3-4 days. If the front lasts for up to three days the rainfall is called *naou;*
- The wadis in the catchment area flow almost every year, but the frequency of the floods varies. Most of the people recalled floods occurring twice or three times or sometimes four times in a wet year;
- If the rainfall starts early (significant rainfall before the 20th of November at the latest) this indicates that the rainy season will be good, as well as longer than normal at this location;
- The onset of winter rainfall is location dependent.

6. Verification of traditional knowledge in rainfall prediction

6.1. Onset of winter rainfall

Traditional knowledge suggests that the earlier the winter rains start the better and longer they will be. It was decided to examine the historical record to see if there is a relationship between the time of the onset of winter rains and their overall amount and duration.

Because irrigation in the area is dependent on wadi flows, it follows that winter rainfalls become significant when they cause wadi flow. Based on runoff plot experiments carried out in different locations in the Marab Hassan catchment, a 5 mm limit was adopted, because it is a threshold at which surface flow is often initiated.

The timing of the first significant rainfall (ie 5 mm or more) was compared with the total seasonal rainfall and with the duration of winter rains. Moving further east and south the rainy season starts later and finishes earlier. This means that the reliable onset of rainfall might also be later and the 5 mm limit as a first significant rainfall might also change.

Figures 7 and 8 show the first and last significant rainfall vs. the total seasonal rainfall for Umm al-Quttayn station. Figure 7 shows a gentle trend that indicates the later the first significant rainfall the lower the seasonal rainfall. The longer the rainy season lasts the higher it is expected the seasonal total rainfall will be, as Figure 7 shows. The Figure shows a well-defined trend of the seasonal total rainfall as a function of the end of the rainy season.

The two Figures lead to the following conclusions:
- The earlier the winter rains start the longer the winter rainy season will be;
- The later the first rain the shorter the rainy season;
- The longer the winter rainy season: the higher the total amount; and the larger the number of significant rainfall events.

Taking the onset of the rainfall as the 10th of November (as shown in the solid line in Figure 7):
- If rainfall starts before 10th of November virtually there is no chance that the seasonal rainfall total will be below 100 mm;
- If rainfall starts after the 10th of November there is a significant possibility of a very dry year. Eight out of twenty-six years have total seasonal rainfall less than or equal to 100 mm, and two of them have total seasonal rainfall of only about 50 mm.

If the onset of the rainy season is taken as 20th of November (the dashed line in Figure 8) and taking the rainfall limit of 200 mm:

Figure 7. First significant rainfall for Umm al-Quttayn station

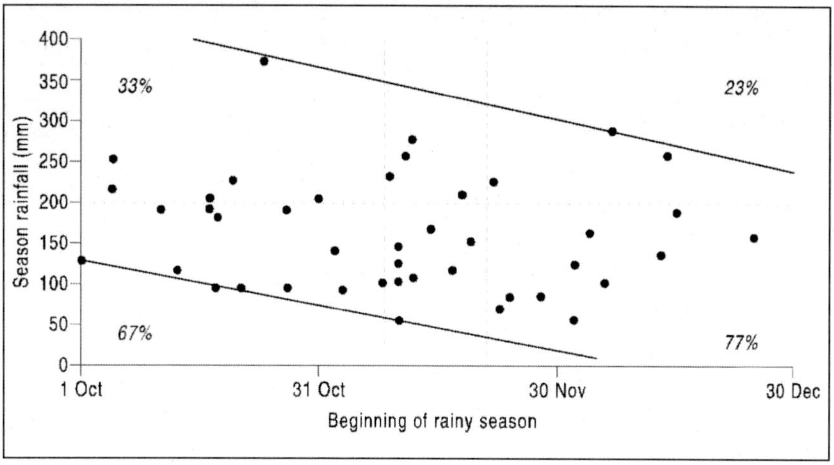

Figure 8. Last significant rainfall for Umm al-Quttayn station

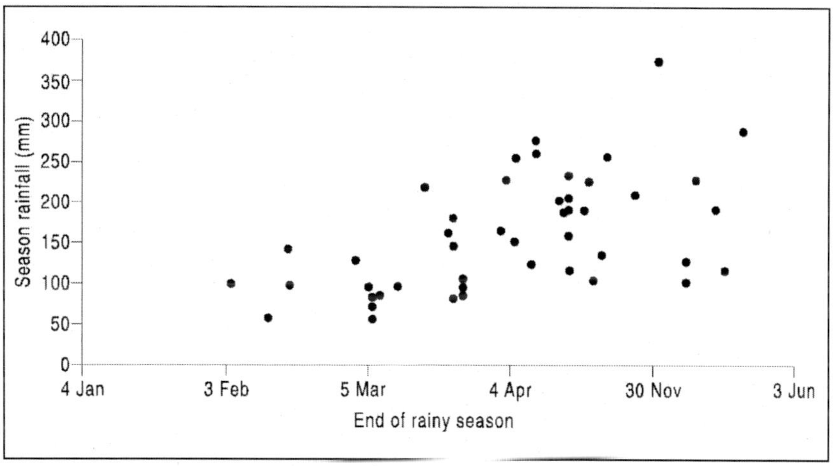

- If the rainfall starts before the 20th of November there is a 33% chance of having total rainfall over 200 mm and a 67 % chance for the rainfall to be less than 200 mm;
- But, if the rainfall starts after 20th of November, the possibility for the rainfall to be above 200 mm is 23 % compared with a 77 % chance of having rainfall less than 200 mm.

The relationships between the number of the rainy days, the number of the rainy days of rainfall higher than 5 mm and the length of the rainy season on the seasonal total rainfall for the same stations have also been examined. The major finding was that the seasonal total rainfall is a function of the number of the rainy days as well as the number of the rainy days with rainfall greater than 5 mm. The trend was very clear for all the stations, examples are shown in Figures 9 and 10 for Umm al-Quttayn station. Obviously, the longer the rainy season the higher the seasonal total rainfall, as illustrated in Figure 11 for Umm al-Quttayn station.

Figure 9. Number of rainy days vs annual rainfall for Umm al-Quttayn station

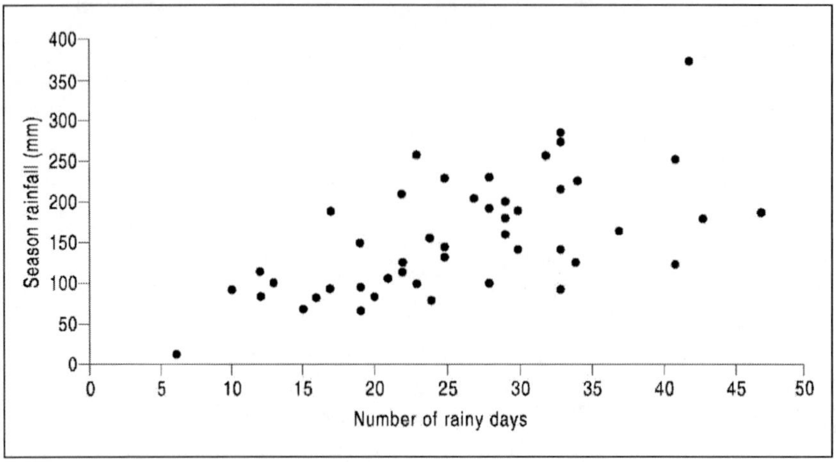

Figure 10. Number of rainy days of rainfall >5 mm for Umm al-Quttayn station

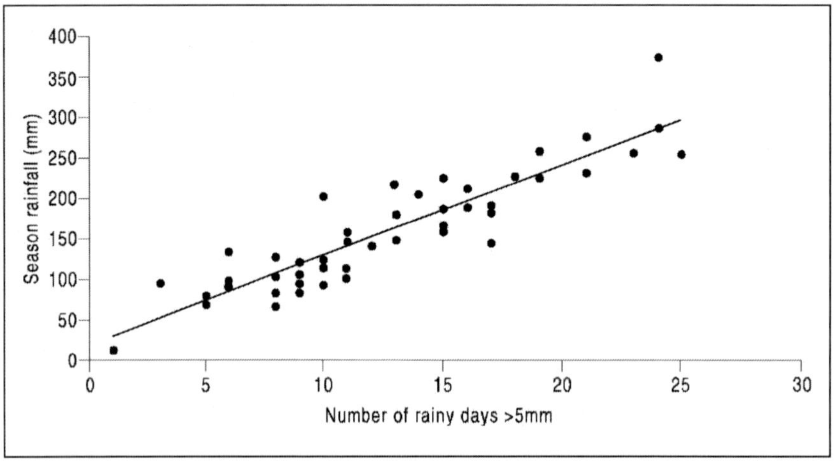

Data about the crop yield in Marab Hassan are limited, but as the current farmer has been cultivating it for almost ten years, qualitative information about yield was obtained from him. At the time of the interview with him he indicated that from 1994 to 2000 they had three good and three bad years. The onset and end of season was checked for the mentioned years and the results shown in Table 7 below.

From Table 7 it is clearly seen that for the good years (the ones with a '*') the rainfall started early (the latest was 8th of November) and ended late in April and May. The length of the rainy season in these years is 5-7 months which is quite good; also the number of rainy days in the season is considered high; of these rainy days almost 30 % received rainfall more than 5 mm, the level likely to cause runoff. In these years the *marab* received floods and the crop yield was good compared with 1994, 1996 and 1999, according to the farmer. The records support what people have said, as with barley the pre-sowing flood is essential, so for the *marab* to receive floods within November is important in order to establish the soil moisture content necessary for barley cultivation.

Figure 11. Length of rainy season for Umm al-Quttayn station (the dots represent the average rainfall for the years in each sample)

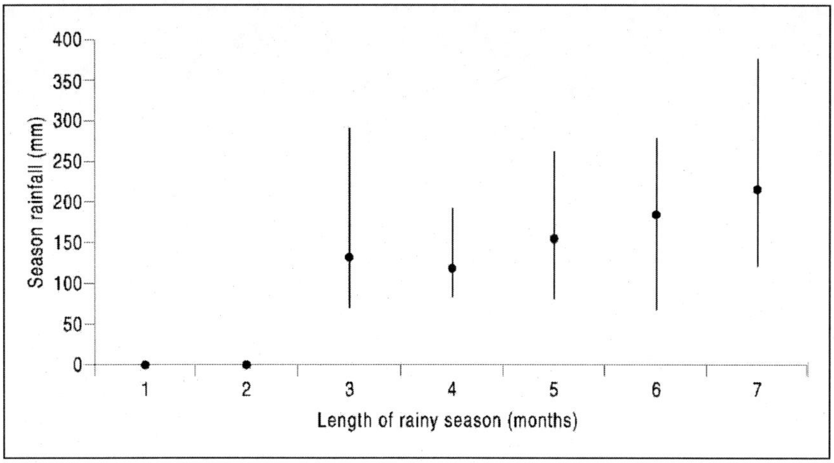

Table 7. Summary of Umm al-Quttayn rainy seasons (1990-1999)

Year	Rain starts	Rain ends	No. of rainy days	days>5 mm	Length of season (month)
1990	14-Nov	3-Apr	37	15	5
1991	10-Nov	13-May	25	11	6
1992	4-Dec	16-Apr	29	15	4
1993	12-Oct	5-Apr	34	10	6
1994	14-Nov	24-Apr	24	5	5
1995*	3-Nov	9-Apr	33	9	5
1996	23-Nov	9-Apr	15	5	4.5
1997"	8-Nov	23-May	23	6	6
1998*	17-Oct	5-May	29	10	7
1999	2-Jan	7-Feb	6	1	1

Table 8. Summary of Umm al-Quttayn station between 1952 and 2000

Rainfall patterns	Dry years	Intermediate years	Wet years
Average seasonal rainfall (mm)	128.5	158.05	168.14
Average duration of seasonal rain (month)	4.3	4.8	4.3
Number of days with significant rainfall events (>5 mm) (day)	10	11	15
Average of total number of rainy days	24	27	27

6.2. Further analysis of rainfall outcomes

Because the winter rainfall periods vary, the preceding analyses have used seasonal rainfall amounts rather than adopting an arbitrary length of the winter rains.

When the predicted seasonal rainfalls (wet, dry, intermediate) are treated as separate groups the significance of the predictions is highlighted. This is illustrated by Figures 12a, 12b and 12c. These 'populations' display different statistics, thus prediction of 'wet'

Figure 12a. Rainfall frequency for Umm al-Quttayn station in dry years (number of dry years is 15, overall average rainfall 1952-2000 is 153.5 mm)

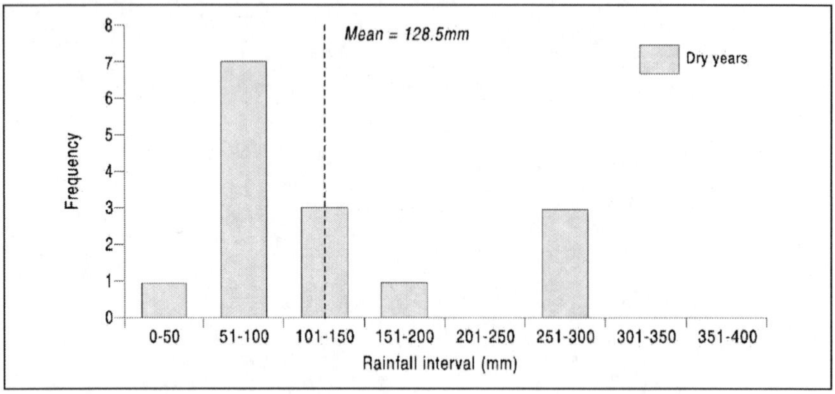

Figure 12b. Rainfall frequency for Umm al-Quttayn station in intermediate years (number of intermediate years is 12, overall average rainfall 1952-2000 is 153.5 mm)

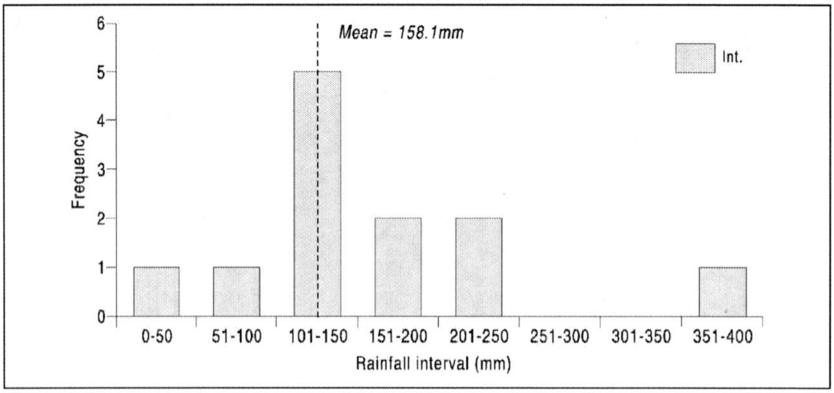

Figure 12c. Rainfall frequency for Umm al-Quttayn station in wet years (number of wet years is 22, overall average rainfall 1952-2000 is 153.5 mm)

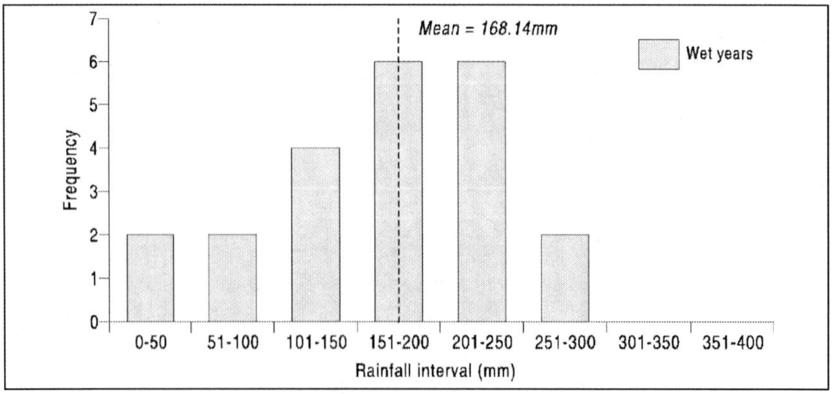

or 'dry' leads to most likely outcomes that are significantly different. Further, such predictions also lead to expectations about the length of rainy season as well as the number of significant events, ie likely growing seasons. Table 8 summarizes the dry, intermediate and wet years' characteristics for Umm al-Quttayn station between 1952 and 2000.

7. Current agricultural practices in the Badia marab: rule of thumb

At present, it is normal for farmers to cultivate without any forecasts of how good the seasonal rains will be. Provided local seeds are available, the *marab* farmers will prepare the land and sow the seeds every year before waiting for October rainfall (Abu-Zanat 2002). This procedure involves considerable expenditure without any assurance that adequate rainfall will occur, so the farmer takes a significant financial risk because the crop could fail.

Only when the local seeds are not available is the decision to sow delayed until a first significant rain has occurred. The timing of this first rain guides the decision on whether or not to sow, and the purchase of non-local seed is only made at this time. This may be due to the higher cost of the seeds, but also depends on the increased confidence in the occurrence of an extended growing season derived from the timing of the first significant rains. This means that the farmers manage their land accordingly (ie according to the onset of the rainfall, as well as the availability of the local seeds). Nevertheless, it has been shown that this predictive 'rule' has considerable validity.

8. Link between traditional and predictive approaches in rainfall forecasting

It has already been shown that using the El-Niño Southern Oscillation Index (SOI) as a rainfall predictor is a very good and reliable approach, especially in such a fragile and harsh environment as the Jordanian Badia. It was also important to test the value of incorporating the 'Traditional Approach' into an overall forecasting methodology. A combination of the 'Traditional Approach' and the global phenomena could be very important for the strategic management aspect in the area. So, in this section the prediction of rainfall using SOI is to be tested against the onset of the rainy season. Table 9 shows the combination of using MAM- SOI and JJA- SOI with the onset of the rainy season ('onset' is here used as meaning the first significant rainfall that is greater than 5 mm).

The 'T' (true) results in Table 9 for MAM- SOI and JJA- SOI predictions mean that the predicted seasonal ARI matches with the actual figures as shown before in Tables 4 and

Table 9. MAM- and JJA- rainfall predictors and onset of the rainy season for Umm al-Quttayn station (1990-2000)

Year	MAM- SOI prediction	JJA- SOI prediction	Onset of rainy season
1990	T	T	T (Nov)
1991	T	T	T (Nov)
1992	F	F	F (Dec)
1993	T	T	T (Oct)
1994	T	T	T (Nov)
1995	T	T	T (Nov)
1996	F	T	T (Nov)
1997	T	T	T (Nov)
1998	T	T	T (Oct)
1999	T	F	F (Jan)
2000	T	F	F (Dec)

6. For the onset of the rainy season, 'T', (true) stands for the season that has started in October or by the end of November at the latest, otherwise false, 'F'.

Table 9 can also be used to develop a decision tree that gives the farmer who cultivates the *marab* advance guidance on whether to sow the seeds or not. The decision tree for barley cultivation in Marab Hassan is shown in Figure 13 and can be summarized as follows:

- From the Figure it is clearly seen that the first step is to check the correlation between MAM- SOI and annual ARI, then between JJA- SOI and seasonal ARI;
- If both MAM- SOI and JJA- SOI give a dry prediction then this is the end and the tree stops here (ie no action and leave the land fallow);
  But if both MAM- SOI and JJA- SOI give wet predictions, then prepare the land for cultivation;
- If the rainfall is predicted to start early this means that farmers have to decide to prepare their land and to sow their seeds, as the prediction has a 73 % probability of being realized;
- If either of them gives a dry prediction and the other gives a wet prediction the next to be checked is the start (onset) of the first significant rainfall;
- If both MAM- and JJA- SOI predict a wet year and this is combined with the early occurrence of significant rainfall, the long wet winter prediction now has a 100 % confidence of success so the cultivation can continue. Wet means a long winter rainy season with a high likelihood of above average seasonal rainfall.

9. Conclusion

It has been shown that using the global climatic indices offers great potential in predicting the winter rainfall in Jordan three or six months ahead, thus giving farmers valuable lead time to plan their activities in the approaching winter months. Also, having an idea in advance about the characteristics of the rainy season is very advantageous in that it can be a guide for the implementation of water harvesting techniques. These approaches will save farmers time, effort and money. As discussed earlier the barley cultivation and the construction of water harvesting schemes in the *marab* area are costly, therefore adopting such schemes needs to be justified by improving the productivity and increasing the net revenue.

Better understanding of the climate, together with statistically reliable forward predictions of the characteristics of the next rainy season offer real opportunities to improve the utilization of water and land resources to achieve real economic benefits. The use of ENSO and other measures allow a 20-30 % improvement in forecasting outcome as well as longer lead times, which allow ground preparation, seed sowing and other preparations.

The traditional approaches to barley cultivation in the *marab* offered useful guidance, especially where the lack of data was a serious problem. This shows that indigenous knowledge can be a good foundation for arid zone management plans, as the most important factor is the people's interest and agreement.

Using readily available data on SOI for 2, 3-month periods offers an early indication of the duration and 'quality' of the winter rains in Jordan. Final confirmation of the prediction is provided by the early occurrence of a significant rainfall. This forecasting methodology allows agricultural decisions to be made earlier and with a much higher degree of certainty than at present. It is anticipated that further work would allow refinement of the forecasting methodology and extend its application both within Jordan and elsewhere.

Figure 13. Decision tree for cultivation

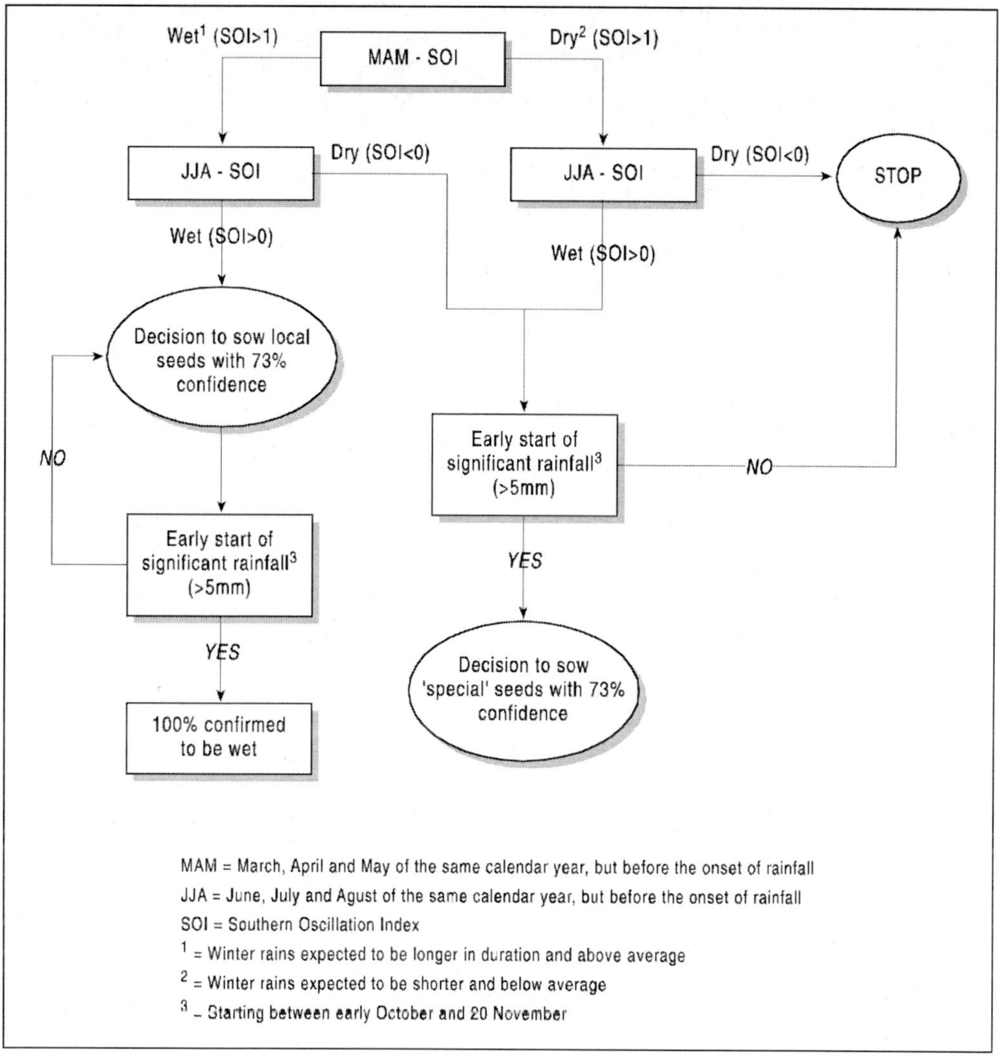

References

Abu-Zanat M (2002). Personal communication. Assistant lecturer at School of Agriculture, University of Jordan, Amman-Jordan.

Allison R, Higgitt D, Kirk A, Warburton J, Al-Homoud A, Sunnaa B and White K (1998). Geology, geomorphology, hydrology, groundwater and physical resources. In: Dutton R, Clarke J and Batikhi A (eds), Arid land resources and their management: Jordan's desert margin. London: Kegan Paul International pp. 21-44.

Beltrando G and Camberlin P (1993). Interannual Variability of rainfall in the eastern Horn of Africa and the indicators of atmospheric correlation. International Journal of Climatology 13:533-546.

De Rojas AE (2000). The climatic impact of La- Niña-related droughts in Entre Rios (Argentina). Drought News Network 12(2):7-8.

Glantz M (2001). Currents of change: EL-Niño's impact on climate and society, 2nd edn. Cambridge: Cambridge University Press.

Ibrahim K (1997). The geology of Al-Bishriyya (Al-Aritayn) area, map sheet no. 3354 II (with special reference to geology at the economic zeolite deposits). Bulletin no. 37, Amman, Geology Directorate, Geological Maping Division.

Millington A, Al-Hussein S and Dutton R (1999). Population dynamics, socio-economic change and land colonization in the northern Badia, with special reference to the Badia research and development project area. Applied Geography 19:363-384.

Nezamosadat MJ (1999). ENSO's impacts on the occurrence of autumnal drought in Iran. Drought New Network 11(2):15-18.

Nezamosadat MJ and Cordery I (2000). On the relationships between ENSO and autumn rainfall in Iran. International Journal of Climatology 20:47-61.

Oosterom EJ, Ceccarelli S and Peacock JM (1993). Yield response of barley to rainfall and temperature in Mediterranean environment. Journal of Agricultural Science 121:307-313.

Osman YZ, Shamseldin AY and Abdo GM (2001). El-Niño-southern oscillation and rainfall variability in central and southern Sudan. International Water Resources Association, Water International 26(2):177-184.

Shatnawi RS (2002). Evaluation of surface water resources of Marab Hassan in the north-east Badia of Jordan. Ph.D. Thesis, University of Wales.

CHAPTER 24

Managing rangelands: learning from institutional experience in order to promote Bedouin participation in the Badia of Jordan

Shaheen Alsirhan
Jordan Badia Research and Development Centre, P. O. Box 902, Jubeiha 11941, Jordan.
E-mail: sirhan96@yahoo.com

The paper[1] examines Jordan's experience of designing and managing rural development and conservation projects including weaknesses in project execution and outcomes that may have resulted from failures of the projects' professional teams to engage effectively, and meaningfully, with the Bedouin communities that the projects are supposed to serve. The implementation of conservation projects can cause problems because, understandably, the main priority of the Bedouin is to protect their own livelihoods and they have little interest in the 'rights' of plants and animals if they feel that such interests may threaten the future of their own families. If conservation work is to succeed, the Bedouin have to be convinced that new income-earning opportunities will be created, through tourism, for example, over which they exert some effective influence.

Keywords: Rangelands, biodiversity, conservation, Bedouin attitudes, institutional experience

1. Introduction

The Badia is the arid-semiarid zone of the Middle East, which receives low rainfall, and is occupied by the Bedouin. These areas constitute large portions of Jordan and the surrounding countries: around 80 % of Jordan; 75 % of Iraq; 90 % of Saudi Arabia; and 55 % of Syria (Alwelaie 1985, Sankari 1993). This study is a contribution to the efforts of the Jordan Badia Research and Development Centre (BRDC) in managing the rangelands through learning from institutional experience in order to promote the local herders' participation.

1.1. Objectives of the study

The key objectives of this study are to reveal: 1) the previous and changing context in which the Bedouin manage the range resources of the Badia, and their attitudes towards those resources; 2) the extent to which communities in different parts of Jordan have been involved in rangeland and biodiversity improvement projects, and how this has changed over time; 3) the extent to which a project's success or failure is associated with community participation; and 4) how previous experience in Jordan (and elsewhere) can help with biodiversity and rangeland protection projects in the north-eastern Badia of Jordan in the future.

1.2. The importance of the rangelands and biodiversity

Rangelands are very significant, from an ecological and physical aspect as well as their social and economic role. According to ash-Shourbagi (1993), plant cover of the rangeland plays a central role in the prevention of soil erosion. Plant cover reduces water run-

---

[1] This paper is based on a thesis submitted for a Masters degree, Department of Geography, University of Durham, UK, 1999. This study was funded by the UK's Darwin Initiative for the Survival of Species (DISS).

off and provides wild animals and domestic livestock with food and refuge. In addition, rangeland acts as a buffer between the desert and farms, villages, and cities. Protecting rangeland biodiversity may benefit medicine and tourism. Herbs such as *Artemisia herbalba, Thymus serpylim* and *Trigonella foenum-graecum* are used as medicine to ameliorate or cure colic, nausea, abdominal pain, headaches, and menstrual discomfort (Brandenburg 1998). Omondi (1994) emphasises that wildlife is a principal attraction of tourism, which is a most valuable resource for any country, for the foreign exchange earned and the jobs provided. The protection of water catchment areas and genetic resources, cultural and natural heritage, and scientific values, are other benefits of biodiversity conservation.

Al-Junaidi (1996) pointed out that the Badia of Jordan is capable of producing feed, which can satisfy around 800 000 sheep for the whole year without any need for additional supplementary feed. al-Junaidi also stated that in 1990, livestock production contributed 32 % of Jordan's agricultural production, 3 % of the gross national product, and 40 % of the national milk requirement. Also, Jordan's Badia is considered an important source of water for the country. The water sources are both underground and surface. The underground water of Jordan's Badia is estimated at 93 000 MCM (al-Junaidi 1996). According to Abu-Zanat (1993) and al-Junaidi (1996) the Badia receives low rainfall, and has high temperatures (around 40 °C in summer) which cause rapid evaporation, thus the opportunities to use surface water are restricted. Nevertheless, surface water is estimated in Jordan's Badia at 126 MCM as a long-term average in a good wet year. There are some dams built to harvest the surface water for agricultural and range purposes. These dams can hold a total of 12 MCM. The average ground slope in the Badia of Jordan is less than 8 % which could benefit water harvesting projects.

Until recent years, social and economic development took place separately from protecting the environment. It became clear later that environmental protection requires economic development and economic development needs environmental protection in order to be sustainable (Mouasher 1997). Jordan has experienced in the Badia a conflict between the needs of two interests. The first is the interest of the Bedouin and livestock, and the second is the interest of biodiversity. The Badia's resources of water, plants, and animals are being put under serious pressure as a result of increases in the population and the growth of livestock numbers. According to Findlay and Maani (1998), Jordan's population rose from about 680 000 in 1952 to 3.45 million in 1990, at an annual rate of 4.3 % per year. Increases in the population led to an expansion of settlement size and numbers, farming activities, and water consumption. The resources of the Badia have been used intensively to support these needs by pumping its water and farming its areas to provide water and food for consumption by the urban city population.

It was mentioned earlier that the Badia of Jordan can satisfy around a maximum of 800 000 sheep for the whole year. However, if the year 1995 is taken as an example, one can imagine what the pressure on the range resources of the Badia can do in one year to resources which can only tolerate a maximum of 800 000 sheep a year when around three million local sheep and goats in addition to around 277 000 legally imported sheep, plus unknown illegal numbers, are kept there. Increasing the livestock numbers leads to overgrazing. The concept of overgrazing in this study includes the heavy grazing by domestic animals of the rangeland's vegetation, leaving nothing behind except unpalatable plants (al-Oun 1997, Oakeley et al 1996). The plant cover deterioration during the second half of last century together with the decreasing amount of feed produced from the Badia will put future generations of livestock owners in a critical position. The projects for improving livestock production will be damaged, agricultural production will be badly affected, and the water and soil resources will deteriorate. All of these effects will prevent people from using the Badia as a source of food (al-Junaidi 1996, ash-Shourbagi 1993).

This study is concerned solely with the human factor of managing the rangeland resources. In order to create a sustainable improvement of the rangeland and to protect biodiversity, it is essential to link the local people with the strategies, objectives, and procedures of the plan.

2. Methodology

The overall aim of this study, as discussed earlier, is to investigate the human aspect of rangeland management and biodiversity conservation in Jordan (and elsewhere with a similar environment). The aim is to suggest trials to promote the herders' participation in improving the range resources in the BRDC area. Mainly, this study is based on an analysis of available reports and publications about rangeland and biodiversity conservation issues. However, as the human dimension of the conservation issues has not been satisfactorily covered in Jordan (in particular, the livestock owners' attitudes and perceptions towards these concepts), it was essential in this study to check these data in the field, from the owners themselves. This was achieved by talking to the owners and observing their activities.

Figure 1. The fieldwork sites (Source: The Royal Jordanian Geographical Centre (RJGC), 1998)

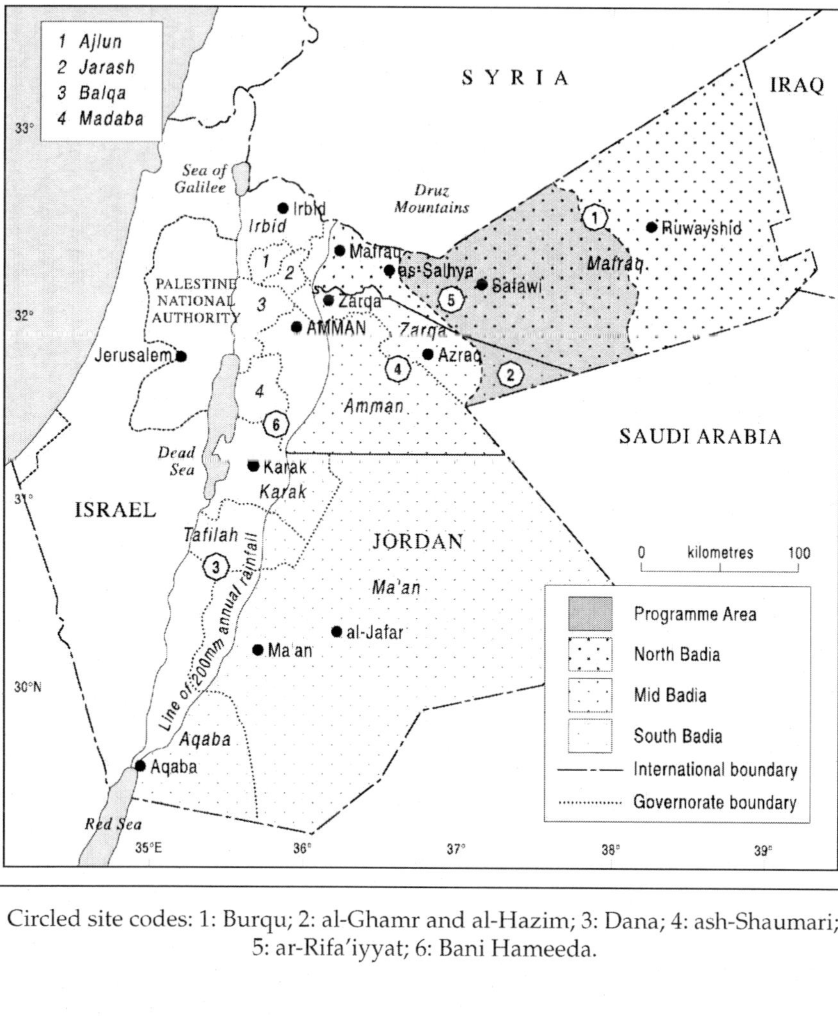

Circled site codes: 1: Burqu; 2: al-Ghamr and al-Hazim; 3: Dana; 4: ash-Shaumari; 5: ar-Rifa'iyyat; 6: Bani Hameeda.

Three types of secondary data were available for this study: 1) information about traditional land and water-use systems in Jordan and in other arid and semi-arid countries with environmental conditions similar to Jordan's; 2) information and data about rangeland management and biodiversity conservation experience in Jordan; and 3) information from governmental and non-governmental organizations. Although these secondary data were very valuable, there was a need to complement them with primary data obtained from informants living in the field study areas. The field study areas were: Burqu, al-Ghamr, al-Hazim, Dana, ash-Shaumari, al-Rifa'iyyat al-Shamaliyya, and Bani Hameeda (Figure 1).

Burqu, al-Ghamr, and al-Hazim are located within or very near the original BRDP area. They are promising areas relatively rich in water and forage making them important resources for both wild animals and domestic livestock. These sites had (and have) access to the technical and logistic facilities of BRDP (now BRDC) was a tremendous advantage when undertaking the fieldwork. The Royal Society for the Conservation of Nature (RSCN) protects Dana and ash-Shaumari; they are both the subject of major conservation area projects. The reasons for choosing both of the nature reserves were: 1) to learn about the RSCN objectives and how they are evolving, 2) to see how RSCN dealt with the local people at the time of establishing those projects and how these relationships were being conducted at the time of the fieldwork, and 3) accessibility of the logistic support required to undertake the field investigations from RSCN's bases at these projects. Other sites in Jordan and the organizations and people involved have been the source of valuable additional information. Discussions were held with personnel responsible for these sites as well as the local beneficiaries. Ordinary activities were observed in the field for a few days at these sites. The researcher was able to visit two sites: ar-Rifa'iyyat ash-Shamaliyya village and Bani Hameeda grazing reserve. Working in these sites was important in order to understand aspects of improving rangeland and livestock production and ways in which the Jordan Co-operative Corporation (JCC) have worked with the local population. It should be mentioned here that the livestock owners in these two sites are not nomadic or semi-nomadic but settled in houses.

This study was based on social methods, both quantitative and qualitative approaches. The main method used in this approach was a survey. It was designed to collect basic data from a small number of people who used resources in the field sites. Because the survey questionnaire was short and precise, it had limited value in revealing the livestock keepers' attitudes and perceptions. This kind of information cannot be obtained from precise questions. It was therefore necessary to engage in semi-structured interviews, observations, and make use of visual materials. On the other hand, during the eleven months spent in Jordan (April 1997 to March 1998), every opportunity was taken to observe the rangeland management, grazing methods, and systems for the protection of biodiversity at the sites included in the research project.

3. Bedouin systems of land and water management

For more than three millennia, Bedouin had a common life style in the Middle East. They moved freely throughout the Badia in the region, searching for water and forage for their livestock. But after World War I, the Ottoman Empire, which had colonized most of the Middle East countries for more than five centuries, was defeated which led to the establishment of Jordan, Iraq, Saudi Arabia, and Syria. In consequence, many changes have taken place in the land and water-use systems in the region.

3.1. Traditional land and water-use systems

The Bedouin traditional land and water-use systems can be understood through discussion of the following terms: nomadism, pastoralism, and the traditional hema system, which also include additional related sub-terms and expressions. Nomadism is a term which refers to human group movements while pastoralism refers to livestock husbandry activities (al-Kasasbeh 1988). As the Bedouin (the inhabitants of the Middle East Badia) used to raise livestock and move regularly in the Badia searching for water and forage for their animals, they were nomads (only a few of them are still nomads) and pastoralists or nomadic pastoralists (Abu-Rabia 1994). Bedouin mobility is in response to the limited resources of the environment (Alwelaie 1985).

Before the 40s and 50s of the last century, artesian wells and trucks were not known to the Bedouin. Water played the main factor affecting Bedouin mobility. Rain was the main source of water. Wells were also dug and barrage dams were built across seasonal streams made by the rain. The Bedouin used to stay deep in the Badia until these wells and barrages dried up then moved to the marginal parts to be close to the adjacent fertile areas which usually receive better rainfall. This mobility in the past was ideal for giving pasture lands time to get some rest and make a comeback, which kept biodiversity in balance (Masri 1991). Besides grazing the pastures, Campbell and Rowe (1998) pointed out that the Bedouin in the northern Badia of Jordan move to the eastern al-Hammad plains in winter in order to prevent mortality amongst new-born lambs in the western parts because of the cold. Similarly, the researcher heard from the Bedouin in the Dana area that they would stay near the top of the mountains in summer and near the low wadis during winter for climatic reasons.

Alwelaie (1985), Masri (1991) and Nesheiwat (1991) reported that another kind of land and water management, which related to the patterns of nomadism and pastoralism, called the hema system, was also wide-spread in the Badia of Jordan and throughout the region. Hema means reserve which is land held and reserved by a group, every member of which has the right to use what is held in common including the range resources. The hema system in Bedouin life was successful due to the strength of group solidarity. Tribal solidarity had to be strong in order to protect, usually, isolated small camping units. Any individual infringing this system would be punished (Shoup 1990).

But is there anything in the traditional land and water-use systems which could be used today to help improve the sustainable use of the Badia grazing resource? It is true that the slow traditional mobility (using the animals to move) is not feasible for the current generation of the nomadic Bedouin but the emphasis should be placed on the opportunity to select some areas which could be divided into two sites. One is concerned solely with summer grazing such as al-Harra of the northern Badia and the other site is concerned solely with winter grazing such as some parts of al-Hammad of the Jordanian Badia. The idea is that the Bedouin move in summer to the summer grazing areas which means leaving the winter grazing areas without grazing while moving to them after the end of summer.

Although the traditional hema system no longer exists, some aspects of it could be shaped and adopted to improve the range resources of the Badia. By distributing some parts of the Badia between the tribes or clans who have traditional claims on these specific areas, this would put an end to the early and over-grazing which have been devastating the range resources. As most of the Badia areas are freely grazed by any Bedouin, this causes a high degree of competition between them, which contributes to early and over-grazing. Re-shaping then re-introducing the hema system could help the Bedouin to manage the range resources. This would give the users of particular areas relief from fearing the competition of other Bedouin which would help them accept ideas of a sustainable use of the range resources. This would also encourage them to protect these areas from other users. However, this will not be an easy target to achieve, taking into account that the land tenure in the Badia is not clear yet, due to the Agricultural Law number 20 (1973) which considers most of the Badia areas to be state lands. Also, most of the Badia is open for grazing by any livestock owner in Jordan which makes the situation more difficult. Therefore, in any attempt to apply such a plan, all concerned parties should be involved.

3.2. Changing land and water-use systems

Numerous changes have taken place in the Bedouin way of life, and its impact on their land and water-use systems. Defeating the Ottoman Empire during World War I was a significant factor for the changes which have subsequently taken place in the Bedouin way of life which have impacted upon their land and water-use. The Bedouin way of life had changed relatively little before and during the Ottoman Empire which lasted for more than five centuries. But between the end of this empire and World War II, the region witnessed the British and French mandates and the creation of Jordan, Iraq, Saudi Arabia, and Syria as separate states. The British and French mandates and the governments of Jordan, Iraq, Saudi Arabia, and Syria regarded the Bedouin societies as states within the regional states. The priority made, with regard to the Bedouin, was to put an end to their political autonomy and encourage them to abandon the nomadic lifestyle (Chatty 1990). Despite the fact that the international borders of the region had never been completely closed until the Gulf War (1990-1991), the national governments have always tried to curtail the long migration and movement patterns. In 1925, Jordan, Syria and Saudi Arabia signed a treaty, by which the Bedouin had to seek permission prior to crossing borders which increased most of the Bedouin trends towards sedentarization as their range of movements became smaller (Shoup 1990).

Dramatic changes have taken place in the traditional management systems since World War II due to a number of factors: the British and French mandates had already ended and countries of the region became independent; more governmental interference in the Bedouin way of life; the introduction of new technologies in the region including forms of communication, boreholes, and farming machines; growth of more alternative employment opportunities; and oil wealth - though not in Jordan. Since the independence of the region's states after World War II, sedentarization has remained a primary focus of the governments (Alwelaie 1985, Chatty 1997).

Even though the Bedouin in the region depended on the range resources of the Badia, they also used the summer grazing in the more fertile areas in the settled parts. They used to be allowed to use fields of sedentary villagers in the fallow areas, which are the fields that are left without crops. In the past, fallow areas were considered common lands and open as pasture to anyone. The owners of these areas (the settled farmers) regarded the fertiliser left behind by the flocks and the assistance that grazing provided in weed control as a form of payment for usage. Furthermore, due to the lack of fodder in the rangelands during the summer months, the Bedouin were forced to look for other alternatives such as moving towards agricultural zones in order to graze the remaining stubble after the harvest and also to get water for herds. This situation created some positive complementary relations with the settled villagers. The villagers provided the nomads with some low cost feed and other needed commodities such as fresh and dry fruits and vegetables whilst, on the other hand, the nomads provided the villagers with dairy products, meat, animal skins, sheep's wool, and goat hair. As a result of these interactions, some nomads started to establish their own communities and to practise agricultural activities in addition to livestock raising (Nesheiwat 1991, Shoup 1990). A further factor that increased sedentarization was the three years drought which hit Jordan and Syria at the beginning of 1958. This pushed some Bedouin to seek wage labour in the urban centres or serve in the military and police forces in both countries as large numbers of livestock died and many of the owners were financially ruined (Chatty 1997, Shoup 1990).

In Jordan, increasing mechanization has accelerated the expansion rate of farming activities. Consequently, the range resources are being heavily over-used. Also, introducing modern ploughs created furrows which turned into run-off gullies which led to soil erosion and to an increase in the run-off water (IFAD 1993). Water sometimes, and in indirect ways, causes deterioration to the rangelands. Nesheiwat (1991) stated that 850 boreholes were drilled in the Badia of Jordan by 1991, 180 of which belong to the government; the rest are private. The government boreholes are used to pump water to the cities and the villages, and to provide some water points for the livestock in the Badia. The private boreholes are used to irrigate farms. The owners also sell water. In addition, there have been other changing factors which can be referred to as population activities. Increasing population has led to increases in demand for food which was secured from both plants and animals. The food requirement for the increasing population has led to carrying out cropping activities in lands not suitable for cropping purposes. Furthermore, the trees of the forests and the shrubs in the Badia are being subjected to a high degree of cutting due to the needs of houses, cooking and heat. Moreover, increasing the population has led to an expansion in building activities which takes place over both the agricultural and rangelands areas.

Raising livestock was presumed a profitable livelihood for the Bedouin until August 1996 when the government removed the subsidy on livestock feeds. According to Rowe (1998), the net profit was JD13.50 per head before and JD1.62 per head after removal of the subsidy. When the subsidy existed, the more livestock the owners had, the more profit they gained. The introduction of the animal feed subsidy after 1971 played a major role in increasing the livestock numbers. Furthermore, with the increase of the human population, the demand for animal products grew and therefore the number of animals also increased. The increased livestock numbers as well as the use of the trucks for transferring the flocks or bringing water, which disturbed the traditional seasonal flock movements, caused early and over-grazing patterns. Due to these patterns of grazing, the palatable and less-palatable plants were eaten before the seed formation stage, leaving unpalatable plants (Abu-Zanat 1993, al-Junaidi 1996, ash-Shourbagi 1993, Nesheiwat 1991).

In summary, after World War II, the Bedouin way of life changed in many ways. There were increases in their trends towards sedentarization besides tremendous increases in the population and livestock numbers. Also, new technologies were introduced into the region including forms of communication, boreholes, and farm machinery which expanded the extent of use of the range resources. Moreover, the number of alternative employment opportunities grew reflecting the expansion of regional oil wealth. As a result of these factors and changes, the range resources were, and have continued to be, over-used which has degraded the resources, reduced the production of plant biomass, and increased soil erosion.

4. Institutional approaches to rangeland management and biodiversity conservation in Jordan

Due to a variety of climates in addition to variety of topography, Jordan has a diversity of flora and fauna. But, as was discussed earlier and due to several factors, the traditional land and water-use systems of the Bedouin have changed. As a result, the natural biodiversity has been badly affected and the rangeland's resources have deteriorated. This situation opened the eyes of a number of national, regional, and international organizations, both governmental and non-governmental, to the need to overcome these conditions. These organizations have been working in Jordan in many fields and on many issues including rural development, improving the agriculture sector, rangeland management, and protecting biodiversity.

4.1. The institutional approaches to rangeland management

The organizations involved in rangeland management and biodiversity conservation are many and so are their activities. Therefore, covering all of them and their activities is beyond the scope of this study, which concentrates on some key organizations and issues. There are a number of institutions involved in rangeland management in Jordan: The Ministry of Agriculture (MoA), The National Centre for Agricultural Research and Technology Transfer (NCARTT), The Jordan Co-operative Corporation (JCC), The Jordan Badia Research and Development Centre (BRDC), The Faculty of Agriculture at the University of Jordan, and The Arab Centre for Studies of Arid Zones and Dry Lands (ACSAD). However, in addition to this list of the institutions there are other institutions which work on the issue of rangeland management, either directly or indirectly, through conducting research and workshops and/or providing consultations, technical aid, and funds. These organizations have been participating in implementing a number of rangeland projects and other activities in Jordan such as:

- Development of the Jordanian part of the Hammad basin. This project was created in 1979 by ACSAD, the Arab Fund for the Economic Development (AFED), and MoA. The objective of this project, with regard to rangeland management, has been to study the range resources of water, flora, and fauna in order to conserve them for the development of the area. Although the project provided the local community with a number of useful services such as constructing some public buildings (Ruwayshid secondary school, for example) and offering the herds some veterinary help, the local herd-owners were not consulted about the establishment of the reserves programme. In consequence, this caused additional high security costs to protect the reserve from the Bedouin (Abu-Zanat 1993, ACSAD 1983, Masri 1995);
- Jordan Arid Zone Productivity (JAZP) (1985-1998). JAZP was a multi-stage project funded by the European Commission. The project aimed to develop appropriate practices for integrated farming and rangeland management in the low rainfall zone (Abu-Zanat 1996);
- The National Programme for Range Rehabilitation and Development (NPRRD). NPRRD was a multi-stage project; funded by IFAD and MoA. It aimed to re-establish the productive capacity of Jordan's rangelands and the historical, social, and economic contribution which were previously made from them. In all phases, beneficiaries of local communities played a central role in the implementation of the project (IFAD 1996);
- Plan of operations agreed upon between the Hashemite Kingdom of Jordan and the World Food Programme (WFP) concerning support to participatory land improvement. This plan was supposed to contribute to Jordan's strategy to improve the agricultural sector and protect the environment (WFP et al 1997);
- al-Lujjoun reserve. In 1981, JCC and a local co-operative society, from local herd-owners, reached an agreement to plant the reserve with fodder shrubs and protect it by the co-operative members for their own benefit. The idea was that those members grazed inside the reserve according to its carrying capacity. According to Tadros and Salem (1993), the implementation of the range reserve increased range productivity by 3-4 times. This led to an increase in milk production, which improved farmers' incomes;
- Ma'in reserve. The reserve was launched by JCC, WFP, and MoA under overall direct management by the Ma'in co-operative society. Technical assistance for the project was provided by the United Nations Development Programme (UNDP) and FAO. The project strategy was to create a new system of range management

and development through herders' participation in all project activities. The stocking rate, within the reserve, increased from 2 500 in 1983 to 20 000 in 1993 for a period of 2-4 months each year (FAO 1995). Development of the area did not face serious obstacles or conflicts between the concerned institutions and the local herders due to prior agreement among the concerned parties.

4.2. Biodiversity conservation

Biodiversity conservation has a recent history in northern Arabia, including Jordan (Chatty 1997). Since it was formed in 1966, the issue of protecting biodiversity in Jordan has been the responsibility of RSCN. It is a non-governmental organization (NGO) created by a voluntary effort when a group of Jordanian hunters noted that many wild animals and plants had disappeared from the country because of the human misuse of the natural resources. This guided them to establish RSCN (Mashaqbeh 1998). The most important project which highly affected RSCN's work was in 1994 when RSCN signed an agreement with the World Bank and UNDP, for a Global Environmental Facility (GEF) loan to ensure the conservation of biodiversity in the Dana and Azraq areas. Besides the Dana and Azraq projects, this agreement allowed RSCN to enhance its capacity through building up its skills and expertise as well as improving its logistic resources (Johnson 1997). The following are reserves established by RSCN as well as some other potential sites seen as the proposed future reserves:

- ash-Shaumari nature reserve: It is located in the east of Jordan with a fenced area of 22 km^2;
- Azraq wetland nature reserve: This reserve is located near the ash-Shaumari reserve with an area of about 12 km^2. It contains a number of pools and puddles. It has a high value for protecting the environment and bird migration. More than 300 species of bird have been recorded beside the mammals Hyena hyena, Canis lupus, Vulpus vulpus, and Canis aureus. Also, other kinds of snakes and fishes have been observed (Abu-Zanat 1993);
- Zubia nature reserve: It covers an area of 31 km^2 located within the north-west part of Jordan. The aims of establishing the reserve were to protect the turpentine-trees and oak woodland and to reintroduce the roe deer (Abu-Zanat 1993, RSCN 1996);
- al-Mujib nature reserve: This reserve is located near the Dead Sea with an area of 220 km^2. Its importance is due to some hot mineral springs and the presence of some endangered species of plants and animals such as the Ibex (Abu-Zanat 1993, RSCN 1996);
- Rum nature reserve: It is located in south Jordan with an area of 510 km^2. It has a great diversity of desert plants and animals (Abu-Zanat 1993);
- Dana nature reserve: The government authorised RSCN to establish Dana nature reserve with an area of 308 km^2 located within the west of Jordan (Swenne 1995, Zandri 1996).

4.3. Findings and discussion

In order to understand more about the institutional efforts related to rangeland management and biodiversity conservation and also to gain current data and information, it was essential to visit a number of sites at which projects have been established and managed by these institutions. Dana, ash-Shaumari, Bani Hameeda grazing reserve, and al-Rifa'iyyat al-Shamaliyya village were visited and several methods were adopted to gain the data needed.

4.3.1. Dana nature reserve

Dana reserve occupies 308 km^2 located within Tafila governorate. Dana has special features that make it the most important reserve among those which are managed by RSCN (Swenne 1995, Zandri 1996). Since establishing the reserve, RSCN has been facing a number of difficulties in linking the local people with the conservation strategy. When the project started, the villagers around the old empty village of Dana in the upper part of the reserve hesitated and resisted the idea of the conservation plan but, a few months later, they started to participate with RSCN's operations in the area. When the reserve started to create jobs for them and when work started in the old village, they became sure that if they did not catch up with the early beginnings, they would miss many opportunities.

However, the researcher's own fieldwork revealed on-going problems. Thirty owners were interviewed in the lower part of the reserve, of whom 21 belonged to the Azazme tribe and nine to the Hweitat. These owners were neglected in the early stages. All of the Azazme interviewed were nomadic (except one owner); they did not have houses and moved within the reserve area. On the other hand, all the owners interviewed of the Hweitat tribe were semi-nomadic (except one). They had houses and lands in villages and cities distributed among the governorates of Amman, Ma'an, and Tafila. This meant that the Hweitat had more options for finding additional resources to generate income in their villages while the Azazme found it very difficult to find other resources for generating income to replace shortages that had resulted from the grazing restrictions which had been imposed by the reserve.

As the topography of the area is difficult, goats were much preferred to sheep and camels. Sheep and camels can only graze in the few flat lands of Dana whereas the mountains are more suitable for goats. Within the 30 households included in the study, only ten flocks of sheep were owned with an average of 17 head each. Also, only three flocks of camels were recorded with an average of 18 head each. All of the owners had goats with an average of 141 each. It is clear that they were not big owners and could be presumed to be poor people. According to Lancaster and Lancaster (1993), the net profit per head of goat in 1993 was JD10. But after the increases in the feed prices in 1996, Rowe (1997) reported that there was a loss, which amounted to around JD18.5 per head. The costs of raising these animals were much more than the revenues earned from their sale. If this situation remains, sooner or later the Bedouin will be forced to search for other ways to generate income.

All of Dana's owners included in the study used the area during each of the previous five years much as they used it before. Also, they did not show any interest in moving outside the area. As they were permanently resident, it was potentially possible for RSCN, or any related institution, to draw up an agreement with them to provide the best management for the resources of the area. Although resident, they still moved inside the reserve, from one mountain or wadi to another. The main reason that made the Bedouin in Dana insist on staying despite the difficult topography of the area was the presence of many natural springs (water) which flow in the area. Over and above the Bedouin's domestic uses, the springs were used to water the livestock and to irrigate some watermelon farms in wadi Feinan. The Azazme cannot leave this area easily because they do not have other tribal lands to go to, as many other tribes in Jordan have. Locally, each household had been given 1.5 dunums for building purposes.

After the increases in the feed prices in 1996, falling livestock prices, and the grazing restrictions made by RSCN, the Bedouin in Dana lived in hardship. In 1997, Rowe conducted a study in the area which indicated that the result of the government's decision to stop subsidising animal feed in 1996 was as follows: 'Barley, the principal feed type, rose in price from JD85 per tonne to JD110 per tonne, while wheat bran rose from JD52 per tonne to JD100 per tonne' (Rowe 1997, 3). Also, sale prices fell from around JD60 for an adult nanny goat in early 1996 to about JD25 in summer 1997. During the fieldwork in Dana, the researcher heard many times from the owners that they had been able to buy a tonne of feed by selling only one goat but more recently they needed to sell 4-5 to get the same tonne. If this situation remained, sooner or later they would have been forced to sell all their livestock. It was found that the average feed amount given to the livestock in the grazing months (March-June) was 1.5 tonne per month while in the non-grazing months (the remaining months) it was 3.84 tonne per month. Thus, according to these figures, grazing played a significant role in reducing the cost of raising the livestock.

After the interviews with the owners, the researcher was able to summarise their viewpoints and then discuss them with the manager of the reserve. In most cases both the owners and the manager had very different opinions with regard to the historical and social rights of the local people (emphasized by the owners) and the biodiversity needs (emphasized by the reserve manager) (see Table 1). There were some misunderstandings, and even mutual hate, between the owners and some of RSCN's employees, especially the rangers.

This total mutual incomprehension is an object lesson for anyone contemplating setting up a project in the Badia (or elsewhere) in which the requirements of biodiversity conservation and range use by owners' herds have to be balanced. Much more discussion and negotiation, leading to mutual comprehension and agreement, is needed before project plans are developed. The researcher argues that most of the problems which are facing RSCN could have been overcome if the local community had been consulted and if RSCN had provided opportunities for the owners before establishing the reserve rather than trying to do so after establishing the reserve. The Azazme inside the reserve complained that they had a low social weight because they did not have highly placed people in the government to help them.

The main challenge facing RSCN then was to reduce the effects of the grazing restrictions on the Azazme's economic cycle in the short-term and involve the people developing the reserve's strategy in the medium and long-terms. It is true that the position of the local people was very critical and the problem was difficult to solve. However, a trial project made by Rowe and Allonby (1998) proved that even though it is difficult, it is not impossible.

This trial project had two objectives; first, to determine the potential growth rates of intensively managed stock and therefore the technical and commercial viability of a larger-scale fattening scheme, and second, to see whether improved stock could be effectively marketed by the trial project, securing optimal prices for livestock from the area and increasing the incentive for owners to sell surplus animals quickly. A total of 37 male goat kids (7-10 months age) were vaccinated, serum sampled, tagged, weighed, and introduced into a pen. The direct benefit of putting the animals in a pen was surely reducing the pressure on the range resources. These animals were fed a mixed ration including berseem, barley, wheat bran, soya, and dried by-products of local tomato and olive processing. Within around two months, the total weight of the 37 animals rose from 509.7 kg in total to 1041.5 kg, faster than before. This project offered the owners a net total profit of JD27.2 per animal.

4.3.2. ash-Shaumari nature reserve

ash-Shaumari is a fenced reserve area occupying 22 km^2 in the east part of Jordan (Clarke 1979). The wild animals reintroduced into the reserve are 165 head of the Arabian oryx (Oryx peucoryx), 15 head of Ostrich (Struthio camelus), and seven head of the Syrian wild ass (Equus hemionus). Also, 130 species of natural plants have been recorded, 134 kinds of birds (most of them are migrant), and 11 different kinds of mammals (Abu-Zanat 1993, Mirza 1997).

Table 1. The viewpoints of the Bedouin and the Dana reserve's manager (Source: Field study conducted by the researcher, 1997)

Herd owners' viewpoints	The comments of the reserve manager
Wild animals inside the reserve get their food from the mountains and the rugged places where the local people and their livestock cannot reach.	Wild animals are forced to use these places due to the high pressure of grazing and hunting, while they used to use all the areas before.
There is no competition between the livestock and the wild animals. The places in which the wild animals are found are rich in forage, which is sufficient and available only for them. Furthermore, one of the Bedouin said that no wild animal has ever been known to have died from hunger before and after they created the reserve. He argued that the forage would be sufficient for all the animals (domesticated and wild) if there was good co-operation between RSCN and the local people. Wild animals, usually, do not need much water in contrast with livestock. Because of availability of springs, there is no big competition between the animals.	Competition is there and so is high pressure, and the locals do not see the change in the wild animal's lives and behaviour over time.
Some tribes from inside Jordan (from south Badia in particular) used to come to graze in wadi Dana during the spring season because of the availability of both forage and water. The local people say that they have no objection to tribes coming from other places because God has given the resources for use by everyone. Many livestock owners who used to come to Dana no longer do because they have settled in other places.	The local people should ask themselves why other tribes are no longer coming, due to high pressure and serious levels of degradation in the area caused by the very high pressure that the local goats are putting on the land.
There is a debate among the local people regarding the importance of protecting biodiversity. Most of them do not agree with the idea that there is any importance of protecting biodiversity, and indeed some of them want to be sure that protecting biodiversity will not have any negative effect on their life.	Local people cannot ever be sure about interaction between domestic and wild animals. Nobody can, but bringing back balance to the ecological system in the area will benefit their livestock because predator animals will have alternative sources of food of wild herbivores instead of their livestock and organising their grazing system will also benefit them in preventing the irreversible degradation of vegetation cover.
Most of those interviewed did not want to draw up an agreement with RSCN to control the grazing because there are no other alternatives to the local grazing areas, and the livestock numbers are greater that any single part of the reserve can support. One of a minority of people who likes the idea of the agreement suggested that more research and discussions with the local people should be done beforehand.	It is very important for them to draw up the agreement with RSCN because RSCN is trying, on one hand, to sustain their presence in the reserve by preventing their degradation of the ecological system, which in the long-term they will benefit from, and on the other hand, RSCN is trying to provide alternative sources of income for them.
There are some lands in Wadi Feinan (Dana) owned or used by some families from all the tribes except the Azazme. The Azazme are excluded because they do not have sufficient social weight.	RSCN is trying to help the Azazme despite their social weight as RSCN feel that they are the most needy group in the area.
All those interviewed agreed that RSCN did not discuss any objectives or plans with them before they created the reserve, and they think that they should have done in order to avoid the conflicts which are happening now.	It is true that RSCN did not discuss the establishment of the reserve with all inhabitants, but as well no modification on the management of the site was done without consulting with them and fully informing them as well.
The reserve benefits some households by offering jobs but harms the majority by driving them to buy more feed with very high prices.	RSCN is not responsible for them buying feed. Low rain and the high grazing pressure is leaving nothing for them to graze, thus buying more feed is essential.
The reserve benefits the government but harms the local people who think that the project's objectives are changing every day.	Everybody has to contribute to the country's policy, RSCN is not always in good relations with the government, and RSCN is not government.
They wish to graze inside the protected area during the dry seasons.	70-80 % of the reserve is still open for living and grazing. RSCN is not trying to stop grazing in the whole reserve, RSCN is trying to implement a grazing regime in which there are grazing areas, controlled grazing areas, and conservation areas to try to sustain the use of the resource.

Around the fenced reserve, ten households were interviewed. All of them were semi-nomadic; having lands and houses in different villages and settlements. Eight owners were from villages or cities located in Mafraq governorate. The other two were, respectively, from Amman and Zarqa governorates. The significance of this is that grazing around the reserve was not restricted to one clan or one tribe. This is an example of the mobility and the free status of grazing in the area which puts more pressure on the resources and limits any sense of group responsibility for maintaining it.

The topography of the ash-Shaumari area (which is Hammada and therefore flat), the big size of the flocks, the need to bring water and feed, and the need for mobility over long distances, meant that all the owners interviewed had trucks. The ownership of cars and pick-ups was restricted to one owner who had one pick-up and one car in addition to two trucks. This owner was thought to be one of the wealthiest traders in Mafraq. It is rare that the herd-owners own cars. They are not as practical as the pick-up and truck to serve them in their daily needs. The Bedouin in the flat hammad desert tend to raise sheep more than goats. All of the owners interviewed in ash-Shaumari had flocks of sheep, with an average of around 390 sheep per owner, while only three owners had flocks of goats, with an average of 26 head per owner, and no owner had camels.

It should be mentioned that within those included in the survey, one owner was considered to be one of the biggest traders in Mafraq (as mentioned above) who was also preparing around 60 000 sheep to be exported to Saudi Arabia; ash-Shaumari is located beside the highway that links Jordan to Saudi Arabia. This flock was excluded from the calculation because it distorted the ownership figures, however it will have had a big impact on rangeland degradation even though they were given supplementary feed. The average quantity of feed bought during the grazing months was 6.2 tonne per month for each owner while in the remaining months (non-grazing months) it was 9.7 tonne per month. This could lead to the conclusion that these owners did not benefit much from the natural forage as the area is small and as the forage was not given time to grow enough before the grazing.

With regard to the status of using the area, not all the ten owners interviewed had used the area continuously for the previous five years. From the ten livestock owners present in 1997, only three owners had come in 1992 and 1993 while four owners had only come in the years of 1994, 1995, and 1996. This significantly indicates that this site did not attract the owners to use it regularly as it is small and did not provide enough feed for the livestock despite the presence of a few shrubs. The main reason attracting some owners to stay longer in the site was the presence of a few plants and shrubs in wadi ash-Shaumari, even though there was not enough to supply the total requirement even in the spring. Despite the fact that water was not mentioned by anyone of those interviewed as an attractive reason to stay at ash-Shaumari, it still played a fundamental role in their decisions. In general, they preferred the rain-water that collected in ponds and dikes. When these ponds and dikes emptied, they obtained free water from an artesian well made by the government in the area or bought it from private wells within the Azraq area.

The reserve manager said that because of the limitation of the reserve's area (22 km^2), it had been facing several ecological pressures, in particular exceeding the carrying capacity, leading to competition on the resources between the reserve animals. Bedouin livestock were totally excluded from grazing inside the reserve. Also RSCN was preparing a plan to increase the reserve size to cover 320 km^2, but they said nothing about any consultation with livestock owners in the area which could lead to conflicts with them in the long-term. The manager of the reserve also said that the livestock owners used to concentrate around the reserve only during winter seasons when wadi ash-Shaumari flows. Those owners, according to the manager, caused pollution when they left the empty plastic feed sacks behind, which were then blown inside the reserve by the wind.

He added that there were very few cases of aggression from the owners by grazing inside the reserve. He even sometimes allowed them to collect firewood inside. He had been thinking about letting the local livestock graze inside the reserve after the seed formation stage. However, research conducted in the area by RSCN showed that it would be better not to let the sheep and goats get inside the reserve; most of them had some diseases that could be transferred to the wild animals in it. However, there might have been a possibility of letting camels graze inside as long they were clear of communicable diseases. The manager did not know the numbers of the carnivores in the area; it would be difficult to count them. These animals do not settle inside the reserve for a long time but use it to hide from the hunters and the local people. Digging holes under the reserve fence is the way that these animals get inside. They attack the young of the herbivores in the reserve. A bulldozer was used to block their holes once or twice a week; however this created more costs managing the reserve. An alternative was being considered, which was to keep the young animals inside a very well built fence on a cement foundation. However, range and wildlife reserves which are based on expensive fencing rather than commitment and participation from the local communities would face more difficulties besides security costs.

The herders interviewed in the area think the reserve is good but not suitable for grazing purposes because its area is small, and because of the presence of the snakes and some carnivores. The reserve does not harm the grazing interests of the owners because of its small size. Also, they do not know how the wild animals get their food and water. They agree that the livestock affect the ability of the wild animals to get food from the rangelands, but God, who provides the living, takes care of both wild and domestic animals and does not leave them without anything to survive on. Also, when God created the world, he gave roles and objectives for every living thing in order to complement each other. Every living thing has a purpose in this life. Nevertheless, conservation negatively affects livestock. There is no benefit from protecting biodiversity, according to the livestock owners, except encouraging tourism. They also think that the old herders cannot give up raising livestock because they do not have other resources to generate income. Therefore, there is no doubt that there was a lack of awareness among the owners with regard to the benefit of protecting biodiversity.

It could be concluded from the discussion above about ash-Shaumari that RSCN was not in a good position to co-operate with the livestock owners to determine the best methods to use the resources. The reasons for that are: 1) the size of the fenced reserve was small (only 22 km^2), 2) the mobility of the owners which confused the manager's efforts to help them create a sustainable plan to use the resources, and 3) the free status of using the area outside the reserve by all the livestock owners in Jordan.

4.3.3. Bani Hameeda grazing reserve

In 1978 and through the efforts of JCC, the local people of the Bani Hameeda villages created a co-operative society to serve the villagers with some agricultural services. The members of this co-operative totalled 297, and owned around 5 000 sheep and 5 000 goats. In 1998, JCC assisted this co-operative to benefit from the Jordanian government-WFP agreement, mentioned above, in order to establish a grazing reserve in an area of 25 km^2. The idea was to grow 900 000 bushes of *Atriplex halimas* within five years at the rate of 180 000 every year. These bushes would be grazed by domestic livestock within 3-4 years of starting the project.

The Bani Hameeda grazing reserve provides a good example of how good organization can help in the management of the resources. The researcher argues that as there are many families and clans who either own or claim to own lands in the BRDC area, a similar approach to Bani Hameeda's could be adopted in a number of these lands mentioned above in that area.

4.3.4. ar-Rifa'iyyat ash-Shamaliyya village

The village of ar-Rifa'iyyat ash-Shamaliyya is located within the north-west part of the BRDC area. A traditional management regime is used in this village similar to the traditional hema system. The reserved area is the rougher/marginal land, which is not easy to plough and farm. Within this difficult land, some spots can be ploughed by donkeys and therefore planted with some crops. The annual rainfall is about 100-150 mm. Although low, the rainfall is sufficient to grow some good forage, which usually lasts for about 4-5 months under normal grazing activities. The reserved area totals some 1.5-2 km^2. The importance of the area is that the local people keep the livestock away from the reserved lands from November to March every year in order that the livestock do not damage the spots planted with wheat and barley, and also to allow the natural forage to grow enough before grazing. In normal seasons they stop buying feed for their animals from March to July every year, and instead they let the livestock graze in the reserve. The informants pointed out that a few years previously some hunters used to come during the weekends, but this practice had stopped. Also, no owner was allowed to come to graze here from outside the village unless he rented it from its owner. Furthermore, the numbers of the herds were not big as the local people depended on other resources to generate income.

A lesson that can be learned from this case is that as the villagers of ar-Rifa'iyyat ash-Shamaliyya are from the same clan and therefore relatives, they do not have problems between themselves in terms of land ownership. This makes them potentially able to organise themselves as a co-operative society. It was mentioned earlier that working with an organised group would be much better in terms of range resource management than working with an unorganised group. The researcher re-emphasises that the above-mentioned approach of Bani Hameeda's could be shaped and adopted in ar-Rifa'iyyat ash-Shamaliyya village.

5. Rangeland and biodiversity in the BRDC area

A number of research studies have been completed since the establishment of BRDC including a base line socio-economic survey, and research on livestock, water, soil, flora, and other resources. They indicate that this part of Jordan has resources of potential value to the area and the country, if they are managed and used in an appropriate and sustainable manner. According to the needs of the area, the kinds of research and studies needed have been prioritised under seven themes: human resources, water resources, ecology, energy and geology, livestock, land resources, and information technology and management (GIS). This research is one of the studies conducted in the BRDC area.

In addition to the secondary data obtained from the BRDC studies and documents, the researcher visited some sites in the BRDC area in order to obtain firsthand information and, therefore, to understand more about the then current position of the rangelands and biodiversity in the area. The sites visited were relatively rich in biodiversity and it is potentially possible to make nature reserves from them. Burqu, al-Ghamr, and al-Hazim were thus visited to gain data including the information about the attitudes of the herd owners toward rangeland management and biodiversity conservation.

5.1. Burqu

The site is important due to the presence of an old dam, which holds winter rain used by the Bedouin for their livestock needs over and above their own daily requirement. The water held in the dam usually lasts around the year. The area of the site is about 950 km^2 divided between al-Harrah and al-Hammad areas. Burqu is a priority site at which RSCN wants to establish a new protected area in Jordan (Budieri 1995). The main problems that face Burqu are overgrazing by sheep, soil erosion caused by the vehicles in the flat al-Hammad, hunting pressure, and the over-pumping of the dam's water by the Bedouin tankers (Budieri 1995, Clarke 1979). Budieri (1995) categorised the Bedouin in Burqu site into three groups: 1) nomads who are totally dependent on the area under study and only move within it throughout the year, 2) nomads who move in different areas and use the productivity of the areas in an opportunistic way, and 3) semi-nomads who have settled in the villages near by and use the area only once a year.

There were 22 owners interviewed around Burqu dam. Within the sample, only eight had tents in the area whereas the rest came to fill water in their water tankers from the dam. The livestock owners tended to own trucks and pick-ups but rarely owned cars. No one interviewed owned a car. On the other hand, four owned pick-ups and 19 owned trucks. The researcher was told that the livestock owners who owned pick-ups in addition to trucks always had other sources of income or they had families in the villages. It seems that pick-ups were more practical when visiting the families inside the villages. However, using water tankers allowed many herd owners to come from far distances such as ar-Rayshah (more than 40 km away from Burqu) in order to collect water. This puts high pressure on the dam's water and disturbs the wild animals in the area. In addition, using both pick-ups and trucks (notably the trucks) causes soil erosion. As no asphalted roads serve the area (except the Amman-Baghdad highway), the livestock owners used numerous numbers of tracks which allowed the wind to erode the soil. All those interviewed had flocks of sheep, at an average of over 600 each, but only 64 % of them had flocks of goats at an average of around 50 each and only one owner had a flock of 30 camels.

One of the owners in this site said that when he had cash money, he could get a tonne of barley for JD120 from the distribution centres but when he did not have cash, the feed price was JD170; it was a sale on credit. This is another indicator of the critical position that the owners then faced. However, increasing the feed prices could effect the rangelands in two ways: in the short-term, it will put high pressure on the range resources as the herders will try to benefit from any single plant in order to reduce the feed costs and in the long-term, as the herders lose more money, they will sell more of their herds which will reduce the pressure on the range resources.

The owners did not use the site constantly. Only around 55 % of those interviewed in 1997 came to the site in 1992, 60 % in 1993 and 1994, while 77 % of them came in 1995 and 1996. In addition, only two of 22 seemed sure to move out of the area while the rest were not sure about their plans. Again, this fact of mobility is the most significant role which makes a conservation of biodiversity project more difficult. Any project will rely on local management by the people who use the resources in the area, but if the people move to other sites (and are replaced by other groups of people), the potential for active local management is reduced.

Water availability was the most significant reason why owners were attracted to Burqu (63.6 % of the owners). Only four mentioned the forage as a reason for them coming. Besides the dam, the sources of water in the area were the rain water, the well of Mutayrah, and some other government wells in Ruwayshid. In the first field visit in July, 100-150 tankers were seen every day filling up with water from the dam whereas only two tankers were seen filling up with water during the second field visit in November as a result of the rainy season in the Badia. It should be mentioned here that any future management plan for the range resources in the area should consider the herders' high degree of dependency on the dam's water.

Almost all of the owners interviewed in Burqu did not think that their herds affected the wild animals' opportunities to obtain food. They thought that the wild animals needed only a little forage and so there was no need to create reserves in the area. With regard to water, they thought that the quantity in the dam was sufficient for both the livestock and the wild animals. Some owners did not know if there was any benefit from protecting biodiversity, and others believed that there were no benefits. Furthermore, they thought that drawing up an agreement between them and BRDC and RSCN to control the grazing would not be possible because of the huge number of the Bedouin who used the site including the fact that it can be used by any Bedouin in Jordan. They heard rumours that the government wanted to create a reserve in Burqu. They were strongly against the idea of protecting the area in this way because they thought this would reduce their access to the dam's water which is considered the best in the area. Unfortunately there are no good alternatives because the water of the government's wells is not healthy for drinking and cooking purposes; many people are suffering now in the area with kidney problems. The Bedouin said if a reserve is proposed, all the users of Burqu would reject it. They said that God makes the forage grow but not the reserve. Also, it is very difficult to make the mobile Bedouin remain in one site for grazing purposes. They will keep moving and searching for water and forage and it would be very difficult for them to keep other herders out of the area. In addition, they would find it difficult to do any work other than raising livestock; it was too late for the old owners to learn new jobs. In order to improve feed availability, some of them see the idea of creating a co-operative to plant grazing shrubs as being good. But the idea of reducing the feed prices sounds much better. Also, they thought that the proposed reserve in the area should be on the international borders because they are 'no-go zones' in any case.

However, in this area, a possible way in which to resolve the problem of over-grazing and biodiversity degradation might be to plant some grazing shrubs as in the JCC project areas previously described. This could reduce the pressure on the range resources after offering the livestock another source of feed and this also could lead to direct control of the hunting activities as the users will be curious about any activity close to the planted shrubs of the expected reserve. On the other hand, in September/October (1998), four specialists from Bird Life International, counted numbers of raptors of international significance (Armstrong 1998, Armstrong et al 1998). Also, they may have discovered a type of partridge not previously recorded anywhere in the world. This would be attractive to bird specialists and thus might create an opportunity for eco-tourism. If eco-tourism emerged in the future, it should benefit the local users in the area financially which may make them more interested in conserving biodiversity. It should be mentioned here that although it may be too late for the old herders to practise new jobs perhaps their children are a good target for any jobs that the expected eco-tourism could offer. But, without previous discussion and acceptance from the herders, any plans or project, the researcher believes, will not be successful.

5.2. al-Ghamr and al-Hazim

The area around al-Ghamr (81 km^2) and al-Hazim (59 km^2), on which there is no literature, are located within the south-western part of the BRDC area. Neither of the sites is regarded by RSCN as being an urgent priority conservation area but they are rich in biodiversity. It must also be stated that the researcher has a personal interest since most of the areas of those sites belong to the as-Sirhan tribe from which the researcher comes. It was hoped that this would facilitate obtaining the information needed. Most of the as-Sirhan do not exploit their land themselves but rent it to other farmers who usually come from Zarqa. These farmers only care about maximising their profit without paying any attention to the conservation of biodiversity. The researcher saw bulldozers removing the plant cover in order to make seasonal farms. This causes damage to the ecological balance in the area, and minimizes opportunities for wild animals to find food. Melons, watermelons, and tomatoes are the crops planted in summer, whereas potatoes and cauliflowers are planted in winter.

A number of sheep owners from the A'nezeh tribe present in Jordan and its neighbouring countries, used to move between these countries, especially Saudi Arabia and Iraq. After the Gulf crisis (1990-1991) between Iraq and the Gulf countries, the Saudi government expelled the Bedouin of Iraqi descent including the A'nezeh. In Jordan now many members of this tribe have lived in Azraq, al-Ghamr, and al-Hazim since 1990. They are not allowed to return to Saudi Arabia. Jordan hosts them as long as they do not have other alternatives. This will be putting additional pressure on the range resources as long the problem of this group has not been solved between Jordan, Iraq, and Saudi Arabia.

In al-Ghamr and al-Hazim, 19 owners were interviewed. The livestock owners, which were met by the researcher in the area, had homes in different villages located in the governorates of Ma'an, Zarqa, Amman, and Mafraq. Once again, this shows that rangelands are common-use lands and open for use by any herd-owner in Jordan which creates more difficulties in any range management and improvement plan. Almost all the interviewees in the area owned trucks (94.7 %), and most also owned pick-ups (68 %) whereas only one herder had a car. Because the sites are located near the border with Saudi Arabia, having pick-ups in addition to trucks could increase the opportunity to obtain additional income from trading between the two countries or smuggling activities. However, as discussed earlier, use of trucks and pick-ups by the Bedouin has made the entire Badia accessible for their herds in a very short time which has caused the deterioration of biodiversity and also has caused soil erosion. On the other hand, bringing water to the livestock has made the herders tend to concentrate in specific places, which causes overgrazing.

As in Burqu, the interviewees owned an average of about 600 sheep. But more people (around 85 %) owned goats, and many more (around 37 % instead of around 5 %) owned camels. Seven of the 19 interviewees owned flocks of camels at a mean of 39 camels per owner. The area is rich in some shrubs which are palatable to camels. The forage and shrubs in the area allow the owners to buy slightly less feed during the grazing months; 11.5 tonnes per month per owner compared with 16.6 tonnes per month during the non-grazing months.

Around 85 % of the owners interviewed in 1997, had used the area regularly during the previous five years. But most of them stayed for only 1-2 months, when they passed through the area moving from north to south or from south to north which may cause confusion for any proposed plan to manage the range resources of the area in the future. The reason that makes the owners prefer not to come to this site or, when they come, not to stay longer, is the large number of farms created and the digging of many wells. Such farms can lead to conflicts with livestock owners. Ten of the interviewees were sure they would leave the area to go south or north, though most of the rest were not sure about their plans for moving.

The presence of the forage and the shrubs was the main reason attracting pastoralists to the area. Water and other different reasons were important but lesser reasons to choose the area. But, if the land-owners maintain their activities of removing the plant cover in addition to over-abstracting the water, this could reduce the herders' interest in the area. The water sources for the livestock in the area are the al-Hazim government well and some other private wells.

The livestock owners in the area do not think that there is competition between wild animals and their livestock for the natural vegetation as the wild animals consume small amount of forage, and in any case there are no big herbivores. Also, they think that wild animals need only a little amount of water. One owner said that he was not aware of any presence of wild animals which, as he thought, referred to the hunting pressures, except some wolfs and hyenas which attack the livestock from time to time. Most of the livestock owners do not think that protecting biodiversity is of any importance and also do not like this idea because it will decrease the opportunities to get free forage. Furthermore, they think that only the rain makes the forage grow; not creating reserves in the area.

Some Bedouin come from Saudi Arabia, and most of the camels in the area belong to them. They come for most of the year because of the availability of shrubs which are palatable to camels. The Jordanian owners do not mind that they come here to graze but they want the same opportunity to go to Saudi Arabia as well, which became very difficult during the 1980s and remained so in the 1990s. One of the Saudi owners said that Saudi Arabia had received a lot of rain in recent previous years and therefore many nature reserves were created along the international borders of the state. It is possible for the Saudi owners to graze inside these reserves but with some conditions regarding the capacity of the forage and water. He added that the Saudi owners are treated very well by the government of Jordan and sometimes better than in Saudi Arabia. Also, he thinks that one of the reasons why Saudi Arabia did not allow livestock owners of neighbouring countries to enter their rangelands was to prevent drug smuggling. They are afraid of drugs entering with these livestock because drugs are causing many social problems in Saudi Arabia.

6. Conclusions and recommendations, with special reference to the BRDC area

It was mentioned earlier that the overall objective of this study is to use what was learnt from the previous discussions in order to promote Bedouin participation in managing the rangelands. Hence, this section will provide the conclusion from these discussions as a step to the better design and implementation of trials in the BRDC area. These trials are intended to be living examples, designed and implemented with the active participation of the Bedouin, to show the local communities the best ways to use their limited resources.

6.1. Conclusions

Before presenting any future trial/demonstration, the following findings should be carefully considered:

6.1.1. Land tenure
It was found that most of the Badia lands, including the BRDC area, are freely grazed by most of the Bedouin of Jordan. This is a problem typical of resource use when no-one feels responsible; 'the tragedy of the commons'. This might be the first obstacle which will face managing the rangelands in the future trials' sites. It will be difficult to determine who best to approach of the users to start any management plan. However, the first step towards resolving this problem could be redefining the lands that the tribes have traditional claims for. In other words, the idea is to make use of the traditional hema system that was mentioned earlier, which will increase the Bedouin sense of responsibility for their resources. However, the social and economic conditions under which hema lands were established have changed radically in recent decades. Thus although the principle may remain, it would not be realistic to assume that the previous hema system could provide a blue print for the future.

6.1.2. The trends of livestock ownership
It was found that the rangelands in the study sites do not provide significant amounts of feed for the livestock and this is associated with the deterioration that has been taking place in the range resources. Also, removing the subsidy of the livestock feed from the government in 1996 hit the herders hard – which resulted in a reduction in the number of sheep. Moreover, it was found that the settled herders tended to own fewer live-

stock than the nomadic herders and that in any case the number of the nomadic herders was decreasing due to sedentarization. Furthermore, the new generation of Bedouin seems to be losing interest in raising livestock. These findings indicate that in the long-term the livestock numbers will decrease which will reduce the difficulties that are facing the management of the rangelands.

6.1.3. Tribal organization

Although the Bedouin traditional organization is still strong, it is not as strong as a few decades ago. It used to be the case that when a shaykh (chief) of a tribe took a decision, the rest of the members of the tribe respected and committed themselves to his 'word'. For many reasons, the power of the shaykhs today is not as strong as before. New generations, in addition to educated people and senior Bedouin army and airforce officers, although they still recognise the shaykhs, have been showing, relatively, different degrees of resistance to the shaykhs' traditional role. This should be put into account when dealing with the local societies. Approaching the local communities should not be only through the shaykhs; the other members also should be consulted and actively involved in discussion about priority and approaches. Through discussion of some cases in the previous sections, it was concluded that working with an organised group is much preferred to working with unorganised individuals. JCC, through implementing a number of successful projects, has tremendous experience of helping local communities organize themselves before starting working with them. JCC's idea is to encourage and help the people to create a co-operative for the targeted community. After creating this co-operative, both JCC and the local people are able to fix the objectives and the methodology to achieve these objectives. JCC had an agreement with WFP, which could help provide the people in the area with many benefits associated with making a co-operative and range reserve.

It should be mentioned here that the idea is not to enforce a specific kind or type of people's organization on them. Any kind or type of people's organization (either formal or informal) would be sufficient in so far as it secures full commitment from all the sections of the community which are involved in implementing the improvement project. It happened once that BRDC and the University of Jordan approached a tribal shaykh in the area in order to carry out an agricultural research project. There were prior discussions between the project's staff which led to the setting of the objectives and methodology to implement the project. BRDC and the University of Jordan relied on the shaykh's word in order to secure full commitments from him to the objectives and methodology. However, as there was no sort of organization, they could not tie that shaykh to his commitments. He cancelled the deal unilaterally, from one side without any understandable reason, which put a lot of investment and effort to waste.

6.1.4. The Bedouin attitudes

The Bedouin, particularly in the BRDC area, are disillusioned because they have been subject to many surveys but they have not seen practical application and development projects. Also, almost all of those interviewed in the study sites did not show any interest in biodiversity and also they agreed with each other that there is no benefit from conservation. They showed fear that conservation of biodiversity will reduce their opportunity to use their traditional lands. In the light of the herders' priorities, set beside the difficulties that they have been facing, protecting biodiversity in their view is considered to be a luxury. It is categorised last in their priorities. It also seems that there is a lack in their awareness of how conservation may help them. This makes creating successful pioneer projects increasingly vital. Rather than discussing theories and concepts with the Bedouin, the researcher came to believe that the time had come to design and implement with them an example of a project which shows the real benefits of managing the rangelands and its impact on their life. Also, another issue needs to be tackled in order to overcome their doubts and resistance. The herders should feel that they are an essential and vital element in the whole process, which is very difficult task but not impossible. This task needs patience and flexibility. Prior discussions are essential, considering them as partners not only receivers. The team, which is supposed to work directly with them, should include some members who are originally Bedouin. These may be experts and/or educated persons who have relevant experience from elsewhere. The experts and educated persons who are originally Bedouin are much more able to understand and deal with them.

6.2. Future trials and demonstrations to integrate livestock owners in managing rangelands in the BRDC area

MoA, JCC, BRDC, and RSCN, in addition to other related institutions, have provided some good examples of integrating the local communities with conservation projects. A number of these projects could be implemented in the BRDC area. Two examples are Tall ar-Rimah and Burqu.

6.2.1. Tall ar-Rimah

It was mentioned earlier that the overall objective of the Jordan Badia Darwin project was to help the local communities, BRDC and RSCN to devise a plan which would benefit the herders while protecting biodiversity. In consequence, three workshops were held in the BRDC Centre (Safawi) for this purpose. A group of herders who belong to a clan from the ash-Sharafat tribe had offered an area, close to the village of Tall ar-Rimah, in order to start a trial as an activity for the above-mentioned conservation plan. The third workshop initiated discussion with those herders for general principles and objectives.

The group has a small area, potentially rich in range resources that could be put under short and long-term plans. To give additional encouragement to the herders, it was planned for the third workshop to discuss with the local communities the benefits of managing the rangelands. As this workshop took place while writing this dissertation in the UK, the researcher suggested, to the people who were in charge of organising these workshops, to link JCC with the whole process in order to enhance opportunities to organise the local people into a co-operative which could be assisted by JCC. In order to show the people an example of a successful project, the researcher took a group of them to see the fattening trial in Dana which was discussed earlier. They had a discussion with the people, who were benefiting from this trial in the lower part of Dana, about its value. The visitors commented to the researcher that they could not see any reason preventing them from applying this trial in their area in Tall ar-Rimah. The approach used in Dana, if the people from Tall ar-Rimah were to adopt it, would be an additional help to the general project of managing the rangelands in their area. It would provide additional benefits for the local herders and for biodiversity. The herders would obtain a higher profit at the same time as fattening some of the young livestock away from the rangelands, which would help protect the range.

Since 2002, a community based rangeland rehabilitation project has started in Tall ar-Rimah area. From the early beginnings, the community participated in defining the problem, selecting the site, selecting the plant species suitable for the area, and choosing the proper grazing method (Al-Tabini et al 2007). The results of the project show that the plant cover has significantly improved and the community participation has been successful.

6.2.2. Burqu

Burqu could be a good place to implement another trial. As mentioned earlier, numbers of raptors of international significance have been discovered in Burqu in addition to a type of partridge that had not previously been recorded anywhere in the world. This will be attractive to bird specialists and also may create an opportunity for eco-tourism. If eco-tourism emerges in the future, it should benefit the local users in the area financially which may make them more interested in conserving biodiversity. But the fieldwork indicates that the livestock owners in Burqu were completely opposed to the idea of protecting Burqu. Therefore, creating a conservation project by BRDC and RSCN (or any other concerned institution) would be very difficult. The first thing to be considered is how to secure a good source of water for the dozens of water tankers, which come to fill up with water from the dam every day. It should be mentioned here that the groundwater in the area is seen by the locals to be of bad quality, causing kidney diseases. However, even though the livestock owners around Burqu did not show any interest in the potential of eco-tourism, helping them to benefit from this is still possible. It should be also mentioned that although it may be too late for the old herders to practice new jobs perhaps their children are a good target for any jobs that the expected eco-tourism would offer. But, without previous discussion and acceptance from the herders, any plans or project, the researcher believes, will not be successful. However, in this area, another possible way in which to resolve the problem of over-grazing and biodiversity degradation might be to plant some grazing shrubs, as in the JCC project areas previously described. This could reduce the pressure on the range resources by offering the livestock another source of feed and this also could lead to direct control of the hunting activities as the users will be curious about any activity close to the planted shrubs of the expected reserve.

In addition to the above-mentioned proposed trials and as a result of the researcher's discussions with Bedouin in the Badia and the points presented at the Darwin meetings, it is clear that there is a need for additional research. For example, there is a need for a better understanding of the current land tenure problems and how they may be overcome. Also, there is a need for an in-depth study of the current situation of Bedouin mobility and whether this will reduce pressure on the rangelands and biodiversity as people settle. Furthermore, there is a need to study the attitudes of the future generations with regard to livestock ownership, and whether this also will ease the problem of biodiversity management by leading to a reduced number of livestock. Having a better understanding of how future generations will deal with livestock could help the institutions mentioned in this study to plan for the future in partnership with those people who retain an interest in rearing livestock.

In the case of the proposed site at Burqu, a technical study of alternative water resources would help reduce the pressure on the dam's water. Also, in al-Ghamr and al-Hazim, there is a need to study the legal position of many wells and seasonal farms as well as their impact on the water quality and plant cover.

In summary, although the problems of balancing the needs of the livestock for access to rangelands with the need for biodiversity conservation are great, there is some reason for optimism. On the one hand, the number of livestock may be declining, and on the other hand, there is some good experience from other projects in Jordan which - in full discussion with the Bedouin in the BRDC area – could be the basis of improved rangeland management which would at the same time enhance both plant and animal biodiversity.

References

Abu-Rabia A (1994). The Negev Bedouin and livestock rearing: social, economic, and political aspects. Oxford: Berg Publisher Limited.

Abu-Zanat M (1993). Maintenance and development of the rangeland in Jordan and its role in combating the desertification. In: Radwan, MS (ed): Workshop on maintenance and development of the rangeland, its role in combating desertification of the Arab World, FAO, AL. Workshop on the rangelands in the Arab World, Amman pp. 364-404.

Abu-Zanat M (1996). The role of the Faculty of Agriculture, Jordan University, in rangeland research and development in Jordan. Workshop on rangeland research and development in Jordan (Workshop documentation), issued by ICARDA, MoA, and GTZ, Amman pp. 172-181.

ACSAD (1983). Studies of al-Hamad basin. Damascus: Arab Centre for Studies of Arid Zone and Dry Lands.

Al-Junaidi M (1996). The development of the natural Jordanian rangeland. Amman: International Engineering and Printing Company Supplies and Services (I.E.C.).

Al-Kasasbeh S (1988). Agricultural settlement in Jordan: The case of Qutrana projects. Ph.D. Thesis, University of Arizona, Arizona.

Al-Oun SS (1997). Livestock marketing in the Jordanian Badia. Unpublished, Ph.D. Thesis.

Al-Tabini R, Al-Khalidi K and Al-Soub H (2007). Community based rangeland rehabilitation project: final report, BRDC.

Alwelaie A (1985). The role of natural and human factors in the degradation of the environment in central, and northern Saudi Arabia. Unpublished, Ph.D. Thesis, University of California, Riverside.

Armstrong AL (1998). Systematic list of birds in the Badia region of Jordan, Jordan Badia Darwin Initiative Project, BRDP, Jordan

Armstrong AL and Leavett R (1998). Some recommendations towards the development of eco-tourism in the Badia region with reference to ornithological interest. Jordan Badia Darwin Initiative Project, BRDP, Jordan.

Ash-Shourbagi M (1993). Rangeland in Arab world: its situation and ways to improve and maintain it. In: Radwan MS (ed): Workshop on maintenance and development of the rangeland, its role in combating desertification of the Arab World. FAO, AL, Workshop on the Rangelands in the Arab World, Amman pp. 19-110.

Brandenburg C (1998). Development anthropology: a primary survey. In: Dutton R, Clarke J and Battikhi M (eds), Arid land resources and their management: Jordan's desert margin. London: Kegan Paul International pp. 247-265.

Budieri A (1995). Burqu preliminary ecological survey. Amman: RSCN.

Campbell DA and Rowe A (1998). Results of a preliminary survey of livestock owners. In: Dutton R, Clarke J and Battikhi M (eds), Arid land resources and their management: Jordan's desert margin. London: Kegan Paul International pp. 189-196.

Chatty D (1990). The current situation of the Bedouin in Syria, Jordan, and Saudi Arabia and their prospects for the future. In: Salzman W and Galaty J (eds), Nomads in a changing world. Naples: Orientale pp. 123-134.

Chatty D (1997). Pastoral tribes in the Middle East and wildlife conservation schemes: The endangered species? Development Anthropologist 15(1&2):12-22.

Clarke JE (1979). A proposal for wildlife reserves in Jordan. RSCN, Amman.

FAO (1995). Community participation in range management for semi-intensive sheep production in the Near East: a case study for Ma'in grazing area in Jordan. Food and Agriculture Organization of the United Nations (FAO), Amman.

Findlay A and Maani M, (1998). Population fertility trends. In: Dutton R, Clarke J and Battikhi M (eds), Arid land resources and their management: Jordan's desert margin. London: Kegan Paul International pp. 197-213.

IFAD (1993). National programme for range rehabilitation and development: pastoral resources and livestock development, annex II. International Fund for Agricultural Development, MoA, Amman.

IFAD (1996). National programme for range rehabilitation and development. Interna-

tional Fund for Agricultural Development, MoA, Amman.
Johnson C (1997). Dana: helping nature helping people. Royal Society for the Conservation of Nature, Amman.
Lancaster W and Lancaster F (1993). Social and economic use of the reserve area: development of the reserve by RSCN and the compatibility or incompatibility between them. BIAAH, Amman.
Mashaqbeh S (1998). Draft report: Darwin workshop no.1, Safawi Field Centre, Jordan. Darwin initiative Badia biodiversity project, CORD, RSCN, BRDP.
Masri A (1991). The tradition of hema as a land tenure institution in arid land management. The Syrian Arab Republic, Rome: FAO.
Masri A (1995). A proposal for developing participatory projects for rangeland improvement in Jordan. ICARDA, Amman.
Mirza O (1997). ash-Shaumari nature reserve. Al-Arab Al-Youm Newspaper, Saturday (25/10/1997), Amman (162):22.
Mouasher A (1997). Dana is a model of sustainable development. RSCN, Amman, al-Reem 60.
Nesheiwat K (1991). Socio-economic aspects of the traditional hema system of arid land management in Jordan. Rome: FAO.
Oakeley R and al-Tabini R (1996). The nature and significance of livestock trading in the north-east Badia of Jordan. BRDP, HCST, Amman.
Omondi P (1994). Wildlife-human conflict in Kenya: integrating wildlife conservation with human needs in the Masai Mara Region. Ph.D. Thesis, McGill University, Montreal.
Rowe A (1997). Pastoralists of the Lower Dana reserve area: attitudes to and prospects for integration into the reserve management strategy. RSCN, Amman.
Rowe A (1998). Bedouin livestock management and socio-economics in the BRDP area. BRDP, HCST, Amman.
Rowe A and Allonby E (1998). Wadi Feinan fattening and marketing trial: financial report and project evaluation. RSCN, Amman.
RSCN (1996). The strategic plan for the Royal Society for the Conservation of Nature (RSCN), Amman.
Sankari M (1993). Maintenance and development of the rangeland in the eastern Arabic countries. In: Radwan MS (ed), Workshop on maintenance and development of the rangeland, its role in combating desertification of the Arab World. FAO, AL, Workshop on the rangelands in the Arab World, Amman pp. 180-297.
Shoup JA (1990). Hima: a traditional Bedouin land-use system in contemporary Syria and Jordan. Ph.D Thesis, University of Washington, Washington.
Swenne A (1995). Rangeland and livestock management. RSCN, Amman.
Tadros K and Salem M (1993). Rangeland resource management at Lajjoun area. ICARDA, Amman.
WFP and MoA (1997). Plan of operations agreed upon between the Hashemite Kingdom of Jordan and the World Food Programme concerning support to participatory land improvement. World Food Programme (WFP), Ministry of Agriculture (MoA), Amman.
Zandri E (1996). Dana nature reserve management plan. RSCN, Amman.

CHAPTER 25

Community-based projects in Jordan's Badia: the Anaqeed Al-Khair case study

Raed Al-Tabini
Jordan Badia Research & Development Centre, P.O. Box 902, Jubiaha, Amman, Jordan.
E-mail: altabini@yahoo.com

William D. Gorman
Department of Agricultural Economics and Agricultural Business, New Mexico State University

James D. Libbin
Department of Agricultural Economics and Agricultural Business, New Mexico State University

Robert Grassberger
Department of Agricultural Economics and Agricultural Business, New Mexico State University

Ismaiel N. Abuamoud
Department of Agricultural Economics and Agricultural Business, New Mexico State University

Salem Al-Oun
Business Administration Department, Faculty of Economics and Administrative Sciences,
Al Al-Bayt University

The Anaqeed Al-Khair project is an agricultural development scheme in the north-east Badia of Jordan, focusing on a cluster of 14 villages with a population of 7 800 people. The project is funded by the Ministry of Planning and implemented by the Badia Research and Development Centre (BRDC) and the Jordan River Foundation (JRF). The project seeks to generate diversified and sustainable income flows for the cooperative members and the community at large within the low rainfall environmental constraints (150-200 mm/yr) that characterize the project zone.

The Anaqeed Al-Khair project has several components: sheep selection and management, honey production, sun-dried tomatoes, forage production, improved sheep wool and goat hair processing. The main obstacles faced by the cooperative are lack of experience in identifying desirable markets, insufficient attention to added value, and lack of a well thought out business plan. In response to the above-mentioned problems BRDC has led the cooperative in overcoming its difficulties, in a number of ways: assistance in finding buyers who offer better terms, establishment of market information systems, facilitating the identification of value-adding technologies and products, analyzing ways to reduce costs of production and marketing, and offering guidelines and methods for the establishment of improved business plans.

Experiences to date suggest that an expansion of activities using the same community-driven methodology is possible, with great potential for replication in other areas of the Badia. The cooperative has decided to concentrate on sheep, honey and sun-dried tomatoes because of their positive income generation, with potential for improvements in marketing. A first formal economic analysis of the results is attempted here so that a statistically valid assessment can be used as the basis for monitoring economic feasibility over the coming years.

Keywords: honeybees, sun-dried tomatoes, sheep, marketing

1. Introduction

At the end of 1998 the Badia Research and Development Programme (BRDP), which later became known as BRDC, changed from a research focus to a policy of research implementation. Based on its early experience as a research agency working in the Badia, BRDC prioritized community-based projects as its focus. Because of financial, political and cultural constraints preventing full-scale project implementation, BRDC instead focused on using pilot projects as a vehicle for gaining the trust and commitment of the local communities and the government.

Past research by BRDC and others leads to the conclusion that desertification and land degradation are largely a function of poor land management practices. The pilot programmes implemented by BRDC seek to teach methods that reduce these impacts and enhance the returns to those who incorporate them. BRDC also recognized that in order to gain acceptance and widespread use, these pilots must be introduced at the grass roots level in the local communities. In contrast to the centralized models used in the past, which brought a plan to the people, this approach seeks to involve the community from the beginning to develop the plan.

The Anaqeed Al-Khair project location is in Jordan's north-eastern Badia, the country's arid and semi-arid zone. The Badia, which makes up approximately 80 % of Jordan (Al-Tabini 2002) and is occupied by Bedouin, may be divided into two major areas based on rainfall and vegetation cover. These are the steppe and the desert. The steppe (home to Anaqeed Al-Khair) has potential for providing animal feed, as it receives an average annual rainfall of 100-200 mm, and provides good vegetation for animal grazing. The desert, on the other hand, receives less than 100 mm of rain annually and has limited grazing resources. The Badia experiences huge seasonal temperature fluctuations. In winter, the minimum mean temperatures may drop to 2-9 °C and snowfall and sub-zero temperatures may occur. The summer maximum mean temperature is around 35-38 °C. Low humidity causes high levels of evaporation or potential evaporation. Rain falls December-March, with great differences in the intensity of showers and storms. Generally described as locally spatial and temporal, maximum annual precipitation ranges from 250 mm in the north-west to 50 mm in the south (Allison et al 1998).

The Anaqeed Al-Khair project is an agricultural development project funded by Jordan's Ministry of Planning and is implemented by BRDC and the Jordan River Foundation (JRF). The project's goal is to improve the local community's socio-economic status. A cooperative was established to ensure that the local community actively participates in the project and that benefits flow directly to its members (Hawamdeh 2004). The cooperative was established late in 2002 with 16 member households. Membership had increased by 2007 to 92 households.

2. The Northern Badia Villages Cluster

The Anaqeed Al-Khair project targeted 14 villages for development (Table 1). These villages were assimilated into one cluster, called the Northern Badia Villages Cluster (NBVC), and then grouped into four sub-clusters according to geographical distribution. The NBVC's population is 7 800 individuals in 1 259 households.

The cluster's population is youthful, with 55 % of the population below age 21. The gender ratio is 49 % female to 51 % male. The work force, those aged 15-65 years, has 80 % of the workers employed in the pubic sector, showing a high dependency on governmental support. In contrast, the private sector provides only 11 % of work opportunities in the Badia, reflecting low levels of investment in the area. The unemployment rate is calculated to be 40 %, which is attributed to the lack of job opportunities and vocational training. The total monthly income of JD189 reflects a high poverty rate when compared to JD337 in all rural areas and JD401 throughout Jordan. The livestock industry is a key income source for NBVC; livestock numbered approximately 38 735 head when the project began (BRDC 2002).

The Bedouin families of NBVC share a perspective on their way of life and cultural identity with the rest of the Bedouin in Jordan. They closely identify themselves with the Badia's environment (the camel as their traditional means of transportation and the *bayt ash-sha'r* (tent) as their dwelling) and regard themselves as the sons of the Badia (Hawamdeh 2004). The identification is not only important as an expression of self-identity; it is also extended to express the individuality of their traditional lifestyle as nomads, even though that lifestyle is shifting to a semi-nomadic and, even, fully settled pastoral or even urban way of life (Millington et al 1999).

The general project goal has been to improve the socio-economic status of the local community. Within this general goal are four objectives: generate diversified and sustainable income flows for the cooperative members and the community; improve local sheep breeds; encourage organic farming activities; encourage local processing of raw materials.

Management has been a key issue. After several community meetings, participants chose a cooperative for the management structure of the project. This approach provides flexibility and assurance that the local community actively participates in the project and that benefits flow directly to them. Although the cooperative was established late in 2002 with just 16 member households, the number of participant households has steadily increased as confidence in the project has grown. The Anaqeed Al-Khair cooperative movement bases its planning and development decisions on members' perspectives and knowledge. Thus, the local community is empowered to explore possible areas of community development, discuss scenarios of future aspirations, and define the direction and shape of social transformation, making that transformation more likely to be sustained (Al-Tabini et al 2005).

The nature of collaboration between the local community and the Anaqeed Al-Khair project implementers reflects concern for the impact that development initiatives have on reshaping the socio-cultural, economic and environmental future of the Bedouin community. The collaboration helps to mitigate the negative perceptions and attitudes that community members may have towards development efforts because the cooperative management structure provides an avenue for community members to share information

Table 1. Northern Badia Cluster villages (Source: BRDC, 2002)

Number	Village	Families	Population	Sheep	Goats
1	Khasha't al-Qinn	154	1081	469	128
2	Umm al-Quttayn	386	2389	1308	1010
3	al-Ma'azula	19	129	686	30
4	al-Mukayfita	286	1966	22888	391
5	Manshiyyat al-Qinn	19	121	113	73
6	Rawdat Amir	120	533	3145	459
7	Mayssa	30	196	2358	413
8	ath-Thilaj	19	69	871	74
9	Umm Husayn	23	173	4580	2221
10	Qasim	78	433	4139	574
11	ar-Rifa'iyyat	104	604	2704	1101
12	Ghadir an-Naqa	6	17	–	–
13	Urayniba ash-Shamaliyya	10	69	–	–
14	ar-Rifa'iyyat ash-Shamaliyya	27	112	–	–
Total		1259	7829	32261	6474

and responsibilities which seem, amongst other advantages, to facilitate income generation. Collaboration between the community and the project team reflects the government's official vision to incorporate multiple participants in the development design process and to support new forms of organizational structures (such as local cooperatives, NGOs and research centres) in its new policies for sustainable development programmes. It also reflects the capability of these organizations to employ effective development designers and implementers who can learn more and adapt effectively through the collaboration process. Finally, collaboration reflects the capacity and competence of each partner, particularly local community members, to play a role in engaging, initiating and adapting local ideas into the design process (Hawamdeh 2004). This has been achieved by introducing the new concept of 'village cluster'. The cluster redefines the relationship between the community members on the one hand and between the community and the developmental organizations on the other, bringing deeper understanding of the conditions and circumstances of residents in the various cluster villages and helping to eliminate reluctance among members of different villages to cooperate together to the benefit of the larger cluster community.

2.1. Role of the cooperative

The Anaqeed Al-Khair cooperative, as a community-based initiative, brings together the individual and community perspective, knowledge, and comprehension to inform the decision-making of development and planning. The engagement of the local cooperative as a key stakeholder is crucial in empowering the local community in researching areas of community development, discussing narratives of future aspirations, and defining the direction and shape of social transformation for better sustainability

From a socio-cultural perspective, the project stimulates the local community's participation by endorsing new grassroots initiatives that reflect socio-cultural relationships and powers through new organizational structures, such as the Anaqeed Al-Khair cooperative. The cooperative then is encouraged and supported to participate at every stage of the implementation process. The plan is that the cooperative will eventually become the long-term manager and owner of the project.

We have recognized the central importance of the market; there is no point in producing even excellent quality goods if nobody wants to buy them. However, researching the market in Jordan has proved to be very difficult. There is little governmental consistency and accuracy in the reporting of imports, exports and other economic indicators (Al-Oun 1998), and when it comes to agricultural products the task becomes increasingly difficult due to varying policies among institutions. Therefore, all of the projects that BRDC has proposed for the larger Anaqeed Al-Khair project are based on the assumption that each individual project is relevant to the Badia community by being related to the existing livestock and agriculture industries. The main aim for each project is to improve the efficiency of regional farming practices to make them more economically viable. To expand sales, BRDC targets both national and international markets. Discussion has taken place among various organizations to create a private company to market Badia products. It has proved difficult to create a private company; it took a few years for the cooperative to establish itself; making all the projects work together has been a big challenge. The idea of establishing a private company within the village cluster for marketing their products is still there, but this will take more times, till the cooperative decides what to focus on as final products. Meanwhile, BRDC has helped the cooperative find international markets for the sun-dried tomatoes through having an agreement with a private company in Amman which works with Spain and Italy. This has given the cooperative a big opportunity to sell the dried tomatoes at a good price.

To better explain the current market, we will discuss each of five production projects individually: sheep selection and management; honey production; sun-dried tomatoes; forage production; and improved sheep wool and goat hair processing.

2.2. Accounting systems – depreciation

The system of project accounting (Tables 4-6) includes a large element for depreciation derived from Tables 2 and 3.

Taking this depreciation into account when considering the profits and losses of the project as a whole is appropriate. However, this should not be confused with the profit and loss 'perception' or, indeed, the reality, of the individual producer or even a group of producers working together in a self-created cluster. In practice, farmers, and especially small farmers or craft-scale producers, don't spend, for example, more than JD 82 000 on an administration building (Table 3). They make do with much simpler facilities. The cluster concept as tested in the Badia has a great value, by encouraging people to work together, in providing economies of scale and in exerting collective pressure on suppliers and at points of sale, and the approach includes the kind of technical expertise and networking strengths of a support organization, in this case BRDC, which have to be costed. But this can be achieved at a fraction of the cost indicated in Table 3.

Therefore, when reading this paper it is important to realize that from the point of view of the producer the balance sheet is more positive than the balance sheet provided here, which is from the viewpoint of the overall cluster programme.

Table 2. Amortized set-up expenses

Items	Quantity	JD / Unit	Units	Total	Life (years)	Annual depreciation
Farm design	1	10 945	each	10 945	30	365
Electrical hookup	1	5 423	each	5 423	30	181
Water hookup	1	4 720	each	4 720	30	157
Total				21088		703

Table 3. Capital purchases (2003-4) & depreciation

Items	Quantity	JD/Unit	Units	Total	Life (years)	SV	Annual depreciation
Land purchase	1	5 000	each	5 000			
Depreciating assets:							
Administration building	1	82 526	each	82 526	30	16 505	2 751
Tomato and honey building	1	52 364	each	52 364	30	10 473	1 745
Pre-fabricated unit	1	3 000	each	3 000	15	600	200
Wool building	1	47 895	each	47 895	30	9 579	1 597
Green fodder facility	1	1 400	each	1 400	10	980	140
Electric switch control room	1	1 948	each	1 948	30	390	65
Fencing	1	5 375	each	5 375	15	1 075	358
Furniture	1	4 447	each	4 447	10	-	445
Computer, printer, telephone	1	1 310	each	1 310	5	-	262
Refrigerators	1	236	each	236	10	47	24
Vehicle shade (car port)	1	2 500	each	2 500	30	1 000	83
Amortized setups	1		each		30		338
Total	1			203 000			8 008

3. The Anaqeed agricultural production projects

3.1. Sheep selection and management

Since 2004, the cooperative has adopted improved sheep feeding, a veterinary regime, selection and culling, and synchronization of ewe pregnancies to result in higher flock lambing and weaning percentages and greater twinning. The project was based on selecting the best sheep in the Badia region, defined as those with high milk productivity and breeding abilities. Indicators in the Badia, based on BRDC research (Campbell et al 1995, Jones 1999), illustrate that good management with typical sheep is a profitable project, though the removal of the government feed subsidy by 1996-7 (Schoen, this volume) greatly reduced profit margins though it also provided an incentive to improve management skills. A rapid increase in the number of project sheep happened because the cooperative with assistance of the BRDC technicians implemented some new and improved techniques in sheep raising, for instance, use of sponges, early weaning, and a better feeding regime.

Gorman et al (2007) reported that an initial purchase of 130 ewes along with 6 rams was made in 2003-4, along with 25 goats. One of the stated objectives of this sheep enterprise was to breed higher quality breeding lambs to be sold into the community and thus increase the quality of sheep production in the region.

Revenue streams from wool, milk, and a few lamb sales yielded gross revenues of JD2 260 in the first year, 2004 (Table 4). However, low first year lamb production in concert with high capital costs netted a loss of JD3 104 in that year. With the purchase of 128 ewes in 2004-5 and the retention of 31 lambs from the previous year, the flock size more than doubled. Revenues from production increased six times over the previous year while operation costs also increased but at a slightly lower rate. However, the added non-cash costs of depreciation led to a small net loss of JD1 516. By year three, lambing rates had improved from 45 % to just less than 70 %, indicating improving management.

During 2005-6 all the goats were sold. Combined revenues with the sale of ewes and lambs accounted for 79 % of the JD18 767 gross revenues that year. Net contribution to profit and general and administrative expenditure (G&A) amounted to JD3 535. However, no lambs born in 2005-6 were retained in the breeding flock. Over the three years Table 4 shows a strong upward trend into profitability.

The Jordanian market for sheep includes locally and internationally produced animals. However, local Awassi sheep maintain a strong price differential over imported sheep from Australia, New Zealand and Sudan (Al-Oun 1998). For instance the prices for imported sheep in 2006 fell to JD60-70 but the Awassi sheep price varied from JD100-120. This demand for local Awassi sheep is due to consumer tastes in Jordan and the Gulf States. Bedouin farmers around the Arabian Badia traditionally maintain Awassi sheep. However, they are facing tremendous problems, such as poor flock management and overgrazing.

Changes in government policies and other regional impacts have adversely affected sheep production costs. Most Jordanian sheep producers have become dependent on imported barley and wheat bran. The price of these feeds has increased as a result of withdrawal of government subsidies. Also adding to the increase in price are local feed shortages caused by regional drought; the four years to 2007 are classified as drought years (Al-Tabini et al 2007). Medicine for animals has become more costly as the Ministry of Agriculture privatized the formerly subsidized services. Shepherding labour also is quite expensive. Most Jordanian farmers rely on foreign labour at an average salary per worker of JD120-150/month. Many large flock owners have given up milking due to the extensive labour requirements and the great distance from labour markets.

Table 4. Income statement for sheep operation (including a few goats)

		01/11/2003 - 31/10/2004			01/11/2004 - 31/10/2005			01/11/2005 - 31/10/2006		
	Units	Amount	JD/Unit	Total	Amount	JD/Unit	Total	Amount	JD/Unit	Total
Revenue from sales										
Wool	kg	130.0	1.0	130	140.0	1.0	140	81.0	1.0	81
Milk	kg	1687.5	0.4	675	1 912.5	0.4	765	3 420.0	0.42	1 436
Manure	units				2.0	50.00	100	0	50	0
Ewes 1	head				65.0	45.38	2 950	68.0	70	4 760
Ewes 2	head				18.0	98.00	1 760	11.0	38.00	418
Culled ewes	head				23.0	29.26	673			
Goats	head				3.0	53.33	160	20.0	89.09	1 782
Goat kids	head				9.0	60.56	545	8.0	30.00	240
Rams	head				4.0	91.25	365			
Bucks	head				2.0	105.00	210			
Lambs	head	26.0	55.97	1 455	87.0	53.83	4 683	150.0	67.00	10 050
Total sales				2 260			12 355			18 767
Cost of goods sold										
Initial inventory				0			0			0
Additions to inventory										
Feed	ton	40	75.0	3 000	100.5	75.0	7 538	87.0	81.13	7 058
Medicine	unit	1.0	117.0	117	1.0	395.0	395	1.0	419.00	419
Ear tags		400.0	0.3	120						
Water	m^3	75.0	2.0	150	285.0	1.0	285	180.0	1.00	180
Farm labour	months	12.0	91.5	1 095	12.0	110.0	1 320	11.0	165.00	1 815
Milking labour (women)	days	120.0	3.0	360	170.0	3.0	510	205.0	2.00	410
Technical supervision	years	0.1	3 600.	300	0.2	3 600.0	600	0.04	3 600.0	150
Electricity	kw	272.7	0.22	60	272.7	0.22	60	180.0	0.22	40
Total cost of production before depreciation				5 202			10 708			10 072
Depreciation				4 035			5 929			5 512
Less lambs retained for breeding herd	head	31.0	147.93	4 586	31.0	26.25	814	25.0	29.23	731
Less lambs retained for breeding herd	head				25.0	78.08	1 952	0.0	72.87	0
Total cost of production				4 651			13 871			14 853
Less final inventory				0			0			0
Contribution to profit and G&A				-2 391			-1 516			3 914
Loss of capital items				713			0	5		380
Net contribution to profit and G&A				-3 104			-1 516			3 535

Mixed feed costs roughly JD14 850 for all 250 sheep for an 18-month period. Medicine and vaccinations cost JD660 for the same number of sheep and time period. Labour, being highly accessible in an area where unemployment is around 50 %, is cheaper in comparison to other regions. All inputs for this project cost JD38 000. Most inputs are bought locally or in neighbouring cities.

The farm has targeted its yearling sales to local farmers in the Badia. The other option is to sell the yearlings on the local market in regional major cities. Milk products are sold at the farm gate to families, cheese-makers and the Tall ar-Rimah dairy factory. In order to escape using a middle-man, milk also can be sold at the farm gate.

The main concept behind this project is to introduce and maintain high quality, healthy local Awassi sheep. Also, through the creation of the farm and its use of best management practices, a model will exist for other farmers to learn new farming techniques.

Most of these assumptions are based on the experience of the BRDP's DFID project (1998-2000). In dealing with demonstration flocks, farmers working with BRDP secured excellent prices and earned JD90 per lamb (sold at age four months in the year 2000) whereas ordinary farmers sold lambs at prices of JD55-60.

3.2. Honey production

The cooperative first acquired bees for honey production in late 2003. Some 50 hives containing honeybee colonies were initially purchased. Of these 50 colonies, 37 were lost after or during the first year of production. The exact causes of the losses were, most probably, cold winter conditions and lack of an adequate food and water supply. The losses are reflected in the 2003-4 income statements as a capital loss and are thus responsible for a large net loss in that year of JD7 659 (Table 5). The consequences of the loss of the colonies are further seen in the 2004-5 income statements when total honey production (and, as a result, revenues) declined by nearly 50 %.

Table 5. Income statement for the honey operation

Year:	01/11/2003 - 31/10/2004			01/11/2004 - 31/10/2005			01/11/2005 - 31/10/2006		
	Quantity	JD/Unit	Total	Quantity	JD/Unit	Total	Quantity	JD/Unit	Total
Revenue from sales	206	10	2 060	109	10.00	1 090	216.0	12.10	2 614
Cost of production									
Initial inventory	0	0	0	0	0	0	0	0	0
Additions to inventory									
Feed	97.5	0.40	39	87.5	0.40	35	48.8	0.40	20
Medicine	1.0	54.00	54	6.0	20.00	120	4.0	15.00	60
Water	2.0	1.00	2	2.0	1.00	2	2.0	1.00	2
Electricity	10.0	0.22	2	10.0	0.22	2	10.0	0.22	2
Jars (0.5 kg)	300.0	0.15	45	300.0	0.15	45	230.0	0.15	35
Jars (1 kg)	100.0	0.25	25	100.0	0.25	25	124.0	0.25	31
Supervisor	3.0	50.00	150	10	300.00	300	1.0	300	300
Labour	122.0	3.62	442	90.0	3.62	326	15.0	3.00	45
Total cost of honey before depreciation			759			855			494
Depreciation			2 743			1 185			976
Total costs of honey			3 501			2 040			1 470
Less final inventory of jars	1.0	13.2	13	1.0	55.50	56	1.0	62.4	62
Contributions to profit and G&A			-1 428			-894			1 206
Loss of capital items (bee colonies)	-37.0	168.4	6 231	-5.0	-126.3	-632	-16.0	-72.1	-1 154
Net contribution to profit and G&A			-7 659			-1 526			52

With the purchase of new colonies for the 2005-6 season, and an increased knowledge of managing the honey production operation as reflected in a lower percentage of hives lost, a positive return to profit and G&A of JD52 was achieved. Table 5 shows a general upward trend into profitability.

The cooperative produces honey in both 1 kg and 0.5 kg jars. The product sells well in local markets under the Badia label. Distributing the honey in major outlets in Amman or other major metropolitan areas where higher prices should be possible could secure a substantial increase in revenue for the cooperative. Focusing on enhancing the brand awareness and quality and continued improvements in management should make honey a sustainable and profitable enterprise for local farmers (Gorman et al, 2007).

3.3. Sun-dried tomatoes

The cooperative started growing a specially selected variety of tomatoes suitable for making sun-dried tomatoes during the 2004-5 fiscal year. The cooperative purchased 450 t of raw tomato products grown by cooperative members. The cooperative dried, packaged and sold 455 kg of the product at JD1.60/kg for a total revenue of JD728. However, operating expenses before depreciation amounted to JD695 resulting in a profit of only JD33 before accounting for depreciation expenses. After depreciation the enterprise showed a loss of JD428 (Table 6) for the year 2004-5.

An analysis of the operation at that time indicated that the cooperative could contract with local farmers to grow the tomatoes cheaper than the growing cost incurred by the cooperative. During the year 2005-6, the cooperative ceased growing tomatoes and contracted all production to outside growers. The cooperative was able nearly to double

Table 6. Income statement for sun-dried tomato operation (1/11/2004-31/10/05)

		1/11/2004-31/10/05			1/11/2005-31/10/06		
	Units	Quantity	JD/Unit	Total	Quantity	JD/Unit	Total
Revenue from sales of sun-dried tomatoes	kg	455	1.60	728	850	1.60	1 360
Costs of production:							
Initial inventory							
Additions to inventory							
Tomato seeds	kg	1.0	36.5	36.5	0.0	36.5	0
Plastic bags	kg	2.0	4.00	8	5.0	4.00	20
Fertilizer and pesticide	unit	1.0	15.00	15	0.0	0.00	0
Salt	kg	250	0.06	15	350	0.06	21
Water	m³	35.0	2.00	70	10.0	2.00	20
Raw product purchases (tomatoes)	tons	450.0	0.4	180	10.0	70.00	700
Labour	day	38.0	5.00	190	40.0	5.00	200
Supervisor	unit	1.0	180	180	0.0	0.00	0
Final inventory				0			0
Total cost of production before depreciation				695			961
Depreciation				461			461
Total cost of production				1 156			1 422
Contribution to profit and G&A				-428			-62
Loss of capital items				0			0
Net contribution to profit and G&A				-428			-62

sales and made only a small loss of JD62 after covering depreciation costs (and a profit of almost JD400 before depreciation). The decision to go to all contract production proved wise and should be continued.

The evidence to date indicates that sun-dried tomatoes can be a long-term profitable enterprise for the cooperative. If the cooperative can further increase its contract grower programme by at least 25 percent without having to increase the contract price of JD70/t, the enterprise will be sufficiently profitable to cover all expenses and make a profit. Expanding the contract-growing programme also increases the income and employment opportunities in the community. There appears to be an opportunity to further enhance profits by concentrating on better packaging, focusing on building a strong brand name and enhancing marketing. These activities should result in larger markets and higher prices. Work on better marketing was part of the plans for the 2006-07 year (Gorman et al, 2007). In 2007 season, the cooperative again contracted a local farmer to plant the tomatoes, to get more output with less cost. Consequently, they sold the produce for a higher profit compared with the previous year. It is difficult to discuss the details since the financial report for the cooperative had not been completed when writing this paper.

3.4. Forage production

In 2003 the cooperative attempted to produce alfalfa but was not successful due to inadequate amounts of water for irrigation. The cooperative then tried barley, but felt this too was an unsustainable practice.

3.5. Goat hair processing: *bayt ash-sha'r* (hair tent) production

The cooperative manufactured two *bayt ash-sha'r* during the 2003-4 fiscal year. This enterprise was intended to use local goat hair fibre resources and provide employment with an anticipated ability of selling the tents to traditional Bedouin families. The enterprise has technically been profitable as the income statement shows a net profit of JD298. However, in this case the actual income statement does not tell the whole story. To date they have been able to sell only one tent at the JD1 500 asking price. The other tent remains in inventory. From a cash flow standpoint the tent activity has resulted in a negative cash flow of JD659 through the 2005-6 fiscal year. The final net outcome will depend upon the price for which the tent is finally sold. However, it appears that quality Bedouin tents can be purchased cheaper in Syria than the cooperative can make them. Since such items move freely across the border, there does not appear to be an opportunity for a profitable tent manufacturing operation and the activity has therefore been discontinued.

4. Results

Working together with BRDC, the cooperative has undertaken a very significant set of production project tests and, which is very unusual in small-scale pilot development programmes of this nature, subjected the early results to a detailed financial scrutiny. This has resulted in two of the projects (growing forage and manufacturing *bayt ash-sha'r*) being dropped respectively because of shortage of water for irrigation and because a cheaper *bayt ash-sha'r* can be manufactured in Syria. The financial scrutiny has also indicated ways in which the performance of the remaining three projects (sheep raising, honeybees and dried tomatoes) can be – and must be – improved in order to make dependable profits and, therefore, to attract a growing interest within the cluster community and beyond. An important part of performance improvement lies in marketing.

Discovering and exploiting market opportunities is a central and never ending task for the cooperative. In practice this will very largely be a task for BRDC because of its expertise and wider network of contacts in Jordan and internationally. Markets are the drivers. Each market will impose conditions on the production system and it is only if these conditions are met that the projects will expand and be sustained.

The cluster programme is also demonstrating that where projects are seen to be both technically within the range of the local communities (with some outside support) and that the products can be marketed, they are often attractive to cooperative members and the surrounding communities and that the activities are subsequently adopted after the cooperative demonstrates success over a trial and error period. These models of technology transfer are vitally important to the sustenance of Bedouin living in their fragile rangeland ecosystems.

The main obstacles faced by the cooperative in marketing their products can be summarized as follows:

- Lack of experience in identifying desirable markets;
- Insufficient attention to value-addition;
- The need to decrease transport costs through more efficient scheduling of larger allotments to larger markets;
- Lack of a well-thought-out business plan based on analysis of input costs and sale prices, advance contracts and new markets.

In response to the above problems, BRDC has led the cooperative in overcoming their difficulties through the following efforts:

- Assistance in finding buyers who offer better terms;
- Establishment of market information systems that will open up more market channels;
- Help in identifying value-adding technologies and products;
- Research of methods for increasing production, which will generate greater income;
- Analysis of ways to reduce costs of production and marketing that will lead to options that can be tested by the cooperative;
- Provision of guidelines on sanitary and phyto-sanitary procedures, grades and standards to meet international trade requirements and generate consumer confidence;
- Provision of guidelines and methods for establishment of improved business plans.

5. Conclusions

- Experience has shown that the larger community is monitoring the project activities and then adopting the successful ideas by watching the experiences of the cooperative;
- The cooperative members and the community can purchase higher quality lambs from the cooperative as breeding stock. The next steps include greater focus on diversification of lamb and goat dairy products in a newly constructed dairy processing facility;
- Experiences to date suggest that an expansion of activities using the same community-driven methodology is possible, with great potential for replication in other areas of the Badia. The cooperative has decided to concentrate on sheep, honey and sun-dried tomatoes because of their positive income generation and potential for improvements in marketing;
- The next step is to improve on this more formal economic analysis of the results so that a statistically valid assessment can monitor economic feasibility over the coming years and shape technical and market improvements.

References

Al-Oun SS (1998). Livestock marketing in the Jordanian Badia. PhD thesis, Durham University.

Al-Tabini R, Al-Khalidi K and Al-Soub H (2007). Community-based rangeland rehabilitation project. Final Report, submitted to USDA-Forest services.

Al-Tabini RJ (2002). An evaluation of the potential of Atriplex nummularia for sheep production in the arid Jordanian rangelands: the effects of defoliation management. PhD thesis, University of Newcastle Upon Tyne.

Al-Tabini RJ and Hawamdeh I (2005). Local community participation at Anaqeed Al Khair project, Proceeding of the international conference on promoting community-driven conservation and sustainable use of dry land agro biodiversity. ICARDA, Aleppo, Syria, 18-21 April 2005.

Allison RJ, Higgitt DL, Kirk AJ, Warburton J, Homoud AS, Sunna BF and White K (1998). Geology geomorphology, hydrology, ground water and physical resources. In: Dutton RW, Clarke JI and Battikhi A (eds), Arid land resources and their management: Jordan's desert margin. London Kegan Paul International pp. 21-46.

BRDC (2002). Socio-economic survey for Um Al-Qotain and Mkaifteh areas. Northern Badia report.

Campbell D and Roe A (1995). Results of a preliminary survey of livestock owners in the Badia Research and Development Programme area. CORD, University of Durham, BRDP report p. 11.

Gorman W, Grassberger R, Abuamoud I, Al-Tabini R and Al Oun S (2007). Profitability Assessment of the Enterprises of the Anakeed Cooperative Community-Based Project, Badia Region, Jordan. Final Report.

Hawamdeh I (2004). Local community participation in livelihoods improvement and sustainability: Anaqeed Al-Khair Project case study. Masters thesis, York University, Canada.

Jones K (1999). Primary veterinary care, training of farmers and paravets. In: Campbell D and Dutton RW (eds), Population and livestock-related studies in the eastern Badia of Jordan. Record of a workshop held under the auspices of the Jordan Badia Research and Development Programme, 22-23 September, 1996, Royal Scientific Society, Jubaiha, Amman, Jordan. Higher Council for Science and Technology pp. 88-92.

Millington A, Hussein S and Dutton R (1999). Population dynamics, socio-economic change and land colonization in northern Jordan, with special reference to the Badia Research and Development Project area. Applied Geography 19(4):363-384.

CHAPTER 26

Community-based rangeland rehabilitation projects: the Tall ar-Rimah case study in north-east Jordan

Raed J. Al-Tabini
Badia Research and Development Centre, Jordan

Hani M. Saoub
University of Jordan, Amman, Jordan

Khalid M. Al Khalidi
Badia Research and Development Centre, Jordan

It is clearly understood and acknowledged that the rehabilitation of arid rangelands here in Jordan, or elsewhere, requires the application of appropriate technology by dedicated workers with skill and patience. The method of furnishing supplemental forage which was demonstrated under this project involves the establishment of nutritious, drought-tolerant shrubs in what could be called 'forage shrub reserves.'

The broad framework of this project aims to enhance the awareness of the Badia people about the importance of rangeland management to their production of livestock, and at the same time allow BRDC, and its supporting associates, to work hand in hand with local communities to demonstrate how certain range management practices can be of direct benefit to the Badia people and the resources on which they depend.

The local community was involved in the project from the starting point of problem definition, possible solutions, site selection, and selection of plant species suitable to the area and the intended use by livestock owners.

A 50 ha demonstration site was prepared, using a variety of water harvesting configurations and digging holes for shrub planting. Three basic types of land treatments, or 'structures,' were demonstrated (with approximate percentages of treated area given for each type): contour furrows, 80 %, crescent-shaped or chevrons (micro-catchments), 10 %; and low rock walls, 10 %.

Statistical analysis indicated that the estimated biomass produced within contour furrows is significantly higher than the production within the crescent and V-shape water harvesting structures. The reason for the difference is related to the production from the contour furrows. It can be concluded that contour furrows are more efficient than other techniques in supporting shrub forage production.

When biomass data were combined for all structures there was a statistically significant difference between *Salsola vermiculata* compared with *Atriplex nummularia* or *Atriplex halimus*. Also, there is a biological difference in biomass production between *A. nummularia* and *A. halimus*, though statistically not significant.

The results obtained from the first botanical survey showed that in the survey which was carried out before starting the project (2002-3), there were only 22 plant species belonging to 12 different families. However, after four years of protection, the number of plant species and families found in the project area were higher; 51 plant species, belonging to 18 families were recorded in the 2005-6 survey.

The project has been able to demonstrate to the government that, with the cooperation of the local Bedu population, rehabilitation of large areas of the Badia steppe is possible. It has also shown the necessity of urgently addressing the issues associated with free access to all areas of land, the effects of drought and the need for preparing drought contingency plans. The total number of sheep, overstocking and issues associated with subsidized feed are other matters that will need to be addressed. The focus of future project activities will be on selected rangeland site management improvement and rehabilitation, with the selected areas managed by Bedu organizations.

Key words: rangeland, community participation, grazing management, plant survey, dry matter production.

1. Introduction

1.1. Rangeland rehabilitation

It is clearly understood and acknowledged that the rehabilitation of arid rangelands here in Jordan, or elsewhere, requires the application of appropriate technologies by dedicated workers with skill and patience. No quick fixes are promised or likely. It has taken thousands of years for these rangelands to reach their present condition, and it will take more than a few dedicated range managers and a few good rainy seasons to turn things around in the Badia. One of the most effective ways of rehabilitating these depleted Mediterranean rangelands is the planting of saltbushes, as shown in many Mediterranean countries over the past 40-50 years. But, there is no miracle solution; planting saltbushes in rangelands or farmland is also subject to a number of constraints in terms of species selection, establishment and management (Le Houe'rou 1993).

The method of furnishing supplemental forage which was demonstrated under this project involves the establishment of nutritious, drought-tolerant shrubs in what could be called 'forage shrub reserves.' The shrubs can be utilized in various ways, at different times of the year depending on what the adjacent range conditions are at any given time and what the livestock requirements are. For example, these forage reserves can be used exclusively during periods of normal seasonal deficits in range forage, or during drought periods. Alternatively, they can be grazed intermittently, while adjacent ranges are providing some, but not fully adequate, feed for livestock.

1.2. Project objectives

The broad framework of the project aims to enhance the awareness of the Badia people about the importance of rangeland management to their livestock production, and at the same time allow BRDC, and its supporting associates, to work hand in hand with local communities to demonstrate how certain range management practices can be of direct benefit to the Badia people and the resources on which they depend. The potential long-term benefits of this effort could be:

- Sustainable, productive use of the range resources;
- Improved socio-economic conditions for livestock owners;
- Enhanced local capacity to manage and preserve a productive ecosystem;
- Increased biodiversity and stability of rangeland ecosystems.

The above long-term benefit can be achieved by the following detailed objectives:

- Demonstrate the feasibility of establishing a forage shrub reserve under the prevailing soil and climatic conditions of the Tall ar-Rimah area;
- Educate the livestock owners and range users of the Tall ar-Rimah area about the importance and implementation of sustainable rangeland management;
- Demonstrate water harvesting and forage production techniques;
- Demonstrate livestock systems responses to the above (phase 2: 2007).

2. Methods and materials

2.1. Implementation with community participation

The local community was involved in the project from the starting point of problem definition, possible solutions, site selection, and selection of plant species suitable to the area and the intended use by livestock owners. This step was carried out by public meetings with landowners, livestock owners and key personnel in the area. The project team consulted with the local community on the following issues:

2.2. Site selection

The site was selected through consultation with the community and with assistance of project experts, so as to ensure that the site is appropriate from both technical and social points of view. The local community has long experience in the area, which is reflected in our case study by recommending areas that supported suitable vegetation cover and where runoff is expected to occur at least annually. Consequently, the project team reviewed the environmental conditions required for the project site with the community, in order to stimulate their views on the selection of the most suitable site and avoid any private agendas.

2.3. Plant species adapted to the area

Rehabilitation of rangelands with appropriate shrubs, forbs or trees can add considerably to feed resources as well as stopping degradation. Among such forages are *Atriplex* species, *Acacia* species, *Salsola* species and *Juncos* species. The *Atriplex* species represent a group of plants well adapted to salinity and drought stress. Several studies have recommended the cultivation of such forages in saline soils and arid areas as good feed resources (Ben Salem et al 1993, Draz 1987, El Shaer et al 1990, Le Houe'rou 1993).

After several meetings with the community, *A. nummularia, A. halimus* and *S. vermiculata* were chosen to be introduced to the site. During the field tour for site selection, local people showed interest in *Salsola* species because they are native to the area and because they are highly palatable. Therefore, the two *Atriplex* species and one *Salsola* species were introduced for comparison.

2.4. The Tall ar-Rimah demonstration site

The Tall ar-Rimah area (Figure 1) is situated north-west of Safawi (Badia area: N32° 17′ 211″ E36° 53′ 916″) and about 70 km east of Mafraq city. One demonstration site was made available to BRDC for the Tall ar-Rimah range project. About 50 ha, situated north-east of the village of Tall ar-Rimah was selected for demonstration of water harvesting techniques and forage shrub establishment and was referred to as the 'Tall ar-Rimah Shrub Reserve.'

The general topography of the Tall ar-Rimah area is gently undulating with scattered hills. As is typical of the rest of the Badia, the climate in the Tall ar-Rimah area is classified as arid. The annual rainfall normally varies from 100-200 mm. Significant rainfall is not expected to occur outside the period from late October to late March, and its timing and intensity can vary widely from event to event and year to year. Most of the rainfall occurs during January and February. Snowfall is not unknown, but is not expected every year, nor in heavy amounts. Mean maximum and minimum air temperatures during January are 13 °C and 3 °C, respectively. The corresponding temperatures are 33 °C and 17 °C in August, the hottest month in this area (BRDP 2001).

Figure 1. Project area at Tall ar-Rimah before starting project activities

The soil texture varies from clay loam at the surface to clay for the rest of the soil profile (Table 1). Soil depths also vary from shallow at the north-west part to deep near the eastern part. Vegetation cover is very sparse, with large areas of bare soil. Native herbaceous plant species common to the area include: *Lolium multiflorum, Lolium rigidum, Phalris canarienses, Medicago scutellata, Lathyrus sativus, Hordeum glucum.*

In spite of the aridity of the area, careful use of water harvesting techniques, coupled with securing cooperation about grazing systems from the local population, has resulted in a very considerable increase in vegetation cover, as seen in Figure 3.

2.5. Water harvesting techniques

The term 'water harvesting' is given to any work done on land surfaces to benefit from rain water, either directly by storing runoff water in the soil profile so that the net runoff volume and hence soil erosion are reduced, or indirectly to store the surface runoff in a reservoir to be used later for different purposes (supplemental irrigation of crops, drinking, or for groundwater recharge). Water harvesting can be simply defined as the collection, storage, and management of rain water and runoff water for useful purposes.

The main factor when designing water harvesting techniques is the ratio between the runoff area (catchments area) and the run-on area (cultivated area or storage area). The runoff area is optimized if it produces enough runoff (adequate runoff coefficient). The run-on area, where water is stored or plants are grown, must have enough storage capacity or adequate water holding capacity to supply the planted crop with water. Thus, the amount of harvested water from a certain area is a function of the amount of runoff that can be produced from that area.

Table 1. Soil texture for different depths in the project area

Location	Sand	Silt	Clay	Textural class
First profile	34.49	28.42	37.09	Clay loam
	18.07	29.87	52.06	Clay
	27.33	21.64	51.03	Clay
	9.94	24.35	65.71	Clay
Second profile	22.24	40.76	37.00	Clay loam
	17.23	33.68	49.09	Clay
	15.49	34.39	50.12	Clay
	21.25	30.32	48.43	Clay

The amount of runoff (for a given duration of time) is calculated by multiplying the amount of rainfall by a runoff coefficient, but not all the runoff is utilized (system efficiency), so the rainfall amount must be multiplied by an efficiency factor.

harvested water = catchment area x rain amount (design) x runoff coefficient x efficiency factor

extra water = cultivated area x (crop water requirement – rain amount [design])

From these two equations, we can get the following:

$$\frac{\textit{crop water requirement} - \textit{rain amount [design])}}{\textit{design rain amount x runoff coefficient x efficiency factor}} = \frac{\textit{catchment area}}{\textit{cultivated area}}$$

The design rainfall is the amount of seasonal rainfall at which the water harvesting system was designed to deliver enough runoff water to fulfil the crop water requirement, or more. However, crop water requirements depend on crop type and climate (a crop will require more water if planted in a hot and dry climate). The design rainfall is set depending on a certain rainfall probability (occurrence and exceedence) level. The selection of a design rainfall is a critical issue because this could lead to either crop failure or destruction of the system by flooding.

There are many water harvesting and soil conservation techniques but they differ according to location, soil type, rainfall, and land use. These techniques depend directly on contour lines (land slope) and soil depth in determining directions, type of technique and spacing.

Finally, the following factors must be considered when designing water harvesting techniques:

- Rainfall distribution during the season;
- Rainfall intensity;
- Soil infiltration rate and land surface characteristics;
- Soil water holding capacity (soil depth and texture);
- Type of crop to be planted (water requirements).

2.6. Land preparation

During the preparation of water harvesting structures and planting of shrubs, the project team concentrated on the involvement of the local community as a labour force. This ensured they learned particularly how these techniques are done, including building terraces and contour furrows, digging holes, planting, and how properly to deal with shrub seedlings.

A 50 ha demonstration site was prepared, using a variety of water harvesting configurations and digging holes for shrub planting. Three basic types of land treatments, or 'structures,' were demonstrated (with approximate percentages of treated area given for each type): contour furrows, 80 %, crescent-shaped or chevrons (micro-catchments), 10 %; and low rock walls, 10 %. The contour furrow is a simple water harvesting technique suitable for rangeland rehabilitation that can be accomplished with an ordinary plough. Contour furrows can be implemented on land slopes 1-8 % with variable soil depth (except very shallow soil). This technique is particularly efficient in intercepting runoff water and redistributing it with a relatively high capacity. The micro-catchments are suitable for slopes up to 8 %, and are not sensitive to slope changes in direction due to their shapes which concentrate the runoff water where the crop is to be planted. Both are highly efficient as water harvesting structures and particularly suitable for range crops, offering the flexibility to plant more than one seedling in each bund. Rock walls help slow runoff velocity, to control the danger of accelerated soil erosion in small gullies, and also to act as water harvesting structures.

2.7. Follow-up observations and management

2.7.1. Plant surveys

Plant surveys are required to measure the changes in the vegetation cover. They are essential to monitoring and evaluation of the rangeland. The range expert on the project used this study as a training tool and as an opportunity for participation with the local community. The community participated in the plant survey, in order to assist in providing knowledge of local names of the plants and their uses. The project team trained a number of community members to help in the survey techniques which then evolved into a community-based plant survey. The results of the botanical survey were reviewed with the community to ensure that the more common plant species (Annex 1) were known to them as well as the less common species that, due to variations in rainfall amounts, might appear less frequently. These steps are designed to enhance community awareness about the local range vegetation and the effect of new or less common plant species in the biodiversity of the plant cover.

2.7.2. Shrub survival rate

Shrubs were counted to determine the survival rates of each species. The project experts and representatives of the local community participated jointly in this process. The local people learned the basis for determining what was considered a living shrub, and gained a better knowledge of how the introduced shrubs performed at the site.

A rudimentary evaluation of the performance of planted shrubs will be reflected in the survival percentages under the different water harvesting methods and levels of initial irrigation. In January 2003, each planted shrub was evaluated and all dead shrubs replaced. The mortality rate associated with the initial planting was documented. Because so many factors can influence initial survival, only mortality by species was documented for future reference. In February 2003, all planted shrubs were checked again. For those shrubs that failed to survive, additional documentation including species and type of water harvesting structure in which the shrub was planted, was recorded. Based on initial expectations 60 % survival is considered the minimum level for 'success,' either for the overall site or for any subset of the plantation (shrub species, water harvesting method, and initial irrigation rate). 70-80 % survival is considered, 'good,' and anything over 80 % would be considered 'excellent'.

Shrub survival is the primary indicator of the success of techniques used in the demonstration site and should be monitored on an annual basis for five years.

2.7.3. Biomass production of shrubs

The biomass production of shrubs, both planted and naturally growing, is estimated using non-destructive sampling, employing the 'reference unit technique' (Al-Tabini 2002). Because of the limited size of the demonstration area and the limited number of shrubs, the preference is to use a non-destructive sampling protocol. The reference unit technique relies on estimating weight from easily obtained plant dimensions such as crown, or branch, area or diameter. Crown or branch volume also works well. The observers must practise estimating the desired indicator of biomass and then samples are collected (clipped, dried, and weighed) to develop the relationship between the ocular estimates of biomass and the dried weight of the biomass. The calibration process is critical to assessing the reliability of the biomass estimates, whether the shrub is planted or naturally occurring. The most direct sampling approach is to estimate the mean forage production, per shrub, for each of the major shrub species present. Biomass per shrub, combined with an estimate of the number of shrubs of a given species, per unit area provides an estimate of the total shrub biomass for the site.

2.7.4. *Evaluation of water harvesting techniques*

The effectiveness of the water harvesting techniques can be evaluated from two perspectives. The first deals with the appropriateness of the spacing of the furrows or water collecting structure. Spacing is estimated based on the water requirements of the plant, design, precipitation, and hill slope runoff efficiency factors. The planting scheme that has been implemented utilizes the current knowledge. Secondary measurements taken as part of the shrub biomass monitoring allows verification that the design supplied a consistent adequate water supply and that biomass production is consistent over a range of site conditions. A second concern deals with the physical effectiveness of the size and durability of the water harvesting structure. Success or failure can be determined by observations of breaching or overtopping. Breaching, or failure of the structure, can be monitored visually and rated as low, moderate, or severe. Low indicates some damage has occurred but the structure is capable of retaining all the water. Moderate indicates the damage is greater and one-half of the water is lost. Severe damage is indicated if the structure is no longer capable of retaining water and allowing it to infiltrate. A fourth observational category can be that the structure is 'undersized'. This would be the case when the structure does not appear to be adequately sized to retain the runoff. Since precipitation is monitored on site, the determination can be made as to whether or not the failure was the result of a larger than expected storm, or if the contributing area estimation was too large. Excessive runoff would be reflected in both overflow and excessive breaching. It is reasonable to assume that the 'contributing area' data collected along with shrub production would yield some insight into the cause of failure.

Local community participation is limited to the testing and evaluation of the introduced technology and monitoring of water harvesting techniques. After the rainy season a small group of people participates in checking the water harvest treatment structures, especially to evaluate how the different types of structures have performed under heavy runoff conditions. Observations are recorded for structures needing repairs and to learn how designs of structures could be improved.

Therefore, it can be seen that the local community participates in collaboration with project staff on this project in three main phases of the project: generation, testing and evaluation of the technology applied.

2.7.5. *Precipitation monitoring*

The information on site description indicates that annual rainfall on the demonstration area is thought to vary from 100-200 mm/yr. Maximum and minimum temperatures are believed to range from 3-33 °C. As part of the demonstration project, a rain gauge and recorder were installed to monitor precipitation. This data is very useful in evaluating overall project success. First, estimates of annual precipitation monitored on-site are compared with the assumed range in precipitation to determine if each study year is normal, wet, or dry relative to expectation. Eventually, the precipitation data collected onsite can be used to document the local variability in rainfall. On an annual basis, measured vegetation response, or biomass production are evaluated relative to the rainfall amount that occurred that year. The same with data used in assessing observed failures or breeches in the water harvesting structures. Although not influenced by the proposed treatments, site-specific precipitation data will be critical in interpreting vegetation responses to the proposed treatments.

2.8. Statistical analysis

Either ANOVA or linear modelling data analysis is undertaken, with the dependent variable being biomass, while independent variables can include species, year, contributing area, slope and water-harvesting technique. Initially, estimates of shrub biomass were obtained by shrub. Individual shrub data were aggregated by species, by water-harvesting technique, and in total for the treated area. Total biomass for the planted area can be compared with estimates of natural shrub production under grazed and non-grazed conditions. Throughout, P = 0.05 was used to define statistical significance. Whenever significant effects of different parameters and the interaction between them were found, mean separations were done using L.S.D and the Tukey test.

3. Results

3.1. Environmental impacts

3.1.1. Rainfall, November 2002-March 2006

During the 2002-3 rainy seasons, from late November to the end of March, Tall ar-Rimah received 163 mm of rainfall, which was 10 % above the long-term average for this area. The rainfall was well distributed over the four-month period, with 11 measurable rainfall events yielding an average of 15 mm per event. This distribution of precipitation created favourable conditions for plant establishment and growth. Table 2 shows the rainfall records for the average of four years. Since the 2002-3 rainy season, the area has been in a drought. The 2005-6 season rainfall was 50 % below the rainfall average. This has a negative impact on natural vegetation cover; more details will be discussed later in this paper. The typical rainy season is November-March each year. However, the highest rainfalls occurred in December and February when they were more than 50 % of the annual average.

In Jordan, some range plants are planted in October or early November. Because of this the plants may stay without rain for two months. This causes negative effects on the plants' survival rate unless the plants are irrigated in the first two months to avoid plant death. This adds more labour and cost.

On the basis of the rainfall patterns, it is recommended that any planting should be late November or early December when the seedlings and the seeds will get enough rain to ensure plant establishment in the soil where it is necessary to have pre-planting moisture.

3.1.2. Performance of the water-holding structures

The three main types of structures used at the Tall ar-Rimah shrub plantation site are contour furrows, and two shapes of micro-catchments, crescent and chevron. For the sake of brevity in this paper, micro-catchments will be referred to as 'micros,' with ad-

Table 2. Annual rainfall distribution from 2002-2006 at the project area

Month	2002-3	2003-4	2004-5	2005-6
November	20	3	17	0
December	63	52	13	8
January	12	19	24	15
February	23	19.5	52	35
March	45	11	3	4
Total	163	104.5	109	62

ditional reference to either type as needed. The contour furrows were prepared with a tractor-drawn implement. The micros were constructed manually with large hoes.

Water harvesting structures were installed in September 2002. After eight months of settling and stabilizing, the average height of the bunds, or ridges, above the original land surface was about 30 cm along the contour furrows and 20-25 cm for the micros. The resulting depth of water-holding depressions averages 30-40 cm, respectively, for the contour furrows and micros. Micros are 3.5-4.0 m wide between ends of bunds. Each wing of the chevron micros is 2.5 m long. The gaps between adjoining micros average 1-2 m. Contour furrows cover 90 % of the water harvesting treatment area and micros cover 10 %. A total of 13 925 m of contour furrows and 0.6 ha of crescent and chevron micros, combined, were installed at Tall ar-Rimah. The total area covered by contour furrows and micros is about 20 ha. The total area under grazing protection is 40 ha. The ungrazed control areas were chosen beside the project area; the land here has similar physical characteristics and the area has no control on grazing activity and no water harvesting in the area.

Other structures installed at Tall ar-Rimah include: two ponds, each about 100 m2 in area, designed to collect heavier runoff events; 1 375 m of low, loose-rock walls to catch sediment and hold soil water for plant growth; and two 0.8 m-high rock diversion walls, one of which is 62 m long and the other 75 m long, built next to a small wadi at the base of the treatment area to direct flood water unto a potential forage production site.

3.1.3. Stability of structures

Contour furrows: Based on field observations made immediately following heavy rains in December 2002 and during the entire winter rainy season, contour furrows performed better than micros in terms of resistance to breaching by heavy accumulations of runoff. The breaching of the contour furrow ridges that did occur was usually at points where water flow was concentrated above the furrows in small gullies, putting added erosional pressure on the relatively loose ridge soil below the furrows. Nothing was observed which would significantly modify either the design or the continued use of contour furrows as applied in this demonstration. Given that the 2002-3 rainy season produced an average rainfall total, that five out of the 11 rainfall events at Tall ar-Rimah exceeded 15 mm, and that contour furrow ridges had not fully settled or sealed between construction and start of the runoff, it can be said that the stability performance of the contour furrows was put to a moderately severe test. It can be concluded that the excellent performance of these structures was the result of a good engineering and contour furrow construction.

Figure 2. *Salsola* sets seeds and grows very fast each season, helping to stabilize the contour furrows

Figure 3. Project area at Tall ar-Rimah showing new plant growth

In the last three rainy seasons, up to and including 2005-6, the contour furrows were stable and required very minor maintenance. The natural vegetation stabilized the ridges, helping to resist breaching. The contour furrows planted with *Salsola*, which sets seeds and grows very fast each season, were the most stable (Figure 2).

Micros: The crescent-shaped and chevron-shaped micros performed well after initial breaching was repaired. Forty percent of the structures were breaching during the first year. More breaching was observed in the chevron micros than in the crescent micros. It appears that there was greater runoff concentration at the apex of the V-shaped catchments than in the curved structures. Location of the micros on, or close to, any existing rills or small gullies increased the tendency for breaching. This should be avoided in the future. Degree of slope could also be a critical factor in successful location of micros; placement on lower slopes with slower runoff offering a better chance of reliable performance. As with the contour furrows, performance of the micros was well tested by the significant amounts of runoff, which impacted these structures prior to their ridges becoming compacted. After they were repaired following initial runoff events, the micros retained water as designed.

Rock walls: Both the low rock walls intended for sediment and soil water catchments and the high rock wall diversion structure showed no weaknesses in stability. The locations were suited to their purposes.

3.2. Vegetation results

3.2.1. Planting period and soil moisture conditions

Shrub seedlings were planted between November 20th and December 20th 2002 without any supplemental irrigation at planting. Significant rains had started at Tall ar-Rimah by November 25th. They provided adequate runoff and soil moisture for the newly planted seedlings. In fact, during the final two weeks of planting many of the empty planting holes were full of water as a result of 60 mm of rainfall between December 18th and 22nd. In the future, if a treatment site has dry soil conditions during the planting period, it is recommended that some irrigation for the new seedlings be applied until rainfall and water harvesting provide enough soil moisture to get the seedlings established. Soil moisture conditions did not require irrigation at Tall ar-Rimah in 2002. No irrigation was applied to the seedlings at Tall ar-Rimah from the period of planting through to the time of writing this paper in 2007, and no irrigation is planned or needed at this site.

Despite the years of drought since the beginning of the project, the shrubs have survived and produced biomass (Figure 3). Al-Farrajii and Hilli (1994) reported that the vived and produced biomass (Figure 3). Al-Farrajii and Hilli (1994) reported that the *Atriplex* shrubs are drought resistant and can thrive well even under severe conditions. In 2005-6, the rainy season was the driest of the four seasons since planting and the shrubs did not receive enough water. Therefore they shed their leaves. Shedding begins with the oldest leaves and progresses toward the apical meristems; small summer leaves often replace large winter leaves. In general, the plants have adapted to drought either by evading drought or by resisting its effects.

3.2.2. Survival rate of planted shrubs

Seedling survival: The most important, and least predictable, performance result in the Tall ar-Rimah demonstration project was the matter of the survival of shrub seedlings after transplantation into this ecologically hostile site in 2002. The seedlings were already older and larger than optimum for this purpose. They were becoming root bound in their plastic tubes. Pruning back of excess foliage, and careful planting at the right time of the year, increased the prospects for seedling survival. Fortunately, the rains came at the right time and were well distributed throughout the rainy season. First year survival of seedlings was an outstanding success for all shrub species planted, and for all water harvesting structures used. Four years from planting the average survival rate for all three shrub species was 88 % (Table 3). This is classified as an excellent result. The shrubs were adapted to the area, and the water harvesting techniques worked well in producing this result.

3.2.3. Forage production of planted shrubs

Forage production, as biomass, was estimated from June 2003 till June 2006. The Reference Unit Method was used (Al-Tabini 2002). Biomass estimates for *A. nummularia, A. halimus* and *S. vermiculata* for different water spreading structures are listed in Table 4.

Statistical analysis indicated that the estimated biomass produced within contour furrows is significantly higher than the production within the crescent and V-shaped water harvesting structures. The reason for the difference is related to the production from the contour furrows. It can be concluded that contour furrows are more efficient than the other techniques in supporting shrub forage production. When biomass data were combined for all structures there is a statistically significant difference between *S. vermiculata* compared with *A. nummularia* or *A. halimus*. Although it is biologically not correct to compare two different plant species, there is a biological difference in biomass production between *A. nummularia* and *A. halimus* even though it is statistically not significant (Table 4).

Figure 4 shows that biomass production has sharply increased every year. It is clearly seen that the increase in biomass production was mainly during the second and third year, constituting 80 % of the total production. This conclusion is in line with other results which suggest that it is important to leave the depleted rangeland from 3-5 years for the rehabilitation process.

Table 3. Survival rate for planted shrubs

Shrub species	No. of seedlings planted 2002	Survival rate June 2003	Survival* rate June 2004	Survival rate June 2005	Survival rate June 2006
A. nummularia	3456	88	92	90	88
A. halimus	3040	85	92	89	87
S. vermiculata	3660	86	88	90	90
Average	10156	86	91	90	88

* Decrease in 'dead' seedlings reported in June is because new foliage growth on some seedlings that were severely desiccated in January showed the plants were still alive

Table 4. Biomass production of shrubs planted in different water harvesting structures, in dry matter (DM)

Estimated biomass from year 2003 to 2006 kg DM/ha					
2006	2005	2004	2003	Water harvesting techniques	Plant species
395	345	196	23	Contour furrows	A. halimus
260	279	186	15	Crescent shape	A. halimus
332	258	133	15	V - shape	A. halimus
329	294	172	18		Mean
452	370	245	24	Contour furrows	A. nummularia
341	314	209	21	Crescent shape	A. nummularia
319	306	151	24	V - shape	A. nummularia
371	330	202	23		Mean
77	45	22	6	Contour furrows	S. vermiculata
60	61	20	5	Crescent shape	S. vermiculata
49	34	14	5	V - shape	S. vermiculata
62	47	19	5		Mean

Figure 4. Shrubs' growth during the period 2003-6

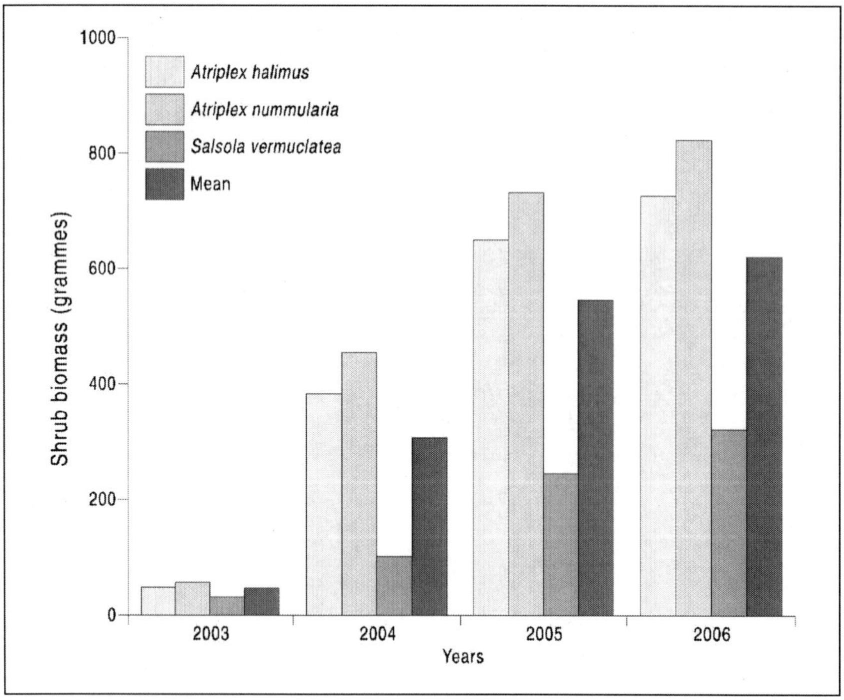

The shrubs were grazed during 2007; the grazing will encourage the shrubs to produce more forage, and to avoid producing wood instead of green edible forage. This pilot grazing appears to have been successful, both socially and biologically, but this will not become fully clear until later in 2008.

3.2.4. Natural vegetation inside and outside the treatment area

Sampling of vegetation was done inside the treatment area with plots distributed randomly in portions of the area that were not disturbed by construction of the water harvesting structures. Outside the treatment area, plots were randomly distributed on grazed rangeland with no disturbance from construction activities. Ocular estimates of the biomass contribution to the vegetation of the metre-square sample plots for each species present were recorded for each plot. All the above-ground plant biomass in each plot was collected, oven-dried, and weighed.

Forage production from the natural herbaceous plants found in the project area during the 2004 survey was estimated as 75 kg/ha inside the reserve compared to 30 kg/ha outside the reserve (Table 5). These differences between the production inside and outside the shrub reserve show the effect that grazing can have on these rangelands and suggest what the potential can be for range forage production under better grazing management. However, the 2005 and 2006 surveys showed that productivity declined to 27 and 16 kg/ha inside the project area. These differences between years are related to the rainfall amount and distribution. For example, the total amount of rainfall in 2003-4 was 104 mm, compared with 109 mm during the 2004-5 season; but there was a great decline in forage production in 2005. From the rainfall data (Table 2) we can see that in the 2003-

Table 5. Natural forage from different plants growing in the site

Year	Natural forage production (kg/ha) Inside	Outside
2004	75	30
2005	27	3
2006	16	3

Figure 5. Number of plants species and families, 2002-6

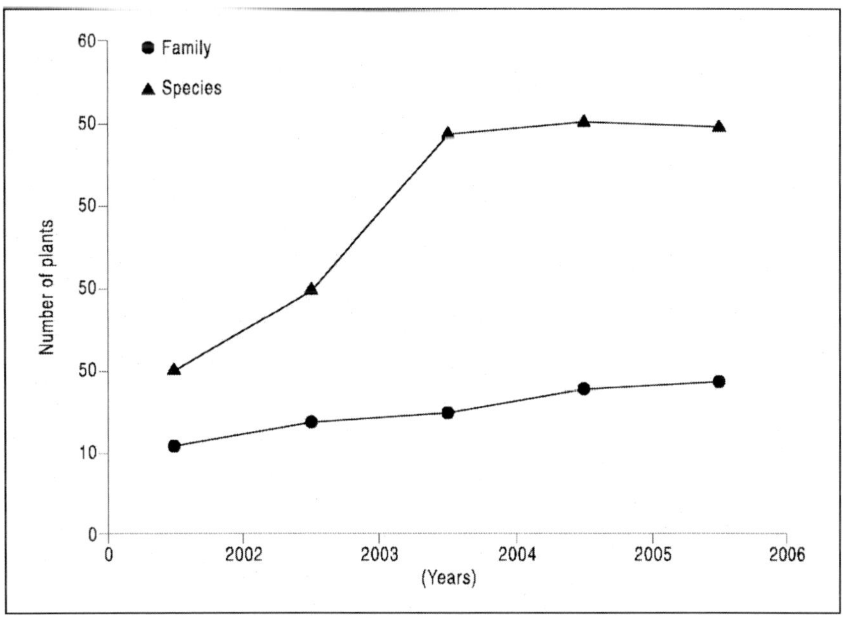

Figure 6. A new plant species, Crocus moabiticus

4 season there were 55 mm of rain in December 2003, while in the 2004-5 and 2005-6 seasons the early rain was significantly low. It appears that early rainfall during December is essential for rapid growth of shrubs and seed germination in the arid dry areas. For the same reason the total number of the plants in 2004 was significantly greater than 2005 and 2006. This result is in line with the explanation which comes from the local community. They said that the early rainfall[1] is very important for the rangeland plants and if the rain comes late the season will be bad for the annuals but good for the shrubs.

The results obtained from the botanical survey showed that in the first survey, carried out before starting the project (2002-3), there were only 22 plant species belonging to 12 different families. However, after four years of protection, the number of plant species and families found in the project area were higher: 51 plant species, belonging to 18 families were recorded in the 2005-6 survey (Figure 5 and Annex 1).

In 2006, a new plant species, *Crocus moabiticus* (Figure 6), in the family Iridaceae, was recorded. This appears to illustrate the importance of managing the vegetative cover and/or the effectiveness of the water harvesting techniques to increase soil moisture. This means that the soil is rich in seeds of different plant species including herbaceous and perennial plants. It is believed that the availability of moisture and protection caused the annual plants to regenerate and produce seeds for new growth (Appendix 2).

During the 2004 survey, the most abundant plant species were as follows: *Schismus arabicus, Sisymbrium Spectulum, Aaronsohnia factorovskyi, Poa bulbosa*, and *Bromus damthonea*. However, during the 2005 survey the most abundant plant species were: *Siedlitizia florida, B. damthonea, S. arabicus, P. bulbosa, Hordeum glaucum steujel*. While in the 2006 survey there was: *Carx stenophylla, Bromus tectorum, S. arabicus, H. glaucum steujel, S. florida*. By 2007 the most abundant plants were *P. bulbosa, S. arabicus, B. tectorum* and *S. florida*. In general it is clear that there are certain plant species that can be considered to be well adapted to the local climatic and soil conditions (Appendix 1).

Contrasting the outside area and the inside area, the major differences between the two zones were the total absence of the grass *P. bulbosa* in the grazed area; and the greater biodiversity in the protected zone including the *Crocus*. The plant community outside the protected area is dominated by *S. florida*, a succulent forb that is relatively unpalatable when green but grazed readily when mature and dry. The outside area is always subjected to severe early grazing which causes low biodiversity and low plant cover compare to the project area (Table 6).

These data support the expected vegetation response to heavy and continuous graz-

[1] The local community called the early rainfall in Arabic harief.

ing: a reduction in biodiversity. Rangelands which are completely and permanently excluded from grazing by livestock could be considered 'unnatural' from an economic viewpoint, and perhaps even from an ecological perspective. However, the goal of improving the biodiversity of range vegetation through better grazing practices and possibly the use of new forage production technologies should remain valid.

3.2.5. Plant production and grazing options

Contour furrows constitute 80 % of the treated area, therefore the stocking rate will be calculated using the forage production from the contour furrows. The average forage production of *A. nummularia* and *A. halimus* is 424 kg dry matter (DM) per hectare, and 77 kg DM/ha for *S. vermiculata*. The total forage production in the site equals the total production of the two *Atriplex* species plus the total production of *S. vermiculata*.

Total production of *Atriplex* species	= 424 kg x 30 ha	= 12 720 kg DM
Total production of *Salsola*	= 77 kg x 30 ha	= 2 310 kg DM
Total forage production in the site	= 15 030 kg DM.	

3.2.6. Atriplex palatability

In general, sheep consume 1.5, 2.0, 2.5 %, respectively, of their body weight in dry matter when grazing low, average, or high digestibility forages. Otsynia et al (1982) reported that the digestibility of browse (shoots) decreased from 84 % to 36 % from April to August while the lignin content increased to 6.9 %. Digestibility of *Atriplex* species averages 59 % in spring and 49 % in summer (El-Aich 1992).

The palatability of chenopod shrubs is moderate to low when grasses are available and increases as the grasses becomes less available (Wilson 1966). Most *Atriplex* spp provide a reasonably high nutritive value for sheep and goats, and are useful as a feed supplement (Rizk 1986). The highest forage values are found during the wet season of the year when plants are green and actively growing (Chatterton et al 1971, Kandil et al 1988, Kandil et al 1990).

To get better utilization of *Atriplex*, it must be supplemented, particularly with a readily available source of energy (Gihad et al 1994). Shehata et al (1988) reported that the digestibility of nutrients were at the highest values ($P<0.05$) in sheep offered *A. nummularia* and barley, whereas the lowest digestibility was attained in sheep offered *A. nummularia* alone. The better digestibility was obtained because of energy from barley (carbohydrate source) which helped the rumen micro-flora in digestion.

Table 6. Plant cover (%) at Tall-ar-Rimah, according to Parker's loop

Plant cover (%) Point	2002[1]	2003	2004	2005	2006
P1	12	52	60	53	23
P2	15	61	43	61	28
P3	21	58	45	54	25
P4	14	56	36	40	12
P5	11	54	39	52	28
P6	9	62	67	48	29
P7	17	59	31	57	36
P8	21	55	34	27	11
P9	22	61	22	11	9
Mean	16	58	44	49	24

[1] Before starting the project.

It has been suggested that in Arab countries where the average rainfall ranges 100-250 mm, attention should be drawn to *Atriplex* shrubs as a supplementary feed rather than the sole feed (ACSAD 1978).

When sheep or goats browse *A. nummularia* or *A. halimus* as a sole feed they do not satisfy their energy production requirements and they may not even be able to satisfy their energy maintenance requirements as animals tend to lose weight (El Shaer et al 1990).

Utilization of *Atriplex* shrubs should not exceed 60 % of available forage in order to ensure regeneration and uniform forage production in the following years (Zaroug 1985).

In conclusion, as the animal feed intake can be measured on the basis of 2.5 % of body weight, and as the average body weight of Awassi ewes is 60 kg, so the daily feed intake will be 1.5 kg dry matter (2.5 % x 60 kg).

Total available forage = 15 030 kg DM.
Proper utilization = 60 %
Forage available to be grazed = 15 030 x 60 % = 9 018 kg DM
Average sheep intake = 1.5 kg
Estimated grazing capacity = 9 018 kg / 1.5 kg = 6 012 sheep days.

On this basis, the following are examples of appropriate grazing options:

- A total of 50 sheep grazing on the site for 4 months (6 012 / 4 x 30 days).
- A total of 200 sheep grazing on the site for 1 month (6 012 / 30 days).
- A total of 429 sheep grazing on the site for 14 days (6 012 / 14 days).

4. Conclusions and recommendations

- Careful planning at each project stage can contribute to achieving a positive outcome:
  - appropriate problem definition and establishment of objectives
  - successful generation of applicable information or innovations
  - appropriate development and use of dissemination mechanisms and their effective implementation
- The level of farmers' participation should be determined by their interests and capacities and by the nature of the work
- Technical scientists contribute new technological information that may provide options for improvement, and methodologies for testing the various options
- Demonstrations of successful rehabilitation of rangeland areas may convince local herders to cooperate in plans to better utilize not only improved areas but also to develop plans for utilization of all range areas. Direct benefits for the herders appeared to be the key incentive
- The project has been able to demonstrate to the government that, with the cooperation of the local Bedu population, rehabilitation of large areas of the Badia steppe is possible. It has also shown the necessity of urgently addressing the issues associated with free access to all areas of land, the effects of drought and the need for preparing drought contingency plans. The total number of sheep, overstocking and issues associated with subsidized feed are other matters that will need to be addressed. The focus of future project activities will be on selected rangeland site management improvement and rehabilitation, with the selected areas managed by Bedu organizations
- The use of native plant species (*S. vermiculata* and *A. halimus*) demonstrated high potential
- It is important to disseminate information to all interested parties, including local communities along with provisional and national governing bodies

- The main lessons learned were both technical and socio-economic and it is clear that they must be considered together in rangeland development programmes
- Full participation of the local people in rangeland development and management will require a major focus on training of technicians and the local people, as well as built-in feedback mechanisms which enable adjustments to be made during project implementation
- Undoubtedly nomadic and other traditional grazing practices are well established in the desert rangelands, sometimes for wise use of these resources, and sometimes not
- The ephemeral and unpredictable nature of desert climates, and therefore the highly variable access to desert forage, support the argument that some forms of nomadic or transhumant grazing are the most efficient ways to harvest this scarce resource. Nomadism is one of the most traditional ways that desert livestock owners can graze wide areas with minimum damage to range vegetation. This practice can continue to be part of grazing management which can combine sound utilization with good conservation of the Badia rangelands.

Acknowledgment

The project is jointly funded by USDA-US Forest Service, USAID-Jordan, Global Environment Facility (GEF) and New Mexico State University. The authors thank Dr. Derick Baily and Rich Philip from New Mexico for their help in data analysis. Special thanks go to Rich Arao, range ecologist consultant, for his big effort in the project and helping to prepare the report on which this paper is partially based.

References

ACSAD (1978). Nutritive and grazing evaluation of Atriplex species for sheep: Part one, nutrition value. Final Report (In Arabic) ACSAD.

Al-Farrajii F and Al-Hilli MR (1994). Halophytes and desertification control in Iraq. In: Squiers VR and Ayoub AT (eds), Halophytes as resource for livestock and for rehabilitation of degraded lands. Netherlands: Kluwer Academic pp. 239-248.

Al-Tabini RJ (2002). An evaluation of the potential of Atriplex nummularia for sheep production in arid Jordanian rangelands: the effects of defoliation management. PhD thesis, University of Newcastle-upon-Tyne.

Ben Salem H and Nefazawi A (1993). Effect of increasing levels of Acacia Cyanophylla intake, digestibility and animal digestion on sheep fed straw based diet. Proceeding 7th meeting of the FAO European sub-network on Mediterranean pastures and fodder crops, Crete, Greece, 21-23 April 1993 118-121.

BRDP (2001). Tal Rimah social survey. Technical Report, Safawi Centre, Jordan.

Chatterton NJ, Goodin JR, Mc Kell CM, RV, P and Rible JM (1971). Monthly variation in the chemical composition of desert saltbush. Range Management 12:289.

Draz O (1987). Fodder shrubs for range improvement: The Syrian experience. In: Kell CM (ed), The biology and utilization of shrubs. San Diego: Academic Press pp. 249-265.

El-Aich A (1992). Fodder trees and shrubs in range and farming systems in North Africa, FAO Animal Production and Health Paper. FAO, Rome, Italy: 1992 No. 102 61-73.

El Shaer HM and Kandil HM (1990). Comparative study on the nutritional value of wild and cultivated A. halimus by sheep and goats. Commonwealth Science and Development Research 29:181-187.

Gihad EA and El Shaer HM (1994). Utilization of halophytes by livestock on rangelands: problems and prospects. In: Squiers VR and Ayoub AT (eds), Halophytes as resource for livestock and for rehabilitation of degraded lands. Netherlands: Kluwer Academic Publishers pp. 77-96.

Kandil HM and El Shaer HM (1988). The utilization of A. Nummularia by goats and sheep in Sinia. International Symposium on constraints and possibilities of ruminant production in the dry tropics, November 5-7, 1988. Cairo, Egypt.

Kandil HM and El Shaer HM (1990). Comparison between goats and sheep in utilization of high fibrous shrubs with energy feeds. Proceeding International Goat production Symposium, October, 22-26, Tallahassee, Egypt.

Le Houe'rou HN (1993). Salt tolerant plants for the arid regions of the Mediterranean iso-climatic zone. In: Leith H and El Massoom A (eds), Towards the ration use of high salinity-tolerant plants. Dordrecht, The Netherlands: Kluwer Academy publisher pp. 403-422.

Otsynia RM, Mckell CM and Van Epps GA (1982). Use of range shrubs to meet nutritional requirements of sheep grazing on crested wheatgrass during fall and early winter. Range Management 23:751-753.

Rizk AM (1986). The phytochemistry of the flora of Qatar. Doha: Science and Applied Research Center, University of Qatar.

Shehata E, El-Sayed IM, Heider A, El-Serafy AM and El-Gallad T (1988). Effect of supplementing Atriplex nummularia with minerals, barley or both on dry matter and water intakes and nutrients metabolism. Proceeding 2nd conference of Agriculture development research, Ain Shams University pp. 111-119.

Wilson AD (1966). The tolerance of sheep to sodium by sheep fed on speices of Atriplex species and Kochia (Bluebush). Australian Journal of Agriculture research 17:155-163.

Zaroug MG (1985). Travel report summary on sheep production project in Jordan. FAO, Rome.

Annex 1. Most common plant species found in the project area, as recorded by Parker's loop

Plant Species	2004	2005	2006
Schismus arabicus	147	56	70
Poa bulbosa	63	103	23
Gymnarrhena micrantha	20	45	0
Malcolomia conringiojes	31	0	39
Siedlitizia florida	26	228	95
Matthiola aspera	32	0	0
Carx stenophylla	8	35	0
Adonis dentata	8	0	0
Bromus damthonea	18	62	41
Sisymbrium spectulum	10	0	0
Hordeum glaucum steugel	8	58	131
Bellevalia desrtorum	30	12	86
Lasiopagon muscoides	24	0	0
Allium trancatum	0	0	0
Silene vulgaris	20	0	0
Aaronsolnia factorovskyi	83	34	32
Hippocrepis costricta	0	67	5

Annex 2. Plant species listed in the project site during the period of study

Family name	Species	2002	2003	2004	2005	2006
Acanthaceae	*Anjrosace maximal*	NA	NA	X	X	X
Apocynaceae	*Eminum spiculatum*	NA	X	X	X	X
Boraginaceae	*Anchusa milleri*	X	X	X	X	X
Caryophyllaceae	*Herniaria hirusta*	X	X	X	X	NA
	Silene arabica Bioss	NA	X	X	X	X
	Spergula fallax	NA	NA	X	X	NA
Chenopodiaceae	*Anabasis syriaca*	X	X	X	X	X
	Noaea mucronata	X	X	X	X	X
	Salsola vermiculata	X	X	X	X	X
	Siedlitizia florida	X	X	X	X	X
Compositae	*Aaronsohnia factorovskyi*	NA	NA	X	X	NA
	Achillea fragrantissima	X	X	X	X	X
	Anthemis hyaline	NA	X	X	X	X
	Anthemis syriaca	NA	NA	X	X	X
	Artemisia inculta	NA	X	X	X	X
	Centaunea postil	NA	NA	X	X	X
	Onoporjon transjanicum	NA	NA	X	X	X
	Filego desrtorum	NA	X	X	X	X
	Lactuca orientalis	NA	NA	X	X	X
	Gymnarrhena micrantha	X	X	X	X	X
	Lasiopagon muscoides	NA	NA	X	X	X
	Matricaria aunea	NA	NA	X	X	X
	Sconzonera pusilla	NA	NA	X	X	X
	Scorzonera pappose	X	X	X	X	X
	Notabasis syriaca	X	X	NA	X	X
Cruciferae	*Cardaria draba*	NA	NA	X	X	X
	Malcolmia conringiojes	NA	NA	X	X	X
	Eruca sativa miller	NA	NA	X	X	X
	Sinapsis alba	X	X	X	X	X
	Sisymbrium Spectulum	NA	X	X	X	X
Geraniaceae	*Erodium hirtum*	NA	X	X	X	X
Iridaceae	*Crocus moabiticus*	NA	NA	NA	NA	X
Labiatae	*Phlomis brachyodon Bioss*	NA	NA	X	X	X
Leguminacae	*Astragalus spinosa*	X	X	X	X	X
	Trigonella stellata	NA	NA	X	X	X
Lilaceae	*Allium truncatum*	X	X	X	X	X
	Gagea reticulate	NA	X	X	X	NA
	Bellevalia desertorum	NA	NA	X	X	X
Malvaceae	*Malva sylvestris*	X	X	X	X	X
Papaveraceae	*Papaver syriacum Bioss*	NA	NA	X	X	X
	Roemeria hybrida	NA	NA	X	X	X
Plantaginaceae	*Plantago spp*	NA	NA	NA	X	X
Poaceae	*Bromus damthonea*	NA	X	X	X	X
	Bromus tectorum	X	X	X	X	X
	Eremopyrum tectorum	NA	NA	NA	X	X
	Hordeum bulbosum	NA	NA	X	X	X
	Hordeum glaucum steujel	NA	X	X	X	X
	Poa bulbosa	X	X	X	X	X
	Poa steudel	X	X	X	X	X
	Schismus arabicus	X	X	X	X	X
	Stipa capensis	NA	X	X	X	X
Polygonaceae	*Rheum palaestinum Feinbr.*	NA	NA	NA	X	X
Raununculaceae	*Adonis dentata*	X	X	X	X	X
Urticaceae	*Urtica pilulifera*	X	X	X	X	X
Zygophyllaceae	*Peganum harmala*	X	X	X	X	X
Total number species tallied		21	23	50	54	51

X= Species present, NA= Species not present

INDEX

In this index page numbers that are in **bold type** refer to Figures, Maps and Tables in the text. Key references are also in bold type and are usually followed by a number of pages dealing with the same topic.